Electron Microscopy and Analysis Group Conference

(EMAG 2013)

Journal of Physics: Conference Series Volume 522

York, United Kingdom
3 – 6 September 2013

Editors:

P. Nellist

ISBN: 978-1-63439-303-4
ISSN: 1742-6588

Printed from e-media with permission by:

Curran Associates, Inc.
57 Morehouse Lane
Red Hook, NY 12571

proceedings.com

Some format issues inherent in the e-media version may also appear in this print version.

Printed by Curran Associates, Inc. (2014)

Institute of Physics
Dirac House, Temple Back
Bristol BS1 6BE UK

Phone: 44 1 17 929 7481
Fax: 44 1 17 920 0979

techtracking@iop.org

Additional copies of this publication are available from:

Curran Associates, Inc.
57 Morehouse Lane
Red Hook, NY 12571 USA
Phone: 845-758-0400
Fax: 845-758-2634
Email: curran@proceedings.com
Web: www.proceedings.com

TABLE OF CONTENTS

PLENARY LECTURES

HOWIE SYMPOSIUM

STEM AND MODELLING

ADVANCES IN ELECTRON MICROSCOPE IMAGING

BIOMATERIALS

NANOMATERIALS

Editorial

It has once again been my pleasure to act as editor for these proceedings, and I must thank all those who have acted as reviewers. I am always struck by the scientific quality of the oral and poster contributions and the vibrant discussions that occur both in the formal sessions and in the exhibition space at EMAG. I am convinced that a crucial part of maintaining that scientific quality is the opportunity that is offered of having a paper fully reviewed by two internationally selected referees and published in the *Journal of Physics: Conference Series*. For many students, this is the first fully reviewed paper they publish. I hope that, like me, you will be struck by the scientific quality of the 80 papers that follow, and that you will find them interesting and informative.

I must also personally thank all the organisers of EMAG2013 for arranging such an excellent meeting. Ian MacLaren, as Chair of the EMAG Group and of the meeting itself, has contributed a foreword to these proceedings describing the meeting in more detail. A particular highlight of the conference was the special symposium in honour of Professor Archie Howie. We all enjoyed a wonderful speech from Archie at the conference dinner, along with some of his electron microscopy-related poetry. I have great pleasure in publishing the conference dinner poems in this proceedings.

I hope you will find these proceedings to be an interesting read and an invaluable resource.

Pete Nellist

Conference committee

Conference chair:	Dr I MacLaren
Programme organiser:	Dr C Ducati
Proceedings editor:	Prof P D Nellist
Trade exhibition organiser:	C Hockey (CEM Group)
Local organisers:	Professor E Boyes, Professor P Gai, Dr R Kröger, Dr V Lazarov, Dr P O'Toole, Dr S Tear and Professor J Yuan
Advanced school organisers:	Dr S Haigh, Dr A Brown
Other committee members:	Mr K Meade, Mr O Heyning, Dr M Crawford, Mr M Dixon and Dr Z Li

Foreword

EMAG2013 was certainly the biggest EMAG conference I can remember in my twenty years of going to EMAG conferences. We had a total of about 180 scientific delegates, not including exhibitors. The conference also benefited from the York conference centre allowing us to integrate the scientific sessions, refreshments and exhibition in one building. This all made for a very vibrant and focused conference. The quality of the presentations was again very high and judging by many of the student talks given, there is plenty of hope for an excellent future for electron microscopy in the UK.

A unique feature of this conference was a chance to organise a session to commemorate the 80th year of one of the pioneers of electron microscopy in the UK, Professor Archie Howie. An excellent symposium was organised in his honour by one of his former students, Professor Pratibha Gai of the University of York. This was dedicated to one field that he made some important initial contributions to: *in-situ* microscopy. The symposium gave a depth and breadth of overview to this field unusual at an EMAG conference, with speakers from the UK, Europe, Japan and the USA and gave a real taste of some of the possibilities in the latest instruments to observe real materials at work at nanometre or even atomic resolution.

Professor Howie himself gave a highly entertaining retrospective of his time working in microscopy from the 1950s until the present in his talk after the conference dinner on the Thursday evening at the excellent York National Railway Museum. We also had excellent plenary lectures from Professor Wolfgang Baumeister on a topic perhaps novel to many at EMAG, TEM in Structural Biology; from Professor Prathibha Gai on atomic resolution *in-situ* studies of chemical reactions in the TEM; and Professor Archie Howie discussing some of the important current open questions in TEM and STEM.

This proceedings therefore presents a snapshot of a broad cross-section of the exciting work going on in electron microscopy in 2013. As always, the papers are ordered thematically according to the sessions organised at the conference. As ever, I have a great debt of gratitude to my fellow EMAG committee members who chaired the various sessions and who, together with the programme chair, Dr Cate Ducati of the University of Cambridge helped to put together such an excellent programme. I also gratefully acknowledge the many hours invested by Professor Pete Nellist of the University of Oxford who then edited this proceedings and ensured that the peer review and revision process ran as smoothly as possible. Finally, I hope that you enjoy reading the contributions made by many to this excellent conference.

Dr Ian MacLaren, Chair of the Electron Microscopy and Analysis Group

Conference banquet poems by Archie Howie

Archie notes that Bede recounts an early English comparison of our life to that of a sparrow, who flies through a mead hall out of a stormy winter's night. The sparrow is briefly exposed to the light, warmth and safety of the hall before it vanishes back into the night and storm. Both poems were influenced by this legend, and the sparrow features in the second.

Brief Life of an Electron

All wires electron clones contain
-Zillions each wait to get the call
To jump to sudden life again
And tunnel through their prison wall.

Then dragged to near the speed of light
They rush through lenses swirled by fields
Into a nano spot with focus bright
Which, scanned around, an image yields.

Pass through the object like a wave
Crashing, caressing, channelling;
There to the quantum rules a slave
Store these events as final fling.

The image builds up as we wait
Our task but to interpret then
Brief life that has this modest fate
So with electrons, so with men!

Journal of Physics: Conference Series **522** (2014) 011003 doi:10.1088/1742-6596/522/1/011003

Heroes of Microscopy

Rejoice at York where EMAG delegates collect!
What icons as role models can they still respect?
Recall that here great Constantine put on Rome's crown
And Alcuin his intellectual roots put down
As poet, cleric and as sharpner of youth's mind.
Here Godricke's eyes and brain the double stars divined!

But York shows too how teams sole efforts magnify
As marks of Minster builders amply testify
Here Viking crews might run amok with axe and sword
But in the boat worked oar and sail in strict accord!
With Prutton and Robards microscopy took root
Gai's JEOL Nanocentre team boosts York's repute!

Past heroes of microscopy we may recall
And wish they'd join us in this railway dining hall.
Bede's little sparrow may be scared to fly in here
But none could block the Flying Scotsman's wild career
Should it steam through with faces pressed against the glass
Glimpse Ruska, Castaing, Cosslett, Crewe in the first class!

Peer review statement

All papers published in this volume of *Journal of Physics: Conference Series* have been peer reviewed through processes administered by the proceedings Editors. Reviews were conducted by expert referees to the professional and scientific standards expected of a proceedings journal published by IOP Publishing.

Addressing Coulomb's singularity, nanoparticle recoil and Johnson's noise

Archie Howie

Cavendish Laboratory, University of Cambridge, J J Thomson Avenue, Cambridge CB3 0HE, UK.

Email: ah30@cam.ac.uk

Abstract. Recent experiments prompt rethinking of the basics of elastic and inelastic electron scattering in electron microscopy. Standard approximations of elastic scattering largely based Bragg's law seem less clearly relevant when individual columns or even single atoms are probed. Phase shift analysis of atomic scattering can provide some checks and insights. The dielectric theory of aloof beam interactions is severely tested by observations of nanoparticle recoil but may be capable of explaining de-coherence effects induced by thermal fluctuations.

1. Introductory challenges

Progress in electron microscopy seems to follow a helical trajectory. Improved instrumentation and pioneering observations provide fresh opportunities and challenges. These may expose the need for radically new thinking or more simply a reworking of earlier ideas that have fallen out of fashion. Bragg reflection concepts, highly relevant in the days of diffraction contrast [1], underpinned the standard approximations for elastic scattering, including diffuse scattering. They reached a high point of Baroque splendour in high resolution phase contrast, where images of crystal defects, simulated by periodic continuation of an artificially very large unit cell involve many hundred diffracted beams [2]. Although it continues to use Bragg concepts such as Bloch waves for image interpretation, scanning transmission electron microscopy (STEM) using high angle dark field mode in reality increasingly provides a quite discordant example where a single column or even a single atom is illuminated in surroundings that are largely irrelevant [3]. Had we started from here it seems unlikely that we would have tackled the problem by imposing some artificial periodicity and then applied Bragg reflection theory! The standard phase shift theory of electron scattering in an atom, recently explored for this situation [4] and further developed in section 2 below, deals more convincingly with the Coulomb singularity in the atomic potential and thus provides a useful check on Bragg methods. Although we may need to use over 200 phase shifts at 100 keV, this is still not excessive compared with the number of Bragg reflections noted above.

Easily accessible and relatively intense plasmon losses in free electron metals like Al attracted attention in electron microscopy about 50 years ago and were used for microanalysis that still looks impressive today. Dielectric excitation theory was then systematically developed for the more general interpretation of energy loss spectra in this valence region including excitation by both penetrating and aloof beams. Then for a period, the interest of microscopists in valence losses evaporated when the more directly interpretable core loss region became available. Recent access to the 1-2eV region including the surface losses of Au and Ag nanostructures has however arrived in a fresh turn of the helix. The startling resurgence that has occurred may merit, as described in section 3 below, the description "plasmonic euphoria".

Instrumentation is now being developed to push electron spectroscopy down towards 10 meV and below where thermal excitation must be considered. Energy gains as well as energy losses will then be observed. Much higher energy loss and gain probabilities can be generated in the visible region in conjunction with pulsed laser illumination. Events linked to energy losses or gains far too small to be measured can also be potentially important in causing electron beam de-coherence and loss of image contrast. In section 4 below the application of dielectric excitation theory to these problems and the effect of thermal excitation on bremsstrahlung emission in this region is investigated.

2. Exact computation of atomic exit waves

2.1. Partial wave theory atomic scattering

An exact solution for the scattered wave emerging from an atom illuminated by an incident plane wave $exp[ikz] = exp[ikrcos\theta]$ is in principle provided in terms of spherical Bessel functions $j_l(kr)$ and $n_l(kr)$ by the long established partial wave theory of electron scattering [5].

$$\psi_{scatt}(r,\theta) = \sum_l\{\exp[2i\delta_l] - 1\}\left(l + \tfrac{1}{2}\right) i^l\{j_l(kr) + in_l(kr)\}\, P_l(cos\theta) \quad (1)$$

This equation gives directly the scattered wave contribution to the exit wave just outside the atom and rigorously describes its free space propagation between there and the far-field asymptotic region where the exact atomic scattering factor $f(\theta)$ emerges.

$$f(\theta) = (2ik)^{-1} \sum_l (2l+1)\{exp\,[2i\delta_l] - 1\}P_l(cos\theta) \quad (2)$$

Using the atomic potential $V(r) = 2mU(r)/\hbar^2$, procedures have been developed for accurate computation of the partial wave shifts δ_l (which are real) but good approximations can be obtained from WKB methods or from the Born integral

$$\delta_l^{Born} = \int_0^\infty kr^2\, j_l^2(kr) U(r) dr \quad (3)$$

When the phase shifts given by eqn. (3) are small and are substituted in eqn. (2) we get a connection to the Born atomic scattering factor generally used in electron microscopy

$$f_B(\theta) = k^{-1} \sum_l (2l+1)\delta_l^{Born}\, P_l(cos\theta) = \frac{1}{4\pi}\int \exp[i\mathbf{K}.\mathbf{r}']]U(\mathbf{r}')d\tau' \quad (4)$$

Here as usual $K = 2k\, sin(\theta/2) = 4\pi\, sin(\theta/2) / \lambda$. The exact scattering factor $f(\theta)$ is complex and therefore significantly different than $f_B(\theta)$ which is real although at 100 keV their magnitudes are quite similar [4].

2.2. Exit wave computations and checks on Bragg scattering at a Coulomb singularity

The partial wave scattering theory just outlined has recently been used to study the scattering of 100keV electrons in different atoms [4] though mostly with extremely small exit radii of only 0.7A. With a larger exit sphere of radius r_e = 1.5A, typical of ionic radii as well as half the near neighbour spacing in many solids, the number of phase shifts required at 100 keV can be $kr_e \approx 250$. For $r > r_e$ the potential was treated as constant as is done in the muffin tin method of solid state physics. The magnitude and phase of the exit wave was computed by adding to the incident wave the scattered wave from eqn. (2) and evaluating the result at points $\rho = \sqrt{(x^2+y^2)}$ in the exit plane (x,y, $z = r_e$). In this computation the Born phase shifts given by eqn. (3) were used since they agreed quite closely with values given by the WKB method. Writing the exit wave as $\psi_e = exp[ikz_e]Fexp[i\eta]$, figs 1 and 2 display for a 100 keV electron the modulus F and phase η as a function of distance ρ from the projected centre of an atom of Au. The focusing effect of the atom in raising the wave magnitude at small ρ as well as the increasing phase shift is evident. Phase grating theory might roughly explain the phase shift but would predict constant unit amplitude which evidently holds approximately only for $\rho > 0.3$A. The small oscillations in this region may be due to ringing at the edge of the muffin tin but could also arise through interference between trajectories converging from opposite sides of the atom.

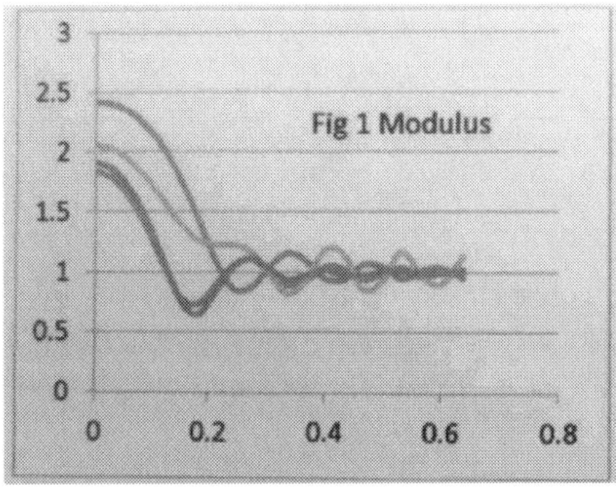

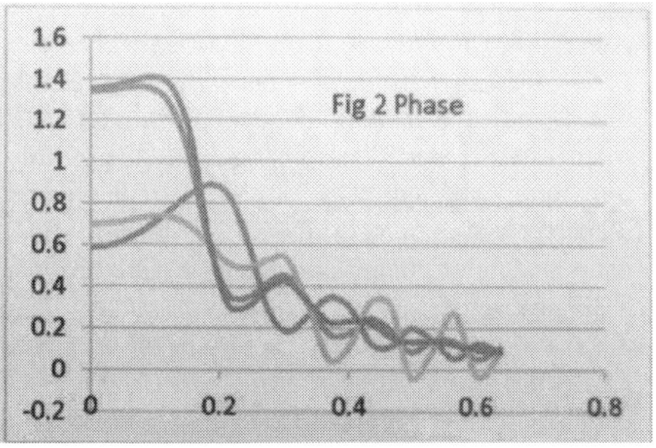

Figs 1 and 2 show the magnitude F and phase η of the exit wave in the exit plane $z_e = 1.5$A as a function of distance ρ in angstroms A from the projected centre of the atom. Light blue denotes the exact computation. The computations based on $f_B(\theta)$ with full phase shifts and reduced phase shifts modulo π are shown in red and green respectively. The dark blue curve is an exact computation for a smoothed but still spherically symmetrical potential defined by low order Fourier components.

For comparison with the exact result (light blue), three other approximations are shown in figs. 1 and 2 starting in red with the Born exit wave computed using the approximation $\{exp[2i\delta_l] -1\} \approx 2i\delta_l$ strictly valid only for small phase shifts. This is the exit wave that would be obtained by using the Born (single scattering) expression of eqn. (4) in the far field and correctly propagating back to the exit plane. It can be seen that this approximation considerably over-estimates the focusing effect. It agrees more closely at larger ρ, although the small oscillations there seem to be in antiphase with the exact theory. The next approximation shown in green is akin to the pseudo-potential method used in band theory where advantage is taken to re-express the phase shifts modulo π, which leaves the exact result of eqns. (1) and (2) unchanged but results in a reduced value for the far-field f_B which can be used to propagate back to the exit plane. In the case considered each of the first five phase shifts could be reduced by π. This approach evidently agrees more closely with the exact results and gets the small oscillations in the right phase. The small changes resulting in $f_B(\theta)$ at small θ values might cause complications however in some EM applications.

More interesting is the third approximation (dark blue) which follows the usual practice of considering only the low order-Fourier components of the scattering potential that might for instance lie in the zero-order Laue zone of a crystal. The phase shifts for such a smoothed, but still spherically symmetric, potential were constructed from eqn. (4) using the orthogonal property of the P_l functions and restricting $f_B(\theta)$ to the small angle region $cos\theta > 0.98$. That the result appears to agree remarkably well both in modulus and phase with the exact computation indicates that essential forward scattering features of the Coulomb singularity are not violated by smoothing the potential in this way. It should be emphasized however that this smoothing is not equivalent to making any of the usual projection approximations either for the potential or for the scattering. It could be interesting to check the effect of these approximations by comparing the above results with those of slice computations taking several steps within the atom to describe the wave spreading effect that is handled automatically in the partial wave theory.

3. Nanoparticle recoil and plasmonic euphoria

The possible effect on small supported clusters of uncontrolled energy and momentum transfers from the electron beam has been a long-standing worry in electron microscopy. In their pioneering work in STEM development, Crewe and colleagues made cine film recordings of the motion of small clusters as well as of single atoms of U and other heavy elements on carbon supports [6]. Their careful analysis suggested however that in most cases this was purely thermal motion and was not induced by the electron beam. Similarly cluster recoil effects in the fast electron wind that might be expected from transfers of linear momentum in Bragg reflection have never been reported though some indirect evidence for torque-induced reorientation was found [7].

On the other hand, kaleidoscopic structural changes were observed more than two decades ago in video studies of metallic nanoclusters under HREM imaging conditions [8]. Beam heating effects were ruled out [9] and these changes seem possibly driven by much rarer events than the momentum transfer in a Bragg reflection or the energy transfer through plasmon excitation and decay. Rather recently however with the much higher and more localized beam currents available in STEM, nanoparticle attraction and recoil induced by placing the probe close to, but just outside, the particle have been observed [9]. This recoil is hard, though perhaps just possible [10], to understand with the usual theory of dielectric excitation by an aloof beam where the transverse forces are usually rather small and nearly always attractive. Coulomb forces due to temporary charging of the nanoparticle could be greater but would also tend to be attractive.

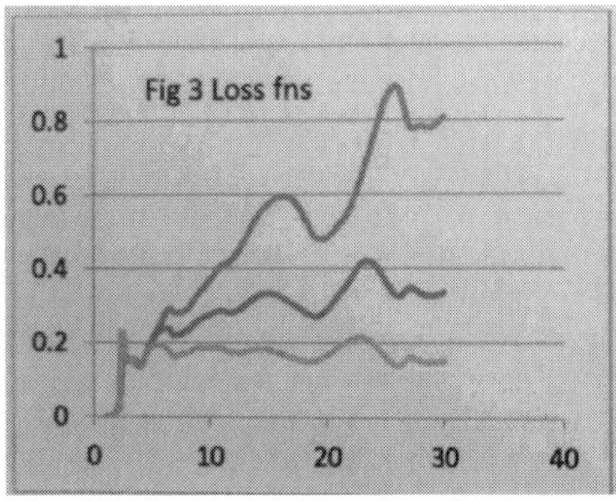

Fig. 3 Dielectric loss functions for Au up to 30eV with loss functions $Im\{-1/\varepsilon\}$ for the bulk (blue), $Im\{-1/(1+\varepsilon)\}$ for the plane in red and $Im\{-1/(2+\varepsilon)\}$ for the spherical dipole mode in green.

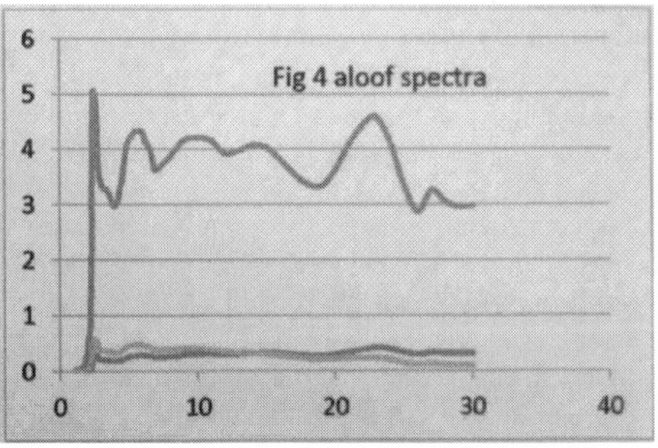

Fig. 4 Aloof beam loss probability P in units of 10^{-6} per eV near a 1.5nm diameter Au sphere for impact parameter b. Blue and red respectively denote the l=1 and 2 modes at b= 1.75nm; green shows the l=1 mode for b = 5.25nm

Figs 3 and 4 indicate however that it may be wise to cast aside plasmonic euphoria and think outside the narrow, sub 5eV spectral window. Substantial valence excitation takes place at energies higher than the sharp 2eV plasmon loss and could for instance result in the ejection of a secondary electron generating much more recoil momentum than is available from photon emission following surface plasmon radiative decay. All these cases call for a more careful treatment of momentum transfer involving the fast electron, the excited atom or cluster and any photon emitted or electron that is ejected [11]. It is also hard to understand the mechanism for the recently observed transfer of angular momentum to a nanocluster from a focused vortex beam. For non-spherical clusters elastic scattering has recently been suggested as an explanation [12] and could perhaps be consistent with the already mentioned detection of Bragg torque momentum transfer. The precise nature of the binding and frictional forces between the nanoparticle and the substrate in different cases is of course a potentially large unknown factor that has been addressed so far mainly by atomic force microscopy.

4. Thermal and laser excitation effects in inelastic scattering and de-coherence

4.1. Energy loss and gain events

Although a rigorous theory of thermal fluctuations in dielectric excitation has been developed [14], the effects are most easily understood, at least on an *ad hoc*, basis for harmonic oscillator excitations such as phonons or plasmons. These excitations have a ladder of states n with energy levels $(n+ ½)\hbar\omega$ and equilibrium thermal occupation probability $p(n) = exp[-n\hbar\omega/kT](1- exp[-\hbar\omega/kT])$. For radiative transitions with $|\Delta n| = 1$ from a state n, the relative probabilities are $n+1$ and n respectively (corresponding to energy loss and energy gain events for a fast electron that stimulates them). When the stimulation is from equilibrium thermal radiation, the difference between these two rates is balanced by the spontaneous decay process $\Delta n = -1$. Summing the two rates indicates that the effect of temperature brings a multiplying factor $F_T = (2n+1)p(n) = coth(\hbar\omega/kT)$ to the total event probability calculated at $T=0$. For $\hbar\omega << kT$, $F_T = 2kT/\hbar\omega$.

Gains as well as losses were first detected many years ago by pioneering transmission electron spectroscopy using thin films of LiF [15]. Creation and annihilation events were stimulated by the electron beam in a thermally excited population of optical phonons of energy $\hbar\omega \approx 0.05$ eV. The film thickness systematically influenced the precise energies of the loss peaks as well as the ratio R of the areas under the loss and gain peaks (which ranges between 5 and 15). Analysis of the results reveals quite good agreement between the observed values of R and the above theory for stimulated transitions of an oscillator which predicts $R = \{\Sigma(n+1)p(n)\}/\{\Sigma np(n)\} = 1 + 1/\{\Sigma np(n)\} = exp[\hbar\omega/kT]$. Possibly because the spontaneous decay process is too slow in comparison with the passage time of the fast electron, it does not seem to make a detectable contribution to the energy gain.

A radically different situation arises in recent experiments using photon-induced near-field electron microscopy (PINEM) [16, 17]. Here, near nanostructures subjected to high intensity pulsed laser excitation, loss spectra are collected in a synchronously pulsed aloof electron beam. Energy and momentum conservation excludes any interaction between the fast electron and the light waves in the far-field region but the observed cascade of energy losses and gains of equal magnitude is well explained through its interaction with the near-field region of the optical pulse emitted by the nanostructure [17-18]. In these experiments the electron is simply acting as a highly localized detector of the nanostructure's response to the pulsed laser field. Because of the enormous intensity of the laser pulse, the contribution of energy loss or gain processes stimulated by the electron itself is by comparison negligible as can be seen from the loss spectra collected when the photon and electron pulses are not in synchrony. It may nevertheless be profitable to try to bridge the gulf between these photon-stimulated and electron-stimulated response situations.

In most cases, the pulsed photon excitation is delivered off resonance to a single mode. If the photon energy $\hbar\omega$ were tuned to the resonance peak of the mode to drive it to a sufficiently high energy level n, the electron-stimulated loss and gain peaks linked to $n+1$ and n respectively might be detectable. Such an investigation would be much easier for a planar or cylindrical structure parallel to the electron beam since the electron can then interact only with evanescent waves at $q_z > \omega/v$ and these cannot be generated by the specimen in response to the laser pulse. In more general nanostructures, emitted waves stimulated by the electron are not cleanly separated from those generated in response to the laser pulse and the two effects must be added. It has even been conjectured that interference effects could be exploited between electron and laser stimulation [19]. To observe such effects however, rather than simple addition of intensities, a sufficiently precise phase correlation between the laser field and the electron wave function would have to be obtained and kept constant throughout the many pulses needed to acquire a spectrum.

4.2 De-coherence processes

Preservation of coherence is vital in the specimen and post specimen region in conventional electron microscopy and in the pre-specimen region in STEM. Various mechanisms of de-coherence were investigated recently [20] for aloof beam interactions where the dielectric excitation theory seemed to offer an explanation for experimental observations [21] in holography. These cases usually involve interference between electron paths separated by some distance d_y as they travel with the same impact parameter close to a dielectric surface. Evanescent waves arising from the dielectric response cause inelastic scattering and there will be immediate loss of coherence when the interaction affects the two paths with significantly different phases i.e. for tangential wave vectors $q_y > 1/d_y$ leading to a change in their relative phase of the paths after the scattering. Figure 5 schematically indicates the situation.

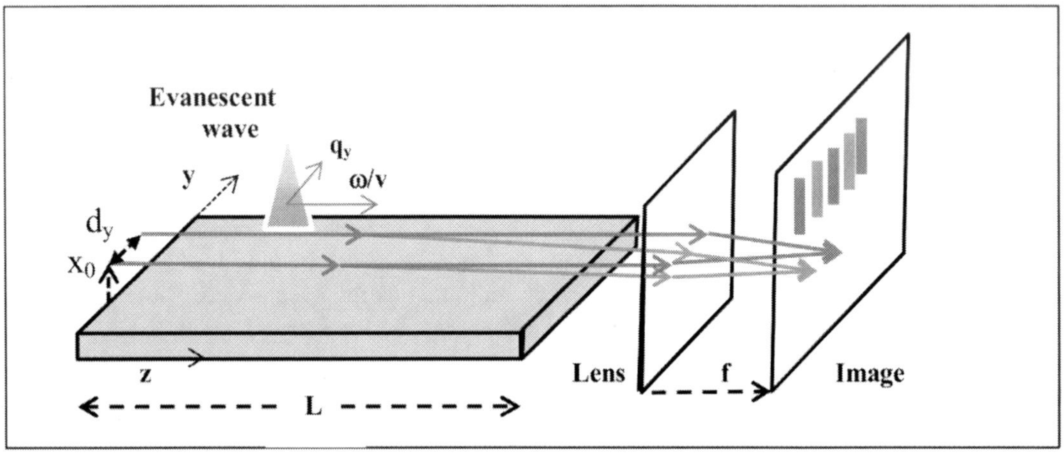

Fig. 5. Schematic drawing for de-coherence events suffered when an aloof beam travels across a conducting plate in two parallel interference paths (red) and is then focused by a lens to form an image with interference fringes. Interaction with the evanescent waves $(q_y, \omega/v)$ stimulated in the plate can cause complete destruction of any phase difference between paths separated by $d_y > 1/q_y$. At much smaller evanescent wave vectors q_y, the phase relation is preserved in the blue paths after scattering but the fringes they later produce will be shifted due to the angle of scattering q_y/k

Recent experiments have [22] have demonstrated de-coherence caused by a random walk accumulation of very small-angle scattering events. The effect was studied with 200keV electrons in an unusual 50cm long, 5mm diameter tube and was found to be proportional to temperature. So far no strong dependence on wall material has been reported but de-coherence increases with distance from the tube axis (and with proximity to the wall). A quantitative explanation was offered in terms of very low frequency magnetic fields emitted by current loops in Johnson noise fluctuations.

Some further insight into this interesting phenomenon should in principle be available by extending the aloof beam dielectric excitation theory already applied to de-coherence in holography. Given the differential expression for the single photon event probability $P(x_0, \omega, q_y)$ that this theory provides in terms of q_y momentum transfer, we can directly calculate the mean square deflection angle $<\theta_y^2>$.

$$\frac{d^2P(x_0,\omega,q_y)}{d\omega dq_y} = \frac{e^2L}{2\pi^2\epsilon_0\hbar v^2} Im\left\{-\frac{1}{\nu+\nu_0\varepsilon}+\frac{\beta^2}{\nu+\nu_0}\right\}\exp[-2\nu_0 x_0] \quad (5)$$

$$\langle\theta_y^2\rangle = \frac{1}{k^2}\langle q_y^2\rangle = \int_{-\infty}^{\infty} q_y^2\, dq_y \int_0^{\infty} \frac{d^2P(x_0,\omega,q_y)}{d\omega dq_y} d\omega \quad (6)$$

In eqn. (5), where $\beta = v/c$, $\nu_0^2 = [q_y^2 + (\omega/v)^2 - (\omega/c)^2]$ and $\nu^2 = [q_y^2 + (\omega/v)^2 - \varepsilon(\omega)(\omega/c)^2]$, the relativistic form [22] has been used since this includes magnetic forces. For the observed 1mm impact parameters x_0, the exponential factor restricts attention to the low frequency range $\omega/2\pi < 10^{10}$ Hz. The dielectric response can then still be described by the Drude expression $\varepsilon(\omega) = 1 - \omega_p^2/\omega(\omega+i\gamma)$ taking the damping constant $\gamma = \varepsilon_0\, \omega_p^2/\sigma(T)$ where $\sigma(T)$ is the DC conductivity at temperature T. Even for high conductivity metals, $\gamma > \omega$ at these low frequencies and so $\varepsilon(\omega) \approx 1 - \omega_p^2/\omega^2 + i\sigma(T)/e_0\omega$. The non-relativistic approximation to eqn. (5) then yields an event probability proportional to $1/\sigma(T)$ which above the Debye temperature is proportional to T due to scattering of carriers by the increased population of thermal phonons. Some of the thermal enhancement effects discussed in 4.1 may thus be included even if losses and gains are not properly distinguished. Initial investigations indicate however that the second (purely relativistic) term in eqn. (5), with a less simple dependence on conductivity, may actually make a bigger contribution. On the basis of work in progress [14] it seems possible that the relativistic theory of aloof beam interactions may be able to account for the Johnson noise observations. It should be noted however that scattering in the x direction normal to the plate is more complex since we can now expect a non-zero mean deflection $<\theta_x>$ together with a fluctuation $<(\theta_x - <\theta_x>)^2> = <\theta_x^2> - <\theta_x>^2$. For electrons traversing a tube and depending on their distance from the wall, there will also be a mean deflection in the radial direction acting like a weak diverging lens with high aberration.

4.3 Touching the warm void

The thermal multiplication factor F_T can cause significant increases in the probability of very low energy bremsstrahlung emission by an electron accelerated in some external potential. At $T = 0$ this process can be viewed as a spontaneous decay between the no longer free quantum states of the electron in that potential. For instance a fast electron propagating through a crystal and described by Bloch waves can lose energy through bremsstrahlung in a process that can be conveniently described as an inelastic interband transition coupled to the emission of a photon [1]. Such coupling, which would be forbidden in free space by energy-momentum conservation, is made possible by the non-free electron form of the electron's $E(\mathbf{k})$ relation in the crystal. For Bloch waves involving only reflections from the zero order Laue zone, we can equally well consider the bremsstrahlung as channeling radiation generated by the undulating motion of the electron between the atomic planes. Similarly, in a uniform magnetic field B, the electron energy states are harmonic oscillator functions with energy levels $E_n = (n+\frac{1}{2})\hbar\omega_L$ and yield spontaneous decay emission at the Larmor frequency $\omega_L = eB/m$ matching the period for the circular orbit $T_l = 2\pi/\omega_L$.

From the classical theory of bremsstrahlung [24] the radiated energy loss rate dE_{rad}/dt is related to the electron's acceleration $a(t)$.

$$E_{rad} = \frac{e^2}{6\pi\epsilon_0 c^3}\int_{-\infty}^{+\infty}[a(t)]^2\,dt = \frac{e^2}{12\pi^2\epsilon_0 c^3}\int_{-\infty}^{+\infty}[a(\omega)]^2 d\omega \qquad (7)$$

Here through Parseval's theorem, the energy loss can also be expressed as an integral over $[a(\omega)]^2$ where $a(\omega)$ is the Fourier transform of $a(t)$. This step provides the spectrum of energy losses. Assuming that these losses are entirely due to single photon emission the loss probability is

$$P_{rad} = \frac{e^2}{6\pi^2\epsilon_0\hbar c^3}\int_0^{+\infty}\frac{1}{\omega}[a(\omega)]^2 d\omega \qquad (8)$$

These equations were used [20] to compute event probabilities P_{rad} for simple trajectories in electron optics including a single 2π, 10 cm radius turn of a 200 keV electron in a mandolin or similar magnetic system. With $a(t) = v^2/r$ we find $P_{rad} = 0.03$ for the emission of a photon of energy $\hbar\omega_L \approx 10^{-6}$eV. With the thermal magnification factor $F_T = 2kT/\hbar\omega$ (arising now from the occupation probabilities $p(n)$ of the photon far-field modes rather than from the Larmor oscillator modes) the expected number of events at room temperature is increased to about 1500. Since these events will be divided equally between energy losses and gains, we can estimate 4 x 10^{-4} eV for the rms energy broadening. Such broadening is too small to be detected but the events might conceivably result in some de-coherence depending on where the bending system is placed in the column.

Unlike the example just considered, the episode of acceleration in most trajectories results in a change in the magnitude or direction of the velocity and eqn. (8) then exhibits a logarithmic divergence at low ω known as the infrared catastrophe. Inclusion of the thermal magnification factor raises this low frequency divergence in P_{rad} to ω^{-1} and creates a $ln(\omega)$ divergence in E_{rad} . In practice it should be possible to avoid any problems by introducing an appropriate low energy or momentum transfer cut off.

Acknowledgements
It is a pleasure to thank both Javier Garcia de Abajo and Alberto Rivacoba for many enlightening discussions on energy loss spectroscopy and electron-photon interactions.

References

[1] Hirsch PB, Howie A, Nicholson RB, Pashley DW and Whelan MJ 1965 *Electron Microscopy of Thin Crystals* Butterworth – London.
[2] Spence JCH 2003 *High Resolution Electron Microscopy* OUP – Oxford.
[3] Pennycook SJ and Nellist PD eds 2011 *Scanning Transmission Electron Microscopy* Springer – New York.
[4] Howie A 2013 http://dx.doi.org/10.1016/j.ultramic.2013.05.002.
[5] Schiff LI 1955 *Quantum Mechanics* McGraw Hill-New York, London.
[6] Isaacson M, Kopf D, Utlaut M, Parker NW and Crewe AV 1977 *Proc. Nat. Acad. Sci.* **74** 1802.
[7] Marks LD and Zhang JP 1992 *Ultramicroscopy* **41** 419.
[8] Smith DJ, Petford-Long AK, Wallenberg LR and Bovin JO 1986 *Science* **233** 872.
Iijima S and Isikashi T 1986 *Phys. Rev. Lett.* **56** 616.
[9] Doraiswamy N and Marks LD 1996 *Surf. Sci.* **348** L67.
[10] Batson PE, Reyes-Coronado A, Barrera RG, Rivacoba A and Echenique PM 2011 *Nanolett.* **11** 3388
[11] Reyes-Coronado A, Barrera RG, Batson PE, Echenique PM, Rivacoba A and Aizpurua J 2010 *Phys. Rev.* **B82** 235429
Batson PE, Reyes-Coronado A, Barrera RG, Rivacoba A, Echenique PM and Aizpurua J 2012 *Ultramicroscopy* **123** 50.
[12] Lloyd S, Babiker M and Yuan J 2012 *Phys. Rev. Lett.* **108** 074802
[13] Verbeeck J, Tian H and Van Tendeloo G 2013 *Adv. Mater.* **25** 1114
[14] Garcia de Abajo FJ (to be published).
[15] Boersch H, Geiger J and Stickel W 1966 *Phys. Rev. Lett.* **17** 379.
[16] Barwick B, Flannigan DJ and Zewail AH 2009 *Nature* **409** 902
[17] Park ST, Lin M and Zewail AH 2010 *New J. Phys.* **12** 123028.
[18] Garcia de Abajo FJ and Kociak M 2008 *New J. Phys.* **10** 073035.
Howie A 2011 *Eur. J. Appl. Phys.* **54** 33502.
[19] Talebi N, Sigle W, Vogelgesang R and van Aken P 2013 *New J. Phys.* **15** 053013.
[20] Howie A 2011 *Ultramicroscopy* **111** 761.
[21] Sonnentag P and Hasselbach F 2007 *Phys. Rev. Lett.* **98** 200402.
Popatov PL, Lichte H, Verbeeck J and van Dyck D 2006 *Ultramicroscopy* **106** 1012.
[22] Uhlemann S, Muller H, Hartel P, Zach J and Haider M 2013 *Phys. Rev. Lett.* **111** 046101.
[23] Garcia-Molina R, Gras-Marti A, Howie A and Ritchie RH 1985 *J.Phys.C* **18** 5335.
[24] Jackson JD 1975 *Classical Electrodynamics(2ndedition)* J. Wiley New York.

In-situ environmental (scanning) transmission electron microscopy of catalysts at the atomic level

P L Gai [1,2,3] **and E D Boyes**[1,2,4]

[1]The York JEOL Nanocentre, [2]Department of Physics, [3]Department of Chemistry, [4]Department of Electronics, University of York, York, YO10 5DD, UK

E-mail address: pratibha.gai@york.ac.uk

Abstract. Observing reacting single atoms on the solid catalyst surfaces under controlled reaction conditions is a key goal in understanding and controlling heterogeneous catalytic reactions. *In-situ* real time aberration corrected environmental (scanning) transmission electron microscopy (E(S)TEM permit the direct imaging of dynamic surface and sub-surface structures of reacting catalysts. In this paper in-situ AC ETEM and AC ESTEM studies under controlled reaction environments of oxide catalysts and supported metal nanocatalysts important in chemical industry are presented. They provide the direct evidence of dynamic processes at the oxide catalyst surface at the atomic scale and single atom dynamics in catalytic reactions. The ESTEM studies of single atom dynamics in controlled reaction environments show that nanoparticles act as reservoirs of ad-atoms. The results have important implications in catalysis and nanoparticle studies.

1. Introduction

Chemical reactions catalysed by solid surfaces in heterogeneous catalysis take place at the atomic level and play a major role in the development of industrial chemicals, energy sources and environmental emission controls. Dynamic active atomic structures at the surface of the solid catalyst under reaction conditions underpin the performance of the catalysts. Insights into the dynamic evolution of structural changes in heterogeneous catalysts under controlled gas and temperature reaction conditions at the atomic level in real time are therefore crucial in the development of efficient catalysts. The direct observation of the surface structure, chemical evolution and the live reaction mode of operation of catalysts under controlled reaction conditions of gas and temperature at the atomic level is therefore a powerful scientific method but extremely challenging.

Previously Gai and Boyes reported the first development of atomic resolution-Environmental transmission electron microscope (atomic resolution-ETEM) which enabled atomic resolution studies of gas-solid reactions at operating temperatures under controlled conditions and for *in-situ* nanosynthesis [1,2]. Highlights of this development included their novel and now widely used ETEM design with objective lens polepieces incorporating radial holes for differential pumping with the regular sample chamber as the controlled reaction environmental cell or reactor (Figure 1) [2]. The direct real time imaging of dynamic interactions between hydrocarbons and oxide catalysts and how the gas molecules move and rearrange the catalyst surface in the selective hydrocarbon oxidation has resulted in the discovery of an atomic scale reaction site mechanism resulting from a simple atomic movement, called the "glide shear reaction mechanism" shown in Figure 2 [1]. During the selective oxidation the glide defects are formed to accommodate the shape misfit between the reduced anion-deficient catalyst surface and the subsurface without collapsing the catalyst structure which is critical to the sustained operation. The authors' ETEM development paved the way to atomic resolution studies of gas (or liquid)-solid reactions at operating temperatures worldwide [3].

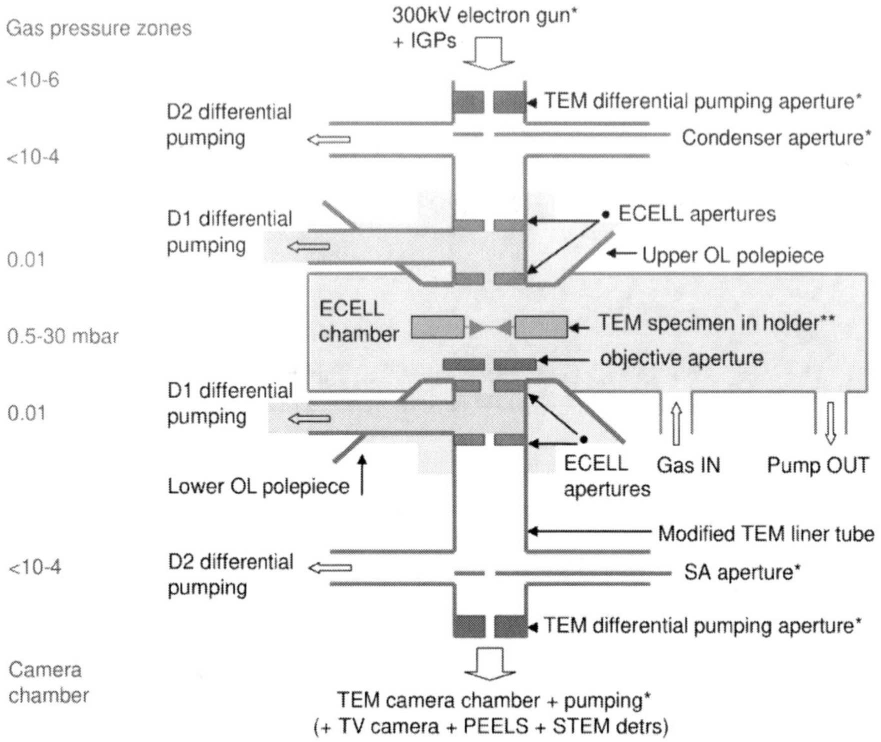

Figure 1: Schematic (not to scale) of key elements of the now widely used atomic resolution ETEM design developed by the authors in the 1990s.

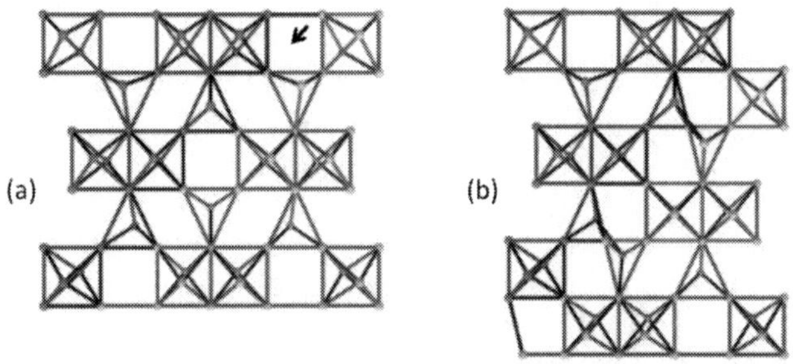

Figure 2 Discovery of glide shear mechanism during selective oxidation of hydrocarbons [1] : (a) fresh catalyst (b) glide shear defect mechanism by simple atomic motion (along the arrow indicated in (a)), without collapsing the catalyst. The glide mechanism is also found in efficient biofuel sythesis [7].

The subsequent development of double aberration corrected ETEM (AC ETEM) utilising a large gap objective lens polepiece has been reported earlier [4,5,6]. The critical role of atomic level defect structures formed by the glide shear mechanism in biofuel synthesis has been demonstrated recently using double aberration corrected-ETEM ((AC-ETEM) [7]. The evolution of atomic scale glide-shear defects occurs at the nanocatalyst surface during activation, creating active sites The direct linear correlation between the surface glide defect density and the improved catalytic performance has offered a new route to energy efficient catalysts for biofuel synthesis [7]. Hydrogen reduction of cobalt oxide catalysts using AC-ETEM has shown advancing atomic scale interfaces of reduced oxides [8].

Recently Boyes et al have reported the development of the ESTEM method for single atom imaging in catalyst reaction studies [9]. With the new AC-ESTEM capability it is now possible to analyse real space STEM Z-(atomic number), especially single atom contrast high angle annular dark field imaging (HAADF) diffraction and spectroscopy, under transient conditions [9].

In this paper we present our original E(S)TEM research into working oxide catalysts and supported nanocatalysts. Supported cobalt based nanoparticles are of interest in a number of hydrogenation reactions including Fischer-Tropsch (F-T) process for synthetic fuel technologies and in environmental control [10,11]. The F-T process contains a series of chemical reactions involving the hydrogenation of CO over solid catalysts to produce various hydrocarbons. More recently cobalt based oxides have been employed for the reactions. Supported platinum nanocatalysts are important in the chemical industry, including in fuel cells, where reactions with hydrogen and oxygen are important, and as part of vehicle exhaust emission control measures [12,13].

2. Experimental Procedures

Co_3O_4 powder was obtained from Aldrich. TEM and diffraction analysis of the powder showed primarily Co_3O_4 [8]. Specimens for conventional EM analysis were made by dispersing Co_3O_4 powder in ethanol which was then treated in an ultrasonic bath for 2 minutes. The suspension was deposited onto holey-C grids. For *in-situ* ETEM studies, bare (unfilmed) Cu grids were used because the C films are known to buckle and react with oxide samples under H_2 at elevated temperatures. The Cu grids were plasma cleaned at 75% power for 5 minutes using a Diener Femto unit to remove the inevitable contamination. After depositing the Co_3O_4 powder, the specimens were further plasma cleaned at 10% power for 30 seconds. Crystals protruding over the edge of the grids were employed in the *in-situ* studies. Model Pt nanocatalysts were prepared by sputtering Pt on carbon supports.

For dynamic real-time *in-situ* studies at the atomic level under controlled gas and temperature conditions, we used a double spherical aberration corrected Environmental (Scanning) TEM (AC-E(S) TEM) developed in-house by Boyes and Gai [4,5,9] in the Nanocentre at the University of York. The AC-E(S) TEM with sub-Angstrom (<100 pm) resolution was developed by modifying a double aberration corrected JEOL 2200FS(S)TEM operating at 200 kV and fitted with a larger gap (HRP) objective lens polepiece and gas tolerant pumping system [4,5,9]. The E(S)TEM development is uniquely equipped to carry out *in-situ* studies of the structure and stability of single atom catalysts and small nanoparticles [9]. The dynamic nature of *in-situ* reactions means that it is rarely possible to obtain a complete through focal series of images. By including in the CTF of each image frame a complete range of spatial frequencies, and by minimising delocalisation effects by operating at close to zero defocus, AC-E(S)TEM allows us to clearly observe single atoms, measure more accurately inter-atomic spacings, analyse crystal structures, and especially external surfaces, from a single image. The wider gap pole piece enables hot stage capabilities and an increased sample tilt range [4,5,9]. Further, AC-E(S)TEM under low electron dose imaging and 'blank' calibration experiments (without the beam and the beam switched on only to record the final reaction end point and compared to *in-situ* data) are used to minimise effects of the electron beam on the sample [1-6, 14,15].

3. Results and Discussion

3.1 Atomic level Reduction of cobalt oxides

The reduction of model Co_3O_4 catalysts has been carried out in H_2 from room temperature (RT) to ~ 450°C from the same area of the sample to directly observe the dynamic phase evolution in the reduction process. Figure 3 shows dynamic AC-ETEM images of the reaction sequence of a Co_3O_4 catalyst crystal at rising temperatures during exposure to H_2 *in-situ*. Figure 3(a) shows Co_3O_4 at RT which is readily identified from other phases because of the presence of a large 0.47 nm {111} atomic plane spacing. The evolution of CoO was observed at the surface of the catalyst from ~200°C and persisted up to higher temperatures of ~ 400 °C. Figure 3(b) shows the presence of CoO at 400 °C in hydrogen with (002) spacings of 0.21 nm. These direct *in-situ* observations have provided new insights into the reduction mechanism from Co_3O_4 to the intermediate CoO. These include the formation of an advancing atomic scale interface between the Co_3O_4 and CoO regions. There are

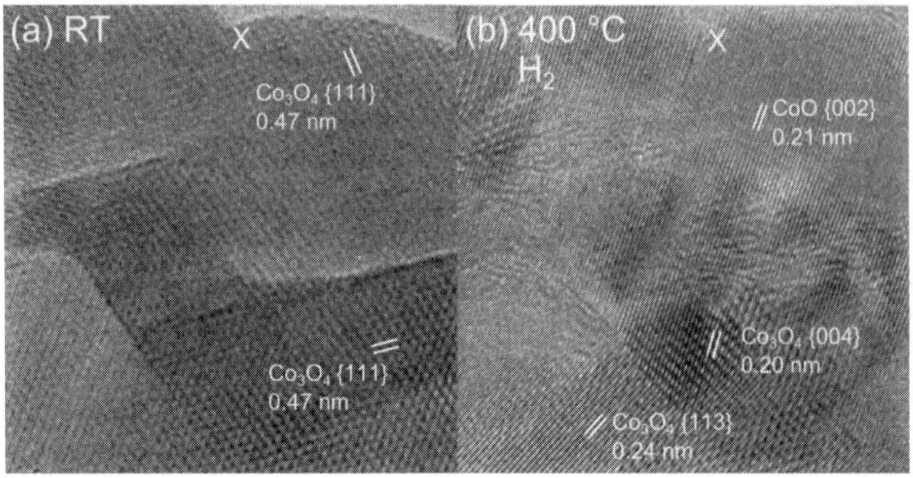

Figure 3. Dynamic AC-ETEM image sequence of the same area (near X) of a Co_3O_4 catalyst in hydrogen gas: (a) Co_3O_4 at room temperature (RT). 0.47{111} lattice spacings are shown. (b) The formation of CoO recorded at ~ 400°C in hydrogen gas. CoO {002} spacings of 0.21nm are shown.

reports in the literature that active sites in the F-T process at low temperatures of 225-250 °C are Co metal. Our *in-situ* observation of CoO at operating temperatures of 250-350 °C has important implications in understanding the catalytic activity of Co-based oxide catalysts in hydrogenation reactions. In our in-situ studies Co is observed only at higher temperatures and higher gas pressures.

3.2 Single atom dynamics

In-situ ESTEM studies were carried out on platinum catalysts supported on carbon in reducing and oxidising atmospheres. Figure 4 (a) displays single atoms and clusters of a Pt catalyst clearly using ESTEM, with the intensity profile of a single atom in (b). The FWHM for the single atom is ~0.11 nm.

A sequence of a small particle (P) undergoing structural rearrangement is shown in Figure 5.

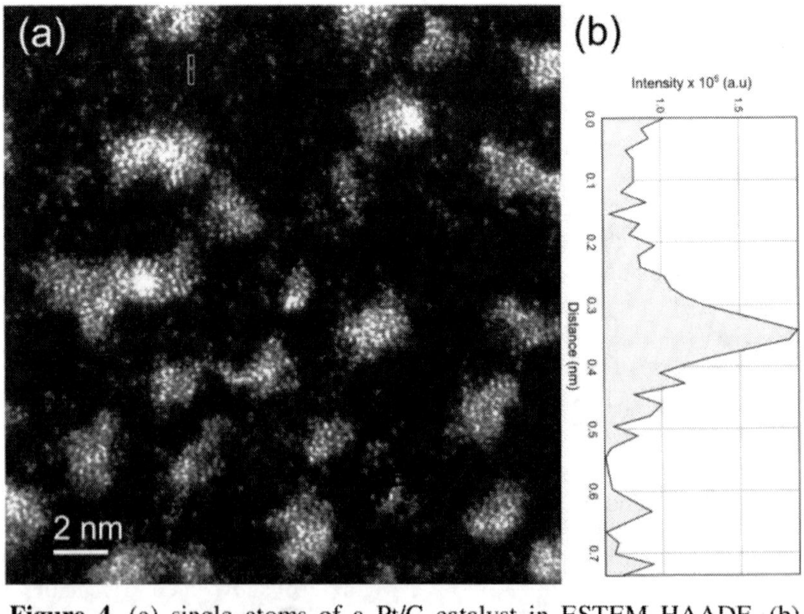

Figure 4. (a) single atoms of a Pt/C catalyst in ESTEM HAADF. (b) Intensity profile of a single atom indicated in (a).

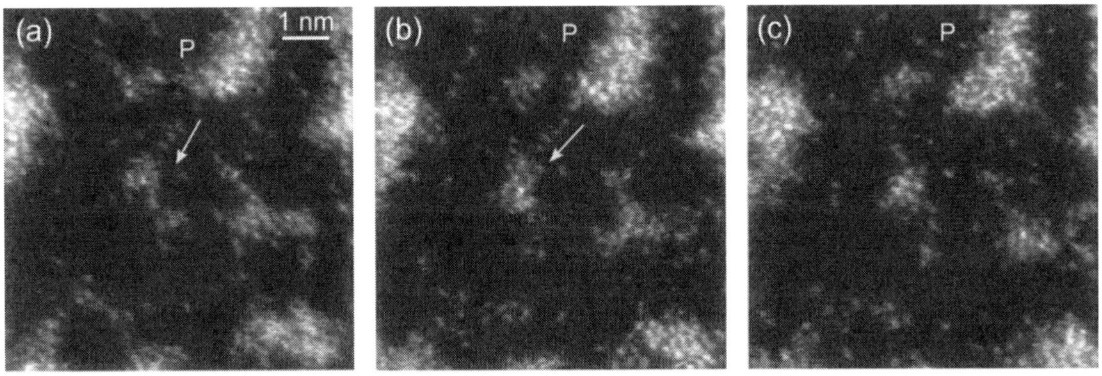

Figure 5. Single atom dynamics: migration of single atoms arrowed in (a) and (b) (e.g. from a cluster/particle P) leading to increased faceting of the particle and the formation of clusters (c).

Figure 5 (b) forms the intermediate stage, indicating the migration of single atoms from the particles (appearing as white dots (arrowed) on the thin amorphous carbon support film. The driving force for the removal of atoms away from the particle is proposed to be minimisation of the particle surface energy, where surface ad-atoms and steps are removed in order to create lower energy facets. The onset of faceting can be seen in (c) and the formation of small clusters .

The reduction-oxidation reactions were carried out at the single atom level and the observations are illustrated in Figure 6. Figure 6 (a) shows single atoms and particles; Figure 6 (b) shows the sample in hydrogen gas at 500 °C. The gas inlet pressure was ~ 200Pa. Fewer single atoms and some faceting of clusters are observed. The hydrogen was then outgassed and oxygen was inserted at ~200^{0}C with the inlet gas pressure of ~ 190 Pa.

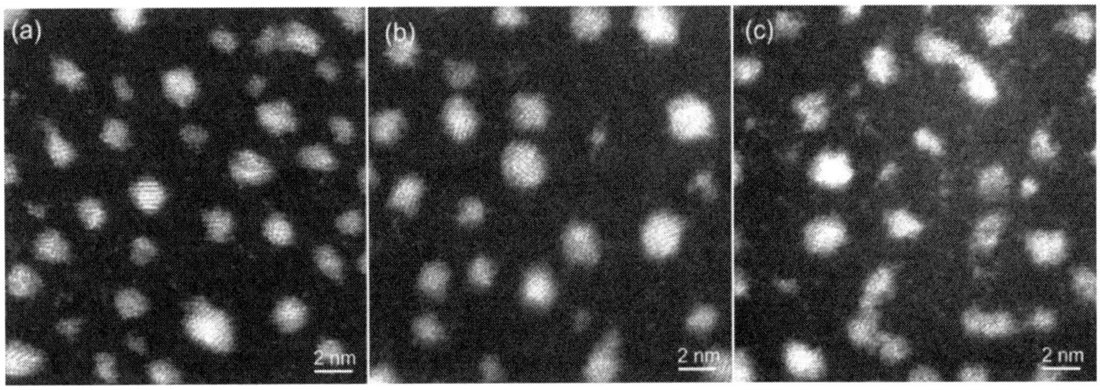

Figure 6. Dynamic ESTEM showing single atom dynamics in Pt/C nanocatalyst in controlled reduction-oxidation reaction environments: (a) RT; (b) Hydrogen gas at ~500 °C showing faceting and fewer single atoms; (c) Oxygen gas at 200°C showing relatively more single atoms indicating that interaction with oxygen inhibits their mobility.

In-situ hydrogen reduction in Figure 6(b) shows the presence of fewer single atoms on the support than in (a), clusters and increased faceting of the nanoparticles. This indicates that the nanoparticles act as a source of potential ad-atoms and clusters. This has important implications in catalysis and understanding the role nanoparticles generally. Recent literature reports based on ex-situ studies indicate excellent catalytic performance in the presence of low coordinated atoms and clusters [16]. Minimisation of surface energy leads to the removal of low coordinated atoms and stable surface facets. This decrease in the low coordinated sites may explain the often observed initial drop in the activity of nanocatalysts [17,18]. In oxygen relatively more single atoms are observed in Figure 6(c) (than in (b)), indicating that the single atom reaction with oxygen prevents their migration.

Conclusions

Using our novel ESTEM development we have demonstrated single atom dynamics in industrially important supported platinum nanocatalysts in reduction and oxidation environments at operating temperatures and atomic level transformations in cobalt oxide catalysts in hydrogen reduction. The results have important implications in catalysis and studies of nanoparticles.

Acknowledgements

We thank the EPSRC (UK) for research grant EP/J018058/1 and L Lari, M R Ward and I Wright for support and collaboration.

References

[1] P L Gai et al. 1995 *Science* **267** 661.
[2] E D Boyes and P L Gai 1997 *Ultrammicroscopy* **67** 219.
[3] P L Gai, E D Boyes et al. 2007 *MRS Bull* **32** 1044.
[4] P L Gai and E D Boyes 2009 *Microsc Res Tech* **72** 152.
[5] P L Gai and E D Boyes 2010 *J Phys* Conf Ser **241**10255.
[6] P L Gai and E D Boyes *Handbook on Nanoscopy* (Springer; eds: D Van Dyke et al.), 2012.
[7] P L Gai and E D Boyes et al. 2009 *Catal Lett* **1** 112
[8] M R Ward, E D Boyes and P L Gai 2013 *ChemCatChem* doi: 10.1002/cctc.201300047.
[9] E D Boyes, M R Ward, L Lari and P L Gai 2013 *Ann.Phys.* (Berlin) **525** 423.
[10] C K Roferdepoorter 1981 *Chemical Reviews* **81** 447.
[11] E Iglesia 1997 *Applied Catalysis A-General* **161** 59.
[12] T Johnson 2008 *Platinum Metals Rev* **52** 23.
[13] M R Ward, T Hyde, E D Boyes and P L Gai 2012 *ChemCatChem* **4** 1622.
[14] P L Gai and E D Boyes 2003 *EM in Het Catalysis*, IOP Publ (London and Philadelphia).
[15] M J Walsh, K Yoshida, A Kuwabara, P L Gai and E D Boyes 2012 *Nanoletters* **12** 2027.
[16] B. Hvolbaek, J K Norskov et al. 2007 *Nano Today* **2** 14.
[17] M J Walsh, K Yoshida, A Kuwabara, P L Gai and E D Boyes 2012 *ChemCatChem* **4** 1638.
[18] N Shiju, K Yoshida, E D Boyes, D R Brown and P L Gai 2011 *Catal Sc and Tech* **1** 413.

Electron Cryomicroscopy: From Molecules to Cells

W. Baumeister
Max-Planck-Institute of Biochemistry, Am Klopferspitz 18, D-82152 Martinsried Germany

E-mail: baumeist@biochem.mpg.de

Abstract. Today's biomolecular electron microscopy uses essentially three different imaging modalities: (i) electron crystallography, (ii) single particle analysis and (iii) electron tomography. Ideally, these imaging modalities are applied to frozen-hydrated samples to ensure an optimum preservation of the structures under scrutiny. Electron crystallography requires the existence of two-dimensional crystals. In principle, electron crystallography is a high-resolution technique and it has indeed been demonstrated in a number of cases that near-atomic resolution can be attained. Single-particle analysis is particularly suited for structural studies of large macromolecular complexes. The amount of material needed is minute and some degree of heterogeneity is tolerable since image classification can be used for further 'purification in silico'. In principle, single particle analysis can attain high-resolution but, in practice, this often remains an elusive goal. However, since medium resolution structures can be obtained relatively easily, it often provides an excellent basis for hybrid approaches in which high-resolution structures of components are integrated into the medium resolution structures of the holocomplexes. Electron tomography can be applied to non-repetitive structures. Most supramolecuar structures inside cells fall into this category. In order to obtain three-dimensional structures of objects with unique topologies it is necessary to obtain different views by physical tilting. The challenge is to obtain large numbers of projection images covering as wide a tilt range as possible and, at the same time, to minimize the cumulative electron dose. Cryoelectron tomography provides medium resolution three-dimensional images of a wide range of biological structures from isolated supramolecular assemblies to organelles and cells. It allows the visualization of molecular machines in their functional environment (in situ) and the mapping of entire molecular landscapes.

Today, there are three categories of biomolecular electron microscopy: (i) electron crystallography, (ii) single particle analysis and (iii) electron tomography. Ideally, all three imaging modalities are applied to frozen-hydrated samples ensuring that they are studied in the most life-like state that is physically possible to achieve. Vitrified aqueous samples are very radiation sensitive and consequently, cryo EM images must be recorded at minimal electron beam exposures limiting their signal-to-noise ratio. Therefore, the high resolution information of images of unstained and vitrified samples must be retrieved by averaging-based noise reduction which requires the presence of repetitive structure. Averaging can obviously not be applied to pleomorphic structures such as organelles and cells (1,2).

Electron crystallography requires the existence of two-dimensional crystals, natural or synthetic, and averaging is straightforward given the periodic arrangement of the molecules under scrutiny. In principle, electron crystallography is a high-resolution technique as demonstrated successfully with a number of structures, in particular of membrane proteins. Often however, the same structures can be studied by x-ray crystallography which tends to be faster and can attain atomic resolution more easily.

In contrast, EM single particle analysis, arguably a misnomer, since it involves the averaging over large numbers of identical particles, has become one of the pillars of modern structural biology. The amount of material needed is minute and some degree of heterogeneity, compositional or conformational, is tolerable since image classification can be used for further purification *in silico*. It is particularly successful in structural studies of very large macromolecular complexes where the traditional methods often fail. In principle, single particle analysis can attain near-atomic resolution but in practice, this often remains an elusive goal.

This may change, however, with the advent of new technology, in particular detectors with improved performance. But even intermediate resolution (subnanometer) structures of very large complexes can provide an excellent basis for hybrid or integrative approaches in which high resolution structures of components and/or orthogonal data, such as distance restraints, are used to generate atomic models.

Two examples will be used to demonstrate the power of EM single-particle analysis in conjunction with data from other sources. The structure of tripeptidylpeptidase II (TPPII), a giant proteolytic complex of 6 MDa has been determined by integrating the crystal structure of its basic element, a dimer of 300 kDa, into the EM density of the holocomplex (3,4). The 26S proteasome, a key player in cellular protein quality control, is smaller (2,5 MDa), but much more complex in its subunit composition; it is built of two copies each of 34 different subunits. It is a labile complex which undergoes rather large conformational changes during its functional cycle. An initial subnanometer structure (5) provided the basis for determining its subunit architecture (6) and, more recently, based upon a massive data set of about 2.5 million individual particles collected by automated image acquisition, an atomic model was obtained as well as first insights into the conformational space (6,7) (Fig. 1).

Electron cryotomography can be used to study the three-dimensional organisation of non-repetitive objects (8,9). Most cellular structures fall into this category. In order to obtain three-dimensional reconstructions of objects with unique topologies it is necessary to acquire data sets with different angular orientations of the sample by physical tilting. The challenge is to obtain large numbers of projections covering as wide a tilt range as possible and, at the same time, to minimize the cumulative electron dose. This is achieved by means of elaborate automated acquisition procedures. Electron cryotomography can provide medium resolution three-dimensional images of a wide range of biological structures from isolated supramolecular assemblies to organelles and whole cells. It allows the visualization of molecular machines in their unperturbed functional environments (*in situ* structural biology) and ultimately the mapping of entire molecular landscapes (visual proteomics) (10).

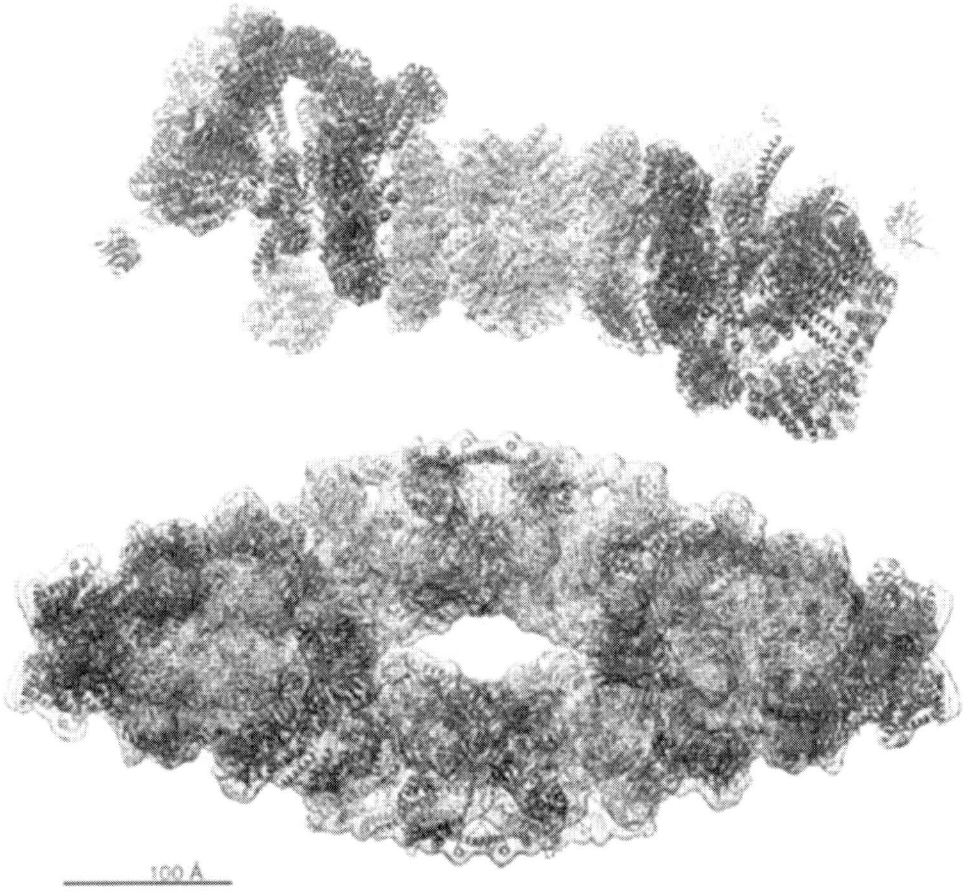

Fig.1: Structure of two large proteolytic complexes, the 26S proteasome (above) and tripeptidylprotease (below) determined by integrative approaches

Until recently the use of electron cryotomography was restricted to relatively thin samples such as prokaryotic cells or the margins of eukaryotic cells. This has changed with the advent of focused-ion beam (FIB) micromachining and developments allowing the application of this technology to samples embedded in vitreous ice. This allows the cutting of 'windows' providing views also of the interior of thicker samples such as eukaryotic cells. By combining the FIB with correlative fluorescence microscopy, it is possible now to navigate large cellular landscapes and to select and target specific areas of interest (11,12).

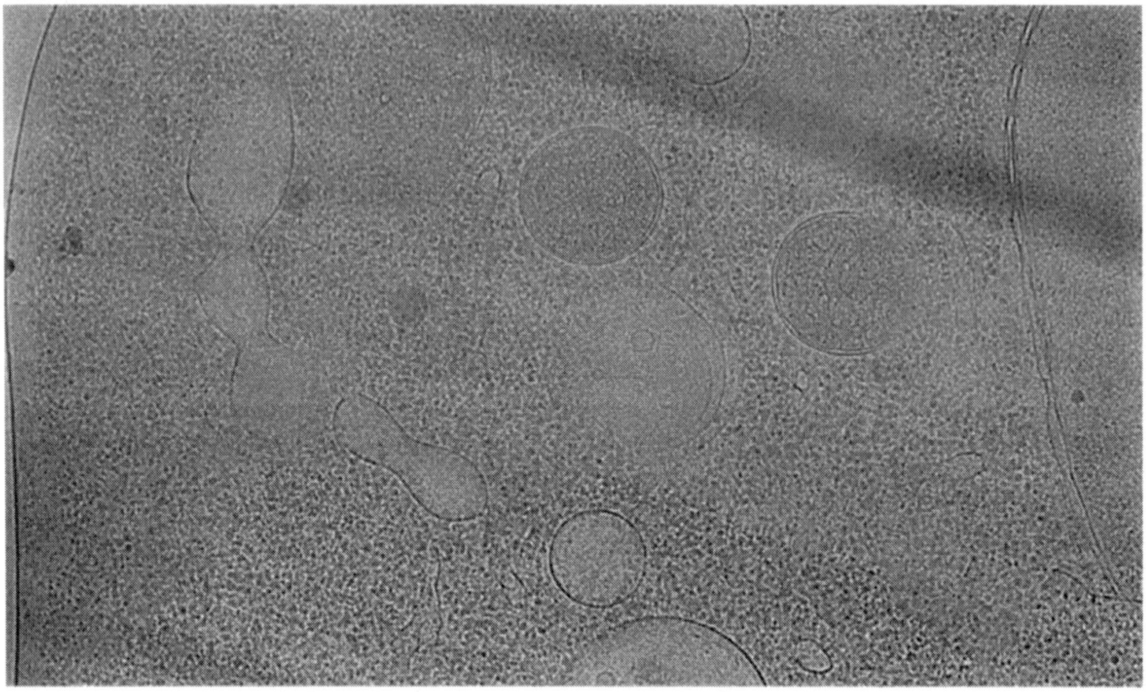

Fig. 2: TEM image of a ,window' cut into an ice-embedded Dictyostelium cell by focused ion-beam micromachining

Given the poor signal-to-noise ratio of the tomograms, it can be challenging to interpret them and take advantage of their rich information content. Image denoising can improve the signal-to-noise ratio by reducing the noise while preserving the features of interest. Segmentation separates the structures of interest from the background and allows their three-dimensional visualization and quantitative analysis. Larger molecular structures can be identified in tomograms by pattern recognition methods using a template structure and once their location and orientation is determined, identical structures can be extracted computationally and averaged. Therefore electron tomography has unique potential to bridge the divide between molecular and cellular structural studies, perhaps the most exciting frontier in structural biology (13).

References

1) *Fitting Kourkoutis, L., J.M. Plitzko and W. Baumeister:* Electron microscopy of biological materials at the nanometer scale. Ann. Rev. Mat. Sci. **42** (2012).

2) *Leis, A., B. Rockel, L. Andrees and W. Baumeister:* Visualizing cells at the nanoscale. Trends Biochem. Sci. **34**,60-70 (2009).

3) *Chuang, C.K., B. Rockel, G. Seyit, P. Walian, A.-M. Schönegge, J. Peters, P. Zwart, W. Baumeister and B.K. Jap:* Hybrid molecular structure of the giant protease tripeptidyl peptidase II. Nature Struct. Mol. Biol. **17**, 990-996 (2010).

4) *Schönegge, A.-M., E. Villa, F. Förster, R. Hegerl, J. Peters, W. Baumeister and B. Rockel:* The structure of human Tripeptidyl peptidase II as determined by a hybrid approach. Structure 20,593-603 (2012)

5) *Bohn, S., F. Beck, E. Sakata, T. Walzthoeni, M. Beck, R. Aebersold, F. Förster, W. Baumeister and S. Nickell:* Structure of the 26S proteasome from *Schizosaccharomyces pombe* at subnanometer resolution. P. Natl. Acad. Sci. USA. **107**, 20992-20997 (2010).

6) *Lasker, K., F. Förster, S. Bohn, T. Walzthoeni, E. Villa, P. Unverdorben, F. Beck, R. Aebersold, A. Sali and W. Baumeister:* Molecular architecture of the 26S proteasome holocomplex determined by an integrative approach. P. Natl. Acad. Sci. **109**, 1380-1387 (2012).

7) *Śledź, P., Unverdorben, P., Beck, F., Pfeifer, G., Schweitzer, A., Förster, F., Baumeister, W.:* Structure of the 26S proteasome with ATP-γS bound provides insights into the mechanism of nucleotide-dependent substrate translocation, P. Natl. Acad. Sci. 110, p. 7264-7269 (2013)

8) *Koster, A., R. Grimm, D. Typke, R. Hegerl, A. Stoschek, J. Walz and W. Baumeister:* Perspectives of molecular and cellular electron tomography. J. Struct. Biol. **120**, 276-308 (1997).

9) *Lucic, V., F. Förster and W. Baumeister:* Structural studies by electron tomography: from cells to molecules. Annu. Rev. Biochem. **74**, 833-865 (2005).

10) *Brandt, F., S.A. Etchells, J.O. Ortiz, A.H. Elcock, F.U. Hartl and W. Baumeister:* The native 3D organization of bacterial polysomes. Cell **136**, 261-271 (2009).

11) *Rigort, A., Bäuerlein, F., Villa, E., Eibauer, M., Laugks, T., Baumeister, W. and Plitzko, J.:* Focused ion beam micromachining of eukaryotic cells for cryoelectron tomography. P. Natl. Acad. Sci. 109, 4449-4454 (2012)

12) *Villa, E., Schaffer, M., Plitzko, J.M., Baumeister, W.:* Opening Windows into the Cell: Focused-Ion-Beam Milling for Cryo-Electron Tomography, in: Current Opinion in Structural Biology (2013), in press

13) *Robinson, C.V., A. Sali and W. Baumeister:* The molecular sociology of the cell. Nature **450**, 973-982 (2007).

Aberration corrected environmental STEM (AC ESTEM) for dynamic *in-situ* gas reaction studies of nanoparticle catalysts

E D Boyes [1,2,3] and **P L Gai** [1,2,4]

York JEOL Nanocentre [1], Departments of Physics [2], Electronics [3] and Chemistry [4], University of York, York, YO10 5DD, UK

Email: ed.boyes@york.ac.uk

Abstract:
Environmental scanning transmission electron microscopy (ESTEM) with aberration correction (AC) has recently been added to the capabilities of the more established ETEM for analysis of heterogeneous nanoparticle based catalysts. It has helped to reveal the importance and potentially unique properties of individual atoms as active sites in their own right as well as pathways between established nanoparticles. A new capability is introduced for dynamic *in-situ* experiments under controlled conditions of specimen temperature and gas environment related to real world conditions pertinent to a range of industrial and societal priorities for new and improved chemical processes, materials, fuels, pharmaceutical products and processes, and in control or remediation of environmental emissions.

1.Introduction

Commercial versions of the ETEM, largely based on the earlier work of the authors [1, 2, 3], are now widely used around the world in both academia and industry, and we have recently introduced [4] the important additional capabilities of aberration corrected environmental scanning transmission electron microscopy or AC ESTEM. The double (Cs, TEM + STEM) aberration corrected JEOL 2200 FS has been rebuilt at York to create the first full function environmental scanning transmission electron microscope (AC ESTEM). It has <0.1nm image resolution in both TEM and STEM modes and already provides a powerful new analytical capability which is being actively developed further. The essential proposition is to add facilities for dynamic in-situ studies of the heterogeneous solid state catalysis of gas reactions, with a hot stage providing controlled high temperatures and a gas environment around the specimen. The challenge has been to combine the new capabilities while retaining or improving the performance of the original full function STEM (and TEM) system with unrestricted imaging and analysis modes, including for Z-contrast STEM, a wide range of electron diffraction modes, and spectroscopy (EDX and EELS). The new system is also a full function ETEM [1, 2, 3]; now also with greatly improved diffraction and the addition of high quality EDX analysis.

2. Experimental procedure, design criteria and initial system operation results

2.1 Single atom imaging and raft analysis

The design goal has been to have all the functions available in all combinations. The proof-of-principle system, now developed further [4], has been used in preliminary experiments to image single atoms and analyse the 3D structure of reacting nanoparticle catalysts under reaction operating conditions. Specifically, single atom imaging and quantitative column analyses [4, 5] have been used effectively with Platinum on Carbon specimens in a hydrogen atmosphere of 2 to 20 Pa and at controlled hot stage temperatures of up to >500°C. Experimental HAADF Z-contrast single atom FWHM peak width resolution has been measured at 0.11+/-0.01nm without contrast enhancement and <0.1nm using it. These experiments show the incidence and movement of individual atoms between, and associated with, larger entities; including rafts, as well as nascent and fully formed nanoparticles with varying amounts of ordered crystallinity. The images, which were previously not attainable, reveal a significant number density (but low mass fraction) of isolated individual atoms existing on the support under real reaction conditions of gas and temperature.

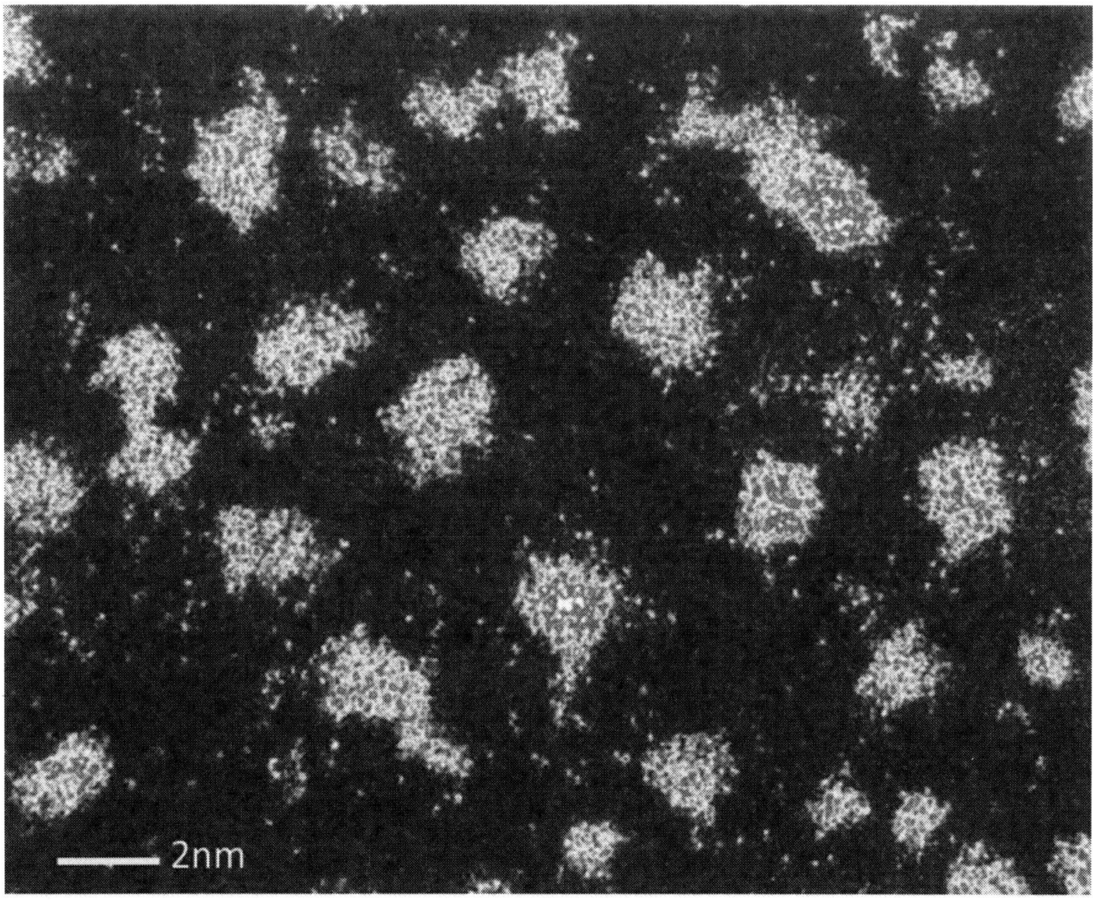

Fig.1: HAADF STEM image of Pt atoms and 1-2 atom thick relatively disordered rafts in an as deposited model Pt on C catalyst in the AC ESTEM with the H_2 gas modifications operational.

The significance of these results is several-fold. Firstly, they reveal migration of individual atoms or in some cases very small clusters between and also located in-between larger units on the support, and indicative of particle growth mechanisms. Secondly, they themselves predominantly have support and gas atoms rather than other metal atom as nearest neighbours. They will therefore inevitably have different energy levels and properties to the same type of atom bound into a nanoparticle crystal structure, or even a raft, and there surrounded with at least some of the nearest neighbours being its mates. This may be a very different environment and may lead either to de-activation or potentially much more active and perhaps differently reacting properties with different reaction type selectivity as well as volume activity. These properties will depend on the nature and strength of the support interaction, as will the site stability of the 'reaction' atom. Thirdly, the geometry of highly dispersed atoms is equivalent to a very high or ultimate surface area catalyst which is at least potentially very effective. Finally, in part the significance of the direct imaging of these species is that although perhaps present as a high number and activity density, and therefore importance, they may form a very small mass fraction and would otherwise remain undetected by other methods of analysis, such as most conventional HRTEM or conventional surface science methods. In essence they are a new component of the reaction pathway to be taken into account in understanding and improving catalyst activity and selectivity by design on a rational basis and better informed by the new data now made available with the development of (AC) ESTEM.

2.2. Nanoparticle development in hydrogen

After heating in hydrogen at 400°C or higher the previous partially ordered rafts, mostly 1-2 atoms thick and with edges always single atom high, are replaced by the ordered high density crystallinity of nanoparticles of Pt, usually with cube form and edges made up of cliffs several atoms high [4] . The particle geometry is extracted from quantitative analyses of Z-contrast HAADF STEM images, using analysis based on the original Crewe work [5] rather than more modern approaches [12]. We analyse reliably only within each image frame because the video signals are normalised within each frame. This can be most easily done where there are individual atoms to act as clear reference signals but also on the basis of the identification of multiple discrete levels. Results are similar to those originally reported by Crewe [5] with linear response, as here, for column heights of up to >10.

2.3. Pumping systems

A key attribute of the new ESTEM instrumentation, like that of the previous ETEM development, is to sustain a controlled gas pressure around the specimen while retaining the full functionality and performance, including <0.1nm imaging resolution, of the core instrument; both for use in the catalyst applications and to allow the parallel use for high vacuum studies of the same piece of high capital investment simply by not turning on the gas supply. Use of a common gas/no-gas microscope configuration ensures minimal need to open the column after initial modifications and contributes to system reliability. The gas pressures for the ESTEM on the JEOL column are achieved with a series of beam line apertures and stages of differential pumping similar in principle to the configuration used for the original ETEM [2] on the 300kV Philips/FEI column but modified in detail to fit the 200kV JEOL system now employed and to provide the novel ESTEM functionalities. Each design is opportunistically adapted to reduce the need to modify the specific core column being used. A key part of both systems is the use of aggressive gas tolerant turbomolecular pumping systems with appropriate new manifolds and with the necessary engineering to ensure full compatibility with <0.1nm image resolution in both TEM and STEM modes. New fixed beam line apertures and differential pumping stages support design gas/vacuum pressures over a 10^{10}x range from up to 100Pa at the sample in the middle of the column down to 10^{-8} Pa in the FEG at the top.

TMPs supplement the ion pumping on the FEI column and replace it entirely on the JEOL one, with a completely oil-free dry pumping system including an additional ion pumped stage before the UHV Schottky FEG. The various pumping modifications for gas have the incidental and slightly paradoxical effect of also improving the basic vacuum of the system (at least in the authors' versions) with the effect of significantly reducing specimen contamination effects in regular high vacuum operations. With the hot stage above 300°C contamination is almost completely avoided anyway. The use of the controlled temperatures of a hot stage, rather than uncontrolled beam-heating, is essential to have calibrated temperatures related to the real world outside the microscope and to reduce as far as practical distorting reaction mechanisms through charged particle effects, as well as to remove contamination with specimens which may initially be less than completely clean. Experiments with aggressively cleaned test specimens suggest most contamination is introduced with the sample rather than being machine derived. The high differential temperatures and associated migration of species across a very locally beam heated sample only add to contamination problems with less than ideally clean samples.

3. Gas pressure calibrations

Gas pressures at the sample have been measured with N_2 gas flow rates through an aperture in the tip of a modified specimen holder and the results calibrated in a separate system held at a succession of independently measured pressures. In the initial proof-of-principle system the manifold pressures were found to be reduced by almost exactly a factor of 100x, or from from mbar to Pa, and have so far been operated mostly in the range 2-20Pa. In the more developed system now being introduced pressures should be able to be increased to 100Pa, to the point with heavier gases the limit is expected to be gas type dependent scattering of the e-beam by gas interactions rather than by the earlier pumping system throughput limitations. Pressures measured at the sample with the test specimen holder built for purpose are calibrated against a gas type insensitive diaphragm gauge in the gas inlet manifold to the microscope.

Compared to the periodic dose methods in surface science UHV analytical methods, the ESTEM can support continuous gas flows and much higher pressures. Somorjai [6] has suggested that the criteria for surface reactions define the 'high pressure' region as being 0.1Pa and higher. This corresponds approximately to a gas supply of 1000 monolayers per second, which should ensure a good level of coverage for a reacting surface. The ESTEM operates with gas supply pressures at least one order of magnitude higher than this and often 100-1000x for up to mega-monolayer gas supply per second. While this is not as high as some industrial reactor processes, which may extend to >100bar, it is clearly much higher than is usual in many other types of analytical systems.

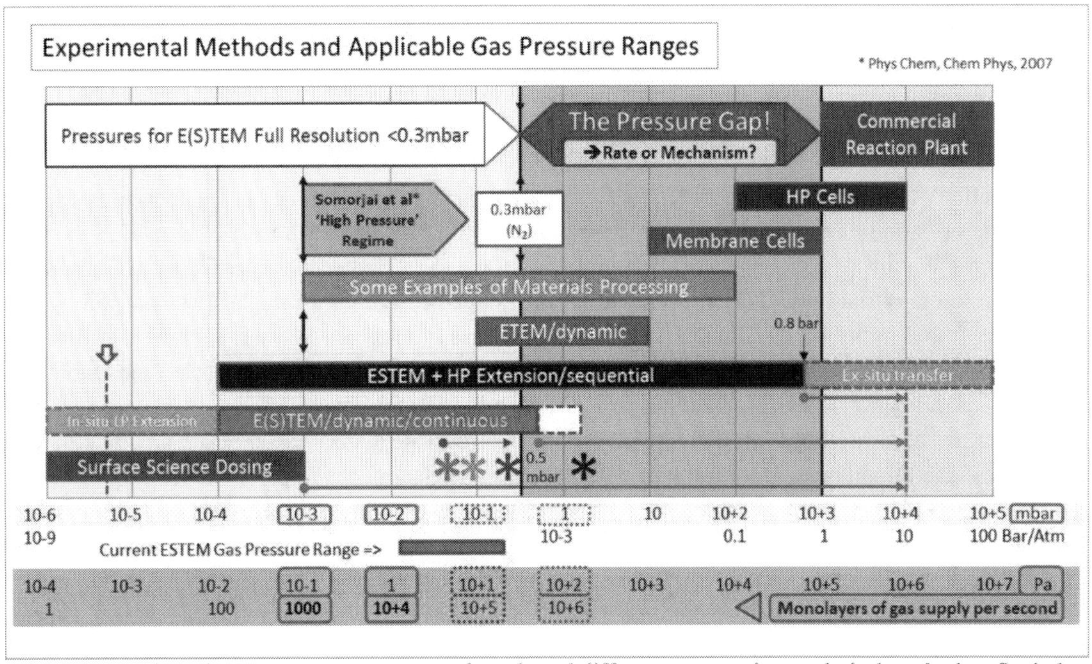

Fig.2. Chart of applicable gas pressure ranges for selected different gas reaction analytical methods. Such data are inevitably somewhat subjective and are intended to be illustrative rather than necessarily accurately definitive for all applications. In the chart E(S)TEM refers to the open version with the gas supply and pumping microscope-based as promoted and used by the authors of this paper, and widely adopted by others for ETEM.

Some of the information about how the ESTEM fits into the pressure landscape is summarised in Fig.2. Higher pressures, to several bar, have been reported for *ex-situ* reaction EM transfer systems [7], window specimen holders [8], and in laboratory and especially synchrotron x-ray analyses.

In the current practice the intent is to use E(S)TEM for real world related gas reaction studies and in part to close the 'pressure gap' through the scientific understanding generated. The applicability of this approach depends on whether the main effect is on reaction rate – when it is OK – or whether the reaction mechanisms are changed significantly and studies at modest pressures are not indicative of results in the real reaction. The sense is that a gas supply to the sample surface of between 10,000 and 100,000 or more monolayers of gas per second currently used in the new E(S)TEM will be sufficient to flood the solid surface and ensure stable reaction conditions, but this will depend on the (often mS or more) residence time of each gas molecule at reaction sites on the catalyst surface.

4. Discussion of technical issues

4.1. Specimen contamination

Outstanding issues include, most importantly, specimen cleanliness and its effect on reaction pathways and energetics. This is more important for dynamic in-situ experiments than for simple characterisations. Pre-cleaning holders, supports and with some care actual samples has been found beneficial as is avoiding e-beam artefacts; especially contamination build-up. A proper solution will require contained preparation and transfer of samples under a protective atmosphere of gas, but not vacuum, without exposure of the actually or potentially reactive samples to air and/or moisture or other contamination. Activation in the microscope of protected catalyst species, as is often used in a real reactor, is a promising pathway. Samples from industrial reactors and even, as we study, ones extracted from real world reaction environments including extended, e.g. 36,000 mile, exposure in a vehicle exhaust system [9], will often be subject either to initial or handling induced material which turns into electron microscopy 'contamination', even if it is itself from the real world operating environment of the catalyst. This may well limit the use of such materials as specimens for *in-situ* studies; although requiring extensive characterisation of the materials in their original forms. In the application it may also lead to some catalyst poisoning deactivation. In this case, and for other reasons, we resort to using model systems better to identify mechanisms and establish operating parameters more clearly than is possible with more complex choices. The simple systems have increased value when related closely to identified issues with real world reaction mechanisms.

4.2. Zone axis analyses

In practical terms an intrinsic problem with using real catalysts as samples is the need for systematic zone axis orientation for the most informative analyses of crystalline nanoparticles of both the active phase and the support if either is so structured. So far most oriented nanoparticles are found in this state due to the difficulties in orienting them in practice, since even with the best stages nanoparticles inevitably move while being tilted. This requires access to apparently simultaneous, but in practice seamlessly multiplexed [10], signals from both real (for position) and reciprocal (for orientation) spaces. Fortunately in STEM – but not in TEM with lens setting hysteresis and therefore inaccurate and slow mode switching – a convergent beam electron diffraction pattern and an image can be recorded – if not optimised – with the same lens settings. Scans are simply switched from field scan to spot analysis with appropriate signal unscrambling; assisted by the different reciprocal space annular zones in which bright field, crystallographic and Z-contrast signals predominantly emerge from the sample exit face. Suitably shaped and sized detectors can be used to pick up the majority signals at the respective radii in reciprocal space. The system has been demonstrated on a dedicated STEM and is in process of being implemented on the York ESTEM. Variations on the theme should allow the beam dose for diffraction to be reduced. This is particularly important for ESTEM which is used as far as possible to study systematic gas and hot stage temperature induced effects. However, like most other forms of electron microscopy, damaging beam-specimen interactions will in the end set the basic limit to the validity of the applications. We need to take all practical steps to minimise the broadly defined 'damage' in gas reaction studies with wide applicability to industrial practices.

4.3. Hot stages

Hot stages for controlled specimen temperatures is an area of rapid development. We continue to use the conventional furnace heating holders, in our case from Gatan [11], because they support use of 3mm disc specimens when required, including ones prepared by evaporation or ion-milling. For high resolution STEM (but not required for TEM) we operate with a replacement Keithley 2614B power supply and independent temperature monitoring. With this type of holder there is inevitably substantial specimen position drift with temperature changes; albeit often predominantly in one direction. With skill and experience, manual or electronically assisted position stabilisation is often possible. A big advantage of these holders is the accompanying thermocouple on the furnace body to monitor temperatures even in the presence of enough gas to modify the readings in the way Pirani gas conductivity gauges work. Recent developments for small amounts of ultrafine (only) powder samples use MEMS technology [13] with much smaller dimension thin film heaters, faster response times and much less reported positional drift with temperature change. The double tile holders required for the most informative zone axis of individual nanoparticles selected for detailed analysis should especially benefit from the new technology when available in this form. However, after a suitable period for position stabilisation it has proved possible using custom electronics for STEM to use the standard Gatan 628 heating holder for sub-angstrom (<0.1nm) TEM and STEM imaging at modest temperatures up to >500°C.

4.4 Parallel chemical studies
Although the AC ESTEM is equipped with a quadropole mass spectrometer this is more for vacuum system maintenance and reaction gas management than reaction product evaluations; in large part due to the very small amount of catalyst being used and the associated furnace and holder environment artefacts or potential artefacts. Current practice [1] is to conduct parallel reaction studies in an *ex-situ* reaction rig with a larger amount of catalyst material, but true parallel in-situ studies of product would clearly be beneficial in addition if not in total replacement.

5. Conclusions
We have constructed the first full function AC ESTEM system and have used it with 0.1nm resolution for single atom, raft and crystalline nanoparticle studies, demonstrating the capability of the facility and compatibility of the same instrumentation for gas studies and parallel high vacuum analyses.

References
[1] P L Gai et al, 1995, *Science*, **267**,661
[2] E D Boyes and P L Gai, 1997, *Ultramicroscopy*, **67**, 219
[3] P L Gai and E D Boyes, 2009, *Microscopy Research and Technique*, **72**, 423
[4] E D Boyes, L Lari, M R Ward and P L Gai, 2013, *Annalen der Physik (Berlin)*, **525**, 423
[5] A V Crewe, J Wall and J Langmore, 1970, *Science*, **168**, 1338
[6] G Somorjai, et al, 2007, *Phys Chem Chem Phys*, **9**, 3501
[7] Various sources: G Parkinson (BP) and L Allard (ONRL) are examples and also Ref 8 et seq
[8] J Creemer et al, 2008, *Ultramicroscopy*, **108**, 993
[9] M R Ward et al, 2012, *Chem Cat Chem*, **4**, 1622
[10] E D Boyes, 2010, *Proc International Microscopy Congress, Brazil*, **I20**, 11
[11] Gatan 628 heating holder powered with Keithley 2614B SMU
[12] J LeBeau et al, 2009, *Phys Rev B*, **79**, 214110
[13] L Allard et al, 2009, *Microscopy Research and Technique*, **72**, 208

Acknowledgements
The authors thank the University of York and the EU through the previous Regional Development Authority, Yorkshire Forward, as well as JEOL Ltd for initial support to set up the York laboratory. We are especially grateful for EPSRC critical mass grant EP/J018058/1 which now supports the facility further development and applications and programme with a view to wider applications and collaborations. Special thanks are due to Ian Wright, Leonardo Lari, Michael Walsh and Michael Ward for invaluable advice and assistance, and to other colleagues for their continuing support of the ambitious world leading project.

Journal of Physics: Conference Series **522** (2014) 012005 doi:10.1088/1742-6596/522/1/012005

Nanocycles of Materials' Transport studied by *in-situ* Electron Microscopy and Diffraction

K Takayanagi[1,4], Y Oshima[2,4], S Lee[3,4], T Tanaka[1,4] and Y Tanishiro[1,4]
[1] Department of Physics, Tokyo Institute of Technology, 2-12-1-H-51, Oh-okayama, Meguro-ku, Tokyo 152-8551, Japan
[4] Core Research for Evolutional Science and Technology (CREST), Japan Science and Technology Agency (JST), K's Gobancho, 7, Gobancho, Chiyoda-ku, Tokyo 102-0076, Japan
[2] Research center for ultra HVEM, Osaka University, 7-1 Mihokaoka, Ibaraki, 567-0047, Japan
[3] Department of Materials Science & Engineering, Tokyo Institute of Technology, 2-12-1-H-51, Oh-okayama, Meguro-ku, Tokyo 152-8551, Japan

E-mail: takayang@phys.titech.ac.jp (K. Takayanagi)

Abstract. A *sub*-50pm resolution electron microscope was applied for *in-situ* studies of ion transport in lithium ion battery and CO-oxidation catalyst. Lithium ions in LiM_2O_4 electrodes (M: transition metals) and titanium ions in Au/TiO_2 catalyst were enabled us to observe individually by ABF-STEM and TEM. Transport of lithium ions was seen, after *in-situ* observations, to induce structural transformation of the electrode materials that causes irreversible cycle of the battery reaction. Transport of the interstitial titanium ions results, due to oxygen gas for CO-oxidation, in modification of Au/TiO_2 interface structure. *In-situ* studies, thus, revealed materials' transformation induced locally by ion/electron transport of the battery/catalyst system at high resolution and high contrast.

1. Introduction

It is of great interest to reveal ionic/electronic process in batteries and in catalysis, so as to manage utilization of their systems realizing the energy saving and resource saving. Ionic/electronic processes induce structural transformation of materials concerned, so that their cyclic process can often be irreversible, which affect efficiency and/or life-time of the system. As for rechargeable batteries, particularly for vehicles, one of issues is long lifetime [1-3]. Charge-discharge cycles change the interface structure between electrode materials and electrolyte irreversibly [4-6], which is thought to deteriorate the battery lifetime. As the shorter charging time causes the faster deterioration, we need nano-scale understanding of the issue. Lithium ion transport, then, is fundamental process at lithium ion battery electrodes. For nano-scale study of the change-discharge process, transmission electron microscopy (TEM) is one of powerful methods [7].

Furthermore, *in-situ* TEM observation could give rich information about local structure change of cathode and/or anode materials during battery reaction, as demonstrated by Huang et al [8-10]. They revealed anisotropic swelling of Si nanowire anode during lithiation by in-situ TEM observation [11]. However, electrochemical measurement such as cyclic voltammetry has not been performed simultaneously with the structural observation. Simultaneous *in-situ* observation is really inevitable for nano-scale understanding and realization of long-life batteries against rapid-charging.

On the other hand, in titanium dioxide (TiO_2)-based catalysis, substrates of titanium dioxide (TiO_2) and other transition-metal oxides are widely as metal nanoparticle catalysts [12]. Gold nanoparticle catalysts for low-temperature *CO oxidation* ($2CO + O_2 \rightarrow 2CO_2$) [12] are especially attractive for applications in air purification and reduction of environmental burdens. The catalytic activity in gas environments depends on the morphology [13,14] and electronic structure [13,14] of contacts between metal nanoparticles and the oxide support. The morphology and electronic structure change *via strong metal-support interaction* (SMSI) [13-17], which causes complication in understanding the catalytic reaction path. The SMSI at Au/TiO_2 caused by the reaction between interstitial Ti ions and adsorbed O_2 molecules [13,14]. *In-situ* electron microscopy, therefore, might give certain information about SMSI, when interstitial titanium ions could be detected directly. In-situ study at high resolution is a mean to find reaction path of CO-oxidation, *via* SIMS.

2. Experimental

In-situ studies were done using an aberration corrected electron microscope achieving a *sub*-50pm resolution, R005 microscope [18]. The microscope is fitted with two aberration correctors for probe forming and for image forming, both of which are operated simultaneously for STEM-TEM imaging. The microscope, equipped with a cold field emission emitter of a tungsten <111> tip, is daily used at accelerating voltages of 300, 200, or 80 kV [18].

For *in-situ* observation of lithium ion transport, we developed a specific specimen holder which has enabled us to observe a current-voltage relation of a $LiMn_2O_4$-nanowire[19]-based lithium ion battery and structure change of the nanowire simultaneously[20]. Structure observations were done by STEM imaging (ABF and/or HAADF) and TEM with diffraction images.

For study of CO-oxidation catalysis, gold nanoparticles deposited on TiO_2, we developed a specimen holder with gas injection nozzles opening at the specimen surface. Specimens of Au/TiO_2 were prepared in a separate vacuum chamber, just before the in-situ observation: A TiO_2 (rutile) crystal was crushed into pieces and annealed at 1170 K for 1 hour at 10^{-4} Pa, and characterized by CL (cathode luminescence). The gas pressure at the Au/TiO_2 specimen could increase/decrease within one second from the base pressure of 2×10^{-5} Pa up to 100Pa.

3. Results and Discussion

3.1 Individual lithium ion imaging in LiV_2O_4 and $LiMn_2O_4$ crystal by STEM

Structure change is discussed in relation to the current-voltage relation of a model LIB. At excess or deficient concentration of lithium ions, spinel electrodes exhibited local disorder. The bulk $LiMn_xO_4$ crystal forms different phases: X-ray analysis [21-22] shows that the crystal is in the cubic spinel phase for x=1, but in the tetragonal phase for x=2. We observed as-grown $LiMn_2O_4$ nanowires by the R005 microscope. High resolution ABF (annular bright field) images shown in figure 1(a) and (b) demonstrate that lithium ions can be detected discriminating from the Mn and/or O columns. In addition, the images clarify the termination of the surface layer. Comparing the observed images with the model structure, the as-grown nanowires had the cubic-spinel structure. Such as-grown $LiMn_2O_4$ nanowires were used for nanowire-based model battery.

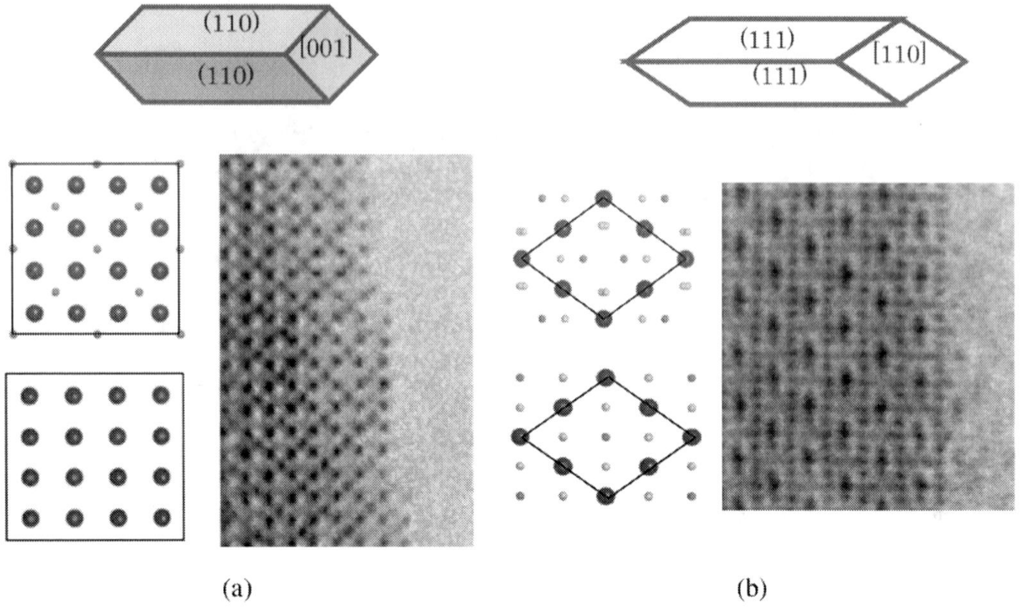

Figure 1. ABF(annular bright field image) of LiMn2O4 nanowire. (a) View from the [010] direction, and (b) view from the [1-10] direction. In schematic models in (a) and (b), green, gray, and red dots indicate, respectively, Mn, O, and Li ions visible in ABF images. In general, the $LiMn_xO_4$ bulk crystal has cubic spinel structure for x=1, and tetragonal structure for x~0: The upper model show the cubic phase, and the below one, tetragonal phase. ABF images in (a) and (b), for the as-grown nanowires, show that the nanowires are in the cubic spinel phase, in consistent with the bulk crystal.

3.2 In-situ observation of nanowire-based lithium ion battery

For simultaneous observation of cyclic voltammetry (CV) and the nanowire cathode (ILE) interface, we fabricated nanowire batteries as illustrated in figure 2: The nanowire electrode has generally thick bundles consisting of short $LiMn_2O_4$ nanowires. Figure 3(a) shows the CV of a "single" nanowire battery. The pair of double current peaks in figure 3(a) corresponds to the 4V battery reaction (vs Li/Li+), where the lithium concentration change between x=1 and x~0 of $Li_xMn_2O_4$. The battery looks like figure 3(b), and a single $LiMn_2O_4$ nanowire looks in TEM images like figure 3(c). In general, $LiMn_2O_4$ nanowires in our batteries maintained the cubic structure during charge process. In the discharge process, the cubic structure changed into the tetragonal phase, which was restored the cubic phase in the next charge cycle.

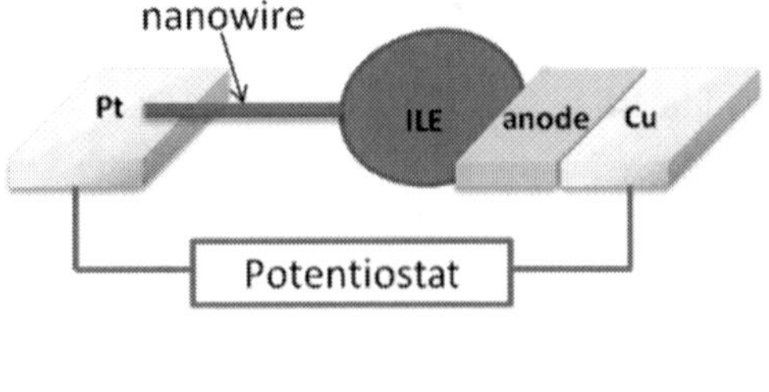

Figure 2. Design of a nanowire-battery. $LiMn_2O_4$ nanowire spans between platinum and ionic liquid electrode (ILE).

It is interesting that the $LiMn_2O_4$ nanowire was completely restored into the cubic phase from the tetragonal phase without fracture. This behaviour is different from the reported case of $LiMn_2O_4$ particles, which were damaged due to the tetragonal formation [23]. The observed reversible phase transformation without fracture supports that $LiMn_2O_4$ nanowires are promising as a cathode material improving the life-time of $LiMn_2O_4$ crystals based batteries.

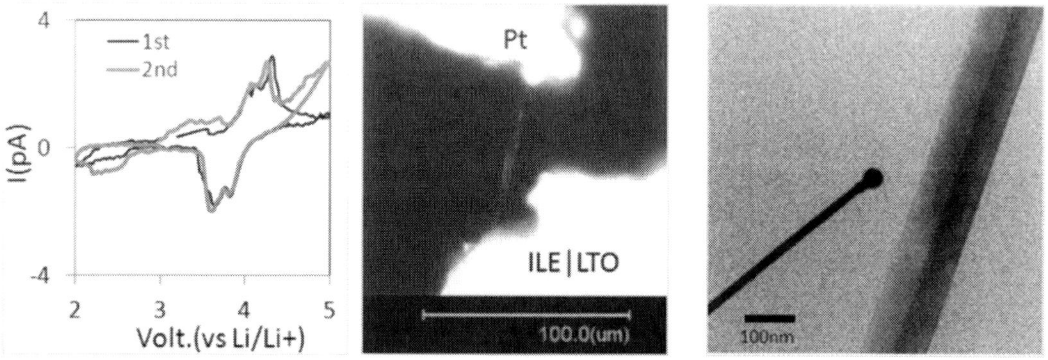

(a) (b) (c)

Figure 3. (a) Typical cyclic voltammogram of the nanowire-battery. Horizontal axis is the voltage vs Li/Li^+ and the vertical axis is the current, I, passing through the $Li_xMn_2O_4$ nanowire. The concentration, x , changes from two (at V=2 for the discharged state) to nearly zero (at V=5 for the charged state). (b) photo-image of a developed $LiMn_2O_4$ nanowire battery. (c) TEM image of a part of the nanowire, which shows that a "single" nanowire only bridges the platinum electrode and ionic liquid electrode (ILE).

3.3 Individual interstitial titanium ion imaging by TEM

Interstitial titanium (Ti) ions can be contained in our prepared Au/TiO_2 specimens. Figure 4(a) illustrates the view of TiO_2 crystal from the [110] direction. Interstitial Ti ions can occupy cation sites at the center of oxygen octahedrons [24-28]. As illustrated in figure 4(c), *titanium ions occupying the iv-site* ("*iv*-Ti ions") can be detected by TEM. In figure 4(b) of a TEM image of TiO_2, a titanium ion is recognizable at the edge, which is adsorbed on the lateral (001) surface. In order to visualize interstitial Ti ions inside the edge, we simulated TEM images of TiO_2 columns with the *iv*-Ti ions, and those without them to derive the intensity given by a "single" *iv*-Ti ions in a column. Intensity calculations of the Ti-columns and Ti-O columns (see figure 4(a)) were agreed with experimental images, and also that a single *iv*-Ti ion gives a definite intensity. Furthermore, the intensity is doubled when two *iv*-Ti ions locate in the same atomic column [29].

High resolution TEM images of wedge-shaped TiO_2 specimens with Au particles were obtained (figure 5(a)). The atomic columns of the Ti-O, Ti-only and O-O are imaged as bright dots, similarly to figure 4(c). We determined thicknesses of the observed Ti-only columns from comparison with the calculation. In order to find "residual" image intensity at the *iv*-Ti sites, a "background-free" image was processed from figure 5(a) by subtracting the calculated image intensity without *iv*-Ti ions: The background-free image is shown in figure 5(b). As marked by rhombus, the residual intensity at the center of each rhombus, tells interstitial Ti ions occupying *iv*-Ti sites. Figures like figure 5(b) gives local distribution of interstitial Ti ions. The *iv*-Ti ions had high density at around Au/TiO_2 periphery, and deficient in the TiO_2. On the other hand, they are rather homogeneous in TiO2 specimens without Au deposit [29].

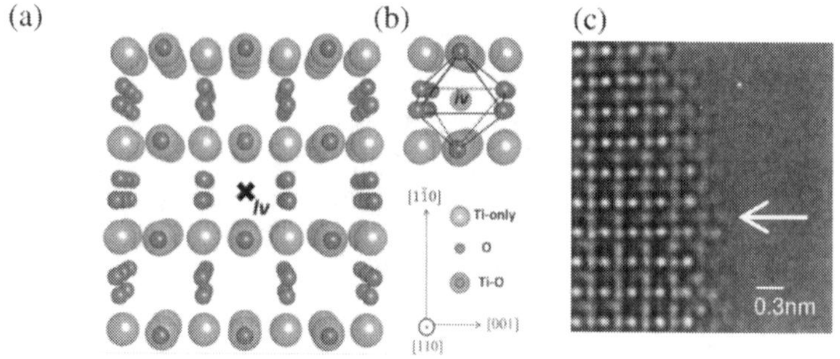

Figure 4. (a) Illustration of TiO_2 crystalline structure, and (b) configuration of an interstitial site of titanium ions, *iv* site. Bright dot indicated by an arrow in (c) is an adsorbed ion, presumably, *iv*-Ti ion on the top surface.

3.4 In-situ study on titanium ions while oxygen gas injection on Au/TiO_2

Following discussions on SIMS, O_2 gases induce diffusion of Ti ions in Au/TiO_2 and TiO_2 (110) surface. In figure 6(a), our Au/TiO_2 specimens were found to have a specific phase at the perimeter of the Au/TiO_2 interface. As shown in figure 6(b), the specific phase grew extensively by exposing to O_2 gas at 100 Pa into a pillar which has a chemical composition of $Ti_{1-x}O_2$ ($x > 0$). Thus, *in-situ* TEM observation gave an evidence of Ti ion diffusion from substrate to surface by exposing to O_2. This observation, together with those in section 3.3, seems to suggest a possible scenario of Au/TiO_2 catalysis: In relation to O_2 gas exposure, interstitial Ti ions accumulate at the perimeter/interface of Au/TiO_2, providing new active sites which promote the O_2 adsorption and dissociation for the CO oxidation through electron donation.

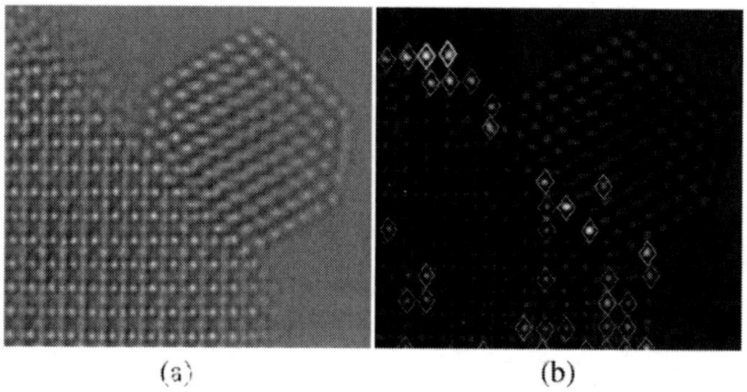

Figure 5. (a) Transmission electron microscope (TEM) image of Au/TiO_2, from [110] direction. (b) the same TEM image as (b), after the subtraction of simulated background intensity for each unit column. The colored rhombus indicates the position and the number of *iv*-Ti ions. Orange, green and light blue represents one, two and three ions, respectively. Yellow rhombus represents interstitial Ti ions at Au/TiO_2 interface / perimeter.

4. Summary

A sub-50pm resolution electro microscope, R005, was applied for in-situ studies using a nanowire lithium ion model battery. Although the $LiMn_2O_4$ nanowire cathode was changed locally into the tetragonal phase while 4V reaction (vs Li/Li+), the cathode was restored to cubic phase without any fracture to continue cyclic charge-discharge process. No fracture while cubic-tetragonal transition is promising for the long-life-time battery with $LiMn_2O_4$ nanowire cathode.

The microscope was also used for study of Au/TiO_2 model catalyst of CO-oxidation. Interstitial titanium ions in TiO_2 substrates with gold nanoparticle (Au/TiO_2) were detected, and are found to be accumulated at Au/TiO_2 periphery in coincident with $Ti_{1-x}O_2$ pillar formation at the periphery.

In conclusion, transport of ions results in structural transformation at nano-micron scale. *In-situ* studies revealed materials' transformation, which is induced by interplay between charge and mechanical balances, at high resolution and high contrast.

References

[1] Tarascon J M and Armand M 2001*Nature* **414** 359
[2] Scrosati B and Garche J 2010 *J. Power Sources* **195** 2419
[3] Pasquier A D, Plitz I, Menocal S and Amatucci G 2003 *J. Power Sources* **115** 171
[4] Aurbach D, Zaban A, Ein-Eli Y, Weissman I, Chusid O, Markovsky B, Levi M, Levi E, Schechter A and Granot E 1997 *J. Power Sources* **68** 91-98
[5] Morcrette M, Rozier P, Dupont L, Mugnier E, Sannier L, Galy J and Tarascon J M 2003 *Nature Mater.* **2** 755
[6] Vetter J, Nov´ak P, Wagner M R, Veit C, M¨oller K-C, Besenhard J O, Winter M, Wohlfahrt-Mehrens M, Vogler C and Hammouche A 2005 *J. Power Sources* **147** 269
[7] Thackeray M, Shao-Horn Y, Kahaian A, Kepler K, Skinner E, Vaughey J and Hackney S 1998 *Electrochem. Solid State Lett.* **1** 7–9
[8] Liu Y, Hudak N S, Huber D L, Limmer S J, Sullivan J P and Huang J Y 2011 *Nano Lett.* **11** 4188
[9] Liu X H, Huang S, Picraux S T, Li J, Zhu T and Huang J Y 2011 *Nano Lett.* **11** 3991
[10] Huang J Y, Zhong L, Wang C M, Sullivan J P, Xu W, Zhang L Q, Mao S X, Hudak N S, Liu X H, Subramanian A, Fan H, Qi L, Kushima A and Li J 2010 *Science* **330** 1515
[11] Liu X H, Zhang L Q, Zhong L, Liu Y, Zheng H., Wang J W, Cho J H, Dayeh S A, Picraux S T, Sullivan J P, Mao S X, Ye Z Z and Huang J Y 2011 *Nano Lett.* **11** 2251
[12] Haruta M, Tsubota S, Kobayashi T, Kageyama H, Genet M J and Delmon B 1999 *J. Catal.* **144** 175
[13] Valden M, Lai X and Goodman D W 1998 Science **281** 1647
[14] Tanaka T, Sano K, Ando M, Sumiya A, Sawada H, Hosokawa F, Okunishi E, Kondo Y and Takayanagi K 2010 Surf. Sci. **604** L75
[15] Williams K J, Boffa A B, Lahtinen J, Salmeron M, Bell A T and Somorjai G A 1990 *Catalysis Letters* **5** 385
[16] Benett R A, Stone P and Bowker M 1999 *Faraday Discuss.* **114** 267
[17] Tauster S J 1987 *Acc. Chem. Res.* **20** 389
[18] Sawada H, Tanishiro Y, Ohashi N, Tomita T, Hosokawa F, Kaneyama T, Kondo Y and Takayanagi K 2009 *J. Electron. Microsc.* **58** 357
[19] Hosono E, Kudo T, Honma I, Matsuda H and Zhou H 2009 *Nano Lett.* **9** 1045
[20] Lee S, Oshima Y, Hosono E, Zhou H, Kim K, Chang H, Kanno R and Takayanagi K 2013 *Nano Lett.* submitted
[21] Thackeray M M, Johnson P J, Picciotto L A, Bruce P G and Goodenough J B 1984 *Mat.Res.Bull.*, **19** 179-187,
[22] Thackeray M M, David W I F, Bruce P G and Goodenough J B 1983 *Mat.Res.Bull.*, **18** 461
[23] Gummow R J, de Kock A and Thackeray M M 1994 *Solid State Ionics*, **69** 59
[24] Elliott S D and Bates S P 2001 *Surf. Sci.* **495** 211
[25] Elliott S D and Bates S P 2002 *Phys. Rev. **B*** **65** 245415
[26] Onishi H and Iwasawa Y 1996 *Phys. Rev. Lett.* **76** 791
[27] Onishi H and Iwasawa Y 1994 *Surf. Sci.* **313** L783
[28] Park K T, Pan M H, Meunier V and Plummer E W 2006 *Phys. Rev. Lett.* **96** 226105
[29] Tanaka T, Sumiya A, Sawada H, Kondo Y and Takayanagi K, *Surf. Sci.* (2013) in press,

Correlated structure-optical properties studies of plasmonic nanoparticles

Emilie Ringe[1], Richard P. Van Duyne[2], Laurence D. Marks[3]

[1] Department of Materials Science and Metallurgy, Cambridge University, 27 Charles Babbage Road, Cambridge CB3 0FS, UK

[2] Department of Chemistry, Northwestern University, 2145 Sheridan Road, Evanston, IL 60208, USA

[3] Department of Materials Science and Engineering, Northwestern University, 2220 Campus Drive, Evanston, IL 60208, USA

Abstract. Interest in nanotechnology is driven by unprecedented means to tailor the physical behaviour via structure and composition. Unlike bulk materials, minute changes in size and shape can affect the optical properties of nanoparticles. Characterization, understanding, and prediction of such structure-function relationships is crucial to the development of novel applications such as plasmonic sensors, devices, and drug delivery systems. Such knowledge has been recently vastly expanded through systematic, high throughput correlated measurements, where the localized surface plasmon resonance (LSPR) is probed optically and the particle shape investigated with electron microscopy. This paper will address some of the recent experimental advances in single particle studies that provide new insight not only on the effects of size, composition, and shape on plasmonic properties but also their interrelation. Plasmon resonance frequency and decay, substrate effects, size, shape, and composition will be explored for a variety of plasmonic systems.

1. Introduction

Gold and silver nanoparticles are the object of intense research because of their fascinating nanoscale behavior. These metals display bright colors when particles of sizes of a few to a few hundreds of nanometers are dispersed in a medium or on a surface. A striking example of this is the yellow and red in medieval stained glass, typically resulting from the incorporation of small silver and gold particles, respectively. The phenomenon responsible for such optical behavior is a localized surface plasmon resonance (LSPR), a coherent oscillation of the electrons in a particle generally smaller than the wavelength of light. This resonance can occur in a variety of materials, however, the ease of fabrication and shape control, stability, well-known surface chemistry, and biocompatibility give Au and Ag a prominent place in the burgeoning field of plasmonics.

The applications of plasmonically active nanoparticles stem from the three wavelength-dependent effects of LSPR: photon absorption, elastic photon scattering, and local enhancement of the electric field at the surface of the nanoparticle. The latter is responsible for the majority of the signal enhancement observed in surface enhanced spectroscopies such as surface-enhanced Raman, fluorescence, and second harmonic generation [1-4]. Photon scattering is most prominent in larger particles and is important in practical sensor implementation and imaging contrast agents [5,6]. Finally, photon absorption leads to plasmon excitation, which subsequently decays through a variety of mechanisms, amongst which many ultimately raise the temperature of the nanoparticle surface, leading to localized heating that can be harnessed to boil liquids or destroy cancerous tissues [7,8].

The above examples are not an extensive list, but they highlight the breadth of applications potentially realizable with plasmonics, given sufficient knowledge and understanding of the underlying resonance phenomenon.

Size, shape, composition, and the surrounding environment provide multiple handles to tune the electron oscillation behavior; the latter being the basis of a number of refractive index sensors [9-11]. It is well known that each of those factors strongly affects the plasmon energy, decay, near-field distribution, and refractive index sensitivity. Mie theory, even in its simplest form (the electrostatic approximation), predicts the importance of shape, composition, and surrounding environment [12]. Many studies have aimed at unraveling the effects of such factors for non-spherical shapes, providing some general rules. However, in the vast majority of these approaches, the many relevant structural (shape, size, composition) and environmental (substrate, surrounding medium) factors have been investigated independently, despite the lack of evidence for their independence. Synthetic limitations, insufficient data size and range, as well as correlation difficulties have hindered progress in understanding such multi-component problems.

In this paper, we discuss the techniques we have developed to surmount many difficulties related to single particle analysis, and present a survey of our recent results on the effects uniquely probed via single particle experiments as well as the interrelation of structural and/or environmental factors on the plasmonic behavior of gold and silver nanoparticles.

2. Methods and methods development

Advances in shape control have led to vast improvements in the yield of anisotropic nanoparticles, as well as their size distribution. However, averaging over the unavoidable shape and size in homogeneities prevents obtaining a detailed understanding of the structure-function relationships; additionally, information about polarization dependence as well as plasmon decay cannot be obtained from a bulk sample. Single particle methods provide unique opportunities to learn about how the minutest structural and environmental factors influence the optical properties of plasmonic particles.

The first correlated, single particle results are attributed to Tamaru *et al.*[13] and Mock *et al.*[14], who coupled optical dark field microscopy results with scanning electron microscopy (SEM) and transmission electron microscopy (TEM) analyses, respectively. Since these early studies on colloidal silver particles, many different experimental approaches have been employed to gain insight on the structure-function relationship in gold and silver nanoparticles, integrating optical absorption techniques, atomic force microscopy, etc; see for instance the recent reviews [15-17] for additional information. These provide a powerful way to observe the effects of structural parameters on optical properties, and, when used in conjunction with statistical analysis, yield predictive rules regarding the effects of size, composition, corner rounding, interparticle coupling, and surrounding environment on the plasmon energy, and lifetime.

Our studies used a combination of optical dark field microscopy and transmission electron microscopy. In dark field microscopy, the sample is illuminated with a hollow cone of white light (possibly polarized), and only the scattered light is collected, creating a dark background as shown in Fig. 1, and providing an excellent signal-to-noise ratio for the particles studied (>50 nm). TEM was chosen as the structural characterization tool because of its superior spatial resolution. We established that optical measurements must be performed prior to electron illumination, as it causes irreversible damage and/or contamination [15]. In practice the experiment is carried out as follows. Nanoparticles are deposited on TEM-compatible substrates (C/Formvar, Si_3N_4, SiO_x, etc), and excess surfactants are removed with an appropriate solvent, attempting not to change the particle morphology (which can happen). Then the spectrum of individual particles is acquired by centering the scattering signal (diffraction-limited spot) on the slit of a spectrometer/detector (CCD), integrating over the height of the spectrum on the CCD. An optical image covering a large sample area is used to number the particles and subsequently serves as a map to retrieve the region of interest once the grid is inserted in a TEM. The entire process allows for the collection of hundreds of correlated single particle data per sample. Obtaining such large data sets is critically important as they enable statistical analysis with relatively small errors, and allow multi-dimensional investigations.

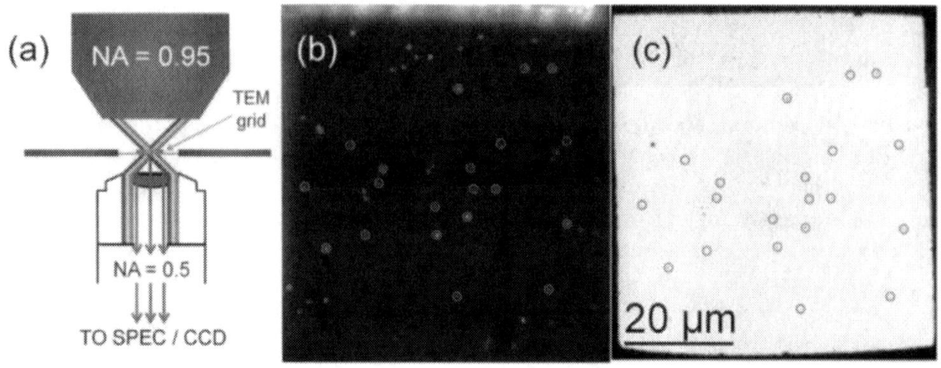

Figure 1. Setup for correlated DFM/TEM measurements. (a) DFM geometry. (b) Black and white optical image used as map for numbering and retrieving single diffraction-limited spots. (c) TEM image of the corresponding area. Reproduced with permission from ref [18].

3. Experimental results: new insight from statistical analysis of single particle data

3.1. Interrelation of size, composition, and dielectric environment on nanocubes

Owing to their high symmetry and flat faces rendering them stackable, and the high electric field enhancement at their sharp corners, plasmonic nanocubes are interesting structures for applications involving device fabrication. Many factors influence their LSPR energy, and to fully investigate and quantify such factors we studied Ag and Au nanocubes of sizes 60-100 nm on two representative substrates: a polymer (Formvar, refractive index n=1.5) and a semiconductor, (Si_3N_4, n~2)[19]. Several effects and interrelations were obtained from analysis of covariance of the data obtained on 175 individual nanocubes (figure 2)[19]. Note that the LSPR FWHM for particles on Si_3N_4 (n~2) could not be determined with acceptable error levels due to etaloning and noise, most pronounced in the blue region (affecting Ag cubes most), hence are not reported.

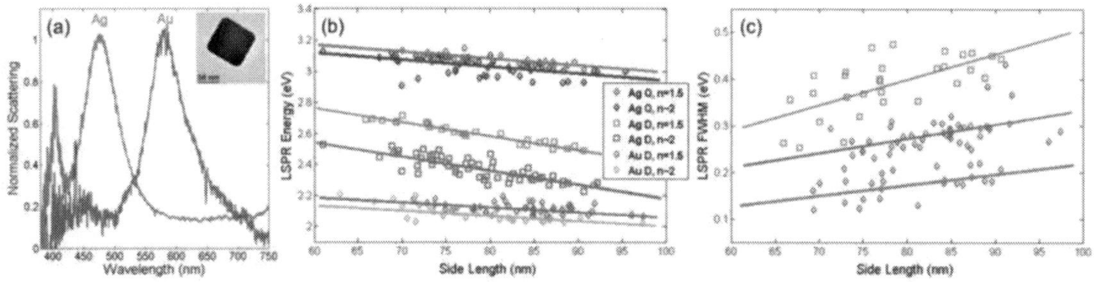

Figure 2. Correlated LSPR/TEM measurements on nanocubes. (a) typical LSPR spectra showing the sharp quadrupole-like resonance in Ag nanocubes, as well as the broader dipole-like resonance in both Ag and Au particles. Interrelation of composition, dielectric environment, and size on the (b) plasmon resonance energy and (c) plasmon FWHM for gold and silver nanocubes on Formvar (n=1.5) and Si_3N_4 (n~2) substrates. The legend applies to both (b) and (c).

As expected from phase retardation effects and radiation damping, the LSPR energy and peak width shifts to the red and broadens with increasing size, respectively, and the LSPR energy for a given size decreases with increasing refractive index of the substrate. Also, in Ag, the quadrupole-like mode (distal) is higher in energy than the dipole-like mode (proximal) for all sizes observed. The interrelations between these factors are more interesting, however, and uniquely addressed using large sets of single particle data. From figure 2, it is immediately evident that the size dependence (the slope) of the energy and FWHM depends on the composition, and, in the case of Ag, mode symmetry, but not on the substrate, as can be proven with analysis of covariance. The LSPR energy size dependence for the Ag dipole (-8.9(5) meV/nm, standard deviation on the last digit in parenthesis) is over twice as large as that of the Ag quadrupole (-4.2(6) meV/nm), which is in turn comparable to that of the Au dipole (-3.9(6) meV/nm). The size dependence of the FWHM also follows a similar trend, with that of the Ag dipole being highest (5.4(1.2) meV/nm), followed by the Ag quadrupole and Au dipole (3.1(8) and 2.3(7) meV/nm). Also, the magnitude of substrate effects is dependent on the mode and composition, but not particle size; the difference in LSPR positions on Formvar and Si_3N_4 are 0.23 and 0.057 eV for the dipolar modes of Ag and Au, respectively; it is 0.053 eV for the Ag quadrupolar mode. While a small change of the size-dependence (different slope in figure 2b) is expected for Au sphere in different dielectric environments [20], this effect was not observed in cubes and other shapes we investigated (decahedra, icosahedra, truncated bitetrahedra; results not shown). This is due to the small range of dielectric constants studied; the effect could only realistically be seen for changes of over a refractive index unit.

Understanding such interrelations between size, composition, and dielectric environment is essential to a holistic design of plasmonic sensors and devices. For example, in the design of a sensor, it is always best to have a narrow peak; if this sensor is based on a random particle array, the quadrupole-like mode of Ag should appeal considerably more than the other modes, as it is sharp and the size dependence of both its energy and width is low, reducing heterogeneous broadening in an ensemble. It should also be possible to interpolate the energy values between those obtained for Formvar and Si_3N_4, providing further predictive power to these finding.

3.2. *Interrelation of size, shape, and corner rounding*

The very first correlated shape-LSPR measurements on nanoparticles pointed out the critical importance of corner rounding on optical properties, and since then, very few studies systematically probed its effects despite the ever-present inhomogeneities in synthesized and fabricated samples. Indeed, in typical solutions of nanocubes and bipyramids, rounding can range from a few to a few tens of nm. Rounding is responsible for a significant amount of scatter in figure 2 and similar plots in the literature. In fact, corner geometry has a major impact not only on the resonance energy and width, but also on the size effects: rounded particles do not have the same size-dependence of the LSPR energy as sharp particles. This conclusion was reached from the statistical analysis of the two large data sets in figure 3, on Au cubes and Ag bipyramids. The equations describing the multiparameter fits, and their associated error (95% confidence level on the coefficients), were found to be

$$\text{Bipyramid LSPR (eV)} = -0.0051(3)*a(\text{nm}) + 0.019(3)*r(\text{nm}) + 2.69(6) \quad (1)$$
$$\text{Cube LSPR (eV)} = -0.0039(15)*a(\text{nm}) + 0.007(2)*r(\text{nm}) + 2.34(12) \quad (2)$$

Where a is the particle size (side length for cube, height of the triangular base for bipyramids), and r is the truncation and rounding as defined in [18,19]. Such well-characterized coefficients provide evidence that rounding is not simply reducing the length of a particle; by affecting the near-field distribution, rounding contributes more than the equivalent change in distance between the tips of a particle, i.e. the effect is not only geometric. Indeed, if it were the case, and truncation would simply lower the distance between poles of opposite charges in the dipolar resonance, the coefficient of r in equation 1 would be negative with a value equal to the coefficient of a), i.e. a fit relating the plasmon energy to the truncated height (a-t) would be best. Clearly, this is not the case as the rounding coefficient is, for bipyramids, about four times the size coefficient. This effect is also observed in computational results on Ag and Au rods based on Mie-Gans theory [18,21]. It is thus important to consider rounding when making predictions of the plasmon behaviour.

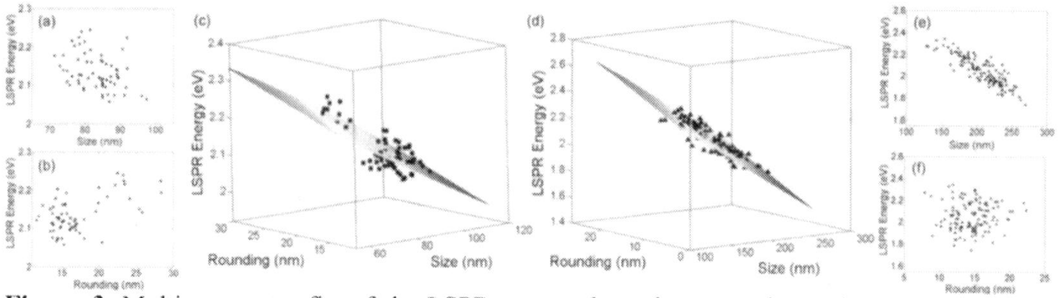

Figure 3. Multi-parameter fits of the LSPR energy dependence on size and corner rounding show similar behaviour in Au nanocubes (a-c) and Ag bipyramids (d-f). Parts d-f adapted with permission from ref.[18]

4. Outlook, opportunities and limitations

Given that structure is a powerful handle to tune the optical response of plasmonic nanoparticles, characterizing structure-function relationships is important in the fundamental understanding of the phenomenon as well as in the design of plasmon-based devices. The high-throughput techniques and statistical analysis presented here yield a more complete and quantitative understanding of such structure-function relationships. The effect of rounding, composition, substrate and size were found to be interrelated, and given the large data sets obtained, this interrelation could be explored, leading to better understanding of size and rounding, as well as quantitative, predictive rules.

Results on single particles can thus be a powerful tool to study materials; however, they are not necessarily the only choice, and clearly not always valid. In the case of small data sets with large variations in multiple components (size, shape, rounding, etc), results may be misleading or have no statistical significance; tests should always be performed on such data to confirm the significance and provide the errors associated with the quantification.

Single particle investigation and structure-function relationships need not be limited to the investigation of the plasmon energy and width correlated with particle shape. In fact, optical-structural correlations can be used to probe a variety of systems, revealing and quantifying other phenomena such as enhancement factor in surface-enhanced Raman spectroscopy[22], polarization-dependent behaviour in anisotropic nanoparticles[13], absorption[23], etc. It is expected that such correlative approaches will provide important clues on the behaviour of nanoparticles, even beyond plasmonic systems. However, introducing new instruments and/or techniques result in additional difficulties related to obtaining a sufficient amount of data to quantify the effects studied; this constraint currently affects refractive index sensitivity, elemental analysis in alloys, and polarization measurements, for example.

Most correlative measurements involve electron microscopy as a structural characterization tool; images are almost exclusively used to measure the size and shape of the nanoparticles. Electron microscopy offers a much larger range of tools, however. The interaction of the electron beam with a metal particle can be used to probe the near field behaviour of plasmon resonances, i.e. the electric field surrounding the particle as a result of electron oscillation at its surface [24]. However, the energy resolution of electron-based techniques, at best around 0.1 eV, is two orders of magnitude worse than that of optical measurements (easily few meV), and instrumentation is vastly more costly. Electron-based approaches are still tremendously useful because of their unsurpassed spatial resolution, an attribute impossible in optical spectroscopy due to the large wavelength of visible light.

Acknowledgments

This work was supported by the NSF MRSEC (DMR-1121262) at the Materials Research Center of Northwestern University.

References

[1] Lakowicz J R, Geddes C D, Gryczynski I, Malicka J, Gryczynski Z, Alsan K, Lukomska J, Matveeva E, Zhang J, Badugu R and Huang J 2004 *J. Fluoresc.* ***14***, 425.
[2] Stiles P L, Dieringer J A, Shah N C and Van Duyne R P 2008 *Annu. Rev. Anal. Chem.* ***1***, 601.
[3] Jeanmaire D L and Van Duyne R P 1977 *J. Electroanal. Chem.* ***84***, 1.
[4] Chen C K, Heinz T F, Ricard D and Shen Y R 1983 *Phys. Rev. B* ***27***, 1965.
[5] Lee K-S and El-Sayed M A 2006 *J. Phys. Chem. B* ***110***, 19220.
[6] Huang X, El-Sayed I H, Qian W and El-Sayed M A 2006 *J. Am. Chem. Soc.* ***128***, 2115.
[7] Neumann O, Urban A S, Day J, Lal S, Nordlander P and Halas N J 2012 *ACS Nano* ***7***, 42.
[8] Lal S, Clare S E and Halas N J 2008 *Acc. Chem. Res.* ***41***, 1842.
[9] Mayer K M and Hafner J H 2011 *Chem. Rev.* ***111***, 3828.
[10] Anker J N, Hall W P, Lyandres O, Shah N C, Zhao J and Van Duyne R P 2008 *Nat. Mater.* ***7***, 442.
[11] Sepúlveda B, Angelomé P C, Lechuga L M and Liz-Marzán L M 2009 *Nano Today* ***4***, 244.
[12] Mie G 1908 *Ann. Phys.* ***25***, 377.
[13] Tamaru H, Kuwata H, Miyazaki H T and Miyano K 2002 *Appl. Phys. Lett.* ***80***, 1826.
[14] Mock J J, Barbic M, Smith D R, Schultz D A and Schultz S 2002 *J. Chem. Phys.* ***116***, 6755.
[15] Henry A-I, Bingham J M, Ringe E, Marks L D, Schatz G C and Van Duyne R P 2011 *J. Phys. Chem. C* ***115***, 9291.
[16] Zijlstra P and Orrit M 2011 *Rep. Prog. Phys.* ***74***, 106401.
[17] Ringe E, Sharma B, Henry A I, Marks L D and Van Duyne R P 2013 *Phys. Chem. Chem. Phys.* ***15***, 4110.
[18] Ringe E, Zhang J, Langille M R, Mirkin C A, Marks L D and Van Duyne R P 2012 *Nanotechnology* ***23***, 444005.
[19] Ringe E, McMahon J M, Sohn K, Cobley C M, Xia Y, Huang J, Schatz G C, Marks L D and Van Duyne R P 2010 *J. Phys. Chem. C* ***114***, 12511.
[20] Myroshnychenko V, Rodríguez-Fernandez J, Pastoriza-Santos I, Funston A M, Novo C, Mulvaney P, Liz-Marzán L M and García de Abajo F J 2008 *Chem. Soc. Rev.* ***37***, 1792.
[21] Prescott S W and Mulvaney P 2006 *J. Appl. Phys.* ***99***, 123504.
[22] Kleinman S K, Ringe E, Valley N, Wustholz K L, Phillios E, Scheidt K A, Schatz G C and Van Duyne R P 2011 *J. Am. Chem. Soc.* ***133***, 4115.
[23] Tcherniak A, Ha J W, Dominguez-Medina S, Slaughter L S and Link S 2010 *Nano Lett.* ***10***, 1398.
[24] Nelayah J, Kociak M, Stéphan O, García de Abajo F J, Tencé M, Henrard L, Taverna D, Pastoriza-Santos I, Liz-Marzán L M and Colliex C 2007 *Nat. Phys.* ***3***, 348.

ETEM observation of Pt/C electrode catalysts in a moisturized cathode atmosphere

K Yoshida[1,2], X Zhang[2,3], N Tanaka[2,3], E D Boyes[4] and P L Gai[5]

[1] Institute for Advanced Research, Nagoya University, Chikusa-ku, Nagoya, 464-8603, Japan
[2] Nanostructures Research Laboratory, Japan Fine Ceramics Center, Atsuta-ku, Nagoya, 456-8587, Japan
[3] Department of Crystalline Materials Science and Ecotopia Science Institute, Nagoya University, Chikusa-ku, Nagoya, 464-8603, Japan
[4] Departments of Physics and Electronics and The York JEOL Nanocentre, University of York, Heslington, York, YO10 5DD, UK
[5] Departments of Chemistry and The York JEOL Nanocentre, University of York, Heslington, York, YO10 5DD, UK

E-mail: ky512@esi.nagoya-u.ac.jp

Abstract. There have been reports of challenges in designing platinum carbon (Pt/C) electrode catalysts for PEMFC. Pt/C electrode catalysts deactivate much faster on the cathode (in moisturized O_2) than on the anode (in H_2). To understand influences of moisture and oxygen on the deactivation of the Pt/C catalysts in proton-exchange-membrane fuel cells (PEMFCs), spherical-aberration-corrected environmental transmission electron microscopy (AC-ETEM) was applied with a high-speed CCD camera. Structural changes of the Pt/C electrode catalysts were dynamically recorded in moisturized nitrogen, oxygen and hydrogen. The mass spectrometry confirmed the moisture content (between 5 to 30 %) of nitrogen driving gas through a humidifier. Coalescence of platinum nanoparticles (D = 3.24 nm) was carefully evaluated in pure N_2 and moisturized N_2 atmosphere. The Pt/C showed considerable structural weakness in a moisturized N_2 atmosphere. Comparable results obtained by AC-ETEM in different gas atmospheres also suggested ways to improve the oxygen reduction reaction (ORR). In this paper, the deactivation process due to moisture (hydroxylation) of carbon supports is discussed using for comparison the movement of platinum nanoparticles measured in moisturized nitrogen and pure nitrogen atmospheres.

1. Introduction

In recent years, proton exchange membrane fuel cells (PEMFC) and direct methanol/oxygen fuel cells (DMFC) [1] have shown great promise as new energy technologies because of their low operating temperature, fast start-up and potential for miniaturization. However, several breakthroughs are still needed to enable widespread use in electric automobiles. PEMFCs and DMFCs typically consist of bipolar plates, electrodes, a membrane separator and an electrode catalyst, and one of the biggest challenges is the synthesis of robust electrode catalysts [2].

PEMFCs use platinum nanoparticles on carbon black (Pt/Carbon) as the electrode catalyst for both the anode and cathode regions [3]. But it is now well established that degradation of the carbon support at the cathode limits the lifetime of Pt/Carbon catalysts and thus the performance of the PEMFC. The main degradation processes of the Pt/Carbon electrode catalyst system have been reported previously - platinum particle sintering, platinum dissolution, carbon support corrosion, carbon mono-oxide corrosion and detachment from the carbon support [4,5]. These degradation processes were indicated by structural changes shown in electron micrographs and through electrochemical measurements. Ex-situ and in-situ microscopy to show the dynamic behaviour of the fuel cell catalyst is very valuable to improve understanding of the degradation mechanisms and thus help to improve robustness. Environmental transmission electron microscopy (ETEM) [6], which was first reported in 1997, has proven to be one of the most efficient tools for in-situ visualisation of the deactivation of heterogeneous catalysts in a reactive gas atmosphere at the nanometer scale [7-9].

Here we report a progressive gas injection system for studying the dynamic behaviour of the heterogeneous catalyst in moisturized environmental conditions. The gas injection system was firstly applied to the Pt/Carbon electrode catalysts of PEMFC, then typical sintering of Pt nanoparticles in moisture is presented.

2. Experimental Procedure

Pt nanoparticles were deposited by sputtering (Ar Ion Sputter E-1030, Hitachi) on a 5 nm amorphous carbon film prepared by chemical vapour deposition (CVD coater NC5 Turbo, Enomoto AV) using methane (CH_4) and ethene (C_2H_4) as precursor gases. Mica sheets (G250-2, Agar Scientific) were used as substrates for the growth of the amorphous carbon film. CVD and ion sputter deposition were carried out at room temperature (288–300 K). Pt/a-carbon was characterised by dynamic aberration-corrected ETEM (Titan ETEM, FEI Company) in nitrogen (N_2, 99.9998%) or moisturized nitrogen gas atmosphere of 10^{-2} Pa. These gases were introduced from cylinders into the sample area. Gas inlet needle valve positions and the pumping rate of the differential pumping ETEM were set to known locations for achieving equilibrium partial pressures between 10^{-5} to 100 Pa in the sample area [10]. The focused electron beam (1800A/cm^2, illuminating a 20 nm × 20 nm sample area) was used to heat the sample and is also considered to create ionized gas molecules around the sample during imaging [11,12].

3. Results and discussion

Figure 1 shows the modified gas injection system. A hot stirrer kept the temperature of the humidifier at 370K. A PTFE boiling stone provide continuous moisture into the driving N_2 gas bubbled through the humidifier. A thermostatic chamber at 473K was used to introduce the moisturized gases to the differentially pumped E-cell through a needle valve. The mass spectrometry confirmed the moisture content of the nitrogen driving gas through the humidifier, the thermostatic chamber and the E-cell. The partial pressures of N_2 and H_2O at the E-cell were 9.17×10^{-3} Pa and 2.90×10^{-3} Pa respectively and the moisture content in the E-cell chamber was estimated to be 23.8 %. Water injection systems have been reported in previous works, such as a window type wet cell by T. Yaguchi et al. [13] and direct injection by one of the present authors [14]. However, those systems could not measure accurately the moisture content at the sample. Our present system enables the differential pumping E-cell to achieve a wet environment and directly measures the moisture content by Q-mass spectroscopy.

We present an application of the WetTEM (moisturised gas environmental transmission electron microscopy) for the Pt/Carbon electrode catalysts. Structural weaknesses of the Pt/Carbon in the cathode atmosphere have been a big issue to improve PEMFC for a long time. Our earlier study had clarified one main deactivation mechanism of Pt nanoparticles sintering due to carbon supports shrinking in an oxygen atmosphere [15]. Pt nanoparticles sintering in pure nitrogen (N_2, 99.9998%) and moisturized nitrogen gas atmospheres were compared to study the influences of water synthesis at the PEMFC cathode.

Figure 1. (a) Humidifier on the hot stirrer, (b) Thermostatic chamber and (c) gas injection and differential pumping line of the ETEM.

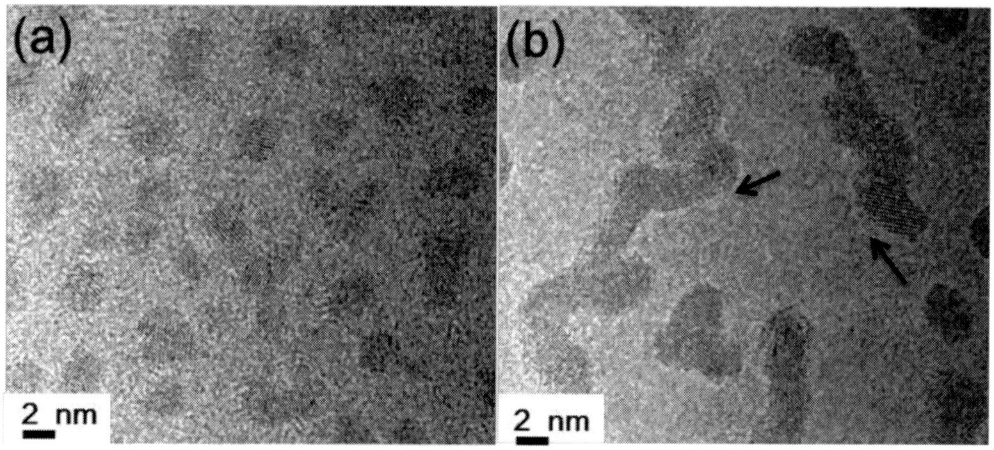

Figure 2. (a) SAC image at 230s of a movie obtained from the Pt/Carbon samples in pure N_2 environment. (b) SAC image at 40s of a movie obtained from the Pt/Carbon in 23.8% moisturized N_2 environment.

Sintering and migration of Pt nanoparticles observed in moisturized N_2 atmosphere was much faster than in a pure N_2 atmosphere as shown in the Selected Area Captured (SAC) images of Figure 2. Fig. 2(b) shows connected Pt nanoparticles which were typically observed in wet condition. The damage and shrinkage of the carbon support is not the reason for the connections because the granular pattern of the carbon support is still seen to survive. White contrast surrounding the Pt connections (arrowed in Fig. 2(b)) also indicates that the carbon films became thinner at the Pt/Carbon interface; presumably because of hydrocarbon desorption. We consider that physical adsorption of water and hydroxylation of the carbon surface is a main reason for the higher mobility of Pt nanoparticles in a moisturized N_2 atmosphere. The present in-situ observations suggest we should induce much stronger trapping sites on the carbon supports for use on the PEMFC cathode.

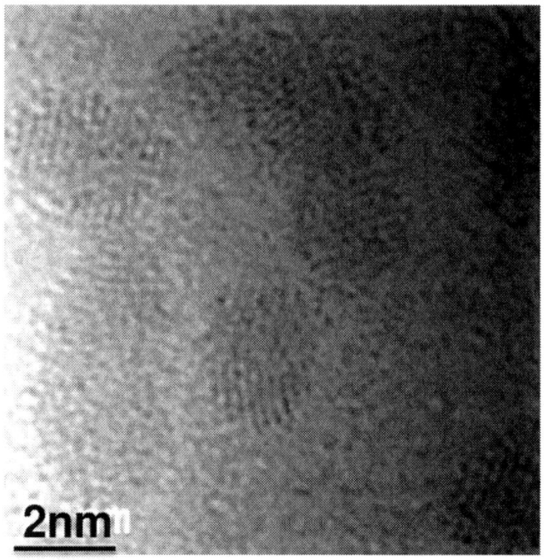

Figure 3. SAC image from a dynamic movie obtained with magnification of 1M times and time resolution of 0.05 s. Image size is 512 × 512 pixel.

Finally, we present the highest magnification SAC image for discussion of future experiments. The SAC image demonstrates the potential of our system for visualising single atoms in ETEM. Such dynamic visualization of single atom behaviour would help a deeper understanding of the mechanisms of deactivation processes for Pt/Carbon catalysts and contribute to development of more robust catalyst electrodes.

4. Conclusion

In this paper we introduce the modified gas injection system, which was newly developed in order to study the deactivation process due to moisture (hydroxylation) of carbon supports. The movements of platinum nanoparticles were successfully visualised on the atomic scale in moisturized nitrogen and pure nitrogen atmospheres.

5. Acknowledgments

Financial support from the Japan Society for the Promotion of Science (B 24710110), the Chubu Economic Federation and Aichi Prefecture are gratefully acknowledged. Author (K.Y.) thanks Young Leaders Cultivation Program of Nagoya University. P.L.G and E.B. thank the EPSRC for support under grant EP/J018058/1.

6. References

[1] S. J. Peighambardoust et al. 2010 *Int. J. Hydrogen Energy* **35** 9349.
[2] B. K. Kakati et al. 2008 *Fuel Cell* **08** 45.
[3] M. Boutonnet et al. 1982 *Colloid and Surfaces* **5** 209.
[4] X. W. Yu et al. 2007 *J. Power Sources* **172** 145.
[5] J. C. Meier et al. 2012 *ACS Catal.* **2** 832.
[6] E. D. Boyes and P. L. Gai 1997 *Ultramicrosc.* **67** 219.
[7] P. L. Hansen et al. 2002 *Science* **295** 2053.
[8] J. M. Howe et al. 2004 *MRS Bulletin* **29** 951.
[9] P. A. Crozier et al. 2008 *Ultramicrosc.* **108** 1432.
[10] K. Yoshida et al. 2013 *Microsc.* doi: 10.1093/jmicro/dft033.
[11] K. Yoshida et al. 2012 *J Electron Microsc.* **61** 99.
[12] A. N. Bright et al. 2013 *Ultramicrosc.* **124** 46.
[13] T. Yaguchi et al. 2012 *J. Electron Microsc.* **61**199.
[14] P. L .Gai et al. 2002 *Micrsc. Microanal.* **8**, 21.
[15] K. Yoshida et al. 2013 *Nanotech.* **24** 065705.

Development of an environmental high voltage electron microscope and its application to nano and bio-materials

N. Tanaka, J. Usukura, M. Kusunoki, Y. Saito*, K. Sasaki*, T. Tanji, S. Muto and S. Arai

Ecotopia Science Institute, Nagoya University, Chikusa-ku,
Nagoya, 464-8603, Japan
*Graduate School of Engineering, Nagoya University, Chikusa-ku,
Nagoya, 464-8603, Japan

*To whom correspondence should be addressed.
E-mail: a41263a@nucc.cc.nagoya-u.ac.jp.

Abstract. Environmental transmission electron microscopy has recently emerged as topics of great interest as well as ultra-high resolution electron microscopy using aberration correctors. Current research in this area has been focusing on dynamic observation with atomic resolution under gaseous atmospheres and in liquids. Nagoya University has been developing a new 1-MV high voltage (scanning) transmission electron microscope, which can be used to observe nano-materials under conditions that include the presence of gases, liquids and illuminating lights, and it can be also used to perform mechanical operations to nm-sized areas as well as electron tomography and elemental analysis by electron energy loss spectroscopy.

1. Introduction

Energy and environmental problems are the important concerns in our future society. Attention must be paid to these concerns from the first step of the development of new materials and devices. To properly assess the development, the test samples need to be analysed under actual usage/reaction conditions. The necessity of environmental and in-situ observation/analysis is increased for structural and elemental estimations[1]. Transmission electron microscopy (TEM) is one of the useful methods for the analysis of nano-materials and nano-devices with atomic dimensions. There are, however, several limitations in TEM such as (1)the requirement of small sample thickness, (2)observation in vacuum, and (3) projected images. As a resolution of the problems, we have been developing a new high-voltage(HV) scanning transmission electron microscope equipped with an open-type environmental cell, large tilting holders, high-sensitivity TV cameras and an imaging filter for electron energy loss spectroscopy(EELS)[2]. It was named HVEM for reaction science(RSHVEM). In this paper, we describe the development and some applications.

2. Instrumental development

Figure 1 shows a front-view of the RSHVEM (JEM-1000K RS). The height of the high-voltage tank is 6.7 m, and the length of the microscope column is 3.6 m. The electron gun uses a LaB_6 thermionic cathode, whose chamber is evacuated independently by two ion pumps to improve the vacuum and increase the life time of the cathode. The three-stage condenser lens system focuses an electron beam to a diameter of less than 1 nm, which enables bright-field and dark-field STEM and elemental-mapping STEM using EELS.

Figure 1. Front-view of RSHVEM.

The most important specification of the RSHVEM is high-resolution observation at 1 MV under gas pressures, for example, 13,300 Pa(=100 Torr). The other specifications are a point-to-point resolution in TEM of less than 0.15 nm, in STEM, less than 1 nm, an energy resolution in EELS less than 1.5 eV, and the use of dedicated sample holders for 3D tomography without the missing edge problems. In the RSHVEM, we adopted the open-type environmental cell in the pole-piece gap with help of differential pumping system. For this purpose, we developed a new side-entry sample holder without a stopper, and enabled the environmental cell to be inserted from the side of the stopper. When the cell is extracted, the instrument can be used as an ordinary HVEM under vacuum conditions. The image recording system for the RSHVEM consists of three CCD cameras for recording low magnification images and electron diffraction patterns (Hamamatsu; 2k x 2k), for high-resolution images (ORIUS; 2k x 2k), for EEL spectra and energy-filtered images(Gatan; imaging filter (GIF); 2k x 2k). Figure 2 shows a high-resolution TEM image of a [110]-oriented cubic silicon carbide(SiC) crystal recorded at 1 MV, where the dumbbell structure composed of silicon(Si) atomic columns and carbon(C) columns is clearly resolved. The separation of the dumbbell is 0.109 nm. For EELS at 1 MV, the zero-loss peak of a width of 0.87 eV was recorded, which proved a good performance of the GIF for spectrometry and a high stability of the total electric system of the RSHVEM. The instrument showed that 200 and 220 lattice fringes of gold are imaged at 11,000 Pa of nitrogen gas[2].

3. Examples showing unique features of the RSHVEM

(1)Copper particles in oxygen and hydrogen

Copper (Cu) particles were prepared by the vacuum deposition of Cu onto commercial silica(SiO_2) powders inside the RSHVEM. A specially designed sample holder with two filaments, one for sample support and heating, and the other for metal deposition, was used[3]. First, in-situ oxidation was performed at 700 C through the introduction of oxygen gas at 1 Pa. Identification of formation of copper mono-oxide was made from lattice images and EELS spectra. Reduction process was also observed dynamically. During the oxidation and reduction processes, the particle shape was similar[2]. Similar observation for gas reaction was made on porous gold catalysts in various gases including CO gas[4].

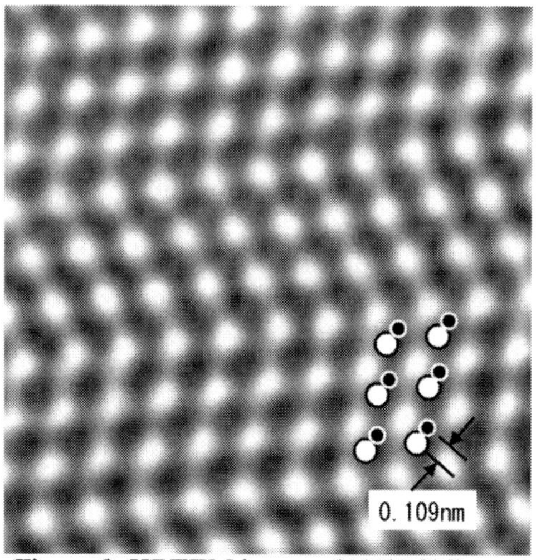

Figure 2. HRTEM image of [110] SiC crystal at 1 MV.

(2)Fracture process of Cu/Si interfaces in hydrogen

The mechanical strength of such interfaces is sensitive to foreign gas atoms which are diffusively supplied from an external environment. In particular, the effect of hydrogen on the interfacial strength is an important issue to study because hydrogen behaves as a typical embrittle element. Using the RSHVEM, we can realize in-situ observation of the fracture process of a semiconductor/metal interface in a LSI device in hydrogen gas.

The experimental TEM sample including a micro-cantilever was a sputter-deposited multilayer prepared via the focused ion beam(FIB) instrument, as shown in Fig. 3. The sample was mounted on a gold wire and attached to a piezo-driven nano-indenter holder (Nanofactory). The hardest SiN layer was pressed with a diamond indenter tip (upper-right) to avoid plastic deformation at the loading point. The atmospheric gas was a mixture of N_2/H_2 gases(95:5), and the pressure was 300 Pa. The in-situ observation revealed that the fracture initiated at the upper edge of the Cu/Si interface as shown in Fig. 3, Simultaneously, we measured the applied load; the measured fracture load of the Cu/Si interface was 75.3 μN. Such quantitative data is compared with calculation using the finite element method(FEM)[5].

(3) Elemental and chemical mapping using a post-column energy filter

We have studied semiconductor interfaces and oxide multilayers using EELS with help of multivariate curve resolution (MCR) method[6]. HVEM is advantageous for small beam broadening in samples, which is related to deterioration of the image resolution of energy filtered TEM images. The data demonstrate that energy-filtered (EF) images with better spatial resolution can be obtained with a shorter recording time using core-loss edges of a lower-energy region compared with those obtained by medium-voltage instruments. Higher voltages increase the mean free path(MFP) for the inelastically scattered electrons, and facilitate the low-loss region to utilize for EF imaging.

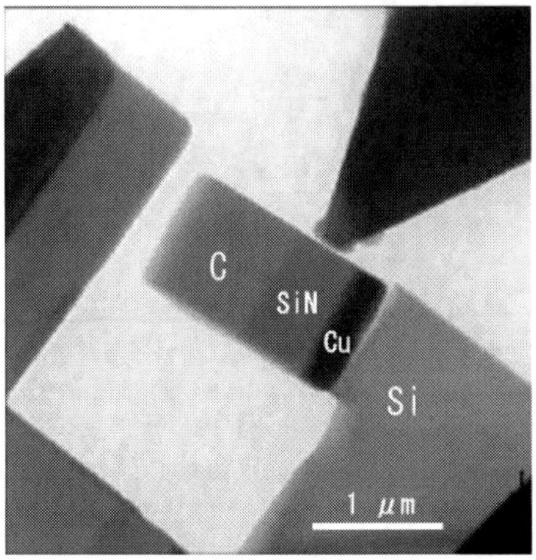

Figure 3. In-situ image of breakage at an interface of Cu/Si in H_2 gas.

(4) Three-dimensional reconstruction in RSHVEM

Other characteristic feature of the RSHVEM is a tomographic imaging without so-called missing edge(or cone) problems. By using a dedicated sample holder with 360 degree rotation. we observed carbon micro-coils(CMC) made from heat-decomposition from acetylene gas with iron or nickel catalysts[7], and studied the growth process from three-dimensional data of the coil. The 3D electron tomography of biological samples is also advantageous in HVEM because of its higher transmittance of electrons. We have observed cells and chromosomes of rat and yeast of a few μm size without slicing[8].

The RSHVEM incorporates scanning-image functions for STEM. Fig. 4 shows a bright-field (BF) STEM images of a yeast cell. We used the budding yeast (Saccharomyces cereviciae) cell to observe the structure of a whole cell. The sample was embedded in epoxy resin with 1 μm thickness after being chemically fixed. Using these images, the 3D reconstruction was successfully realized[9]. HV-STEM tomography has shown the

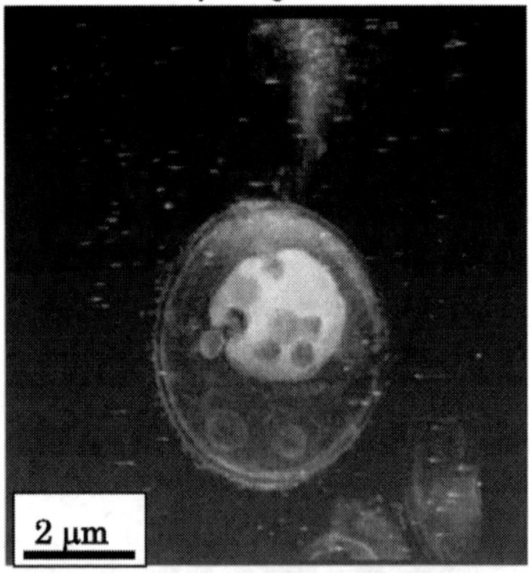

Figure 4. BF-STEM tomographic image of a yeast cell.

potential to visualize a whole cell structure clearly. This is because STEM avoids the deterioration of images caused by inelastic scattering in membrane films, embedding resins and gas atmospheres.

4. Trail of '*in-place* observation' for reducing the effects of irradiation

One of the most important problems to be considered in environmental electron microscopy is the effect of irradiation by ions of atmospheric gases produced by electron irradiation inside the cell. This effect increases with increasing gas pressures. Here we propose 'in-place observation' or 'in-vestigio' observation as one of the solutions to this problem, where electron beam is timely switched off during the chemical reaction between samples and gases by beam blanking. The RSHVEM is equipped with a reliable blanking system. The blanking is performed with electrostatic deflectors above the third condenser lens. We have succeeded in performing the 'in place observation' of (200) lattice fringes of a gold film on the TV screen of the present instrument at 1 MV without a large amount of sample drift[2].

5. Conclusion

In the present paper, the characteristic features of a new HVEM for reaction science were described as well as its experimental data, particularly focusing on mechanical operation in hydrogen atmosphere, 3D tomography by using STEM mode at 1 MV. Finally, we emphasize the importance of 'in-place observation', where the electron beam was smartly blanked during chemical reactions on various kinds of advanced materials to minimize the effects of irradiation of electrons and ions caused by incident electrons.

Acknowledgements

The authors would like to sincerely thank Profs. S. Hirano, T. Matsui of Nagoya University and Prof. Mori of Osaka university for their efforts to realize the new TEM/STEM. The instrument was completely installed with the best performance by JEOL Ltd., and in particular, Dr. S. Ohta and Mr. M. Ohsaki are acknowledged for their efforts. We would also like to thank Dr. K. Murata of NIPS and Dr. Y. Takahashi at Kansai University, for allowing his data in advance. The development of the RSHVEM was fully supported by the Ministry of Education, Science, Culture and Sport (MEXT) in Japan .

References

[1] Gai P L and Boyes E (2003), 'Electron Microscopy for Heterogeneous Catalysis', (IOP Press, Bristol).
[2] Tanaka N, Usukura J, Kusunoki M, Saito Y, Tanji T, Muto S, Arai S (2013), Microscopy, **62**, 205.
[3] Kamino T, Yaguchi T, Konno M , Watabe A, Marukawa T, Mima, T, Kuroda K, Saka H, Arai S, Makino H, Suzuki Y and Kishida K.(2005), J. Electron Microsc., **54**, 497.
[4] Fujita T, Guan P, McKenna K, Lang X, Hirata A, Zhang L, Tokunaga T, Arai S, Yamamoto Y, Tanaka N, Ishikawa Y, Asao N, Yamamoto Y, Erlebacher, J and Chen M, (2012), Nature Mater. DOI: 1-:1038/NMAT3391.
[5]Takahashi Y, Arai S, Yamamoto Y, Kondo H, Tanaka N(2012), Scr. Mater, to be submitted.
[6] Muto S, Yoshida T and Tatsumi K (2009), Mater. Trans, **50**, 964.
[7] Motojima S, Kawaguchi M, Nozaki K and Iwanaga H (1990), Appl. Phys. Lett., **56**, 321.
[8] Yamasaki J, Mutoh M, Ohta S, Arai S, and Tanaka N(2013), Microscopy, to be submitted.
[9] Murata, K., Esaki, M., Ogura, T., Arai, S., Yamamoto, Y. and Tanaka, N.(2013), Microscopy, to be submitted.

In-situ Reduction of Co_3O_4 in H_2 using Environmental HRTEM

M R Ward[1,2], **E D Boyes**[1,2,3], **P L Gai**[1,2,4]

[1] Department of Physics, University of York, York, YO10 5DD, UK
[2] York JEOL Nanocentre, University of York, York, YO10 5DD, UK
[3] Department of Electronics, University of York, York, YO10 5DD, UK
[4] Department of Chemistry, University of York, York, YO10 5DD, UK

E-mail: `mrw502@york.ac.uk`

Abstract. We present an *in-situ* study of the reduction of Co_3O_4 to CoO. Co based catalysts are promising for Fischer Tropsch process reactions in which heavy fuels such as diesel are synthesised from CO_2 and H_2. Co_3O_4 is the precursor and must be reduced by H_2 to Co. Although there have been many studied of this process, both *ex-situ* and *in-situ*, there is no consensus on how Co_3O_4 transforms to CoO including how defects influence the transformation mechanism. Our results show that dislocations are formed between Co_3O_4 and CoO regions and a fuse-wire like transformation is observed in crystals above 15 nm diameter.

1. Introduction

The commercialisation of synthetic diesel has received much attention since the 1980s [1, 2]. One source of synthetic diesel is from the Fischer-Tropsch process. The Fischer-Tropsch process involves many reactions including hydrogenation and gasification but it allows non-oil resources such as gas, biomass and coal to be converted into liquid diesel fuel. The reaction responsible for the conversion is the Fischer-Tropsch reaction: $(2n+1)H_2 + nCO \rightarrow CH_{2n+2} + nH_2O$

The Fischer-Tropsch reaction requires a catalyst otherwise the yield of desired product would be low [3]. Ni, Fe and Cu based catalysts were found to be catalytically active in this reaction but they generally had poor yields of heavier hydrocarbons such as diesel at low temperatures [3].

Co based catalysts are now the preferred choice for the production of diesel using the Fischer-Tropsch process [4, 5]. The Co takes the form of nanoparticles which may be loaded onto a support such as Al_2O_3. The catalysts are usually prepared using a catalyst precursor in the form of Co_3O_4 [2, 6]. Ultimately, the reduction of Co_3O_4 determines the activity of the Co catalyst.

It is well known that fcc Co_3O_4 can transform to an intermediate fcc CoO phase before reducing to either fcc or hcp Co [7, 8, 9]. The fcc phase of Co is believed to dominate at temperatures above 350 °C [10]. The role of defects in the reduction process to CoO is not clear and how it effects the reduction to Co. Here we present aberration corrected environmental TEM (AC-ETEM) study of the *in-situ* reduction of Co_3O_4 in H_2

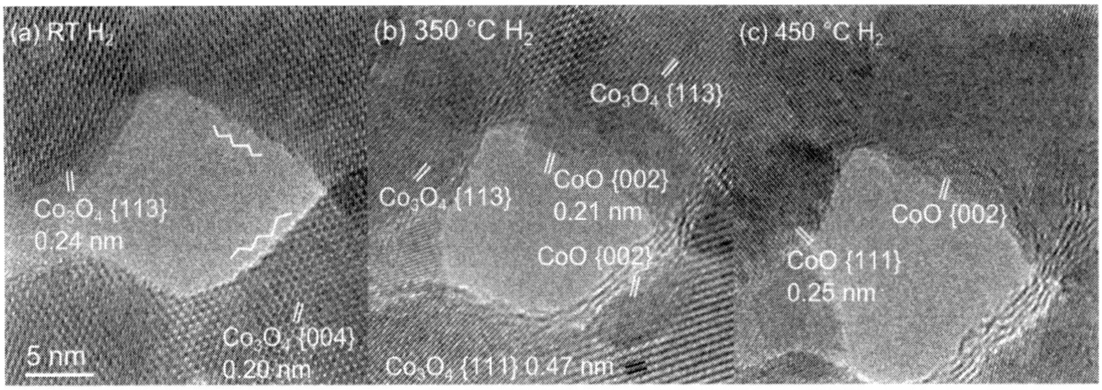

Figure 1. HRTEM reduction sequence of the reduction of Co_3O_4 to rocksalt CoO with (a) showing the Co_3O_4 crystal at room temperature (b) 350 °C and (c) 450 °C. The wavy lines in (a) indicate the presence of stepped Co_3O_4 surfaces which is where the CoO formation is initiated as shown in (b) at 350 °C. At 450 °C the entire parent crystal has transformed

2. Method

The Co_3O_4 powder used for the *in-situ* studies originates from Aldrich. For *in-situ* work (S)TEM specimens were prepared by suspending approximately 20 mg of the powder in ethanol and depositing 5 μl onto an empty Cu TEM grid. C films were not used because they become unstable at high temperatures and can react with H_2.

The microscope used for these studies was a double aberration corrected JEOL 2200FS with extra large pole-piece gap that allows the use of the Gatan 628 heating holder. Extensive modifications were made to the microscope to allow gas to be injected into the specimen chamber which are described elsewhere [11, 12]. Aberrations were corrected using a reference sample (Au nanoparticles on 2.5 nm amorphous C).

Regions of the Co_3O_4 were imaged before the gas was added for reference. The temperature was increased in 50 °C increments up to 550 °C. The regions of interest were not exposed to the electron beam during the heating steps. A landmark on the Cu grid far away from the regions of interest was used as a locator when the sample drifted during heating. At each temperature, images were recorded.

The images featured in this paper were all taken *in-situ* in H_2 at the temperatures indicated.

3. Results

Figure 1 shows a HRTEM reduction sequence of an area of Co_3O_4 at room temperature, 350 °C and 450 °C. The majority of the Co_3O_4 crystals contained a mixture of stepped surfaces and flat surfaces such as those shown in frame (a). The contrast of the similarly aligned grains in the figure is caused by the slight difference in orientation (a few °) resulting in different phase contrast in the HRTEM image.

We observed that CoO formation began at temperatures as low as 200 °C [13] but in general we found that the majority of crystals did not form CoO regions until 350 °C as is the case here. We observed that the CoO formation generally begins at stepped surfaces or at corners as shown in Figure 1(b). The CoO region cuts further into the Co_3O_4 crystal like a fuse-wire as the heating continues and by 450 °C the entire crystal had transformed to CoO. Despite heating to 550 °C we did not observe Co in this experiment. We believe this is because of the low pressure used in this experiment [13].

In the widest part of the crystal, which was about 50 nm wide, dramatic contrast variations

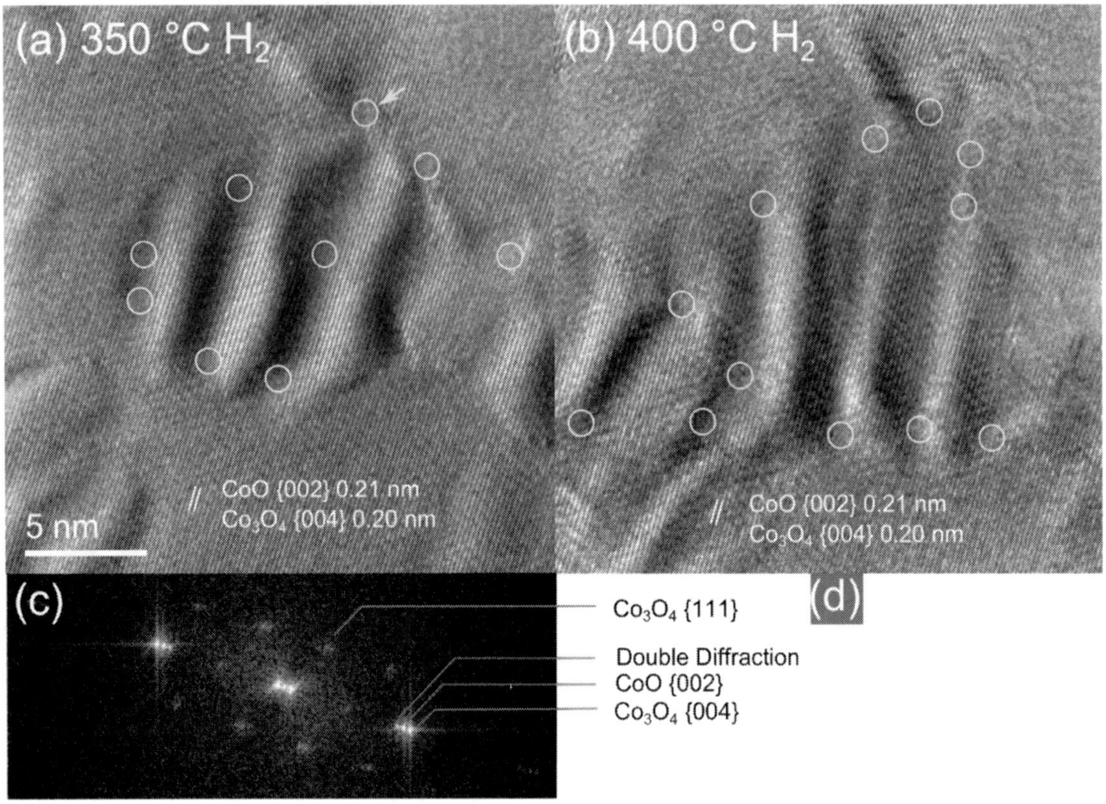

Figure 2. HRTEM images of a Co_3O_4 crystal approximately 50 nm wide showing dramatic contrast variations in the centre of the crystal at (a) 350 °C and (b) 400 °C in H_2. (c) is the FFT of (a) and (d) shows an example of a defect shown in the filtered image of (a). Defects are highlighted in the images with circles and the defect in (d) is highlighted with an arrow in (a)

were observed. Examples of this contrast is shown in Figure 2. Such oscillations in intensity have been attributed to the presence of defects and the strain associated with them. The HRTEM images in Figure 2 show that the contrast changes between 350 °C and 400 °C. The FFT indicates the simultaneous presence of CoO and Co_3O_4 as shown in Figure 2(c). There is an additional reflection on the opposite side of the Co_3O_4 $\{0\,0\,4\}$ reflection which is caused by the double diffraction of Co_3O_4 and CoO. This suggests that in the case observed in Figure 2, the Co_3O_4 to CoO interface is observed parallel to the electron beam unlike in Figure 1 where is it perpendicular.

By Bragg filtering the FFT (selecting one or more set of reflections) it was deduced that the contrast changes observed in the centre of the crystal at the two different temperatures is related to the presence of dislocations. In the filtered image (an example is shown in Figure 2(d)), the dislocations appear as forks. By including more reflections in the filtered image, most of the dislocations were deduced as having Burgers vectors of type $\vec{b} = {}^1/_2\langle 1\,1\,0\rangle$ [13]. Dislocations were also found close to Co_3O_4 to CoO interfaces [13]. We currently do not know the nature of the defects (e.g. edge or screw) but our preliminary data indicate that they play a role in the reduction.

4. Conclusions

We have demonstrated that defects form in the reduction of Co_3O_4 to CoO and mobile interfacial boundary forms between Co_3O_4 and CoO in large crystals. We believe the defects originate from the difference in lattice constants of Co_3O_4 and CoO [13].

Further work is required to deduce the nature of the defects and how they influence the formation of Co. Higher pressure runs are required and ideally a double tilt heating holder is required for deducing the Burgers vector of defects in more zone axes. Environmental HAADF-STEM [12] could also be useful since the contrast originating from the specimen is not dependant on phase contrast unlike HRTEM images which are often non-intuitive.

Acknowledgments

M. R. Ward thanks the EPSRC for PhD funding.

References

[1] Rofer-DePoorter C K 1981 *Chemical Reviews* **81** 447–474
[2] Schulz H 1999 *Applied Catalysis A: General* **186** 3 – 12
[3] Steynberg A, Espinoza R, Jager B and Vosloo A 1999 *Applied Catalysis A: General* **186** 41 – 54
[4] Iglesia E, Soled S L and Fiato R A 1992 *Journal of Catalysis* **137** 212 – 224
[5] Iglesia E 1997 *Applied Catalysis A: General* **161** 59 – 78
[6] Colley S E, Copperthwaite R G, Hutchings G J, Terblanche S P and Thackeray M M 1989 *Nature* **339** 129–130
[7] Lin H Y and Chen Y W 2004 *Materials Chemistry and Physics* **85** 171 – 175
[8] Bulavchenko O, Cherepanova S, Malakhov V, Dovlitova L, Ishchenko A and Tsybulya S 2009 *Kinetics and Catalysis* **50**(2) 192–198
[9] Potoczna-Petru D and Kępiński L 2001 *Catalysis Letters* **73**(1) 41–46
[10] Dinega D P and Bawendi M G 1999 *Angewandte Chemie International Edition* **38** 1788–1791
[11] Gai P L and Boyes E D 2009 *Microscopy Research and Technique* **72** 153–164
[12] Boyes E D, Ward M R, Lari L and Gai P L 2013 *Annalen der Physik* **525** 423–429
[13] Ward M R, Boyes E D and Gai P L 2013 *ChemCatChem* 2655–2661

Low-pressure ETEM Studies of Au assisted MgO Nanorod Growth

Linus D. L. Duchstein, Christian D. Damsgaard, Thomas W. Hansen, Jakob B. Wagner*

Center for Electron Nanoscopy, Technical University of Denmark, DK-2800 Kgs. Lyngby, Denmark

*e-mail address for correspondence: jakob.wagner@cen.dtu.dk

Abstract. Environmental transmission electron microscopy (ETEM) studies MgO nanorod growth from Au catalyst nanoparticles in a controlled gas atmosphere have been performed, in order to elucidate the mobility of Au surface atoms and the configuration of the Au/MgO interface. MgO nanorod growth is driven by the electron beam and found to be strongly dependent on the gaseous environment in the microscope and electron beam current density.

1. Introduction

Nanometer sized Au particles supported on MgO (100) surfaces have been used as model catalyst for CO oxidation at room temperature [1,2]. The exact morphology and size of the particles are strongly linked to the activity of the reaction as shown theoretically by McKenna *et al.* [3]. Environmental TEM is an excellent tool to study individual nanoparticles under conditions similar to their working environment [4]. However, the highly energetic electron beam might influence the sample in undesirable ways. The focussed electron beam has earlier been shown to cause growth of MgO nanorods on the MgO support under high dose irradiation in high vacuum [5,6]. In the present work, the presence of residual gas in the microscope column is shown to enhance the MgO nanorod growth significantly. The growth of MgO nanorods are found at relatively low beam currents compared to earlier work[6]. In order to exploit the opportunities of ETEM on such beam sensitive systems a more systematic approach has to be considered.

2. Experimental Details

MgO smoke particles are produced by combustion of a Mg metal ribbon to form close-to-perfect cubes exposing MgO (100) surfaces. The MgO cubes are collected on a TEM grid by placing the grid in the smoke for a few seconds. Au nanoparticles are deposited on the cubes by sputter-coating for 1s using a Cressington 208 HR Sputter Coater forming 2-6nm epitaxially oriented Au nanoparticles, see inset Fig. 1a.

The samples are investigated in an FEI Titan 80-300kV ETEM equipped with a differential pumping system that allows samples to be exposed to a gaseous environment. Further specifications of the microscope can be found in [3]. Using the differential pumping system rather than the original microscope pumping system consisting of ion getter pumps and a liquid nitrogen cold trap results in a slight degradation of the vacuum. The residual gas in the column is found to be predominantly water by means of mass spectrometry. The samples are studied in this residual gas in the pressure range 10^{-5}Pa to 10^{-4}Pa. All experiments were carried out at room temperature.

3. Results

Figure 1 shows a TEM image of the prepared sample together with a size distribution of the formed Au particles. The sputter coating parameters chosen resulted in nanoparticles with a size below 4nm, giving a realistic model catalyst system.

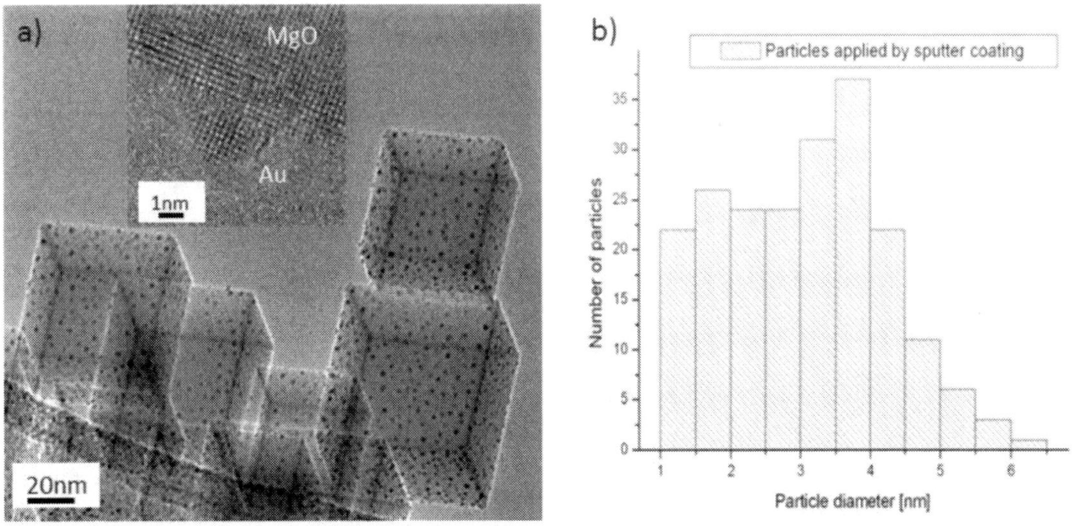

Figure 1. a) TEM micrograph of MgO supported Au nanoparticles. Inset: HRTEM micrograph of MgO supported Au nanoparticle b) Particle size distribution of the Au particles.

MgO and Au are both cubic crystals with a slight mismatch (~3%) in the lattice distance (a_{Au}=4.08Å, a_{MgO}=4.21Å). The Au nanoparticles are epitaxially oriented as seen in the inset in Fig. 1a. When the samples are illuminated by the electron beam, nanorods grow at the Au/MgO interface. Eventually the Au particles disappear and the nanorods dissolve in the electron beam. Figure 2a shows stills from a video capturing the growth and dissolution process of the nanorods. The measured nanorod length during growth and dissolution is presented in Fig. 2b.

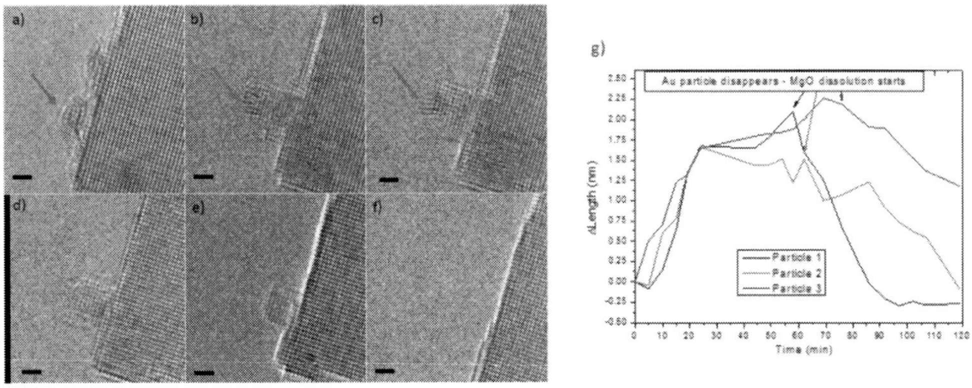

Figure 2. a)-f) HRTEM micrographs as stills from an image sequence showing Au-assisted growth of a MgO nanorod. When the Au particle disappears, the nanorod dissolves. The red arrows indicate the Au nanoparticle. P=10^{-4}Pa, j=1A/cm^2. The scalebars are 1 nm. g) Tracking the MgO nanorods grown at P=10^{-4}Pa and j=1A/cm^2.

A detailed view of the Au/MgO interface during growth can be found in Fig. 3. The time interval between each image is 3s with the particle/nanorod interface highlighted to illustrate the dynamics of the kinked interface.

Figure 4 summarizes our findings on the effect of the electron dose and water vapour pressure on the sample. The images are acquired after approximately 30min of electron irradiation in each case. In general higher electron beam current density (Figs 4c and d) increases the structural changes of the MgO surface. Increasing the (water) pressure in the microscope affects the mobility of the MgO species and results in increased roughness of the MgO (100) surfaces. The dashed lines in Fig. 4 indicate the layer that underwent structural change during electron irradiation.

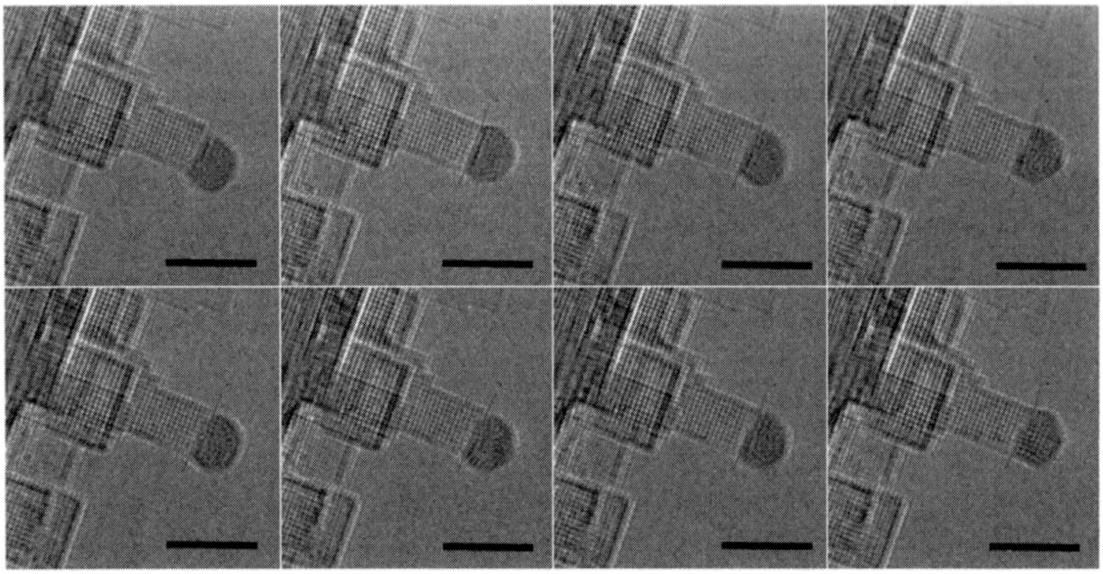

Figure 3. HRTEM micrograph of one nanorod and the attached Au particle in a series of stills separated by 3s. $P=10^{-4}$Pa, j=1A/cm^2. Scalebars are 5nm.

4. Discussion

The growth and dissolution observed in Fig. 2 have been quantified in order to illustrate the dynamical mechanism. After initial growth the nanorods reach a point of steady state where the mobility of the Au particle increases.

The shrinkage of the Au nanoparticles observed prior to the final dissolution of the MgO nanorods indicates Ostwald ripening of the Au nanoparticles. The diffusion barrier of Au atoms at the surface is relatively small as is the case for most metals and metal oxides [3]. Without Au as a collection agent for the MgO species the nanorods dissolve due to the increased mobility of MgO at the tip.

The longer nanorods typically have a diameter comparable to the growth particle at the tip. Even though the Au particles are highly mobile and in principle can act as collection point over a larger area resulting in wider features at the MgO surface, the competing dissolution of the MgO favours nanorods with diameters similar to the Au nanoparticles. It is evident that the Au nanoparticle is highly mobile, moving perpendicular to the growth direction of the nanorod in discrete atomic steps dictated by the epitaxial relationship between Au and MgO. Figure 3 clearly shows the mobility of the Au particle and the Au/MgO interface illustrating the Au/MgO interface to act as a collection point (due to negatively charged metal particles) where the highly mobile MgO species are trapped and recrystallize in rods.

The structural changes on the MgO surface depends on the electron beam current density and the pressure. The apparent roughness of the MgO surfaces after 30min. electron irradiation ranges from monolayer roughness to 3-4nm as indicated in Fig. 4. At low pressure ($P=10^{-5}$Pa) and relative low electron dose rate (10^{-15}A/nm^2) the surface mobility of MgO species is observed to be relatively small.

Increasing the electron dose or the pressure by leaking in water vapour, increases the mobility on the surface resulting in the formation of kinks and steps. The presence of water species in the surrounding environment influences the charge transfer in the system changing the overall energy landscape and thereby increasing the probability of MgO species getting trapped at the Au/MgO interface.

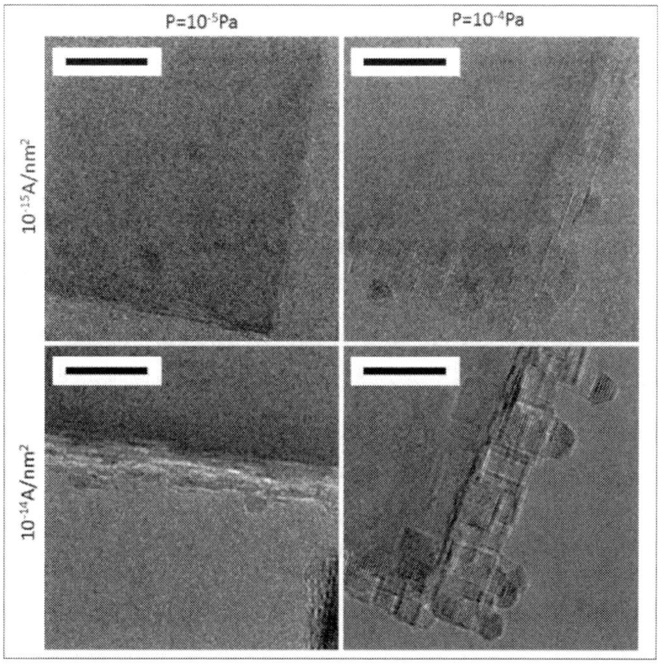

Figure 4. Au/MgO exposed to different electron dose rates at different pressures for approximately 30min in each case. Scale bars are 5nm.

5. Conclusion

The present results of Au nanoparticles on MgO under non-UHV conditions in an electron beam suggest that the mobility of MgO species and Au ability to trap such is strongly dependent on the gaseous surroundings and the electron beam current density. Quasi one-dimensional structures of MgO are grown and stabilized by the Au nanoparticles. As soon as the Au disappears, most likely due to Ostwald ripening, the MgO nanorod dissolves leading to an overall lower surface energy of the MgO cube.
The findings strongly suggest that the electron beam has a severe effect on the observations and materials behaviour especially when gases are involved. A slight increase in pressure in the ETEM compared to conventional TEM strongly enhances the electron beam induced mobility of MgO species at the MgO (100) surfaces.

References

[1] R Meyer, C Lemire, S K Shaikhutdinov and H-J Freund 2004 *Gold Bulletin* **37**, 72.
[2] M Haruta and M Date 2001 *Appl Catal a-Gen* **222**, 427.
[3] K P McKenna, P V Sushko and A L Shluger 2007 *The Journal of Physical Chemistry C* **111**, 2823.
[4] T W Hansen, J B Wagner and R E Dunin-Borkowski 2010 *Materials Science and Technology* **26**, 1338.
[5] P M Ajayan and L D Marks 1989 *Physical Review Letters* **63**, 279.
[6] A G Nasibulin, L Sun, S Hämäläinen, S D Shandakov, F Banhart and E I Kauppinen 2010 *Crystal Growth & Design* **10**, 414.

In-situ EBSD Phase Transformation and Recrystallisation

H S Ubhi[1], J Parsons[2], N Othen[2], S Campbell[3], R Poole[3] and A Gholinia[4]

[1]Oxford Instruments Analytical Ltd., High Wycombe, Bucks, HP12 3SE, U.K.
[2]Gatan UK, 25 Nuffield Way, Abingdon, Oxon, OX14 1RL, U.K.
[3]MAHLE Engine systems UK Ltd., Riccarton, Kilmarnock, KA1 3NA Ayshire, U.K.
[4]School of materials, University of Manchester, Grosvenor Street, Manchester, M1 7HS, U.K.

Corresponding author: A Gholinia (ali.gholinia@manchester.ac.uk)

Abstract. With the advent of in-situ heating stages that can fit into SEM's and the combination with EBSD, it is now possible to directly observe phenomenon such as phase transformations and recrystallisation at high spatial resolution and to link these processes to microstructural parameters.
This presentation will report some results from preliminary in-situ EBSD heating experiments conducted in an SEM on the transformation of ausenite to ferrite in a plain carbon steel and recrystallisation in bronze alloy strip cast on a steel substrate. The microstructural changes observed during these experiments will be reported in terms of EBSD maps, grains size and crystallographic texture that evolves during a) a heating cycle from ferrite to austenite and cooling to ferrite and b) the recrystallisation microstructure for bronze and steel during isochronal heating.

1. Introduction

In-situ heating stages for SEM's with conventional wire wound heaters are too large to be titled to 70 degrees or allowing shorter working distances. Consequently, for EBSD analysis most heating studies have been performed in the past [1,2], using dedicated Scanning Electron Microscopes (SEM) and heating stages. In addition, such stages may also lead to a lot of sample drift during analysis and also usually restrict the user to low heating rates.

A new heating stage developed by Gatan UK, shown in Figure 1, was used in this study. This stage has a very compact design allowing it to be used in almost any SEM. The compactness of this stage is due to its patented doped-Si wafer heater. It can be used in combination with EBSD analysis, where the stage can be tilted to 70 degrees while maintaining working distances as short as 12mm. It also has very stable high heating rates and sample drift is minimal. Furthermore the control thermocouple position allows very tight control and accurate sample temperature measurements. The heater also has a rapid heating rate, such that its maximum temperature of 950 °C can be reached in a few seconds.

The EBSD analysis was performed using the Oxford Instruments EBSD system. Heating above 400°C results in infra-red (IR) radiation which can disrupt EBSD analysis. However, this problem can be eliminated by using an IR filter in the EBSD detector, as implemented in the Oxford Instruments NordlysMax detector. Dynamic measurements such as in-situ heating are preferably done at high EBSD acquisition speeds in order to effectively capture rapid events that occur during heating.

Figure 1) In-situ heater at 70° stage tilt during EBSD analysis sample heated to 750°C.

2. Results and Discussion

The new heater was used to study the recrystallisation behaviour of a deformed bronze strip cast on a ferritic steel substrate system and the phase transformation of austenite to ferrite in a low carbon steel sample.

2.1. In-situ heating study of recrystallisation

It is usually very difficult to establish the nucleation sites of recrystallisation from the results of conventional ex-situ post-mortem studies nor is it easy to show the effect of interface such as steel and bronze on the recrystallisation.

The results from a successful in-situ heating with EBSD analysis are shown here from a steel-bronze system. A small sample size of about 2mm x 2mm x 2mm was used to have a very rapid response with good temperature control from the heater. The chamber was purged with nitrogen to reduce effects of surface oxidation.

Figure 2a shows the sample heated to 520 °C where there is evidence of nucleation sites for recrystallisation in bronze. It is apparent that as expected the nucleation starts at grain boundaries in the bronze and not at the steel interface [3].

Figure 2b shows that at 680 °C the steel goes through a sudden recrystallisation, whereas the bronze is partially only recrystallised. The bronze starts to recrystallise earlier than the steel substrate, however, the recrystallisation rate is much slower in bronze as compared to the steel.

2.2. In-situ heating study of phase transformation

Figure 3 shows the phase transformation from austenite to ferrite in a series of EBSD IPF and phase coloured maps, for a cooling rate of 1 °C per minute between 945 – 895 °C and 895 – 880 °C. Upon reaching 895 °C the temperature was held constant for a period of 6 minutes before cooling was resumed. EBSD data was acquired at 20kV and 870Hz at 1.5μm step size, where each map took 6 minutes. The average hit rate or indexing percentage for the maps was above 80%.

The relatively fine austenite grains of about 20μm are seen to transform to large ferrite grains of about 200μm. (The scale bar in the maps is 200μm)

The high speeds used for the EBSD analysis allowed this investigation to capture the details of the rapid dynamic events during the phase transformation in order to study phase boundary and texture analysis, which will be reported elsewhere.

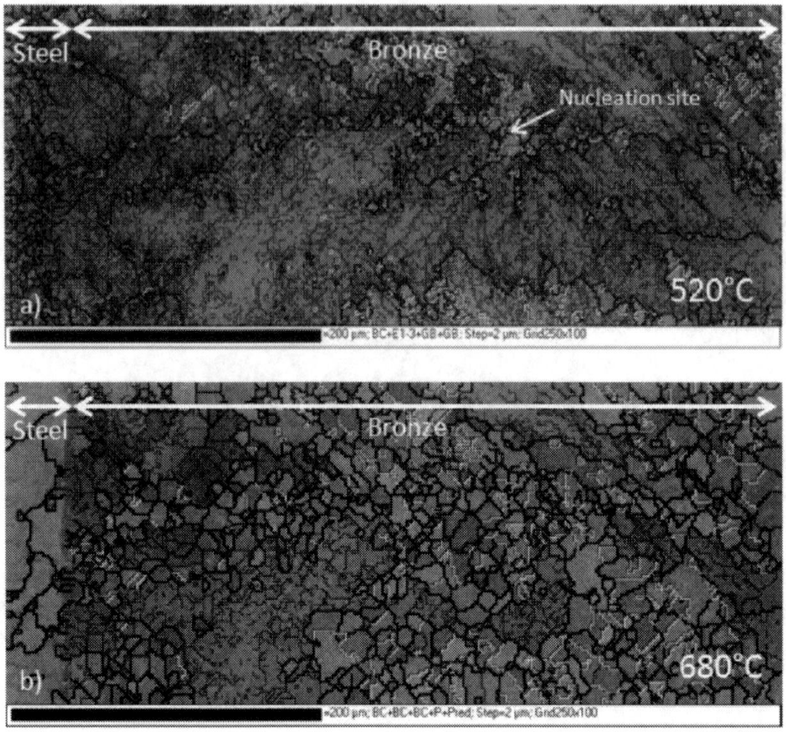

Figure 2) In-situ heating EBSD maps from bronze on a steel substrate in Euler colours where thick dark lines show high angle boundaries, thin dark lines show low angle boundaries and light lines show twin boundaries at a) 520 °C and b) 680 °C.

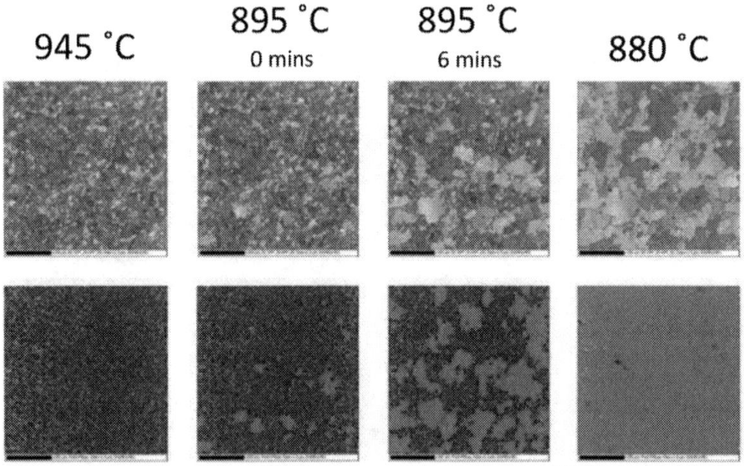

Figure 3) In-situ cooling EBSD maps are shown from 945 °C to 880 °C (for a cooling rate of 1 °C per minute between 945 – 895 °C and 895 – 880 °C). Upon reaching 895 °C the temperature is held constant for a period of 6 minutes, before cooling is resumed. The top row shows the IPF colour maps and the bottom row phase maps where austenite is blue (dark) and ferrite is red (light).

References

[1] G.G.E. Seward, et al., Scanning, Vol.24, 232-240 (2002).
[2] M.A. Meyer, et al., Microelectronic Eng., Vol. 64, 375-382 (2002).
[3] A. Gholinia et al., Materials Science and Technology, Vol.26, 685-690 (2010).

Journal of Physics: Conference Series 522 (2014) 012012 doi:10.1088/1742-6596/522/1/012012

Expunging Diffraction Contrast: EELS and HAADF in STEM

L. M. Brown[1]
Cavendish Laboratory, J. J. Thomson Ave., Cambridge CB3 0HE

E-mail: lmb12@cam.ac.uk

Abstract. In 1975 the Scanning Transmission Electron Microscope promised to deliver point-by-point information, both structural and chemical, relating to problems of grain boundary segregation, nanostructured phases, glassy materials, and heterogeneous catalysts. Diffraction contrast, so important in TEM, for a while continued to play an important role. But the full utilisation of Crewe's ADF technique, developed for single atoms, but now applied to crystals, required the elimination so far as possible of diffraction contrast. Archie Howie recognised that elimination of Bragg scattered beams, whilst retaining incoherently scattered electrons at larger angles, could be achieved by modifying Crewe's annular detector. High Angle Annular Dark Field, HAADF, is now the most widely used imaging mode in STEM. 'HAADF' might with justification stand for 'Howie's Adaptation of Annular Dark Field'!

1. Introductions

The 'Howie-Whelan' equations for diffraction contrast are extraordinarily successful. The concept of 'diffraction contrast', that is, the observation in the image plane of the microscope of intensity variation due to electrons scattered out of the objective aperture by diffraction, owes its origin to the earliest interpretation of transmission electron micrographs [1]. Until then, contrast had been thought of as 'absorption contrast', like 'mass-thickness' contrast in X-ray images. Such a notion cannot produce the striking effects of thickness fringes, bend contours, stacking-faults and dislocation images. The development by Mike Whelan of contrast from stacking faults, using the simplification of only one diffracted beam multiply scattered back and forth from and into the primary beam, but suffering an abrupt phase change at the stacking fault, led him to a formulation of the contrast using Bloch waves. This he extended to the situation where the phase change is a continuous function of beam traversal. The resulting equations were disseminated in Summer School lectures in 1959. In 1960 Archie, in the course of writing his thesis, produced 'intuitively' the famous coupled equations directly written for the primary and diffracted beam amplitudes scattered in an imperfect crystal. At this point, Archie and Mike decided to collaborate to produce the magnificent papers [2] showing compelling agreement between theory and experimental details of dislocation images. The arrival in Cambridge in 1959 of Hashimoto enabled the three men to include absorption, especially 'anomalous' absorption, into the equations, to produce a relatively complete theory [3]. On this basis, many details of dislocations in many different materials could be studied, especially including extended nodes in networks,

[1] To whom any correspondence should be addressed.

dislocation loops in quenched and irradiated metals, dipoles and forest interactions, etc. The equations, universally called the 'Howie-Whelan equations', now form a topic in most undergraduate lecture courses and in many standard texts.

In 1970 Archie and I both attended a summer school organised by Prof. Valdrè [4] and heard a series of lectures given by A. V. Crewe. It seemed to us that the future of electron microscopy lay in developing the ability to perform chemical analysis in the microscope, based on point-by-point acquisition of images simultaneously with spectra. Archie once said that once you have done something really fundamental – he was thinking of the development of diffraction contrast – it is difficult to change gear! However, under his leadership, the group nevertheless progressively overcame many technical and conceptual difficulties. It educated students, many of whom now lead the field and who apply STEM to a very wide range of problems. Ultimately, immediately after Archie and I retired, the group formed the nucleus of the SuperSTEM unit at Daresbury, which continues many of the lines Archie initiated. Studies of defects using diffraction contrast, once such a major activity, now play a minor role.

This paper is intended to provide a readable outline account of some of the developments along the way, for the benefit of new students who must wonder how the subject became what it now is!

2. Outline developments:

In addition to graduate lectures on electron microscopy, an important activity of the group was the weekly seminar. This was initiated under the banner of the 'Metal Physics Group' by Prof. Hirsch. It included both advanced presentations, often by visitors, but also by research students. Attendance at the lectures and the seminar was nearly obligatory. The seminars were held on Wednesday afternoons, to coincide with the one day of the week when there were no undergraduate laboratories in progress, and before the usual undergraduate supervisions began. This arrangement eliminated possible excuses for non-attendance. After about eight months' work, first year students presented an outline of 'progress so far' – a daunting experience, used in part to assess whether they would progress to be enrolled as a graduate student for a Ph.D. Informal discussion was strongly encouraged, and often new ideas emerged and critical responses which had a palpable effect on the quality of the work.

Every year graduate lectures were offered on electron microscopy and spectrometry, including practical work. The lectures were given initially by me, then for many years by Archie, then by successive generations of postdocs. They were attended usually by ten to fifteen students from other departments as well as from physics. A look at the schedules over the years reveals many well-known names: Peter Ward, Mike Stobbs, Alan Winter, Bernard Buxton, Yuan Jun, John Rodenburg, Andrew Bleloch.

Something of the intellectual progress can be gleaned from the fact that in 1985 the name of the group was changed from 'Metal Physics' to 'Microstructural Physics', and then in 2000, just before it was disbanded, to 'Nanostructural Physics'.

A research group in a department dedicated to science education is unlikely to have the resources to develop and construct entirely new instruments. The policy must be to purchase them, and then to add such modifications as seem timely and effective in the pursuit of research objectives. The table shows in outline how the group developed STEM after the first instrument arrived and the Physics Department (Cavendish Laboratory) moved to a new building. In addition to the line shown here, Archie used the acquisition of two front-line high resolution microscopes to investigate glassy structure, epitaxial thin films, and other microstructural problems. I continued investigations of defect structure in fatigued and irradiated metals, but added segregation studies at grain boundaries and in diamond.

3. Road to HAADF.

Early experience with dedicated STEM exposed us to many unfamiliar problems. High resolution analytical techniques included microdiffraction and EELS. Both yielded novel results rather quickly. But the acquisition of data was cumbersome, subject to errors such as drift of beam and specimen, as well as noise. As outlined in the table, significant

Table 1

YEAR	DEVELOPMENT	COMMENTS
1974/75	VG HB5 STEM ARRIVES: FIELD EMISSION GUN, TOP ENTRY SPECIMEN LOADING; INSTALLED IN NEW CAVENDISH	GRIGSON COIL SERIAL MICRODIFFRACTION ACQUISITION; MAGNET SCAN FOR SLOW SERIAL EELS COLLECTION; ADF DETECTOR
1974	ARCHIE COMMENCES STUDIES OF IN SITU EPITAXY IN VACUO	LECTURES ON EPITAXY AND DEFECTS BY J. W. MATTHEWS, M. J. STOWELL, W. M. STOBBS
1976	DRIFT TUBE ENABLES FAST HYSTERESIS-FREE ELECTROSTATIC SCANNING FOR EELS	PHIL BATSON (GLASGOW SUPPORT) COLLIEX STARTS ENERGY FILTERED IMAGING
1977	ARCHIE COMMENCES CATALYST STUDIES	
1977	EDX DETECTOR INSTALLED: X-RAY ANALYSIS OF CATALYSTS AND OTHER PARTICLES STARTS	LECTURES BY RUFUS RITCHIE ON PLASMONS AND INELASTIC SCATTERING
1977	PHIL GASKELL ARRIVES	
1979	CL DETECTION BY TAPERED SILVER TUBE	PENNYCOOK
1979	DIFFERENTIAL PHASE CONTRAST - SPLIT DETECTOR	JOHN CHAPMAN (GLASGOW SUPPORT)
1979	H.A.A.D.F.	HOWIE,TREACY,PENNYCOOK
1980	DENNIS McMULLAN ARRIVES	
1980	VIRTUAL OBJECTIVE APERTURE - CRAVEN DESIGN EFFECTIVELY REDUCES X-RAY BACKGROUND	CRAVEN GOES TO GLASGOW
1980	PHYSICS OF ELECTRON AND PHOTON DETECTION	LECTURES BY PENNYCOOK
1980	LECTURES ON STRUCTURE OF GLASSY MATERIALS	P. H. GASKELL, C. S. CARGILL
1981	HB501 ARRIVES FOR SURFACE STUDIES	CAROUSEL SPECIMEN LOADING - IMPROVED VACUUM AT SPECIMEN
1983	SECONDARY ELECTRON DETECTOR	IMESON/McMULLAN
1983	MICRODIFFRACTION CAMERA	RODENBURG/McMULLAN
1985	COINCIDENCE COUNTING	BLELOCH/MULLEJANS
1987	McMULLAN SPECTROMETER	CCD/EELS/ENERGY FILTERED MICRODIFFRACTION
1988	SP501 FOR BEAM WRITING	
1990	HOWIE BECOMES HEAD OF DEPARTMENT	
1991	STM PURCHASED	HOWIE
1992	ESEM PURCHASED (MP AIDS POLYMERS AND COLLOIDS GROUP GET STARTED)	DONALD
1994	PROF. RON BURGE ARRIVES	
1995	SPH. ABERRATION	KRIVANEK/DELBY

	CORRECTION	
2000	VOLTAGE STABILISER FOR SPECTROMETER	McMULLAN
2000	FUNDING SECURED FOR SUPERSTEM AT DARESBURY	GOODHEW, KIELY, BLELOCH, BROWN, BRYDSON
2001	RETIREMENT OF HOWIE AND BROWN	GROUP MERGES WITH PCS GROUP

improvements were made in the first few years. In addition, the group lacked experience in high vacuum and field emission. A crucial development was Archie's bold venture into studies of 'supported' catalysts, usually heavy metal particles of nanometre size dispersed in a matrix of much lower atomic number. He set Mike Treacy the task of looking at these. The problem soon encountered was the variable scattering due to diffraction effects from particles and matrix. A version of the annular detector pioneered by Crewe was provided with the microscope. The detector was positioned on the electron exit end of the specimen cartridge, giving an effective lower collection angle of about 12mr, with the upper angle limited by the bore of the cartridge to about 50mr. The resulting images were very poor, scarcely visible particles on a background made patchy by variations in thickness as well as diffraction contrast. Archie recognised that increasing the outer collection angle would at least partially eliminate the diffraction contrast. But how could one do this neatly? The cartridge bore seemed an impermeable obstacle.

As shown in Table 1, at the same time, Pennycook was developing methods for the collection of cathodoluminescence (CL) from single defects. He had designed and used successfully a tapered internally reflecting silver tube, the narrow end of which could be placed close to the source of light in the specimen, and the wider end of which could be matched to a quartz light guide, thence to a photomultiplier detector. With Treacy, he recognised that the CL collector could be modified by placing a small annular glass scintillator close to the specimen, thereby avoiding the bore of the cartridge. Fig. 1 shows the configuration. The inner angle of collection could be increased to about 100mr, and the outer angle to 400mr, essentially infinite so far as scattering theory is concerned. The resulting images were much improved, now useful, enabling size distributions of the particles to be obtained. The image could be further enhanced by dividing the HAADF signal by the energy-filtered low angle signal to eliminate some of the variation due to thickness in the substrate. This image is a version Crewe's 'Z-contrast' image, now made useful for crystalline specimens. Although HAADF stands for High Angle Annular Dark Field, it seems appropriate to refer to it as Howie's Adaptation of Annular Dark Field. Fig. 2, reproduced with permission from Treacy's thesis [5], shows images of Pd particles in a charcoal substrate. In Fig. 2(a) is a bright field image, energy filtered to remove losses greater than about 5eV. The thicker substrate at the top of the image transmits few zero-loss electrons. Fig. 2(b) shows the ADF image, which clearly suffers from confusing diffraction contrast and mass thickness contrast from the substrate. Fig. 2(c) shows the HAADF image, now showing the particles with greater and much more reproducible contrast, although residual substrate scattering is present. Further images of Pt on alumina show similar effects.

The principle of HAADF can be understood simply. Atomic motion, either zero point motion or thermal agitation, has the effect of producing a diffuse background in the diffraction pattern, but more importantly, of reducing the intensity in the Bragg peaks. When the amplitude of the motion exceeds the spacing of the Bragg planes, coherent interference between the scattered waves is lost. The atoms scatter independently, or incoherently. The higher order Bragg reflections, scattered to larger angles from the more closely spaced Bragg planes, are most affected. The trick of HAADF is to eliminate from the image those beams from the lower order reflections. If the scattering is truly incoherent, it is caused by Rutherford scattering from the atomic nuclei, with intensity proportional to the square of the atomic number, Z. Crewe's Z-contrast is achieved by dividing the HAADF signal by the inelastic signal, proportional very approximately to the number of electrons.

As first emphasised by Pennycook and Boatner [6], because they are formed from incoherently scattered electrons, HAADF images display intensity much less dependent upon foil orientation, thickness, and microscope defocus than images resulting from diffraction contrast or phase contrast. Atoms of higher atomic number than the background appear bright. It becomes possible to interpret the images intuitively. It is largely for this reason that HAADF imaging has become so popular. However, as pointed out clearly by Treacy, Howie and Pennycook [5], the Bragg scattering is never totally suppressed. For example, long wavelength phonons contribute to the root mean square atomic vibrations, but leave the relative motion of atomic near neighbours highly correlated. The effect of long range strain is to change the background intensity in the image, and the effect of electron channelling down atomic columns enhances coherent scattering and localised intensity. A review by Nellist [7] is recommended.

In spite of these complications, HAADF is by far the most utilised imaging mode in STEM. It is compatible with EELS, and transmits over an angular range large enough to acquire useful convergent beam diffraction patterns. A search through Google throws up 100,000 references of all sorts, still much less than a similar search for diffraction contrast, which yields nearly 6 million. It is fair to say that without HAADF, STEM would be far less widely practised than it is.

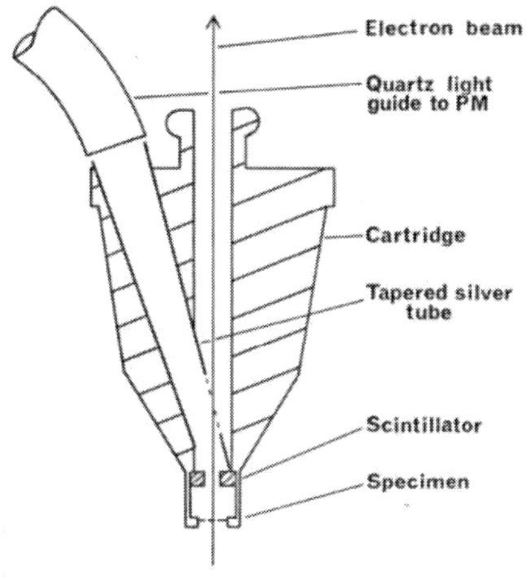

Fig. 1. The first high angle annular dark field detector [5]

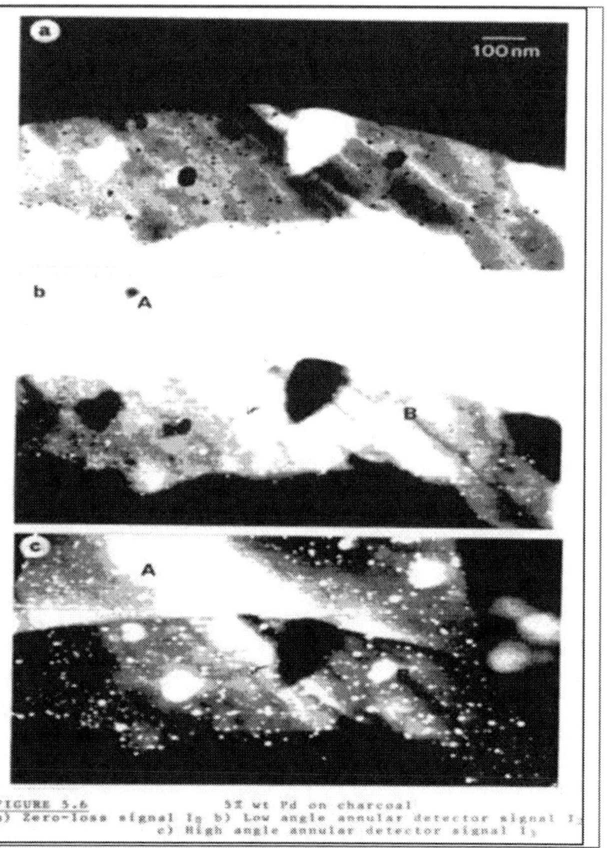

Fig. 2. Pt catalyst particles supported in graphite film. (a) zero loss image (b) ADF image (c) HAADF image [5]

4. Development of EELS

The early acquisition of spectra by scanning the magnetic field in the spectrometer was slow, noisy, and suffered from hysteresis. The rule was: the same spectrum must be obtained three times, and the features observable in all three were probably real! Many visitors played a role in improving the system, starting with Phil Batson's design of the 'drift tube' which enabled fast electrostatic scanning, and ending with Dennis McMullan's design for the parallel acquisition of spectra using CCD detection and optical coupling to the spectrometer output. Early work by Howie and his students confirmed that the spatial resolution achievable is very great, fundamentally limited by the impact parameter [8]. This permits atomic resolution for core transitions greater than about 100eV. At about the same time, it was demonstrated that lattice fringe contrast is preserved in images using only electrons which have lost energy to plasmon excitations with energies around 25eV, although the resolution is degraded by the increased scattering angles [9]. This latter observation promoted much discussion as to which comes first as the electron traverses the specimen, the inelastic scattering or the elastic diffraction! But the impact parameter for the plasmon is large enough that either way contrast is preserved. An extraordinary development of these ideas is the recent intense interest in the excitation of valence losses – mainly plasmon losses – in particles of nanometre size. It is possible to excite the losses when the probe is positioned not inside the particle, but near it (the so-called 'aloof' mode of excitation) and to observe energy losses due to various resonance modes. Howie has reviewed the fully relativistic theory [10, 11] which can treat edges and corners of particles, as well as interfaces. The interested reader is recommended to view recent spectacular recent images of excitations in triangular metallic nanoparticles [12]. This is a far cry from the diffraction contrast of nearly 60 years ago! Over that period, Archie and his group have been leaders in the remarkable progress in electron microscopy, witnessed by successive EMAG conferences.

5. References

1. Hirsch, P. B., Horne, R. W., and Whelan, M. J. 1956 *Philos. Mag.* **1** 677-684, also Whelan, M. J., Hirsch, P. B., Horne, R. W., Bollmann, W., 1957 *Proc. Roy. Soc. Lond.* **A240** 524-538.
2. Howie, A., Whelan, M. J., 1961 *Proc. Roy. Soc. Lond.* **A263** 217-237 & **A267** 206-230.
3. Hashimoto, H., Howie, A., Whelan, M. J., 1962 *Proc. Roy. Soc. Lond.* **A269** 80-103
4. Crewe, A. V., 1971 in *Electron Microscopy in Material Science,* U. Valdrè, Ed., Academic Press, N. Y.
5. Treacy, M. M. J., Howie, A., and Pennycook, S. J., 1980 *Electron Microscopy and Analysis 1979* T. Mulvey, Ed., Conference Series No. 52, Inst. of Physics, Bristol and London; p 261; also Treacy, M. M. J., 1979, *Electron Microscopy of palladium and platinum catalysts,* University of Cambridge Ph. D. thesis.
6. Pennycook, S. J., and Boatner, L. A., 1988 *Nature* 336 656-567
7. Nellist, Peter D., 2011 Ch. 5 in *Aberration-Corrected Analytical Transmission Electron Microscopy,* Rik Brydson, Ed., John Wiley and Sons, Chichester.
8. Craven, A. J., Gibson, J. M., Howie, A., and Spalding, D. R., 1978 *Philos. Mag.* **38** 519-
9. Craven, A. J., and Colliex, C., 1977, *Electron Microscopy and Analysis 1977* D. Misell, Ed., 271 – 274.
10. Howie, A., 1999, Interpretation of spatially resolved valence loss spectra, Ch 6 in *Topics in Electron Diffraction and Microscopy of Materials* , P. B. Hirsch, ed., Inst. of Physics, Bristol.
11. Howie, A., Aizpurua, J., García de Abajo, F. J., and Rafferty, B. K., 1999, *Jpn. J. of Electron Microscopy,* **48**, 673-679.
12. Nelayah, J., Kociak, M., Stéphan, O., García de Abajo, F. J., Tencé, M., Henrard, L., Taverna, D., Pastoriza-Santos, L., Liz-Marzan, L. M., and Colliex, C., 2007, *Nature Physics,* **3** 348-353.

6. Acknowledgments. The summary of the development of diffraction contrast in the introduction here is based on an account by Prof. Sir Peter Hirsch (e-mail dated 2007). The table of history is derived from group seminar lists. The dates are accurate only to within a year or two. I am grateful to Prof. Howie for comments, and to the organising committee and Prof. P. L. Gai for the invitation to speak at EMAG 13.

A methodology for the extraction of quantitative information from electron microscopy images at the atomic level

P L Galindo[1], J Pizarro[1], E Guerrero[1], M P Guerrero-Lebrero[1], G Scavello[1], A Yáñez[1], B M Núñez-Moraleda, J M Maestre, D L Sales[2], M Herrera[2] and S I Molina[2]

[1]Departamento de Ingeniería Informática, Escuela Superior de Ingeniería de Cádiz, Universidad de Cádiz, Campus Río San Pedro, s/n, Puerto Real, Cadiz 11510, Spain
[2]Departamento de Ciencia de los Materiales e I.M. y Q.I., Facultad de Ciencias, Universidad de Cádiz, Campus Río San Pedro, s/n, Puerto Real, Cadiz 11510, Spain

E-mail: pedro.galindo@uca.es

Abstract. In this paper we describe a methodology developed at the University of Cadiz (Spain) in the past few years for the extraction of quantitative information from electron microscopy images at the atomic level. This work is based on a coordinated and synergic activity of several research groups that have been working together over the last decade in two different and complementary fields: Materials Science and Computer Science. The aim of our joint research has been to develop innovative high-performance computing techniques and simulation methods in order to address computationally challenging problems in the analysis, modelling and simulation of materials at the atomic scale, providing significant advances with respect to existing techniques. The methodology involves several fundamental areas of research including the analysis of high resolution electron microscopy images, materials modelling, image simulation and 3D reconstruction using quantitative information from experimental images. These techniques for the analysis, modelling and simulation allow optimizing the control and functionality of devices developed using materials under study, and have been tested using data obtained from experimental samples.

1. Introduction: methodology overview

The extraction of quantitative information (such as strain, composition and morphology) from electron microscopy images is a challenging task. It has recently been shown an excellent agreement between experiments and theory in HAADF-STEM [1-3], demonstrating that current models of image formation are adequate in the modeling of both low and high-angle scattering, including thermal diffuse scattering, which dominates Z-contrast images.

Over the last decade, a set of tools for electron microscopy HAADF image analysis, nanostructure modeling and HAADF image simulation has been developed, giving rise to a methodology that enables us to obtain quantitative results for the understanding of materials at nanoscale.

The methodology first defines a 3D strained model using information coming from two sources: growing conditions and electron microscopy images. While growing conditions may give us detailed information about substrate composition, and expected nanostructures density, composition, size, shape, etc., more precise information may be obtained from the analysis of high-angle annular dark field (HAADF-STEM) electron microscopy images using both classical (EELS, strain mapping, etc .) and advanced techniques(compositional analysis with atomic resolution). The second step is the determination of a relaxed 3D atomistic model. This can be accomplished by calculating the displacement fields in the model, and applying these displacements to atomic coordinates of crystal positions in the bulk, or can be directly calculated using molecular mechanics. The final step is the simulation of electron microscopy images that can be compared with experimental ones in order to assess the accuracy of our model that might be refined in an iterative process. In the following paragraphs we will discuss in detail these steps.

Finally, let us note that at the development stage, computationally intensive studies are usually carried out using high performance computing. The development of compatible, portable, scalable, and user-friendly computational tools running into Graphical Processing Units (GPUs) will make possible its distribution to other universities, research centres and specialized companies.

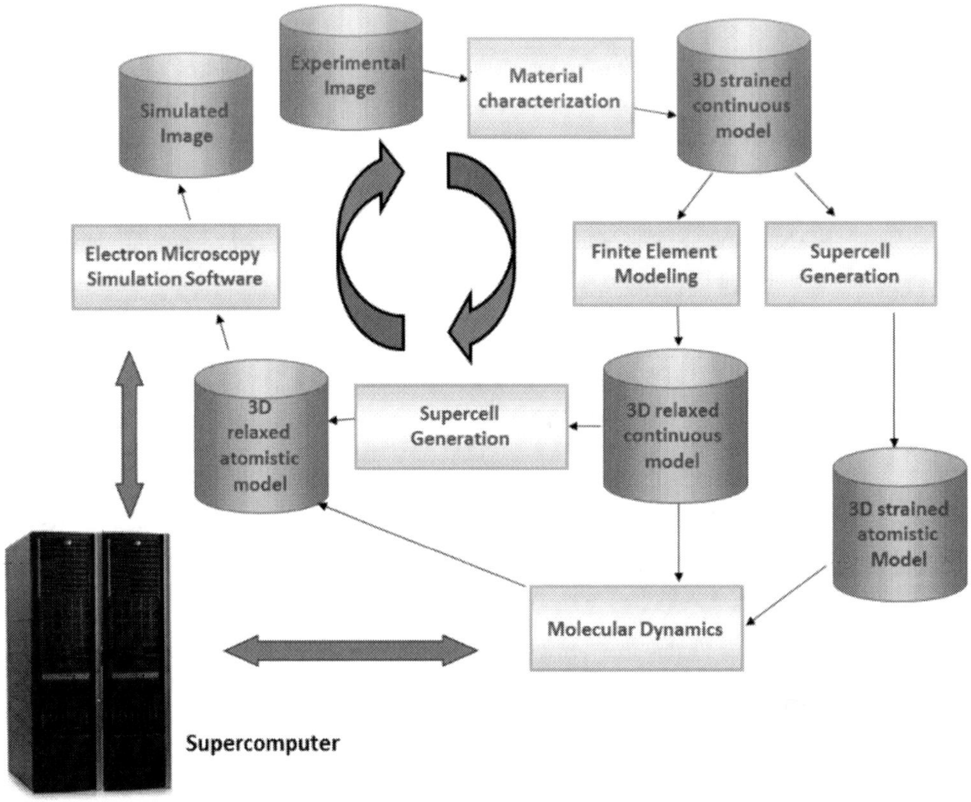

Figure 1. Schematic diagram of the methodology developed at the University of Cadiz for the quantitative extraction of information from electron microscopy images

2. Material characterization: strain and composition analysis techniques for 3D solid modelling

Two main techniques have been developed for the analysis of electron microscopy images. The first one, under the name Peak Pairs, is an algorithm for the determination of strain from high-resolution images . The second one is a technique to perform column-to-column compositional analysis of materials from the analysis of intensities in atomic-column resolved HAADF-STEM images. Both techniques have been implemented as plug-ins for the Digital Micrograph under the names PPA and qHAADF respectively, and are distributed by HREM Research, Inc.

2.1. Peak Pairs

Strain mapping techniques constitutes a well-established numerical image processing technique that allows, from the analysis of high resolution electron microscopy images, the determination of the displacement of lattice fringes in a strained material with respect to a perfectly regular lattice measured in a reference (relaxed) material. By measuring the position of peaks in the image, it is possible to precisely determine distortions of the lattice with respect to a given reference.

Peak Finding techniques work in real space and for strain analysis were first introduced in 1993[4] and then developed by different researchers [5-7]. In 2007 the Peak Pairs [8] algorithm was proposed, where from a noise filtered image, pairs of peaks were detected along a predefined direction and distance in the affine transformed space. Since this transformation reduces potential errors in detecting adjacent maxima in heavily distorted areas and the procedure works without user's intervention, Peak Pairs has become a useful tool for strain mapping [9,10].

PPA is the acronym of Peak Pairs Analysis, the plug-in for Digital Micrograph that implements the Peak Pairs algorithm, available from HREM Research Inc. Its key features are (a) Peak location at sub-pixel resolution using 2D filtering (Bragg, Wiener,...) and cubic interpolation techniques, (b) Robust Peak detection in noisy images using a small reference region, (c) Local strain tensor determination at atomic-column level, (d) Local representation of strain at the atomic-column level using color scales superimposed on HREM image, (e) Automatic peak intensity measurement and (f) Image distortion correction due to a projector lens system (CTEM) or a scanning system (STEM).

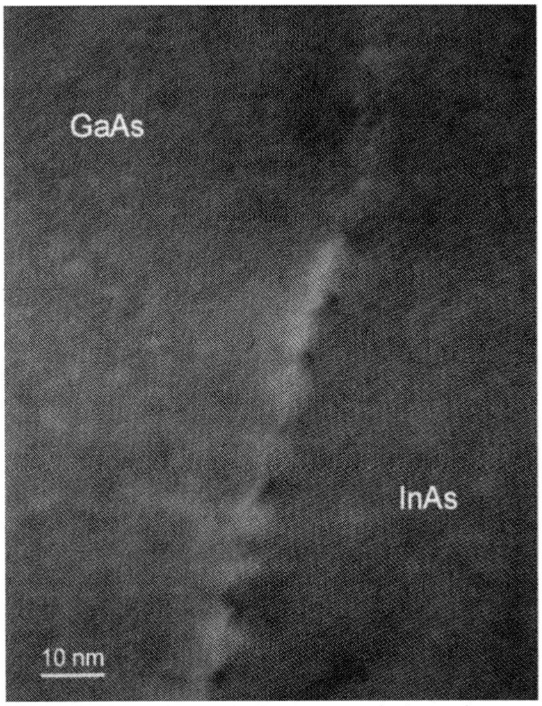

Figure 2. High-resolution transmission electron microscope (HRTEM) image of an InAs/GaAs (001) interface taken with a JEOL 2000 EX TEM at 200 kV. Point resolution = 0.21 nm.[9]

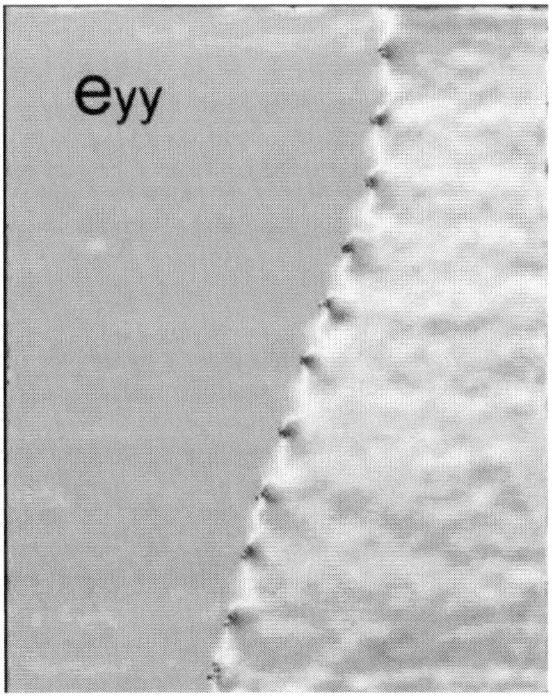

Figure 3. Components of the strain tensors exx , eyy , exy and eyx showing the presence of two well-distinguished materials corresponding to GaAs and InAs and a misfit dislocation network just at the InAs/GaAs interface.[9]

2.2. Colum-to-column compositional mapping (qHAADF)

In [11] a phenomenological method to determine quantitatively the composition of a material with Atomic column spatial resolution was described. The method is based on the analysis of local integrated intensities of aberration-corrected Z-contrast images. A set of reference samples of known composition is used to quantify the relationship between the experimental Z-contrast intensities and the thickness and composition of the analysed material. The method has been applied to different alloys such as InAsP[11] and InGaAs[12] and has also allowed to determine maps of Bi content (expressed as number of atoms) in each atomic column of a $GaAs_{1-x}Bi_x$ epitaxial layer[13].

qHAADF is the acronym of Quantitative HAADF, the plug-in for Digital Micrograph that implements this technique, available from HREM Research Inc. It performs column-to-column compositional analysis of materials from the integrated intensities of an atomic-column resolved HAADF-STEM image, and its key features are (a) Gives a map of integrated intensities and/or composition in materials with atomic-column spatial resolution, (b) Experimental HAADF images are used to obtain the fitting parameters to determine composition, (c) Automatic finding intensity peaks and noise filtering tools, (d) Compositional segregation profiles can be obtained from epitaxial layers, (e) Counts atoms in each atomic column and (f) Locates interstitial atoms in materials.

3. Finite Element Analysis and Molecular Dynamics for 3D atomistic modelling

Several methods have been used to simulate the strain distribution in nanostructures, being the finite element method (FEM) the most extended. Among its advantages we can mention lower computing and time requirements, allowing the use of arbitrary shape structures with anisotropic properties and non-uniform composition.

3.1. Finite Element simulations to determine atomic positions

Finite element method (FEM) modelling may be used to calculate the stress and strain in a three-dimensional defining a model taking into account shape, composition, boundary conditions and elastic parameters of the materials. The approximate shape may be determined from electron microscopy images taken at different conditions, and is usually defined visually using a CAD tool using combinations of several volumes having different geometric shapes. Composition modeling should take into account segregation and diffusion across different materials, and different models may be considered, such as the Muraki model for segregation. When available, composition maps are preferred, such as those coming from EELS maps used frequently to estimate the composition in nanowires. Boundary conditions are defined using symmetries, periodicities, or fixing them against rigid body shift in a desired direction, as it is usually done on the substrate. The basic mesh distribution for finite element calculation is usually calculated using the Delaunay algorithm, as it can be applied to all geometry objects. Anisotropic behaviour and values at 300 K are assumed for elastic constants of known materials, and Vegard's law is usually applied to estimate values at different compositions in the case of alloyed materials, such as ternaries or cuaternaries, for which elastic constants are unknown. Finally, the displacement, strain and stress fields may be calculated by solving anisotropic elastic theory equations over the domain of continuum mechanics. The displacement field may then be applied to atomic coordinates of crystal positions in the bulk in order to determine atomic positions of the unstrained crystal [10].

3.2. Molecular Dynamics atomic position determination

Molecular Dynamics (MD) is a computer simulation technique allowing atoms and molecules to interact for a period of time under known laws of physics. The major advantage of this techniques as compared to the previous one(FEM) is that can MD may be used in the presence of defects, including point (vacancies, interstitials), linear (dislocations) and planar (grain boundaries, stacking faults) defects. This possibility is crucial for a better knowledge of their mechanical properties and therefore of a great technological interest.

Once all rules have been set up about how atoms interact with one another, initial positions and velocities of all system particles are defined. We can use two different techniques: methods minimizing the potential energy (often called *energy minimization techniques*) and methods modeling the system behavior with time propagation. The most usual model in semiconductor simulation breaks energy up into two parts: covalent and non-covalent terms. Covalent terms describe the deviation of bond lengths (stretching energy), bond angles (bending energy) and torsion angles (torsion energy) away from equilibrium values, while non-covalent terms include forces between all non-bonded atoms. It has been shown that FEM and molecular dynamics give similar results.[14]

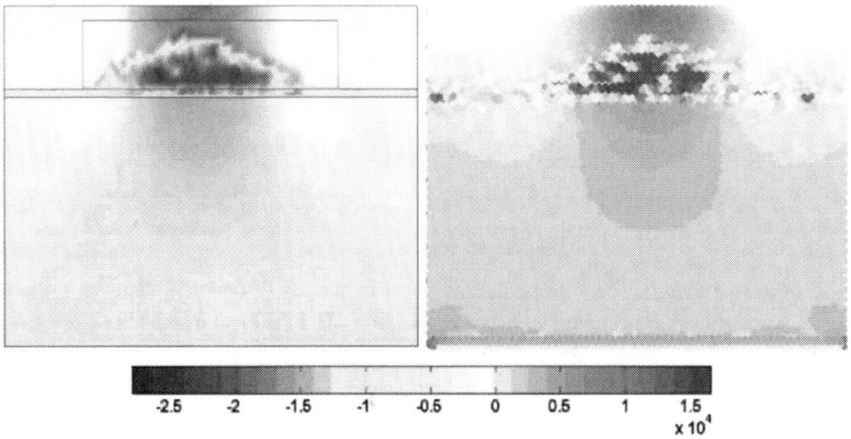

Figure 4. σ_{xx} component of the stress tensor (in bar): (left) Finite element model and (right) Molecular Dynamics (right)

4. SICSTEM: a simulation software for HAADF-STEM images

High angle annular dark field scanning transmission electron microscopy (HAADF-STEM) is a powerful tool to quantify size, shape, position, and composition of nano-objects with the assessment of image simulation. A parallel software called SICSTEM[15] has been developed that can afford HAADF-STEM image simulations of nanostructures composed of several hundred thousand atoms in manageable time.

The input to SICSTEM software is a supercell described as a set of (x, y, z) coordinates for atom positioning, its composition, site occupancy, and Debye–Waller factor and the characteristics of the microscope beam energy, third and fifth order objective aberrations, objective aperture, detector angles, etc. At each desired position of the specimen, an incident focused probe wave function is generated, and then transmitted and propagated across the specimen taking into account phase grating and TDS until it reaches the exit surface. Finally, the electron intensity is integrated on the detector. In HAADF-STEM simulation, the calculation of each pixel requires a complete multislice simulation thus requiring more than $1x10^6$ multislice calculations for a 1024x1024 image. Nevertheless, each pixel in the scanning process can be calculated independently from each other, making easy the division of tasks in parallel. The software has been designed to be able to generate not only one-dimensional line scans or two-dimensional images, but also focal series, very useful in three-dimensional HAADF-STEM research.

In order to test the accuracy across variations in thickness, the contrast ratio of Ga/As peaks has been determined for thickness values in the range of 0–20 nm. Figure 5 illustrates a comparison with Ishizuka's Fast Fourier Transform (FFT) multislice approach, where the contrast ratio of As/Ga peaks as a function of specimen thickness is shown.

We run the SICSTEM software in order to get a HAADF-STEM simulated image of an InP capped InAsP /InP nanowire. The composition for the test nanowire assumes that the maximum composition in the nanowire is 100% pure InAs. The image was simulated considering a 100 kV dedicated VG

Microscope HB501UX STEM having a probe size of 0.85 Å. The number of atoms in the supercell was 422184. The simulated image of the nanowire is shown in Figure 6. This image has been obtained at the supercomputer of the University of Cadiz, a cluster composed of 320 cores and a peak performance of 3.75×10^{12} floating point operations per second. The processing time was 78 h for a 560x1024 pixels resolution image.

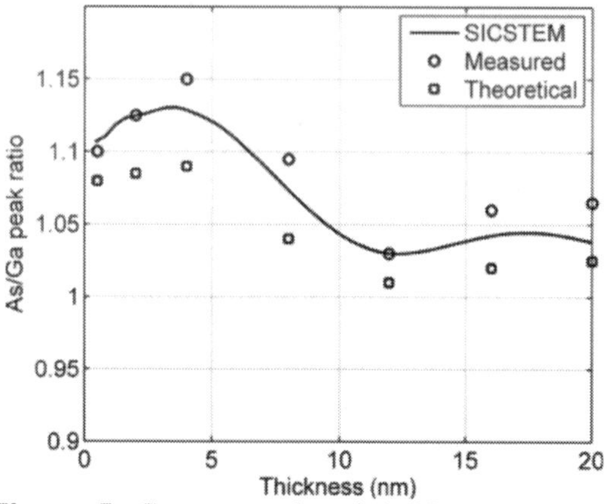

Figure 5. Contrast ratio of As/Ga peaks as a function of specimen thickness. Dots and squares represent the contrast ratios calculated using Izhizuka´s FFT multislice code and the theoretical estimates respectively while the line represents the As/Ga peak ratio obtained by SICSTEM software.[15]

Figure 6. HAADF-STEM simulated image of an $InAs_xP_{1-x}$ nanowire (560 x1024 pixels) using SICSTEM software.[15]

5. Electron Tomography techniques for 3D atomistic reconstruction

Electron tomography (ET) can be used to determine the shape, and therefore the properties, of an object. It deals with retrieving 3D information from 2D projections of the object under study. While ET has been used widely in biology, its application to the nanoscale world is relatively recent because of the limitations in resolution and precision aberration of electron microscopes. Tomography 3D reconstruction can be used to estimate spatial parameters of the object of interest as its dimensions and spatial distribution. Despite a variety of reconstruction algorithms exists and good quality reconstructions can be achieved in theory, in practice the quality of the reconstruction depends strongly on the quality of the input data. In fact, the real imaging process supplies noisy images and consequently no suitable input data for the direct application of such algorithms. In the field of electron tomography, some of the problems arise from geometrical distortions suffered from the image collected by the CCD. In addition, there are other factors that affect the final reconstruction, as the unknown real tilt axis inclination, the actual value of the tilt angles and the actual specimen position with respect to the holder.

5.1. Needle-shaped samples reconstruction using quantum dots position as fiducial markers

The alignment of tomographic series is a mandatory step in order to obtain reliable reconstructions. The alignment can be carried out in several ways, depending on the amount of user interaction needed. At atomic scale, classical tomography reconstruction can be replaced by Tomography Through Points (TTP) reconstruction, that is, the 3D information can be obtained by applying image processing techniques to the projections and by using geometrical models instead of the classical integral approach.

5.2. Reconstruction of 3D atomic positions in simulated graphene

The marker-based model for alignment allows a direct application in all those cases in which markers can be easily detected. Therefore, a tomographic approach for the quantification of the graphene corrugation is a good choice. Indeed, a single graphene layer has a simple geometric configuration (honeycomb regular structure) and the atoms can be located with great accuracy [16]. It means that a kind of data simplification can be performed in order to optimize the whole reconstruction process. The main points of the proposed approach are (a) treating each atom as a marker, it is possible to obtain a tomographic reconstruction of the structure, (b) using a small angles range to obtain the tomography, focusing problems are avoided and more homogeneous images can be obtained, (c) exploiting an algebraic approach, the reconstruction speed can be improved without loss of precision and (d) estimating the positions of the atoms in 3D space instead of using the whole image, as it is done in classical tomography, problematic issues such as missing wedge or projections noise can be avoided or at least minimized.

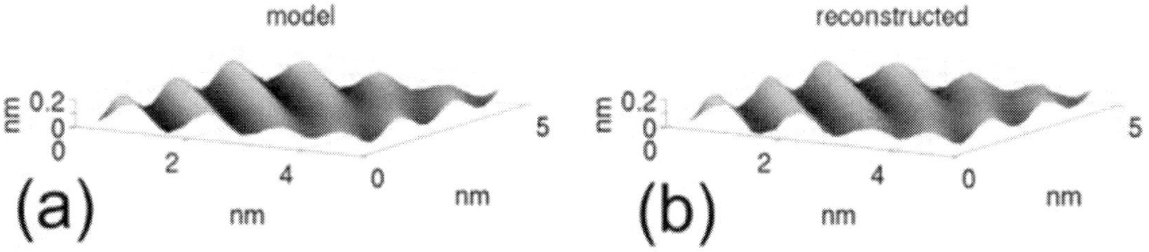

Figure 7. Original model(a) and reconstructed(b) graphene corrugation using Tomography through Points

5.3. Discrete tomography – the ADART algorithm

DART is a heuristic Discrete Tomography technique, capable of computing reconstructions of high quality from a small number of projections. The algorithm is very effective for binary images, but it can also be used to reconstruct images that contain more than two gray levels[16]. In DART, the update operation is restricted to pixels in the boundaries across gray levels in the reconstructed object.

ADART [17] is an algorithm based on DART that is capable of computing high quality reconstructions from substantially fewer projections than required for conventional continuous tomography. ADART reconstruction errors are significantly reduced with respect to DART. This reduction is remarkable for binary images, especially in noisy environments, and is nearly constant and independent of noise level for multiple grey-level images.

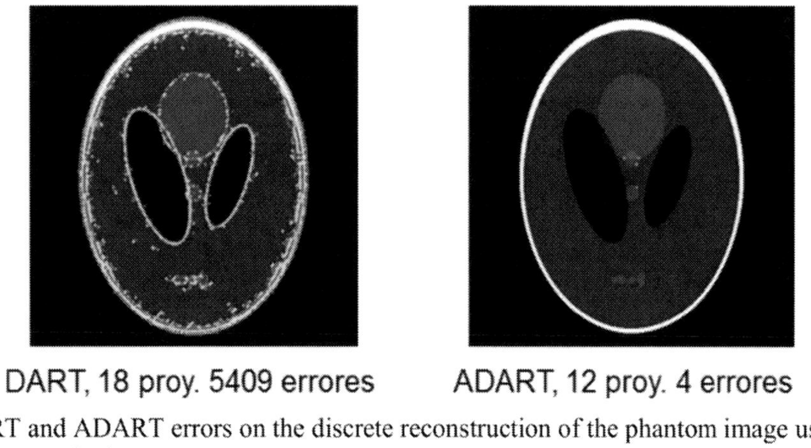

Figure 8. DART and ADART errors on the discrete reconstruction of the phantom image using 18 and 12 projections respectively.

6. Acknowledgements

This work was supported by the Spanish MINECO (projects TEC2011-29120-C05-03, MAT2010-15206 and CONSOLIDER INGENIO 2010 CSD2009-00013) and the Junta de Andalucía (PAI research groups TIC-145, TEP-946 and TEP-120).

References

[1] Allen L J, D'Alfonso A J, Forbes B D, Findlay S D, LeBeau J M and Stemmer S, 2012, *Journal of Physics: Conference Series* **371** 012009

[2] LeBeau J M, Findlay S D, Allen L J and Stemmer S 2008 *Phys. Rev. Lett.* **100** 206101

[3] LeBeau J M, Findlay S D, Wang X, Jacobson A J, Allen L J and Stemmer S 2009 *Phys. Rev. B* **79** 214110

[4] Bierwolf R, Hohenstein M, Phillip F, Brandt O, Crook G E and Ploog K 1993 *Ultramicroscopy* **49** 273

[5] Bayle P, Deutsch T, Pilles B, Lancon F, Marty A and Thibault J 1994 *Ultramicroscopy* **56** 94

[6] Jouneau P H, Tardot A, Feuillet B, Mariette H and Cibert J 1994 *J. Appl. Phys.* **75**, 7310

[7] Rosenauer A andGerthsen D 1999 *Adv. Imaging Electron Physics* **107** 121

[8] Galindo P L, Kret S, Sánchez A M, Laval J Y, Yáñez A, Pizarro J, Guerrero E, Ben T and Molina S I 2007 *Ultramicroscopy* **107** 1186

[9] Galindo P L, Pizarro J, Molina S I and Ishizuka K, 2009 *Microscopy and Analysis* March 23

[10] Sales D L, Pizarro J, Galindo P L, Garcia R, Trevisi G, Frigeri P, Nasi L, Franchi S and Molina S I 2007 *Nanotechnology* **18** 475503

[11] Molina S I, Sales D L, Galindo P L 2009 *Ultramicroscopy* **109** 172 (Erratum: 2009 **109** 1315).

[12] Hernández-Maldonado D, Herrera M, Alonso-González P 2011 *Microsc. Microanal.* ***17*** 5780

[13] Sales D L, Guerrero E, Rodrigo J F 2011 *Appl. Phys. Lett.* **98** 101902

[14] Núñez Moraleda B M, Galindo P L, Yáñez A, Guerrero E, Guerrero-Lebrero M P, Molina S I 2009 MSM Oxford, Póster P2-13.

[15] Pizarro J, Galindo P L, Guerrero E, Yáñez A, Guerrero M P, Rosenauer A, Sales D L, and Molina S I 2008 *Appl. Phys. Lett.* **93** 153107

[16] Scavello G, Pizarro J, Maestre J F, Molina S I and Galindo P L 2012 *Appl. Phys. Lett.* **101** 213106

[16] Batenburg K J and Sijbers J 2007 *14th IEEE Int. Conf. Image Process., San Antonio, TX*

[17] Maestre-Deusto F J, Scavello G, Pizarro J, and Galindo P L. 2011 *IEEE Trans. Imag. Proc.* **20(8)** 2146

EELS modelling of graphitisation

C R Seabourne[1], R Brydson[1], A P Brown[1], C D Latham[2], M I Heggie[2] and A J Scott[1]

[1]Institute for Materials Research, SPEME, Faculty of Engineering, University of Leeds, Leeds, LS2 9JT, UK
[2]Department of Chemistry, Faculty of Engineering and Physical Sciences, University of Surrey, Guildford, GU2 7XH, UK

E-mail: c.r.seabourne@leeds.ac.uk

Abstract. The impact of graphitisation-type processes on the carbon *K*-edge ELNES is explored for model systems using the CASTEP density functional theory code. For *c* lattice direction expansion, contraction of spectral peaks occurs between 20 and 27 eV above the edge onset, for *a/b* lattice dimension expansion spectral peaks are heavily compressed in terms of energy separation and are shifted towards the edge-onset, consistent with a $1/a^2$ relationship. For a nanotube model system, it is shown that for higher curvature, an additional feature was observed in the spectrum ~5 eV, arguably consistent with 'fullerene'-type character.

1. Introduction

Graphitic materials are now amongst the most characterized and written about systems in science, and the range of systems is considerable; nanotubes [1], fullerenes [2,3], potentially tunable cone-type structures [4,5], the wonder-material graphene and doped derivatives [6-10], and further discoveries regarding the folding and other unusual behaviours of the material [11]. Graphene is ideal for study in the electron microscope due to its inherent thinness, avoiding effects such as dechanneling in atomic columns in the specimen [12], though careful microscope parameters must be chosen so as to avoid knock-on damage [13]. A similar approach can be adopted when studying bulk graphite in the microscope, to obtain high resolution information. This high level information can include the electron energy-loss spectrum (EELS) of graphite. This work focuses on the core-loss region (carbon *K* edge), in particular using density functional theory (DFT) modelling, but one of the authors of this paper has also considered multiple scattering (MS) as a technique to model the edge [14].

DFT modelling is used to address two particular challenges regarding the carbon *K*-edge energy-loss near edge structure (ELNES) and the potential impacts of graphitisation, amorphisation and irradiation damage in nuclear reactors or the electron microscope; lattice expansion/contraction in the *a/b* and *c* directions, and the influence of curvature on the *K*-edge ELNES, as considered by using a series of nanotube examples. This work should allow us to relate experimentally observed changes in the carbon *K*-edge ELNES (upon for example irradiation or heating) to structural changes as modelled by the systems in this work. This may be particularly helpful if it is difficult to measure lattice parameters or curvature of nanoscale carbons experimentally.

2. Density functional theory methodology

For all experiments, the CASTEP DFT code was used for ELNES predictions [15,16]. It is important when using a DFT code to properly converge relevant parameters to ensure accurate results. This was carried out by converging parameters directly against the ELNES results themselves, thus following a methodology previously outlined in the literature [17]. In this instance, the key parameters for convergence are the basis-set size, as defined by the kinetic energy cut-off value, and the density of *k*-point sampling in reciprocal space, in these cases using a Monkhorst-Pack type grid [18].

A bulk graphite crystal was chosen as the convergence system. For the kinetic energy cut-off, an initial value of 400 eV was chosen, and upon doubling that value to 800 eV, there was virtually no change in the predicted, unbroadened ELNES result (the carbon *K* edge for a single atom in the standard unit cell), thus this parameter was considered to be converged. A similar process was carried out for the *k*-points grid, which was doubled from 12×12×4 to 24×24×8. In this case, barely any changes were observable by visual inspection to the predicted spectra, but a further mathematical analysis was carried out. For a range of points from -2 eV to 50 eV, the modulus percentage change between the two results was found at each data point on the energy axis. This percentage change was then averaged across all the data points, and found to be less than 10 %, confirming only small changes occur, thus the parameter is converged.

Table 1: Summary of converged parameters for carbon *K* edge calculations, using the CASTEP DFT code. The *c* lattice parameter was subsequently optimised to a value closer to experiment.

Structural information of model convergence system	Kinetic energy cut-off / eV	Density of *k*-points in reciprocal space / $Å^{-1}$
$a = b = 2.46$ Å, $c = 6.80$ Å $\alpha = \beta = 90°$, $\gamma = 120°$	800	0.018

Where the predicted spectra were then subject to broadening, an energy-dependent final state lifetime based scheme was used, as presented previously [19].

Following initial convergence experiments, a Broyden, Fletcher, Goldfarb, Shanno (BFGS) geometry optimisation was carried out with a generalized gradient approximation Tkatchenko Scheffler (GGA-TS) correction [20], leading to optimised cell dimensions of; $a=b = 2.454$ Å, $c = 6.673$ Å.

3. Results

3.1. *Expansion / contraction of lattice c dimension*

From a geometry optimised starting point of $c = 6.673$ Å, with the $a=b$ lattice parameter fixed at 2.454 Å, the *c* dimension was varied by ± 10 % and ± 5 %. This led to a total of five predicted carbon *K* edge spectra, for expansion / contraction of the *c* dimension, Figure 1. Contraction physically speaking might represent behaviour under pressure; expansion is a simplified approach to understand trends in for example the intercalation of cations between graphite layers (although we do not seek to account for the resultant change in chemistry). Calculations were carried out in the ground electronic state. In each case, there were two inequivalent atoms in the unit cell, whose spectra are simply averaged. It is assumed their chemical environments are sufficiently similar so as to make adjusting for an absolute energy-onset difference not relevant.

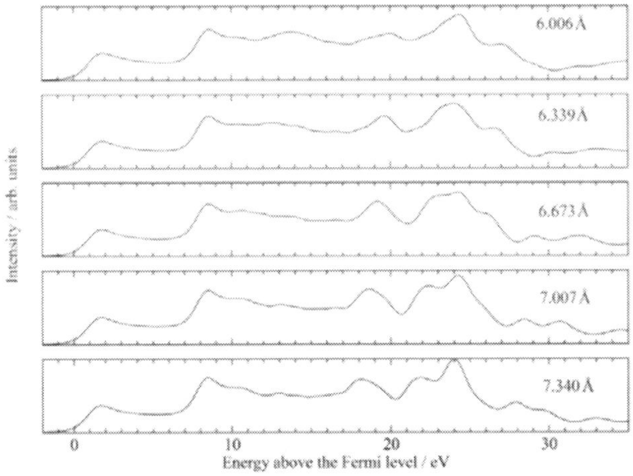

Figure 1. Predicted carbon *K*-edge ELNES for graphite using the CASTEP DFT code, carried out in the ground electronic state. The *c* axis lattice parameters are labelled in each case, for constant $a=b$. The results are shown with a 0.4 eV Gaussian broadening.

We observe contraction of peaks between 20 and 27 eV (above the edge onset) as the *c* axis parameter increases (spectral peaks move closer together), and arguably some loss in intensity between 13-16 eV. Also a significant peak develops for smaller *c* values at ~27 eV.

3.2 Expansion / contraction of a/b lattice parameters

As in section 3.1, the *a/b* lattice parameter was varied from 2.454 Å by ± 5 % and ± 10 %, with the *c* lattice parameter kept constant, Figure 2. Contracted cells are obviously under considerable strain considering the standard C-C bond distance of 1.42 Å.

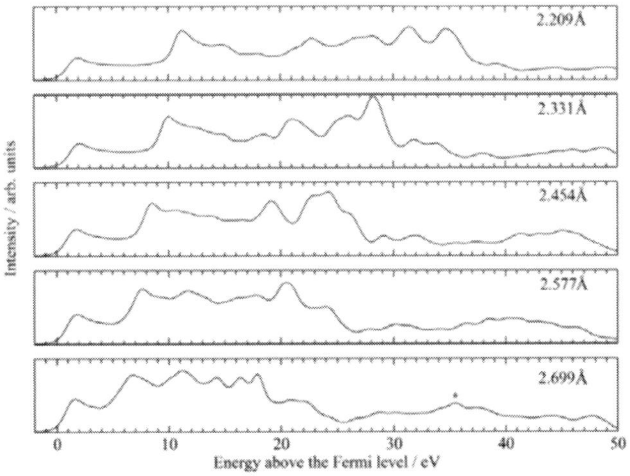

Figure 2. Predicted carbon K-edge ELNES for graphite, using the CASTEP DFT code, carried out in the ground electronic state. The *a/b* lattice parameters are labelled in each case, for constant *c*. The results are shown with a 0.4 eV Gaussian broadening.

Based on previous work, various peak separations vary as a function of $1/a^2$ [21,22]. Plotting the separation in energy between the first two peaks versus $1/a^2$ leads to a linear fit, with a R^2 regression coefficient of 0.995, similarly the energy separation from the π^* peak (the first in the spectrum) to the peak labelled with an asterisk (or its equivalents for the different lattice parameter values) plotted versus $1/a^2$ is also a linear fit, regression coefficient of 0.98, and there are other similar examples.

3.3 Nanotube radius of curvature

Single-wall nanotubes were constructed as $n \times m$ ($n=m$ armchair, diameter = $0.783(n^2 + nm + m^2)^{1/2}$ Å). The C-C bond length was 1.42 Å. Inequivalent positions gave similar spectra.

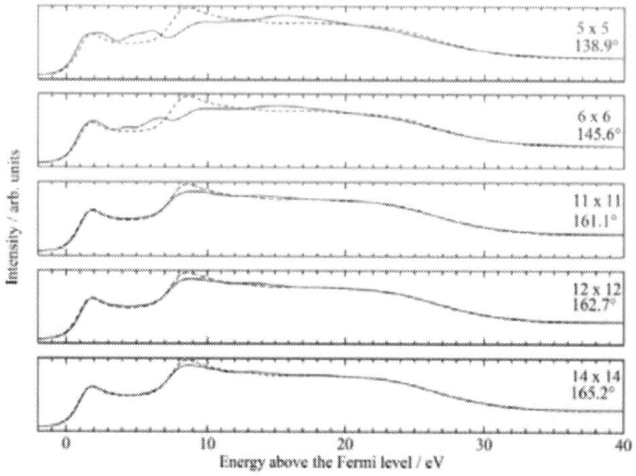

Figure 3. Predicted carbon K-edge ELNES for various labelled nanotube sizes (also labelled are angles between adjacent hexagons, which for graphene is 180°), using the CASTEP DFT code, carried out in the ground electronic state, with inequivalent positions not averaged due to similarity. The results are shown with the addition of energy-dependent final state lifetime broadening [19]. Shown on each plot is a dashed line, which is a predicted spectrum for bulk graphene (with a 25 Å vacuum to simulate the monolayer).

For increasing nanotube radius of curvature the spectrum begins to approach that of graphene [23]. When the system is however 'more curved', for example for the 5×5 nanotube (diameter 6.78 Å), an additional peak is observed at ~5 eV above the edge onset – this may be something that is indicative in experimental spectra, indicating curvature in a specimen, or perhaps fullerene character [24].

4. Conclusions

Modelled carbon *K*-edge ELNES show that curvature may lead to an additional feature at ~5 eV above the edge onset and that for *a/b* lattice expansion peaks are compressed closer together towards the Fermi level. These are structural trends useful when analysing carbon *K*-edge ELNES from irradiated samples for example, where it is not necessarily trivial to measure lattice parameters or 'curvature'. Future work may consider multi-walled nanotubes and displacement from AB graphite stacking.

5. References

[1] Iijima S and Ichihashi T 1993 *Nature* 363, 603
[2] Kroto H W et al. 1985 *Nature* 318, 162
[3] Howard J B et al. 1992 *Carbon* 30, 1183
[4] Krishnan A et al. 1997 *Nature* 388, 451
[5] Hage F S et al. 2013 *Nanoscale,* submitted
[6] Novoselov K S et al. 2004 *Science* 306, 666
[7] Novoselov K S et al. 2005 *Proc. Natl Acad. Sci. USA* 102, 10451
[8] Geim A K and Novoselov K S 2007 *Nat. Mater.* 6, 183
[9] Geim A K 2009 *Science* 19, 1530
[10] Ramasse Q M et al. 2013 *Nano Lett.* ASAP, dx.doi.org/10.1021/nl304187e
[11] Heggie M I et al. 2011 *J. Nucl. Mater.* 413, 150
[12] Egerton R F 2009 *Rep. Prog. Phys.* 72, 016502
[13] Gass M H et al. 2008 *Nat. Nanotech.* 3, 676
[14] McCulloch D G and Brydson R 1996 *J. Phys.: Condens. Matter* 8, 3835
[15] Clark S J et al. 2005 *Z. Kristallogr.* 220, 567
[16] Gao S P et al. 2008 *Phys. Rev. B* 77, 115122
[17] Seabourne C R et al. 2009 *Ultramic.* 109, 1374
[18] Monkhorst H J and Pack J D 1976 *Phys. Rev. B* 13, 5188
[19] Moreau P et al. 2006 *Phys. Rev. B* 73, 195111
[20] Tkatchenko A and Scheffler M 2009 *Phys. Rev. Lett.* 102, 073005
[21] Daniels H et al. 2007 *Phil. Mag.* 87, 4073
[22] Craven A J 1995 *J. Micros.* 180, 250
[23] Suenaga K and Koshino M 2010 *Nature* 468, 1088
[24] Zhang Z et al. 2011 *Carbon* 49, 5049

Two-dimensional Object Functions and Three-dimensional Illumination Functions: their Validity, Interaction and Utility

L. Jones[1], P. D. Nellist
Department of Materials, University of Oxford, Parks Road Oxford OX1 3PH, UK

E-mail: lewys.jones@materials.ox.ac.uk

Abstract. In high-angle annular dark-field scanning transmission electron microscopy (HAADF-STEM) the data is generally interpreted as the convolution of the sample's sharply-peaked object function with the intensity of the real-valued point-spread function (PSF) of the illumination. As a single image is itself necessarily two-dimensional (2D) it is typically assumed that the object function and PSF can also be accurately described as 2D; this is the so-called 2D-object imaging assumption. Here the validity of these two-assumptions is evaluated using experimental HAADF STEM focal-series. It is found that the contrast contained in each image of the focal-series is accurately described by the convolution of a focus-invariant 2D object with a 2D optical-transfer function (OTF) which describes the illumination at the focus value used for imaging.

1. Introduction

The detector geometry in the high-angle annular dark-field (HAADF) imaging mode in the scanning transmission electron microscopy (STEM) causes the recorded signal to be regarded as being incoherent. Because of this, it is the intensity distribution of the focussed STEM probe that most closely dictates both the recorded contrast and resolution. This is highlighted by the so-called two-dimensional (2D) object function model used to describe the imaging of on-axis crystalline samples:

$$I(\boldsymbol{r}) = |P(\boldsymbol{r})|^2 \otimes O(\boldsymbol{r}) \tag{1}$$

In this model the image data, I, that is recorded as a function of real-space probe position, $\boldsymbol{r}$, is described by a convolution between the intensity (modulus squared) of the 2D real-space probe-function, $P(\boldsymbol{r})$, and a 2D representation of the object, $O(\boldsymbol{r})$, where the object the sample is represented by a collection of sharply-peaked points at the atomic column positions and is sparse otherwise.

While this model describes the contrast in individual HAADF images, it is becoming increasingly common to record through-focus image series for either optical-sectioning [1] or image reconstruction [2] type experiments. We now have to consider the third-dimension (either real-space z-direction or equivalently defocus), and whether it remains reasonable to assume that the probe and/or object behaves two-dimensionally. The first of these is easy to address as from a consideration of the geometric optics of the highly-convergent STEM probe we can immediately see that this varies greatly with defocus [3], such that $P(\boldsymbol{r},z)$.

[1] To whom any correspondence should be addressed.

The two remaining scenarios to consider are then:

$$S(\mathbf{r},z) = \begin{cases} |P(\mathbf{r},z)|^2 \otimes O(\mathbf{r}) & \textit{Object behaves 2D} \\ |P(\mathbf{r},z)|^2 \otimes O(\mathbf{r},z) & \textit{Object behaves 3D} \end{cases} \quad (2)$$

The first of these assumes that the object function can be treated as being independent of defocus, and the second that focus-dependent information appears in the object function. To determine which of these two regimes operates we need then to evaluate the "apparent thickness" of the object function and focal-series allows us to do this.

2. Method

To interpret the z-response of the sample we must first understand the defocus-response of the illumination. Fortunately the 3D form of the STEM probe is well understood and can be readily simulated. Probe simulations were performed in the MatLab™ programming environment matching the experimental conditions (200kV acceleration and a 22 mrad objective aperture).

Next, to capture the entire profile of the contrast response a HAADF-STEM focal-series should be recorded with a total defocus-range such that the first and last images are just out of focus (this can equivalently be achieved with fixed optics and a piezo scanning stage). Here focal-series were recorded from [100] oriented MgO smoke cubes, where sample thickness can be inferred from the cube width. Within this image-stack then the laterally resolved contrast would begin at near-zero, peak when in-focus and then deteriorate again. To quantitatively measure the contrast transfer performance of the images in the series the common technique of Fourier transform (FT) inspection is useful. With lattice-resolved images of crystalline samples, transforming each image in the series into reciprocal-space allows the periodic image components (laterally resolved contrast) to be observed as FT-spot. To quantify the contrast we can then simply integrate the modulus in the area immediately adjacent to each FT-spot and track this through defocus. This can be repeated for every observable spot in the FT and after fitting of say a Gaussian profile curve, a defocus –spread, or –persistence is measured from this fit. Interpolating this defocus-spread across all reciprocal space and finding the minima then gives the ultimate depth-sensitivity exhibited by the experimental data [4].

3. Results and Discussion

First the expected contrast transfer performance of the simulated illumination:

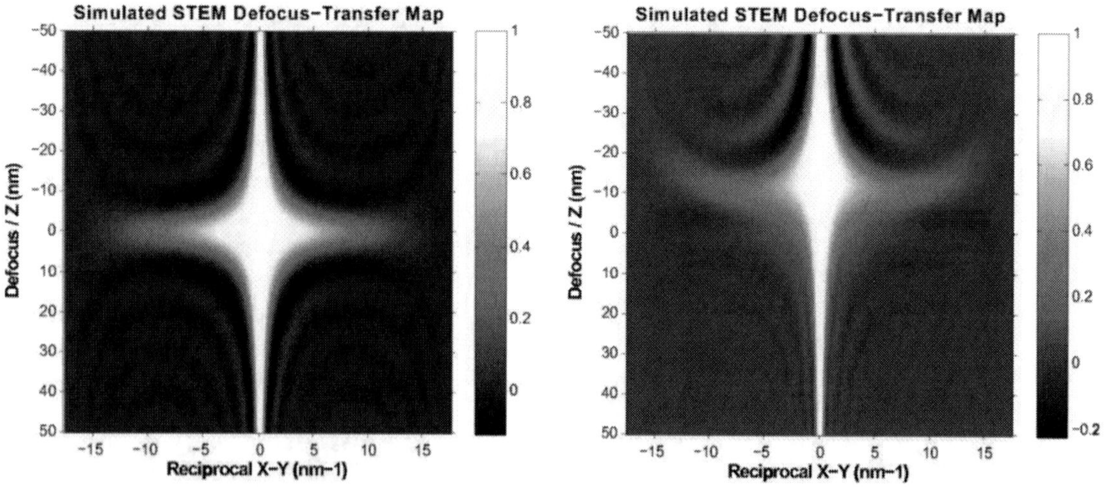

Figure 1. Simulated OTFs for two aberration conditions. Left: aberration free, and right: the same but with +50 μm spherical aberration ($C_{3,0}$). Horizontal axis shows reciprocal nanometres (spatial frequency) while vertical axis shows real-space defocus.

In the perfect (diffraction limited only) probe the band of peak information transfer is flat with respect to defocus and centred about zero defocus. Upon the introduction of positive spherical aberration this both shifts and sweeps towards negative defocus (familiar to many operators C_s must be balanced with defocus). In this investigation only the defocus width of the maximal transfer band is needed and can be measured with again the same Gaussian fitting. Further examples of such simulations, including discussion of the OTF shapes stemming from other common non-round aberrations can be found in [5].

After transforming the experimental data-cube to Fourier-space and thresholding (to say a 10% intensity level) only those voxels with usable contrast are revealed, Figure 2.

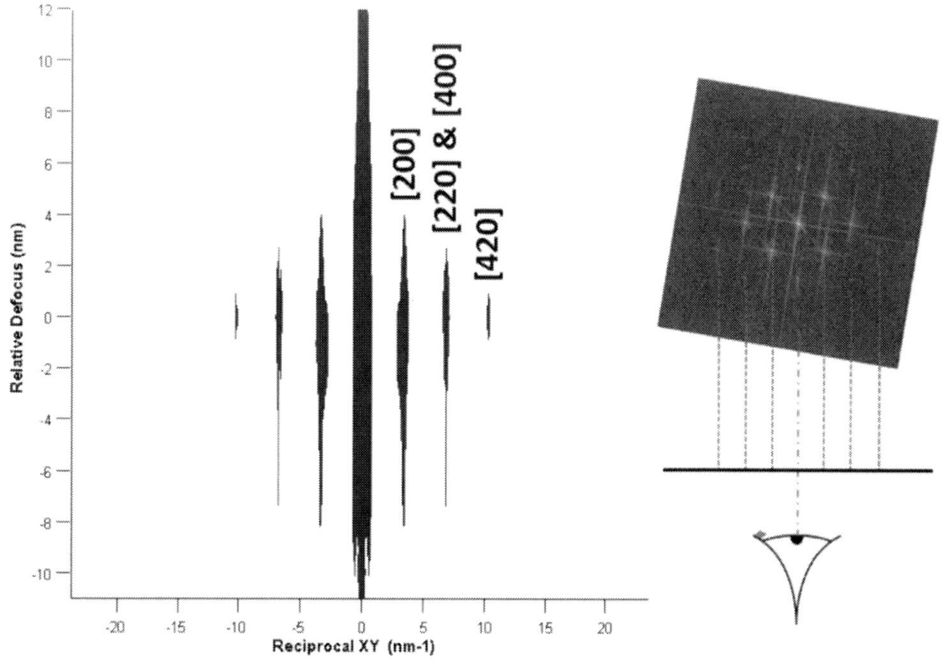

Figure 2. Side-on 3D-visualisation of the lattice contrast in an experimental focal-series. Vertical rods correspond to the intensity iso-surface encapsulating FT-spot intensities. Inset shows the viewing direction of the side-on stack.

Using this representation the locations of FT-spot rods can be readily identified and analysed.

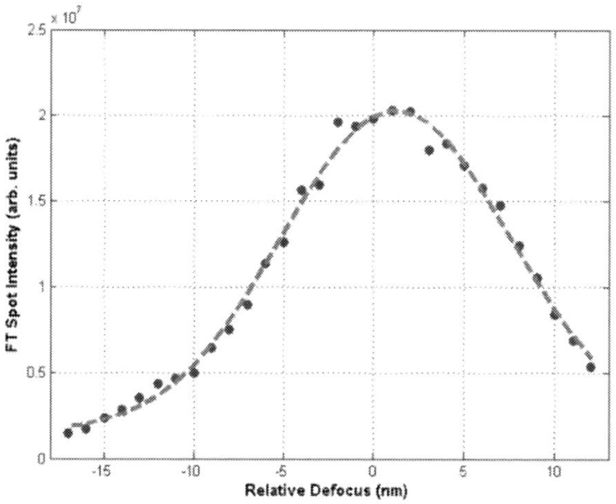

Figure 3. Integrated intensities (blue dots) from the immediate area around an example [200] FT-spot (3 pixel radius) across a 30-frame experimental focal-series. Sample was ≈83 nm thick. Dashed line indicates the fitted Gaussian profile through the data used to measure the defocus spread / persistence. Note that at the start of the series, the FT-spot intensity is near zero indicating no resolved image contrast.

Now the ratio between the FT-spot defocus-spread (contrast-persistence) and the OTF z-response gives the 'normalised thickness' this is the contribution to the defocus-spread in the convolution in equation (2). Plotting this as a function of the real (measured) thickness we get Figure 4.

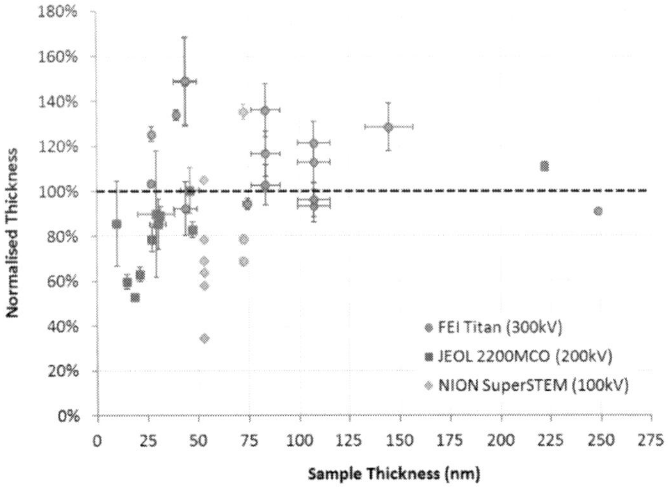

Figure 4. Plot of the calculated 'normalised thickness' versus the measured real thickness (red squares). The dashed line represents the 100% datum where experimental objects behave perfectly 2D. Additional results from different voltages and objective aperture sizes are shown (circles and diamonds) indicating the generality of the result.

Now we find that for a wide range of sample thicknesses all appear approximately 2D. The reader may verify this by comparing again the contrast-persistence in Figure 3 of around 15 nm with both the true sample thickness of ≈83 nm, and the OTF profile in Figure 1 (also ≈15 nm).

4. Conclusions

- With increasing interest in focal-series for optical-sectioning or image reconstruction the 2D object-function assumption and imaging model merits evaluation.
- Experimental focal-series facilitate such an investigation, measuring the defocus-persistence of FT-spots (an indicator of resolved contrast).
- Comparison with simulated probe-OTFs allows the additional z-blurring from the sample to be investigated equivalent to the apparent sample thickness in the convolution model.
- We find that for the thickness range investigated the MgO samples appear two-dimensional and we confirm the validity of the 2D object function assumption for simply projecting crystals.
- This conclusion then suggests that it may be possible to retrieve the 2D sample object function from the over-determined 3D focal-series, and an image reconstructed free from the effects of residual illumination aberrations.

5. References

[1] G. Behan, E.C. Cosgriff, A.I. Kirkland, P.D. Nellist, *Phil. Trans. A* **367** (2009).
[2] L. Jones, P.D. Nellist, *Microscopy and Microanalysis* **18** (2012).
[3] A.R. Lupini, N. de Jonge, *Microscopy and Microanalysis* **17** (2011).
[4] L. Jones, P.D. Nellist, "Testing the Accuracy of the Two-dimensional Object Model in HAADF STEM", *Micron* (in review) (2013).
[5] L. Jones, P.D. Nellist, "Three Dimensional Optical Transfer Functions in the Aberration Corrected Scanning Transmission Electron Microscope", *Journal of Microscopy* (in submission) (2013).

Acknowledgments

The authors thanks Dr Armand Béché for help with the experiential imaging. This research was supported by the EU-FP7 Grant Agreement 312483 (ESTEEM2) and the UK EPSRC.

An in-house developed annular bright field detection system

Leonardo Lari[1], **Ian Wright, Daniel Pingstone, Jonathan Steward, Daniel Gilks and Vlado K Lazarov**
Department of Physics, The University of York, Heslington, YO10 5DD, York, UK
and
The York-JEOL Nanocentre, York Science Park, Heslington, YO10 5BR, York ,UK

E-mail: leonardo.lari@york.ac.uk

Abstract. Annular bright field (ABF) detectors have been developed in the last few years allowing the direct imaging of low-Z atoms from oxygen down to hydrogen. These types of detectors are now available as standard attachments for the latest generation of top-end electron microscopes. However these systems cannot always be installed in previous generation microscopes. In this paper we report the preliminary results of an in-house implementation of a ABF detection system on a CEOS aberration corrected JEOL 2200FS STEM. This has been obtained by exploiting the standard BF detector coupled with a high vacuum compatible, X-ray tight and retractable shadowing mechanism. This results in the acquisition of near zero-angle scattered electrons with inner collection semi-angle from 2.0 mrad to 23 mrad and outer semi-angle in the range from 3.0 mrad to 35 mrad. The characteristics and performances of this ABF detection system are discussed.

1. Introduction

In the last decade, probe aberration corrected scanning transmission electron microscopy (STEM) has proven to be one of the leading methods for atomic scale investigation of crystalline materials. The advantage of these methods lies mostly in the high angle annular dark field (HAADF) imaging which features the so-called Z-contrast, where the intensity of the image is proportionally related to the power of two of the atomic number. This property makes HAADF images readily interpretable without the need of extensive image simulations [1]. One of the deficiencies of the HAADF imaging is inability to image light elements, *e.g.* oxygen and other low Z elements, due to their low electron scattering power. Recently, annular bright field (ABF) detectors have been developed allowing direct imaging of low-Z atoms, as oxygen in strontium titanate [2]. Starting initially by using bright field detectors in conjunction with a beam stopper [3] ABF detectors are now available as standard attachments for the latest generation of top-end electron microscopes, and have been demonstrated to be able to directly image lithium columns in LiV_2O_4 using a JEOL R005 [4] and more recently hydrogen columns in VH_2 using a JEOL ARM microscope [5]. However, these systems with a dedicated detector cannot always be installed in previous generation microscopes. In this work we present preliminary results obtained by an ABF detection system in-house designed and developed which can provide a relatively cheap solution for implementing an ABF detector on earlier generation STEM equipped with a BF detector.

[1] To whom any correspondence should be addressed.

Journal of Physics: Conference Series **522** (2014) 012016 doi:10.1088/1742-6596/522/1/012016

2. Methods

A beam shadowing mechanism was designed to block the central part of the standard JEOL bright field (BF) STEM detector of a JEOL 2200FS double CEOS™ aberration corrected (S)TEM operating at 200kV and equipped with an in-column Omega type electron energy filter. Figure 1a) shows its location in the port above the BF detector located in the camera chamber of the microscope. This port is normally used to place a post omega filter annular dark field detector. The beam stopper (BS) was made in the departmental workshop using Al for the parts to go inside the camera chamber, and steel to interface the beam stopper to the microscope column. An Al rod holds an off-axis beam stopping disk. The rod rotates around its axis and has two rest positions one centred on the electron optics axis (*ON* position, Figure 1b) and at 90 degrees (*OFF* position, Figure 1c) where is retracted out of the electron beam path. The BS disk has a diameter of about half of the size of the BF detector entrance aperture. An additional mechanism was fitted to allow fine centring of the BS respect to the BF detector axis. Protection from X-ray generated by the beam stop is provided by a cylinder of Ta mounted around the Al rod at the internal side of the camera chamber wall as indicated in Figure 1c). X-ray leakage trough the chamber was tested using a Thermo Scientific™ Mini 900X radiation monitor.

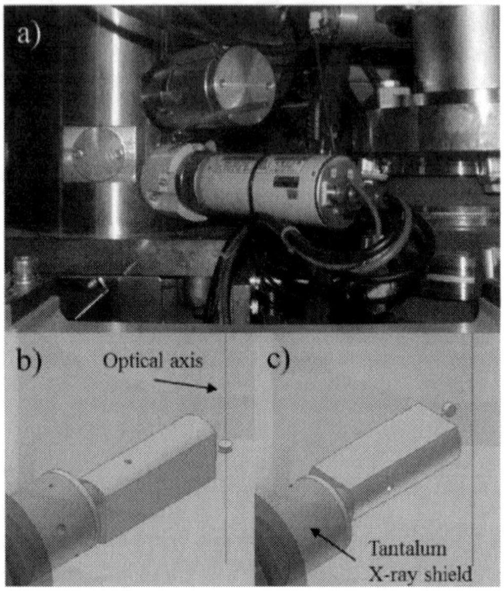

Figure 1. a) location of the ABF beam stopper above the bright field detector in the camera chamber; b) beam stopper *ON* position; c) beam stopper *OFF* position

Two TEM samples were prepared for measuring the ABF detector acceptance semi-angles and for testing its performance. Si [110] and $SrTiO_3$ [110] crystals were mechanically polished and ion milled to electron transparency using the cross sectioning methodology as described in [6].

3. Results and discussion

The response of the BF detector as a position of the electron beam is shown in Figure 2a). The electron beam was allowed to scan an area encompassing the BF detector entrance aperture in STEM alignment mode. A scanning time of 19msec and the smaller condenser aperture (12mrad convergence

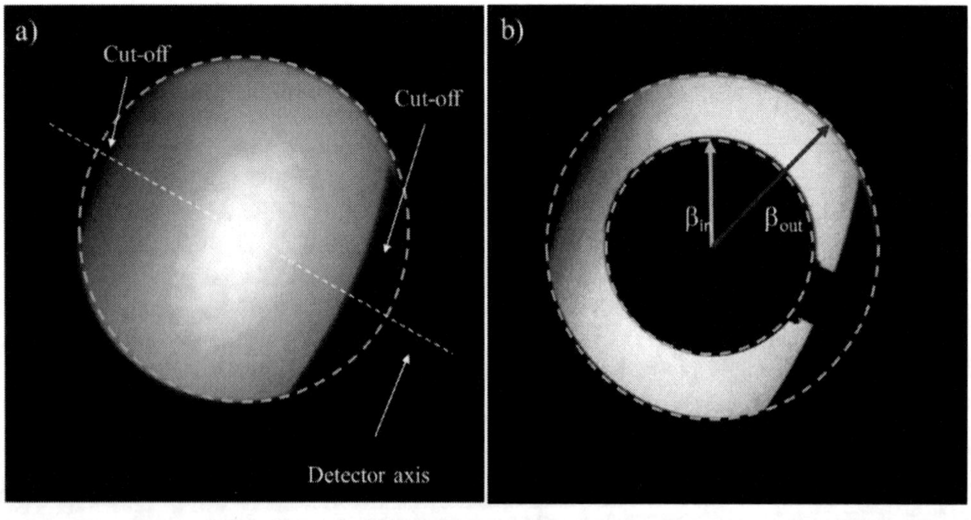

Figure 2. a) Large field of view BF STEM image showing the BF detector response; b) BF detector configuration showing inner and outer collection semi-angles.

semi-angle) was used in order to avoid saturation of the BF detector. During the image acquisition the beam stopper was kept in the *OFF* position. As indicated by the two arrows inset in Figure 2a) there are two cut-off areas perpendicular to the detector axis. Figure 2b) shows the ABF configuration with the beam stopper centred respect to the BF detector. Using the Si [110] sample the inner collection semi-angle β_{in} and the outer one β_{out} were measured from Kikuchi lines spacing for each of the fifteen camera length setting available in STEM mode for our microscope. The results are listed in Table 1. Larger camera length values produce very narrow collection ranges contained within the central region of the BF detector hence the imaging is coherent and very close to normal BF conditions.

Table 1. List of ABF acceptance semi-angles versus nominal camera length settings

Camera length	β_{in} (mrad)	β_{out} (mrad)
4 cm	23.4	38.7
5 cm	20.0	33.0
6 cm	16.6	27.3
8 cm	11.1	18.3
10 cm	9.5	15.6
12 cm	7.8	12.8
15 cm	7.0	11.6
20 cm	6.2	10.3
25 cm	6.1	10.1
30 cm	7.7	12.7
40 cm	6.3	10.4
50 cm	5.2	8.6
60 cm	5.0	8.3
80 cm	4.5	7.4
100 cm	3.1	5.1
120 cm	2.5	4.1
150 cm	2.0	3.2

β_{in} inner semi-angle; β_{out} outer semi-angle

However, shorter camera lengths (from 4 cm to 10 cm) enable larger shadowing of the central part of the BF detector (above 10 mrad) which provides angular acceptance range close to the standard ABF values as used in [5]. Figure 3 a) shows a HAADF image of the $SrTiO_3$ sample acquired along the [110] zone axis. In this projection three distinctive atomic columns are present, O and Ti single columns and Sr/O mixed column. The HAADF image was acquired in the range 85-170 mrad. Brighter columns correspond to the columns occupied by Sr atoms. Scanning instabilities are visible as streaks running at about 45 degrees with respect to the horizontal. In the same acquisition conditions (50 cm camera length) an ABF image was acquired as shown in Figure 3b). Probe deconvolution using a maximum entropy approach [7] was applied to the collected HAADF images as shown in Figure 3, whereas the ABF image (Figure 3b) was filtered using the smooth function of Gatan™ Digital Micrograph.

A full series of simulations of $SrTiO_3$[110] were produced using QSTEM [8] for the range of camera length settings available. The aberration coefficients obtained by the CEOS™ stem alignment software at the beginning of the microscopy session were used as calculation parameters, nominally tri-fold astigmatism 22nm, axial coma 15nm, spherical aberration coefficient 1.1 μm, four-fold astigmatism 660nm, star aberration 316 nm, and five-fold astigmatism 80μm. The multislice simulations were performed using the frozen phonons approximation. The best fit for the acceptance semi-angle used 5.2-8.6 mrad was obtained for a sample thickness of 360 slices corresponding to 99 nm, as shown in the inset of Figure 3b). Deviation from a perfect match with the experimental image can be due, besides the noise, to the portion of the signal cut off as shown in Figure 2. On the left side there is a reduction of the outer collection semi-angle of about 15% while on the other side (without considering the rod where the blanker is attached) the reduction is of 25%. Nevertheless this problem could be solved by introducing an annular aperture instead of a central beam stopper. This would practically imply the use of the lower camera length settings.

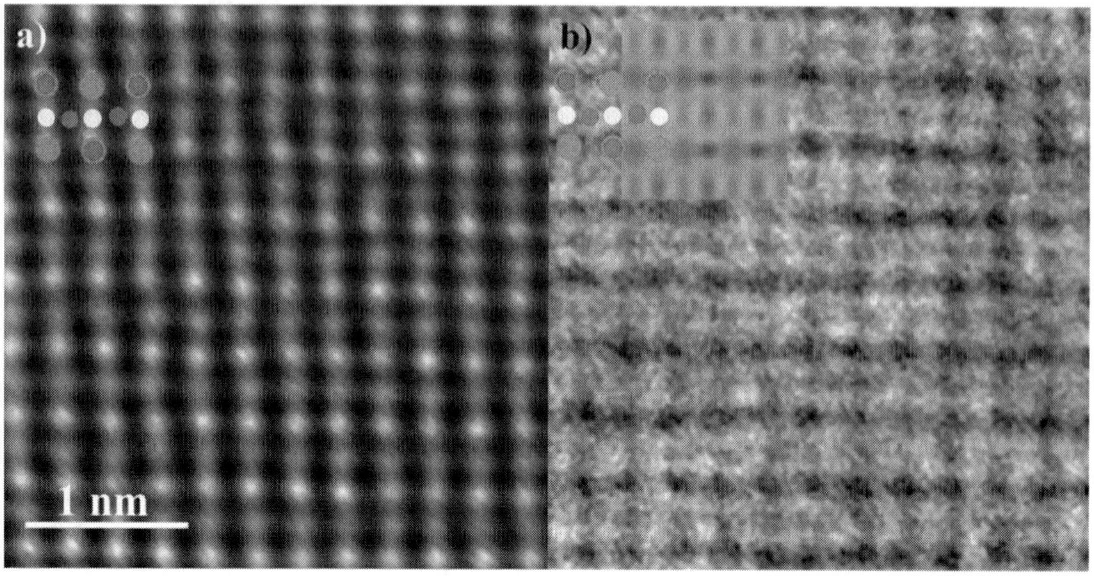

Figure 3. a) HAADF STEM image of $SrTiO_3$ along [110] zone axis; b) ABF image from same area as in a) with collection semi-angles (β_i=5.2 mrad and β_o=8.6 mrad). Atoms are colour coded as following: yellow (Ti), green (Sr) and red (O).

4. Conclusions

In this paper we have shown the preliminary results of an in-house developed annular bright field detection system using a beam stopper coupled with the existing standard STEM bright field detector. The bright field detector collection area was found to be irregular with a reduction of collection semi-angle along the detector longitudinal axis of 15% and 25% respectively on each side. Useful camera lengths setting for appropriate shadowing of the central bright field region was found only for nominal camera lengths below 10cm. Our findings indicate that an annular aperture coupled with the bright field detector, instead of a beam stopper, would better define the ABF collection semi-angles to be fed into the image simulation software.

Acknowledgments

The authors would like to thank the University of York Physics department workshop technicians for their help with this project.

References

[1] Nellist P D and Pennycook S J 2000 *Advances in Imaging and electron physics* **113** 147
[2] Okunishi E, Ishikawa I, Sawada H, Hosokawa F, Hori M, and Kondo Y 2009 *Microsc. Microanal.* **15** 164–165
[3] Findlay S.D., Shibata N, Sawada H, Okunishi E, Kondo Y and Ikuhara Y 2010 *Ultramicroscopy* **110** 903–923
[4] Oshima Y, Sawada H, Hosokawa F, Okunishi E, Kaneyama T, Kondo Y, Niitaka S, Takagi H, Tanishiro Y and Takayanagi K 2010 *Journal of Electron Microscopy* **59** 457–461
[5] Findlay S D, Saito T, Shibata N, Sato Y, Matsuda J, Asano K, Akiba E, Hirayama T and Ikuhara Y 2010 *Applied Physics Express* **3** 116603
[6] Lari L, Lea S, Feeser C, Wessels B W, and Lazarov V K. 2012 *J. Appl. Physics* **111** 07C311
[7] Grillo V. 2011 *Microscopy and Microanalysis* **17** 1292
[8] Koch C T, "Determination of core structure periodicity and point defect density along dislocations", PhD thesis, Arizona State University (2002)

A practical approach to quantify the ADF detector in STEM

D. S. He and Z. Y. Li*

Nanoscale Physics Research Laboratory, School of Physics and Astronomy, University of Birmingham, Birmingham B15 2TT, UK

*Email: z.li@bham.ac.uk

Abstract. We present a practical approach to quantify the annular dark field (ADF) detector in scanning transmission electron microscope (STEM). The non-uniform response of the detector as a function of the beam current is investigated. The brightness and contrast of the preamplifier have been taken into account to find the black level of the detector. The efficiency map is obtained.

1. Introduction

Quantitative analysis of annular dark field (ADF) image in scanning transmission electron microscopy (STEM) is useful in the three dimensional characterisation of the nano-object [1-3]. Here, the performance of the ADF detector is crucial and the quantification procedure is necessary for the quantitative analysis of ADF-STEM. So far, several procedures have been reported [3-5]. These procedures made use of additional calibrated instruments, either the dynamical signal analyzer [4] or Faraday cup [5], but none had taken into account of the non-uniformity of the detector in quantification [3-5].

In this work, we present a practical approach to characterise the performance of the ADF detector. By varying the beam dose, we identify a linear response region of the detector and find the black level of the detector. We then obtain the efficiency map of the detector.

This work is conducted on the JEOL2100F, with CEOS probe corrector, located in the University of Birmingham. The JEOL ADF detector is a P47 phosphor scintillator disk with a hole in the center, whose inner and outer diameters are 3 and 8 mm respectively. The brightness and contrast of the detector can be controlled by the software. The charge coupled device (CCD) camera installed is Gatan 2k × 2k UltraScan1000 with the conversion factor of 0.266 electron/count provided by the manufacturer.

2. The electron beam dose measurement

In the present study, the CCD camera is used to measure the e-beam dose. The response of the camera is shown in Figure 1: (a) is an image taken with 7s exposure time when the specimen was removed from the field of view and the beam was spread evenly across the CCD sensor. At this exposure time, the CCD is at about half of its full dynamic range. (b) is the intensity line profile showing that the CCD has a low background level. (c) is the count rate integrated over the entire CCD, after background subtraction, as a function of the exposure time. The dashed line is a linear fit with the function of $y = ax$ for exposure time less than ~13s. The linear response of the CCD camera enables us to work out the beam current by dividing the total number of electrons reaching the CCD (counts × 0.266) by exposure time. Table 1 shows typical values of the beam current obtained in this way, with emission current of 138 μA, for different condenser aperture sizes and spot sizes. Here aperture 4 and 3 are the smallest and the second smallest aperture of the microscope. The CCD camera saturates at ~65000 counts per pixel.

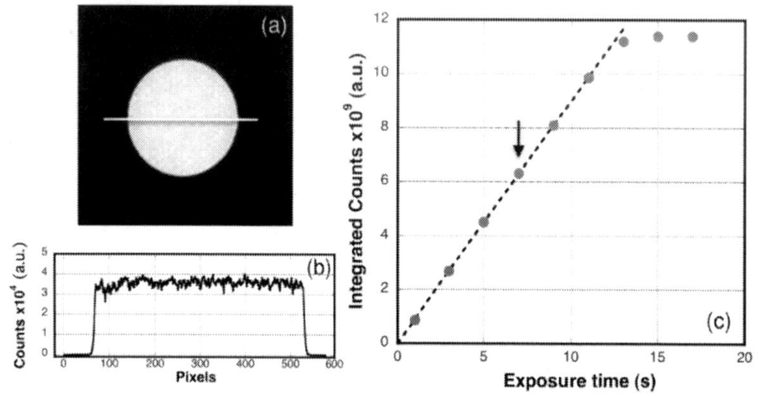

Figure 1 The linear response of CCD camera. (a) An image of the CCD taken when the specimen was removed from the field of view. (b) The line profile intensity indicated in (a). (c) The counts integrated over the entire CCD as a function of exposure time.

Table 1 The beam current for the typical microscope setting

Aperture	4					3				
Spot size	10C	9C	8C	7C	6C	10C	9C	8C	7C	6C
Beam current (pA)	5.6	8.4	16.5	24.5	50.1	11.5	17.9	35.1	52.9	108.3

3. The performance of the ADF detector as a function of the beam dose

Fig. 2 (a) shows the response of the JEOL ADF detector by focusing the electron beam directly on it. It is clear that the response of the detector is non-uniform. For the same beam current, the upper left region near the inner boundary has the highest output count and the lower right region near the outer boundary has the lowest output count. The output counts at three locations, indicated respectively by the dot, square and diamond in Fig. 2 (a), were investigated as a function of beam current. Here, the beam current was varied by changing aperture size and spot size as calibrated in Section 2. The results are shown in Fig. 2 (b) with the same symbols. In general, for each location, there are three operating regions: a region that has zero counts when the electrons reach the detector (cut-off region); a region where the output count is linear to the incident beam current (linear region, fitted with lines) and a region where the count is saturated (saturation region). Operating in both the cut-off and the saturated regions leads to distortion of the linearity of the data, which should be avoided.

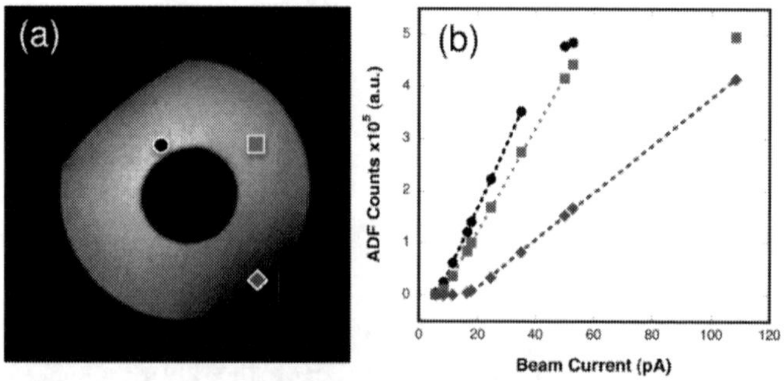

Figure 2 (a) An example of the scanning of the ADF detector. (b) The highest, intermediate and the lowest response of the detector as a function of beam current.

4. The influence of the brightness and contrast setup

The setting of the brightness and contrast of the preamplifier can change the output count of the ADF detector. Two important parameters are considered: the X-intercept and slope (as indicated in Figure 3 (a)). It is found that the X-intercept represents the electron dose that must be exceeded to generate counts. The X-intercepts and slopes differ from one position to another on the detector. Figure 3 (b, c) and (e, f) show the slope and X-intercept as a function of the brightness and contrast respectively, with dashed curves showing fits. The fitting functions used are $y = a$ in (b) and (d), $y = ax+b$ in (c) and $y = be^{ax}$ in (e) and (f), where a and b are constants and x and y are variables. Figure 3 (b) shows that the brightness has little influence on the slope and has a linear relationship with the X-intercepts shown in Figure 3 (c). As indicated in the dashed circle in Fig. 3 (c), the intercepts of three locations all reach zero at brightness of about 2900, showing that every electron contributes proportionally to the output counts. With this setup, the count read out can be linked to the number of electrons arriving at the particular location of the ADF detector.

The relationship between contrast and the slope, as well as contrast and the X-intercept, is not linear. As shown in Figure 3 (e) and (f), the slopes increase exponentially as a function of contrast while the X-intercepts decrease exponentially against the contrast. This is because that the X-intercept is equal to the offset level of the detector divided by the slope as indicated in Figure 3 (a). However, the relative slopes (the relative conversion factors) stay unchanged as the contrast setup is varied (Figure 3 (d)). If the offset was zero, i.e. at brightness of 2900, the relative slopes would be the same because there is no cut-off region, thus the count-current curves would pass through the origin. As a result, the relative conversion factor can be mapped directly by scanning the entire detector area. Alternatively, the slope must be extracted from fitting to the linear operating region for each location of the detector.

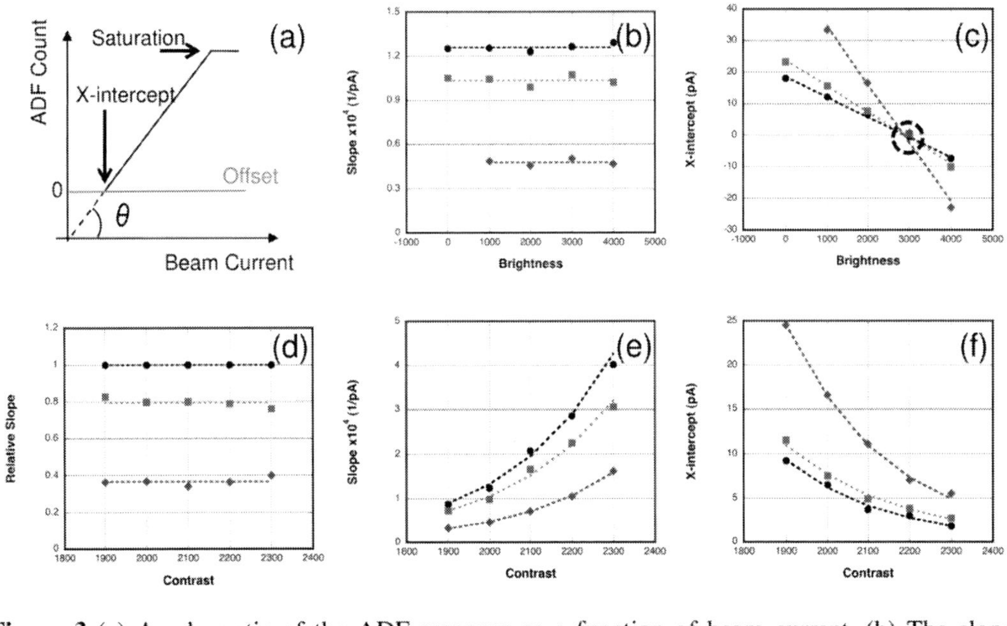

Figure 3 (a) A schematic of the ADF response as a function of beam current. (b) The slope as a function of brightness at contrast = 2000. (c) The X-intercept as a function of brightness at contrast = 2000. (d) The relative slope as a function of the contrast at brightness = 2000. (e) The slope as a function of contrast at brightness = 2000 (f) The X-intercept as a function of contrast at brightness = 2000. In (b-f), the dots, squares and diamonds are from the highest, intermediate and lowest efficiency locations indicated in Figure 2 (a).

5. The efficiency map

The conversion factor (the number of electrons to ADF counts) has been mapped point by point as follows: (1) the response was mapped at different beam currents and aligned using a correlation coefficient; (2) at each position, after being averaged over 10×10 pixels windows, the linear range was fitted with a linear function; (3) the slope of the fitting parameters was obtained and normalised to the largest slope. The obtained detector map would be slightly distorted if the beam was not perfectly aligned. In Figure 4 (a), the geometrical distortion of the mapping has been corrected, assuming the detector has a perfectly round shape. The circularly averaged response of the ADF detector is shown in Fig. 4 (b) where it can be seen that for most parts of the detector the efficiency is between 0.6-0.8. It is worth mentioning that the map in Fig. 4 (a) shares similar characteristic detector features as the map shown in Fig. 2 (a), which is obtained by scanning the detector using a fixed beam current. The non-uniformity of ADF detectors is common when compared to other commercially available microscopes, though with varying degree [3-5]. This is likely a result of current design geometry.

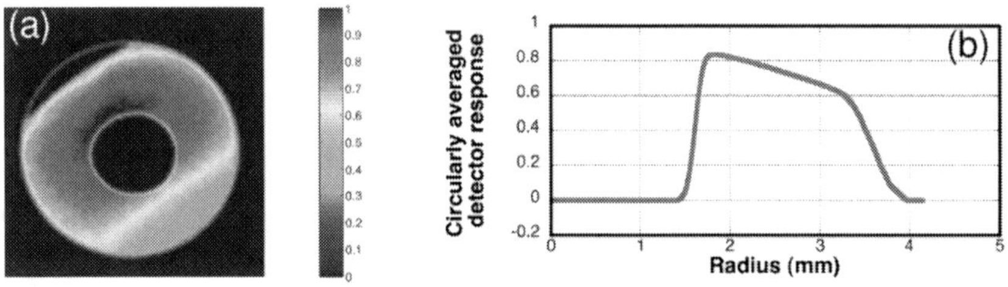

Figure 4 (a) The efficiency map of the detector. (b) The circularly averaged efficiency.

6. Conclusion

We have described a practical approach to quantify the ADF detector. Through a detailed analysis of detector response rate, we obtain the efficiency map which has taken into consideration the non-uniformity of the detector. This method can be applied to most commercially available microscopes which are not fitted with a Faraday cup. By analysing the response of the detector as a function of beam current, we find that though the detector response is not uniform, it is linear within certain beam current regions. By varying the brightness and contrast of the preamplifier, we find that different regions of the detector share the same black level (at brightness=2900). The brightness and contrast do not change the relative efficiency of the detector. Such an efficiency map would be useful when the absolute electron scattering cross-section is needed for quantitative comparisons between experimentally recorded and simulated images [7].

7. Acknowledgements

We acknowledge the financial support from the EPSRC. The STEM instrument employed in this research was obtained through the Birmingham Science City project. DSH acknowledge the studentship from the University of Birmingham and the China Scholarship Council.

References

[1] Li Z Y et al. 2008 Nature **451** 46.
[2] Young N P et al. 2008 Physical Review Letters **101** 246103.
[3] Katz-Boon H et al. 2013 Ultramicroscopy **124** 61.
[4] LeBeau J M and Stemmer S 2008 Ultramicroscopy **108** 1653.
[5] E H et al. 2010 Jounal of Physics: Conference Series **241** 012067.
[6] JEOL Instruction Manual 2007 (EM-24560)
[7] Findlay S D and LeBeau J M 2013 Ultramicroscopy **124** 52.

How flat is your detector? Non-uniform annular detector sensitivity in STEM quantification

K E MacArthur[1], L B Jones, P D Nellist
Department of Materials, University of Oxford, Parks Road Oxford OX1 3PH, UK

E-mail: katherine.macarthur@materials.ox.ac.uk

Abstract. Recording HAADF STEM data on an absolute scale for image quantification is becoming increasingly common. A particular challenge with this method is that most image simulation programs model the detector as being circularly symmetric and exhibiting uniform detection sensitivity across its entire active region. For a real detector this is rarely the case; it then becomes vital to understand how far one's detector deviates from the ideal. Here we investigate a collection of detector maps recorded using hardware from each of the current major manufacturers. Using these maps we compare their different asymmetries and any non-uniformities in their sensitivity. To facilitate this we define the parameters; 'flatness', 'roundness', 'smoothness' and 'ellipticity', evaluate each hardware with respect to these and rank them.

1. Introduction

Quantification of high-angle annular dark-field scanning transmission electron microscopy (HAADF STEM) data has been growing in interest in recent years [1,2]. Typically the images (recorded in counts) are normalised by the count-rate equivalent to the entire STEM probe being incident on the detector. This detector calibration for HAADF STEM requires the beam to be rastered over the detector to produce a 'detector map'. This map then yields both the number of counts in the detector hole (vacuum level - equivalent to the D.C. offset) and the gain of the amplifying electronics, also referred to as detector sensitivity. Knowing this offset and gain, the data can then be expressed as a fraction of the incident beam intensity, by simply subtracting the D.C. offset and dividing by the gain, facilitating direct comparison with image simulations. The scaled images can then be expressed in units of 'fractional beam intensity' allowing for local variation in sample composition to be observed, when thickness is known[3], or variations in thickness to be measured where composition is fixed[4].

A particular challenge with this method is that most image simulation programs model the detector as being circularly symmetric and exhibiting uniform detection sensitivity across its entire active region[5]. Here we investigate a collection of detector maps recorded using hardware from each of the current major manufactures. Working from the premise that no detector is ideal or uniform we set out to characterise how asymmetric and non-uniform these detectors are in their response. To facilitate this we define the parameters; 'flatness', 'roundness', 'smoothness' and 'ellipticity', evaluate each hardware with respect to these and rank them. We also present some important methodological

[1] To whom any correspondence should be addressed.

subtleties in the mapping procedure including sample occlusion, detector saturation and map-distortion resulting from post-specimen optics. Considering all these parameters and effects we show how any detector or mapping-method deficiencies can contribute to errors in experimental image quantification.

2. Method

To determine the dark-field 'count-rate' that corresponds to 100% of the STEM probe's current one must first map the detector's sensitivity. This map is produced by forming focused probe at the detector plane either, for example using a confocal or diffractive mode, which is then rastered across the detector. The integrated ADF signal is read out with probe position, similar to acquiring a normal STEM image, to produce the sensitivity map. If the probe current used for imaging would saturate or damage the detector, then it is acceptable to drop the current by a known ratio by adjusting the mapping dwell time or probe forming aperture size (amplifier brightness and contrast settings must remain unchanged between mapping and imaging).

These maps allow the entire visible surface of detectors to be inspected and analysed for their symmetry and manufacturing quality. All the comparisons derived in this paper are directly from the detector maps shown, Figure 1, relying on the following definitions:

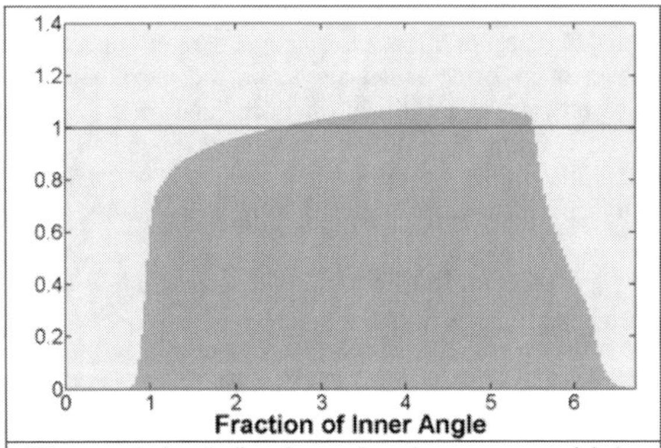

'Flatness' is defined as the detector sensitivity with respect to scattering angle (radially) after averaging azimuthally. Expressed here in units of normalised standard deviation.

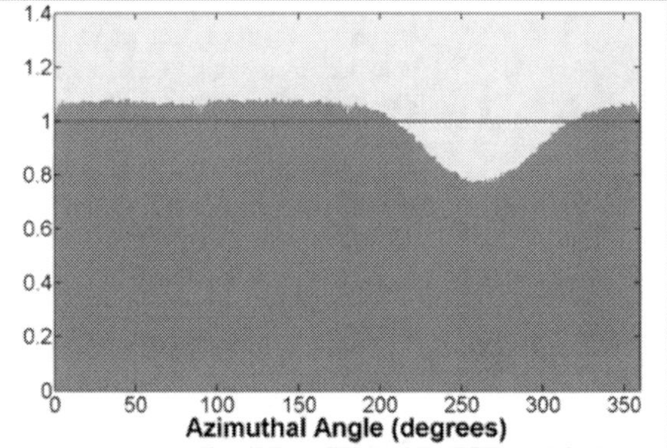

'Roundness' is a measure of the consistency of the detector sensitivity around the detector (azimuthally) after averaging radially. Expressed in units of normalised standard deviation.

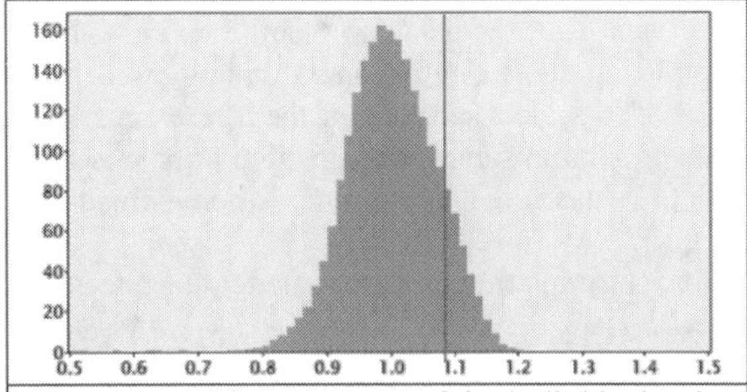

'Smoothness' is a measure of the individual point sensitivity in comparison to all pixels. This is measured as the full-width-quarter-maximum of the active region in the normalised histogram.

'Ellipticity' is defined as the deviation from an ideal circular shape expressed as the percentage of the major over the minor diameters of the inner angle opening.

Finally, the angle ratio is simply defined as the outer angle divided by the inner angle.

3. Results and Discussion

The definitions described above were applied to each of the six detector maps, the results of which are summarised below. For all parameters, apart from the angle ratio, the value present in the table is the measure of deviation from the ideal cases; therefore a larger number means a more asymmetric detector. To calculate the final value for ranking an arithmetic average was carried out of all the parameters, except angle ratio.

Manufacturer	Company 1			Company 2	Company 3	Company 4
	Detector A	Detector B	Detector C	Detector D	Detector E	Detector F
Detector Map	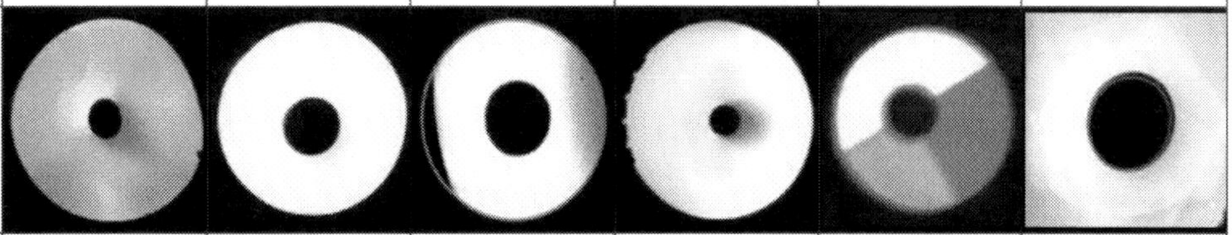					
Angle Ratio	5.47 x	3.09 x	2.89 x	**5.91 x**	2.90 x	3.50 x
'Flatness'	8.9 %	**6.8 %**	24.9 %	10.4 %	9.7 %	14.5 %
'Roundness'	8.2 %	5.4 %	10.6 %	5.4 %	28.1 %	**2.4 %**
'Smoothness'	30.0 % *	**15.1 %**	16.3 %	18.0 %	87.2 %	23.2 % *
'Ellipticity'	19.6 %	4.8 %	8.9 %	**0.5 %**	4.3 %	13.3 %
Average	16.7 %	**8.0 %**	15.2 %	8.6 %	32.3 %	13.4 %

Figure 1. Scaled sensitivity images of the six detectors studied, grouped by manufacturer. For each type of non-uniformity, the percentage deviation from perfection is shown. For each performance metric the most uniform detector is highlighted in bold. Lastly, an overall "non-uniformity" score is tabulated.

As well as the hardware effects already compared there are several alignment and methodology factors, present in the maps above, which will also greatly affect quantification methods. Firstly, the map of detector A demonstrated two main problems; sample occlusion and post specimen optics. The presence of any sample, in this cases holey carbon film, has the potential to disturb or distort the beam during the map and will also affect the gain measurement used for later quantifications. The post specimen optics effect is apparent from the 3-fold symmetrical distortion present in the map, resulting from the post-specimen hexapole correctors present in the TEM/STEM system. The corrector lenses will not only affect the detector map but will also distort, thereby providing radial non-uniformity to, the electron flux the detector will see during imaging.

Secondly detector F demonstrates evidence of 'burn in' from saturation. This is due the detector being saturated with too large a dose of current prior to the map being acquired at a lower current level. The image therefore contains residual counts which will affect quantification results by increasing the average detector sensitivity measured.

Lastly detector E demonstrates an example of a quadrant detector. This necessitates four separate readouts and therefore four separate amplifying electronics. In this example the quadrants have not been properly gain normalised during the experiment, resulting in such an intensity variation in the final detector map. With correct normalisation between the quadrants, however, it is possible to combat some of the hardware imbalances.

4. Conclusion

- The largest detector, physically, is detector D with the maximum angle ratio, providing signal to noise benefits from collecting over a greater range of angles.
- The flattest detector is Detector B. This detector will collect scattering to different angles most fairly and will be the most accurate for composition mapping studies.
- The roundest detector is Detector F. This detector would be the best for avoiding imaging artefacts such as small shifts in column positions.
- The smoothest detector is Detector B, suggesting it may have the best manufacturing quality.
- The least elliptical detector is Detector D. This is essential for accurate inner-angle measurement for reliable comparison with simulation.
- The largest collection-angle range detector (inner- to outer-angle ratio) is Detector D. For a fixed inner-angle this larger ratio means a bigger outer-angle and improved total electron collection giving the best signal to noise ratio possible.

We have also demonstrated the importance of mapping procedure in the quantification process.

5. References

[1] A. Singhal, J. Yang, J. Gibson, Ultramicroscopy 67 (1997) 191–206.
[2] J.M. Lebeau, S. Stemmer, Ultramicroscopy 108 (2008) 1653–1658.
[3] A. Rosenauer, T. Mehrtens, K. Müller, K. Gries, M. Schowalter, P.V. Satyam, S. Bley, C. Tessarek, D. Hommel, K. Sebald, M. Seyfried, J. Gutowski, A. Avramescu, K. Engl, S. Lutgen, Ultramicroscopy 111 (2011) 1316–27.
[4] H. Katz-Boon, C.J. Rossouw, C. Dwyer, J. Etheridge, Ultramicroscopy 124C (2012) 61–70.
[5] S.D. Findlay, J.M. LeBeau, Ultramicroscopy 124 (2013) 52–60.

6. Acknowledgments

The authors gratefully acknowledge Dr Dogan Ozkaya, Johnson-Matthey, Dr Tim Pennycook, SuperSTEM Laboratory, and Dr Armand Béché, EMAT Antwerp, for their help in acquiring detector maps. They would also like to thank EPSRC and Johnson-Matthey for their financial support. The research leading to these results has received funding from the European Union Seventh Framework Programme under Grant Agreement 312483 - ESTEEM2 (Integrated Infrastructure Initiative–I3).

High resolution electron exit wave reconstruction from a diffraction pattern using Gaussian basis decomposition

Konstantin B. Borisenko and Angus I. Kirkland

Department of Materials, University of Oxford, Parks Road, Oxford, OX1 3PH, UK

E-mails: konstantin.borisenko@materials.ox.ac.uk and angus.kirkland@materials.ox.ac.uk

Abstract. We describe an algorithm to reconstruct the electron exit wave of a weak-phase object from single diffraction pattern. The algorithm uses analytic formulations describing the diffraction intensities through a representation of the object exit wave in a Gaussian basis. The reconstruction is achieved by solving an overdetermined system of non-linear equations using an easily parallelisable global multi-start search with Levenberg-Marquard optimisation and analytic derivatives.

1. Introduction

Restoration of the exit wave from a diffraction pattern (diffractive imaging or ptychography) has been recognised as a promising technique in electron microscopy capable of providing diffraction resolution limited data and eliminating the need for an objective lens [1]. In addition, recording a single diffraction pattern (DP) requires a considerably lower radiation dose compared to recording a ptychographic series of diffraction patterns or a focal or tilt series of images which is particularly important for radiation sensitive materials and for atomic resolution tomography [2,3].

Currently, the most common algorithm used for restoration of the exit wave from a DP is an iterative numerical algorithm based on the original ideas due to Gerchberg and Saxton [4]. However, in general the restoration is an ill-posed problem of finding a $2n$-parameter complex wave function from n pixel diffraction data, which in principle can lead to multiple solutions. To avoid this problem the solution space is usually downscaled so that the complex exit wave sought has the same total number of pixels as the input DP. An alternative approach is to obtain DPs from overlapping regions in a ptychographic scheme to provide additional redundancy for the successful restoration [5].

In the present work we test a direct approach for converting a single DP into the corresponding exit wave using a representation of the exit wave in a Gaussian basis. This approach does not require scaling, greatly reduces the number of variables needed to find the solution and allows analytic calculation of the DP and derivatives needed for the restoration process.

2. The exit wave in a Gaussian basis

2.1. The diffracted wave in one dimension

For simplicity of derivation we first consider the exit wave with unity amplitude and phase shift $\phi(x)$ in one dimension as:

$$\Psi(x) = \exp(i\phi(x)) \tag{1}$$

In case of coherent plane wave illumination the Fraunhofer diffracted wave, $D(k)$, is a Fourier Transform (FT) of $\Psi(x)$:

$$D(k) = \int_{-\infty}^{+\infty} \exp(i\phi(x))\exp(-i2\pi kx)dx \tag{2}$$

Using the Euler identity the diffracted wave can be rewritten as:

$$D(k) = \int_{-\infty}^{+\infty} (\cos(\phi(x)) + i\sin(\phi(x)))\exp(-i2\pi kx)dx = \int_{-\infty}^{+\infty} \cos(\phi(x))\exp(-i2\pi kx)dx$$

$$+ i\int_{-\infty}^{+\infty} \sin(\phi(x))\exp(-i2\pi kx)dx \tag{3}$$

We now represent $\phi(x)$ as a sum of Gaussian functions:

$$\phi(x) = \sum_n a_n \exp(-\frac{1}{2}(x-b_n)^2 c_n^{-2}) = \sum_n G_n(a_n, b_n, c_n, x) \tag{4}$$

The FT of a Gaussian function in its general form is:

$$\int_{-\infty}^{+\infty} a\exp(-\frac{1}{2}(x-b)^2 c^{-2})\exp(-i2\pi kx)dx = \cos(2\pi kb)ac\sqrt{2\pi}\exp(-2\pi^2 k^2 c^2)$$

$$-i\sin(2\pi kb)ac\sqrt{2\pi}\exp(-2\pi^2 k^2 c^2) = \cos G_{1t}(k) - i\sin G_{1t}(k) \tag{5}$$

For simplicity of derivation, we consider the case with n=2 where $\phi(x) = G_1 + G_2$:

$$D(k) = \int_{-\infty}^{+\infty} \cos(G_1 + G_2)\exp(-i2\pi kx)dx + i\int_{-\infty}^{+\infty} \sin(G_1 + G_2)\exp(-i2\pi kx)dx = \mathrm{Re}(k) + i\,\mathrm{Im}(k) \tag{6}$$

The observed diffracted intensity is then calculated as $|D(k)|^2 = \mathrm{Re}(k)^2 + \mathrm{Im}(k)^2$.

Using identities for the sine and cosine of a sum of angles the real part of $D(k)$ becomes:

$$\mathrm{Re}(k) = \int_{-\infty}^{+\infty} \cos(G_1)\cos(G_2)\exp(-i2\pi kx)dx - \int_{-\infty}^{+\infty} \sin(G_1)\sin(G_2)\exp(-i2\pi kx)dx \tag{7}$$

The imaginary part of the $D(k)$ is then:

$$\mathrm{Im}(k) = \int_{-\infty}^{+\infty} \sin(G_1)\cos(G_2)\exp(-i2\pi kx)dx + \int_{-\infty}^{+\infty} \cos(G_1)\sin(G_2)\exp(-i2\pi kx)dx \tag{8}$$

Using first order approximations for the sine and cosine function which are valid for small phase shifts (small a coefficients in the basis Gaussians) and small overlap between the basis Gaussians ($|b_1 - b_2| > c_1$ or c_2) $\sin(G(x)) \approx G(x)$ and $\cos(G(x)) \approx 1$. Ignoring the delta function at k=0 it can be shown that the first order approximation of the FT in this case is:

$$D(k) \approx \sin G_{1t}(k) + \sin G_{2t}(k) + i\cos G_{1t}(k) + i\cos G_{2t}(k) \tag{9}$$

The FT of the exit wave represented as a sum of n Gaussians to first order can be then easily calculated as:

$$D(k) \approx \sum_n \sin G_{nt}(k) + i\sum_n \cos G_{nt}(k) \tag{10}$$

2.2. The diffracted wave in two dimensions

A symmetric Gaussian function in two dimensions has the general form

$$G(a, b_x, b_y, c, x, y) = a\exp(-\frac{1}{2}((x-b_x)^2 + (y-b_y)^2)c^{-2}) \tag{11}$$

Using the expressions obtained for the FT of a general Gaussian in 1D the FT in 2D is:

$$\int_{-\infty}^{+\infty}\int_{-\infty}^{+\infty} a\exp(-\frac{1}{2}((x-b_x)^2+(y-b_y)^2)c^{-2})\exp(-i2\pi kx)\exp(-i2\pi ky)dxdy$$
$$= ac^2 2\pi\exp(-2\pi^2 c^2(k_x^{\ 2}+k_y^{\ 2}))(\cos(2\pi k_x b_x)\cos(2\pi k_y b_y)-\sin(2\pi k_x b_x)\sin(2\pi k_y b_y)$$
$$-i\sin(2\pi k_x b_x)\cos(2\pi k_y b_y)-i\cos(2\pi k_x b_x)\sin(2\pi k_y b_y)) = \mathrm{Re}(k_x,k_y)+i\,\mathrm{Im}(k_x,k_y)$$
$$= \cos\cos G_{1t}(k_x,k_y)+i\sin\cos G_{1t}(k_x,k_y) \quad (12)$$

By analogy to the first order approximation for the DP as a sum of 1D Gaussians the DP of an exit wave represented as a sum of 2D Gaussians in a first-order approximation can then be written as:

$$D(k_x,k_y) = \mathrm{Re}(k_x,k_y)+i\,\mathrm{Im}(k_x,k_y) \approx \sum_n \sin\cos G_{nt}(k_x,k_y)+i\sum_n \cos\cos G_{nt}(k_x,k_y) \quad (13)$$

The observed DP intensities can then be rewritten as

$$D^2(k_x,k_y) = \sum_n \cos\cos G_{nt}(k_x,k_y)^2+\sum_n(\cos\cos G_{nt}(k_x,k_y)\sum_{m\neq n}\cos\cos G_{mt}(k_x,k_y))$$
$$+\sum_n \sin\cos G_{nt}(k_x,k_y)^2+\sum_n(\sin\cos G_{nt}(k_x,k_y)\sum_{m\neq n}\sin\cos G_{mt}(k_x,k_y)) \quad (14)$$

The general expression for the partial derivative of the diffraction pattern intensities computed using the above first-order approximation with the general parameter p (p = a_n, b_{xn}, b_{yn}, c_n) of an individual contributing basis Gaussian n in the exit wave is therefore:

$$\frac{\partial D^2(k_x,k_y)}{\partial p_n} = \frac{\partial\cos\cos G_{nt}(k_x,k_y)^2}{\partial p_n}+2\frac{\partial\cos\cos G_{nt}(k_x,k_y)}{\partial p_n}\sum_{m\neq n}\cos\cos G_{mt}(k_x,k_y)$$
$$+\frac{\partial\sin\cos G_{nt}(k_x,k_y)^2}{\partial p_n}+2\frac{\partial\sin\cos G_{nt}(k_x,k_y)}{\partial p_n}\sum_{m\neq n}\sin\cos G_{mt}(k_x,k_y) \quad (15)$$

In practice, the accuracy of the simulated DP can be increased by using higher order approximations for the sine and cosine functions in addition to more Gaussian functions in the basis to represent the exit wave. A straightforward, although slow, implementation of this strategy is to use a numerical FT to compute the DP from an exit wave represented as a sum of Gaussians and also to use numerical derivatives. In this approach the exit wave can be restored by solving an overdetermined system of non-linear equations using an appropriate algorithm.

3. DP simulation and exit wave restoration

To test the approach, an exit wave from a model of a 0.41 nm diameter carbon nanotube fragment with a total of 60 atoms was simulated at 80 kV using the multislice algorithm [6]. The noiseless DP intensities computed from the FT of the exit wave and weighted to remove the central beam were subsequently used as input for the reconstruction code with the Levenbergh-Marquardt optimisation method implemented in MATLAB to solve the system of non-linear equations. These reconstructions are shown in Figure 1.

4. Results and discussion

The solution of the non-linear equations often depends on the initial approximation of the function and it was found that in the present case a scan of initial values is required to provide a high quality restoration. This final restoration shows good agreement with the input exit wave as shown in Figure 2. The computations of the first order approximation with analytic derivatives work well to quickly refine the initial estimate of the exit wave. However, for a better quantitative agreement higher order approximation and larger basis sets with multiple Gaussians were needed. In this particular example we have found that two Gaussian functions per peak were adequate to obtain the restoration shown in Figures 1 and 2, although more Gaussians are needed to reproduce the phase shifts more accurately.

It is well known that Gaussians are appropriate for descriptions of the atomic potential, and therefore, for modelling exit wave shifts. A sum of a number of differently shaped Gaussians is often satisfactory to produce meaningful results that require fitting of the atomic potential as, for example, in quantum chemistry. Moreover, the method described here can be extended to a general object where the amplitude of the exit wave is not unity and other functions, such as Lorentzians, which have analytical FTs and better reflect the behaviour of atomic electron density in the vicinity of the nucleus, can be used to better represent the exit wave.

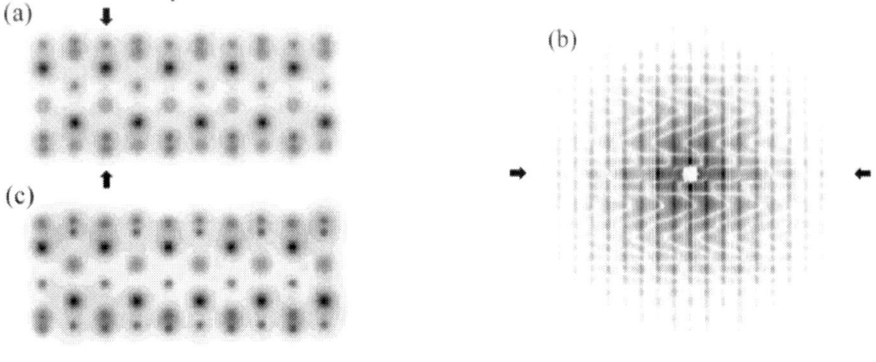

Figure 1. Original exit wave phase (a), logarithm of the square root of the corresponding input diffraction pattern (b) and the restored exit wave phase (c) all shown in inverted contrast to improve presentation.

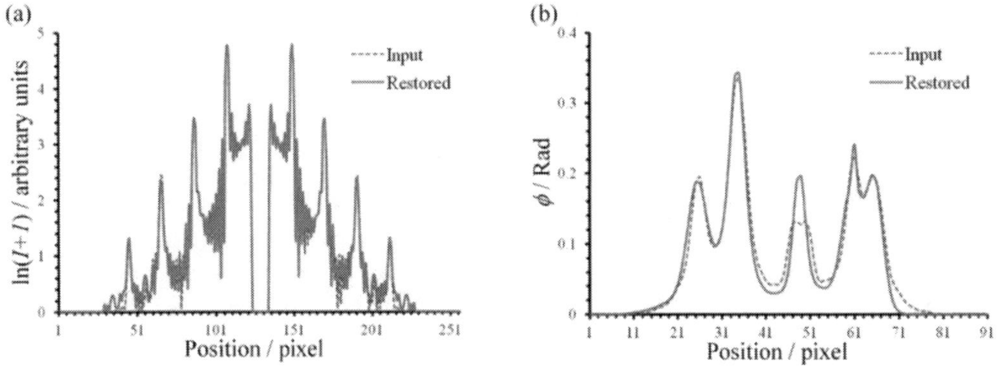

Figure 2. (a) Comparison between the input and restored diffraction patterns intensities, I, along the line defined by the arrows in Figure 1(b); (b) comparison between the input and restored exit wave phases, ϕ, along the line defined by the arrows in Figure 1(a).

5. Acknowledgements
Financial support from the European Union under the Seventh Framework Program under a contract for an Integrated Infrastructure Initiative (Reference 312483-ESTEEM2) is gratefully acknowledged.

6. References
[1] Humphry MJ, Kraus B, Hurst AC, Maiden AM and Rodenburg JM 2011 *Nature Comm.* **3** 730
[2] Morgan AJ, Martin AV, D'Alfonso AJ, Putkunz CT and Allen LJ 2011 *Ultramicroscopy* **111** 1455
[3] Borisenko KB, Moldovan G, Kirkland AI, Van Dyck D, Tang HY and Chen FR 2012 *J. All. Compounds* **536** S94
[4] Gerchberg RW and Saxton WO 1972 *Optik* **35** 237
[5] Maiden AM and Rodenburg JM 2009 *Ultramicroscopy* **109** 1256
[6] Kirkland EJ 1998 *Advanced Computing in Electron Microscopy* (New York: Plenum Press)

Journal of Physics: Conference Series **522** (2014) 012020 doi:10.1088/1742-6596/522/1/012020

Application of Low Energy STEM with the In-lens Cold FE-SEM

Y. Orai, T. Sunaoshi, S. Okada, T. Ogashiwa, H. Ito, M. Konno

Hitachi High-Technologies Corporation, 882, ichige, hitachinaka shi, Ibaraki, Japan

orai-yoshihisa@naka.hitachi-hitec.com

Abstract. We have developed a high resolution STEM imaging method using accelerating beam voltages of 30 kV and less. In this study, we examined and determined the low accelerating voltage parameters necessary for obtaining lattice resolution in STEM images, using the cold field emission scanning electron microscope (CFE-SEM), Hitachi SU9000. We also investigated and optimized the objective lens conditions for optimum lattice fringe imaging. The STEM images and associated Fourier transform image obtained at 30 kV show the Si (222) lattice fringes and reflection spots, corresponding to the d-lattice spacing of 0.157 nm. The STEM images and associated Fourier transform image at 15 kV show the Si (111) lattice fringes and reflection spot, corresponding to the d-lattice spacing of 0.314 nm.

Introduction

In the past, the transmission electron microscope operating at an accelerating voltage of 200 kV or higher has been used extensively for structural characterization of semiconductor, metallic and ceramic materials. As the dimensions of IC semiconductor devices shrink down to 20 nm (and smaller) it is critical to produce thin specimens of less than 20 nm in thickness. This is necessary to avoid overlapping individual structures and therefore blurring the STEM image of an individual device. Additionally, inorganic carbon materials (primarily carbon nanotubes and graphene) and organic polymeric materials are being widely investigated as viable post-CMOS device materials. The abovementioned materials share the STEM specimen problems of electron beam-induced structural disorder (due to knock-on damage) and a reduction of STEM image contrast due to the small ratio of elastic to inelastic scattered electrons and the narrow angle of the scattered electrons. Low energy microscopy reduces beam-induced damage and enhances scattering contrast, but also reduces resolution as the electrons' wavelength increases. Accordingly, it is necessary to determine methods for reducing specimen damage and improving contrast without loss of resolution. To specifically address that need, research institutions, academia, and private industry have collaborated to develop and commercialize a novel low energy FE-SEM/STEM. [1-4].

The imaging resolution of a 30 kV FE-SEM with a STEM function has recently been improved, thanks to a new lower spherical aberration objective lens, and a high stability and coherency (ΔE) electron gun. The cold field emission scanning electron microscope (CFE-SEM) with an in-lens type objective lens, Hitachi SU9000, provides all the above mentioned benefits and made possible the development of our high resolution STEM technique. Through using this new method it is possible to routinely achieve lattice resolution, showing the lattice fringes of the graphite (002) plane with a spacing of 0.34 nm [5].

In this study, we describe the observation conditions for obtaining lattice resolution enhancement in STEM imaging at an accelerating voltage of below 30 kV.

Experimental method

The SU9000 CFE-SEM's accelerating voltage range is from 500V to 30 kV, making it possible to attempt lattice imaging below 30 kV without special equipment. We selected a single-crystal silicon specimen to illustrate how our low accelerating voltage STEM technique would apply to the analysis of current semiconductor devices. The sample was thinned by an Ar^+ ion milling technique, producing a final sample thickness of about 20 nm.

Theoretical resolution is limited by electron diffraction and aberration effects. For every SEM, there is an optimum beam half angle for minimum theoretical resolution. In the SU9000 FE-SEM's case, this optimum beam half angle is approximate 10mrad at 30kV (maximum accelerating voltage). Normally this optimum beam half angle is selected when observing the real objective image because the image resolution is best at this condition. On the other hand, for lattice image observation is necessary to set the beam half angle to $\alpha_i > 2\theta_B$. In order to observe the Si (111) lattice fringes (d=0.314 nm) at accelerating voltages of 30 kV and 15 kV, the beam half angles needed to be set to $\alpha_i > 22.5$ mrad and $\alpha_i > 32.5$ mrad respectively. Therefore, in order to obtain the lattice image of Si at 30 kV and 15 kV, we adjusted the objective aperture, 2nd condenser lens, and objective lens to produce a larger beam half angle than the one for optimum resolution.

The objective lens spherical aberration coefficient (Cs) increases as the amount of defocus between STEM structural imaging and STEM lattice imaging also increases. The optics of the SU9000 need a great amount of defocus for optimizing lattice fringe imaging and clearly presents a significant challenge. For this reason, it is difficult to observe the structural and lattice fringes information of the specimen simultaneously. However, it is possible to reduce Cs by reducing the focal length which results in a short working distance. Therefore, we made an experimental sample stage shown in Fig.1 to reduce the distance between the objective lens and the actual sample. This resulted in a reduction of Cs from about 2 mm to 1 mm.

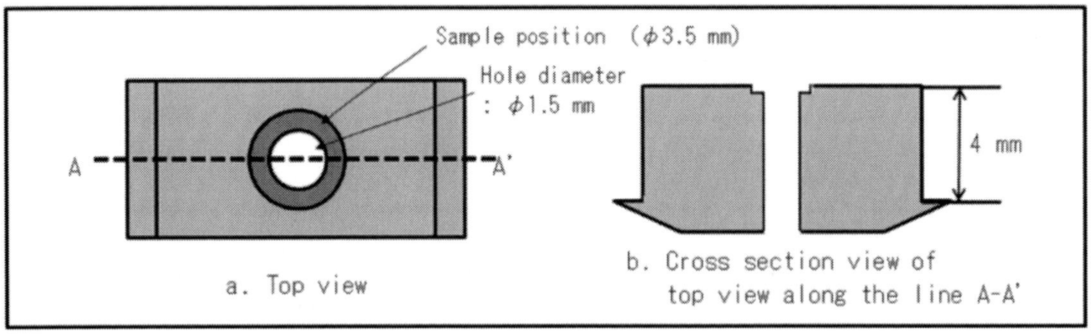

Fig.1 Schematic images of the experimental sample stage

Results and Discussion

Fig.2 shows a comparison of image resolution and amount of defocus using standard optics and optimized optics. We observed BF-STEM images of graphite lattice fringes in 3 positions. The working distance and Cs according to computation are 3.0mm, 2.0mm at sample position (a), 2.7 mm, 1.7 mm at (b), 1.8 mm, and 1.0 mm at (c). Sample position (a) is the "standard" STEM observation position and sample position (c) is upper limit at an accelerating voltage of 30kV setting. We calculated the amount of defocus difference (Δf) between the lattice and in focus conditions by noting the value of the lens' digital analog converter (DAC) at each focus point. The amount of defocus difference is reduced as the sample position rises. The ensuing reduced Cs at sample position (c) (25 nm defocus) makes it possible to clearly observe the structural image and lattice image simultaneously.

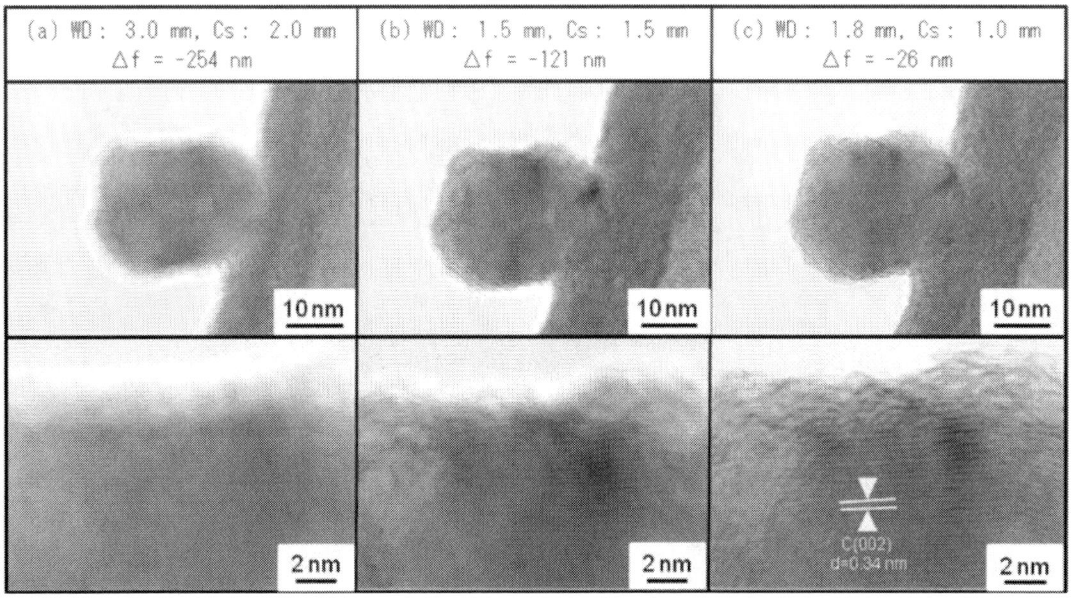

Fig.2 BF-STEM lattice images of graphite at an accelerating voltage of 30 kV

Next we investigated the image resolution when changing the amount of defocus. We observed the BF-STEM images of single crystal Si<011> by a through-focus method at an accelerating voltage of 30 kV. Fig.3 shows the widest lattice image and narrowest single crystal Si<011> lattice image (with inset Fourier transform image) which could be observed by the SU9000. Amount of defocus (Δf) was based on using (a) as 0-reference. As Δf was changed, the BF-STEM image at Δf = -341 nm provided the best result showing the Si (222) plane, which has a spacing of 0.157 nm. The inset Fourier transform image confirms the Si (222) reflection spot. The BF-STEM and Fourier transform images at Δf = -223 nm show the Si (220) plane, which has a spacing of 0.192 nm. This result emphasizes that high resolution imaging of less than a 200 pm is possible without aberration correctors, and creates new possibilities and applications for low energy STEM.

Fig.4 shows a high resolution BF-STEM image with its inset Fourier transform image, observed along the Si<110> zone axis at an accelerating voltage of 15 kV. The BF-STEM image shows the Si (111) plane, which has a spacing of 0.314 nm and Fourier transform image shows the Si (111) reflection spot, corresponding to 0.314 nm. These results reveal that we can observe lattice images at below 30 kV by using a high-performance in-lens CFE-SEM such as the SU9000. This method has the potential for high contrast visualization, less beam damage, and high resolution imaging on the latest semiconductor devices and carbon materials.

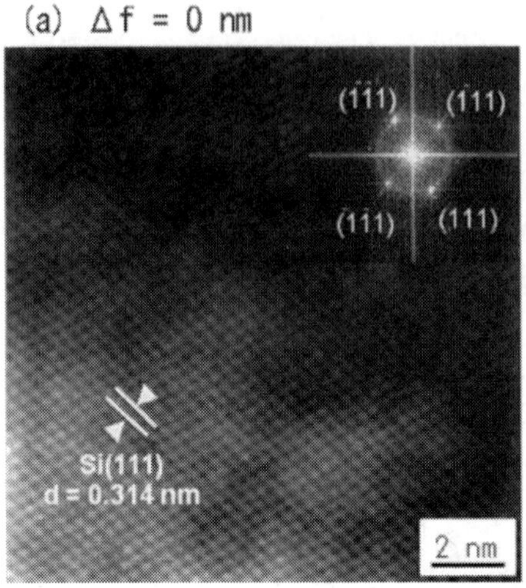

Fig.3 Observation of Si lattice imaging by through-focus method at an accelerating voltage of 30 kV.

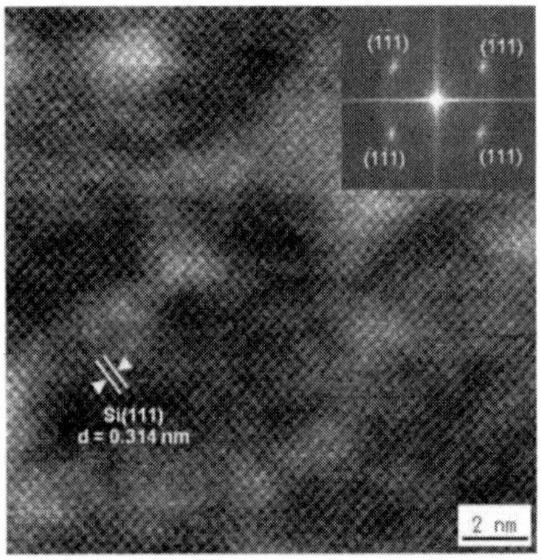

Fig.4
High resolution BF-STEM image with inset Fourier transform image observed along the Si<110> zone axis at an accelerating voltage of 15 kV.

Conclusion

We reported a method for STEM image observation of lattice fringes in silicon, with a minimum lattice spacing of 0.157 nm, by using an in-lens type FE-SEM without aberration corrector at an accelerating voltage of 30 kV. We also observed lattice fringes in silicon (Si (111) lattice spacing of 0.314 nm) at an accelerating voltage of 15 kV. These results reveal the potential for high contrast visualization without loss of resolution for the latest semiconductor devices and any carbon-based materials with minimal radiation beam damage.

References

[1] K.Suenaga, et al : Nature vol.468 (2010), 1088-1090
[2] T. Sasaki, et al : Microsc. Microanal. 18 (Suppl 2), (2012), 1514-1515
[3] U. Kaiser, et al : Ultramicroscopy 111 (2011), 1239–1246
[4] O. Kamimura, et al : Ultramicroscopy 110(2) (2010), 130–133
[5] Vinh Van Ngo et al., Microscopy Today, volume 15, issue 02, (2007) p12-16

Compressed Sensing Electron tomography using adaptive dictionaries: a simulation study

A.AlAfeef[1], P.Cockshott[1], I.MacLaren[2], S.McVitie[2]

1 School of Computing Science, University of Glasgow, Glasgow, G12 8QQ, United Kingdom
2 School of Physics and Astronomy, University of Glasgow, Glasgow G12 8QQ, United Kingdom

E-mail: a.al-afeef.1@research.gla.ac.uk

Abstract. Electron tomography (ET) is an increasingly important technique for examining the three-dimensional morphologies of nanostructures. ET involves the acquisition of a set of 2D projection images to be reconstructed into a volumetric image by solving an inverse problem. However, due to limitations in the acquisition process this inverse problem is considered ill-posed (i.e., no unique solution exists). Furthermore reconstruction usually suffers from missing wedge artifacts (e.g., star, fan, blurring, and elongation artifacts). Compressed sensing (CS) has recently been applied to ET and showed promising results for reducing missing wedge artifacts caused by limited angle sampling. CS uses a nonlinear reconstruction algorithm that employs image sparsity as a priori knowledge to improve the accuracy of density reconstruction from a relatively small number of projections compared to other reconstruction techniques. However, The performance of CS recovery depends heavily on the degree of sparsity of the reconstructed image in the selected transform domain. Prespecified transformations such as spatial gradients provide sparse image representation, while synthesising the sparsifying transform based on the properties of the particular specimen may give even sparser results and can extend the application of CS to specimens that can not be sparsely represented with other transforms such as Total variation (TV). In this work, we show that CS reconstruction in ET can be significantly improved by tailoring the sparsity representation using a sparse dictionary learning principle.

1. Introduction

In materials science, Electron tomography (ET) is considered an effective technique that provides indispensable information for the study of particles and structures in the Nanoworld. The 3D morphology of nanostructures can be provided by reconstruction of an aligned set of 2D TEM images acquired around a single or double tilting axis. The reconstruction step in ET is performed using an image reconstruction algorithm such as weighted back projection (WBP) or simultaneous iterative reconstruction technique (SIRT). SIRT algorithm provides reconstruction of a relatively higher Signal-to-noise ratio than those obtained using WBP. However, the reconstruction still suffers from elongation, blurring and artifacts due to the missing wedge limitation in ET. The number of projections that can be recorded is limited by the maximum tilt angle (typically $\pm$ 72) above which projections can not be acquired. A segmentation step is usually needed to overcome such artifacts and to distinguish between artifacts and main components in the reconstruction. This segmentation process is often performed manually leading to potentially subjective and time-consuming interpretation of the data.

The reconstruction process of ET can be explained by considering a 3D object as a set of 2D sections, the reconstruction of that object can then be created by reconstructing each 2D section from the corresponding 1D projection. An object may be reconstructed from its projections by solving a system of linear equations which can be modelled as:

$$Px = b \tag{1}$$

Where P is the discrete Radon transform that converts the measurements (1D projections) b into object domain (2D imaged object) x. To create a reconstruction (tomogram), the unknown vector x needs to be calculated from measurements b which is not straightforward because of the underdetermined nature of the reconstruction inverse problem due to the limited number of projections. Also the reconstruction will be further degraded with the presence of noise and alignment errors.

It is well-known that the quality of a tomographic reconstruction can be enhanced by including additional prior knowledge about the specimen throughout the reconstruction process as in SIRT and recently CS-based approaches. CS has been shown to be powerful technique to reducing the artifacts that arise from ET reconstruction. The key prior knowledge employed in CS is that the signal is sparse in a transform domain, meaning that it can be approximated in a more compact form. This sparsity assumption can then be formulated as a problem of simultaneously minimising a cost function consisting of a data consistency term and one or more sparsity constraints in an appropriate transform domain (e.g., wavelets and TV). It should be noted that, in order for the CS algorithm to be successful and effective, an object to be reconstructed need to be compressible in a certain domain (for example, the TV can be used for reconstruction, if the object under study can be described as a piecewise constant). However, many samples do not satisfy such constraints, and therefore CS reconstruction can not be effective. Recently, an algorithm (referred to as K-SVD) for training a dictionary for sparse signal representation was proposed [1]. K-SVD is a signal representation approach which, from a set of signals, can derive a dictionary able to approximate each signal with a sparse combination of the atoms. The goal in this work is to make use of the K-SVD algorithm for designing sparsity dictionaries and utilizing these dictionaries for CS recovery in ET.

2. CS Recovery with Adaptive Transform using K-SVD

In [2], Candes *et al* , discovered important results that formed the theoretical basis of compressed sensing as currently studied, where an exact recovery of the SheppLogan phantom was obtained from 22 radial samples of its discrete Fourier transform. This work kicked off the field of compressed sensing theory and lead to successful application to other Inverse Problems such as Magnetic Resonance Imaging (MRI) [3] and recently ET as in [4, 5, 6] and [7]. The reconstruction in ET based on CS can be expressed in terms of minimizing the cost function in Eq.(2)

$$\arg\ \min_{x} \left\{ \|Px - b\|_2^2 + \lambda \|\Psi x\|_1 \right\} \tag{2}$$

where Ψ is a predefined sparse transform (such as wavelets and total variation). While these transforms, especially total variation that have been proven to work well for the general class of ET samples [5, 7], they are not optimized for each ET application. In EM, significant prior information exists about the object being imaged and this prior information can be used to design more efficient sparsity dictionaries for CS ET. The K-SVD [1] algorithm aims for learning adaptive transforms (dictionary) Ψ and a sparse matrix Ω, such that for any signal x, there exists a sparse linear combination from Ψ that minimizes the representation error by solving:

$$\arg\min_{\Psi,\Omega} \|x - \Psi\Omega\|_2^2 \ \ s.t \ \ \|\omega_i\|_0 \leq v, \forall i \tag{3}$$

Journal of Physics: Conference Series **522** (2014) 012021 doi:10.1088/1742-6596/522/1/012021

Where ω_i are columns of Ω, and v is the desired sparsity. In this work, the K-SVD algorithm will be used for designing adaptive sparsity dictionaries for a restricted class of ET images and utilizing these dictionaries for CS recovery. The sparsity in the proposed method is enforced on overlapping patches of the reconstructed image x and the dictionary is directly adapted to the image leading to higher reconstruction fidelity. For reconstruction in ET, the proposed algorithm can be formulated as in Eq.(4)

$$\arg\min_{\alpha,\Psi,x} \sum_{ij} \left\{ \|B_{ij}x - \Psi\omega_{ij}\|_2^2 + \lambda \|Px - y\|_2^2 \right\} \; s.t \; \|\omega_{ij}\|_0 \leq v, \; \forall i, j (4)$$

Here, B_{ij} is a mapping operator that extracts a square patchs of size $\sqrt{n} \times \sqrt{n}$ from image x; $\Psi\omega_{ij}$ is the sparse approximation of the patch, with the maximum number of non-zero values $\leq v$; Ψ is the trained dictionary; Ω is the set $\{\omega_{ij}\}_{ij}$; and v is a positive constant.

The first term in Eq.(4) measures the quality of the sparse approximations of the image patches with respect to the dictionary Ψ. The second term in the cost forces data fidelity in image domain. In one step (dictionary learning), x is assumed fixed, and the dictionary is learned with the sparse representations of the image patches. In the other step (reconstruction update), Ψ and Ω are fixed, and x is updated to satisfy data consistency. The reconstruction update step involves a least squares problem that can be solved using the corresponding normal equation and employing the conjugate gradient method that is initialized with a zero-filled Fourier reconstruction for x. v is the sparsity threshold in the K-SVD algorithm: higher v produces fewer atoms to approximate a patch in the sparse coding step.

3. Results

To evaluate the quality and accuracy of proposed reconstruction algorithm, the CS test phantom (CS-phantom) proposed in [8] is used (Figure 1-a). CS-phantom is tailored for testing the accuracy and properties of CS solvers in the noise-free domain. For comparisons, we performed the reconstruction using a WBP method (Figure 1-b), a TV-based method (Figure 1-c), and proposed approach (Figure 1-d) using simulated data set consisted of 27 projections. Peak Signal to Noise Ratio (PSNR) in dB is showed for the reconstruction in (Table 1). The reconstruction scheme outlined in this work is shown to produce substantially better reconstruction both visually and in terms of PSNR, as compared to the WBP and TV-based method.

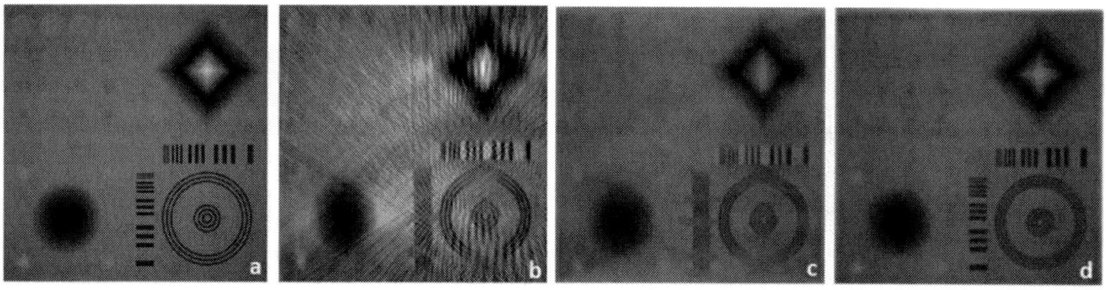

Figure 1. Reconstructed phantom images, (a) CS-phantom used for the simulation of a tilt series from $-65°$ to $65°$ with an increment of $5°$. (b) WBP (c) CS TV based algorithm (d) proposed method.

Table 1. PSNR of reconstructions.

Technique	CS-phantom
WBP	8.20 dB
CS-TV	21.17 dB
CS-Dictionary	30.5 dB

4. Conclusions and Discussion

A novel sparse reconstruction technique that incorporates prior information through dictionary training is introduced for CS in ET. Reconstruction results illustrate that the proposed technique can yield significantly improved image quality compared to commonly used sparsity transforms in CS ET. Future work will explore other interesting properties of the dictionary that may provide even more advanced reconstruction for highly under sampled ET.

4.1. Acknowledgments

This research was supported by a Lord Kelvin Smith Scholarship of the University of Glasgow.

References

[1] Aharon M, Elad M and Bruckstein A 2006 *Signal Processing, IEEE Transactions on* **54** 4311–4322
[2] Candes E J and Tao T 2006 *Information Theory, IEEE Transactions on* **52** 5406–5425
[3] Lustig M, Donoho D and Pauly J M 2007 *Magnetic Resonance in Medicine* **58** 1182–1195 ISSN 1522-2594
[4] Saghi Z, Holland D J, Leary R, Falqui A, Bertoni G, Sederman A J, Gladden L F and Midgley P A 2011 *Nano Letters* **11** 4666–4673
[5] Leary R, Saghi Z, Midgley P A and Holland D J 2013 *Ultramicroscopy*
[6] Monsegue N, Jin X, Echigo T, Wang G and Murayama M 2012 *Microscopy and Microanalysis*
[7] Goris B, Van den Broek W, Batenburg K, Heidari Mezerji H and Bals S 2012 *Ultramicroscopy* **113** 120–130
[8] Smith D and Welch E 2011 Non-sparse phantom for compressed sensing mri reconstruction *International Society for Magnetic Resonance in Medicine 19th Scientific Meeting-ISMRM* vol 11 p 2845

Journal of Physics: Conference Series **522** (2014) 012022 doi:10.1088/1742-6596/522/1/012022

A novel 2D and 3D method for automated insulin granule measurement and its application in assessing accepted preparation methods for electron microscopy

J Mantell[1], D Nam[2], D Bull[2], A.Achim[2] and P.Verkade[1,3]

[1]Wolfson Bioimaging Facility, University of Bristol, Medical Sciences, University Walk, Bristol BS8 1TD, UK.

[2]Visual Information Laboratory, University of Bristol, Merchant Venturers Building, Woodland Road, Bristol BS8 1UB, UK.

[3]Schools of Biochemistry and Physiology and Pharmacology, University of Bristol, Medical Sciences, University Walk, Bristol BS8 1TD, UK.

j.mantell@bristol.ac.uk

Abstract. Transmission electron microscopy images of insulin-producing beta cells in the islets of Langerhans contain many complex structures, making it difficult to accurately segment insulin granules. Furthermore the appearance of the granules and surrounding halo and limiting membrane can vary enormously depending on the methods used for sample preparation. An automated method has been developed using active contours to segment the insulin core initially and then expand to segment the halos [1]. The method has been validated against manual measurements and also yields higher accuracy than other automated methods [2]. It has then been extended to three dimensions to analyse a tomographic reconstruction from a thick section of the same material. The final step has been to produce a GUI and use the automated process to compare a number of different electron microscopy preparation protocols including chemical fixation (where many of halos are often distended) and to explore the many subtleties of high pressure freezing (where the halos are often minimal, [3]).

1. Introduction

Insulin is secreted into the blood stream by beta cells inside Islets of Langerhans of the pancreas. After its biosynthesis, insulin is processed and packaged in granules, surrounded by a membrane. Quantitative measurements of the amount of insulin in the cell have physiological importance, so it is important to be able to take measurements from images, ideally in an automated way. In TEM the granules usually appear as dense cores often surrounded by a halo and enclosing membrane. However the sample preparation route can strongly affect the appearance, such that the cores vary in intensity and texture, (and so can be confused with other cellular organelles such as mitochondria), and some granules have no surrounding halo or only partially visible membranes (Figure 1). All of these factors make accurately segmenting granules a challenging task. We have developed a novel method based on active contours which we have shown to be superior to other mathematical methods previously explored [1] and does not require an expensive software package to run [2]. Indeed we have now built a (soon to be) freely available GUI which runs through MatLab. We first validated our algorithm against manual measurements. Per image, the automated method was not substantially faster but it was then applied to a large dataset where samples were prepared in the same laboratory for electron microscopy using a large set of different sample preparation protocols. For this analysis we have measured in excess of 10,000 granules and have found that there are clear differences between chemically fixed (CF) and high pressure frozen (HPF) samples, and also subtle variations depending on the post-HPF freeze substitution recipes. Furthermore, in order to achieve true volume measurements we have extended the method to 3- dimensions (3D) by applying it to an image z-stack from a tomographic reconstruction dataset.

a)

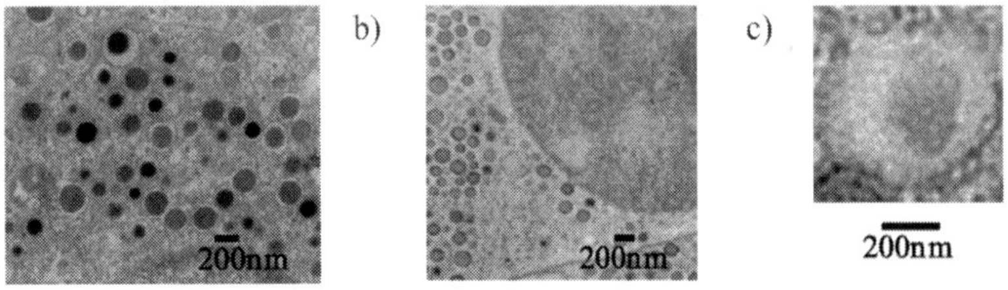

Figure 1. Varying appearance of secretory granules a) CF b) HPF c) Incomplete membrane

2. General description of the algorithms used for core and halo segmentation

2.1 Core segmentation

Due to the complexity of the image, the core segmentation process is broken down into 3 steps (Figure 2).The first step uses a locally scalable region-based level-set active contour with a shape reguliser to provide a good pre-segmentation of the cores. Other unwanted cellular components segmented at this stage, are discarded if they are not round enough or are too large. The cores are then accurately segmented by using a dual level-set active contour. The algorithms have been published elsewhere [1].

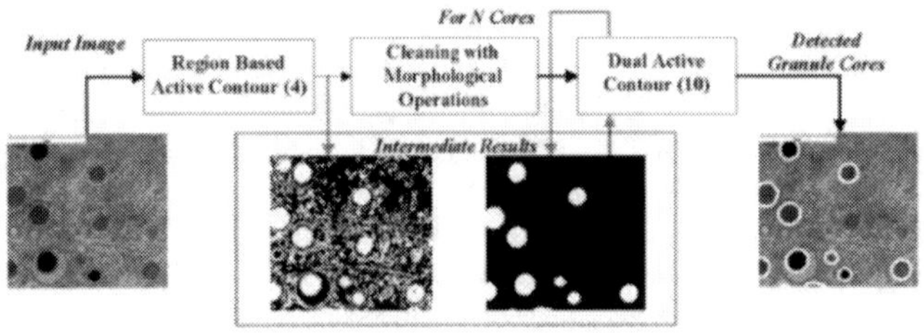

Figure 2: Granule core detection algorithm using section of TEM image 200 x 200 pixels (pixel size is 5nm)

2.2 Segmentation of granule membranes

The initial active contour segmentation result (segmented cores), and the original TEM image are starting points for membrane segmentation. The core image is slightly dilated and then removed from the image. The granule membrane is then sampled by projecting outwards from the core in all directions until the first non zero pixel is found. Morphological operations, sampling and application of a vector field convolution snake, are needed to fill gaps in incomplete membranes. The application of a further sliding band convergence filter is used to verify when cores have no halo which is often the case for the HPF samples. The work flow is shown in Figure 3.The mathematical equations were presented elsewhere [4].

2.3 Extension to 3D

The beta cell tomogram (z-stack) slice images are quite different from their conventional 2D counterparts as they are reconstructed from a series of tilt images from a 300nm section and also the reconstruction algorithm plays a part. We selected the most commonly used weighted back projection in the IMOD software package [5]. This leads to an image with relatively poor contrast when compared to the conventional images from 70nm sections. The poor contrast has the greatest effect on the appearance of the granule halos. An initial segmentation is first performed on each slice to get a good estimate of the granule core locations; a second refining segmentation is then done on each granule using a dual region-based active surface, first for the core. For the halo we use a dual Bayesian active surface. A novel boundary prior and gradient term have been incorporated in a dual active surface, to address the challenges faced with halo segmentation. This method has been shown to be superior to other active surface models [6]. We show a completely encapsulated granule core and halo in Figure 4 manually segmented in (a) and (c) and using our algorithms in (b) and (d). Note that since the granule is segmented in each slice it does not look completely smooth in the vertical plane.

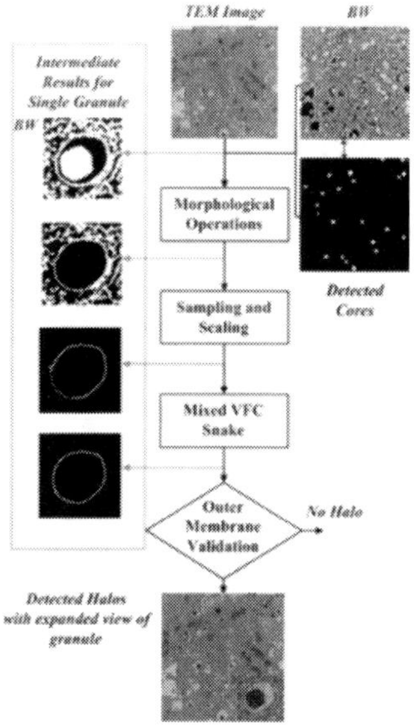

Figure 3: Membrane segmentation algorithm

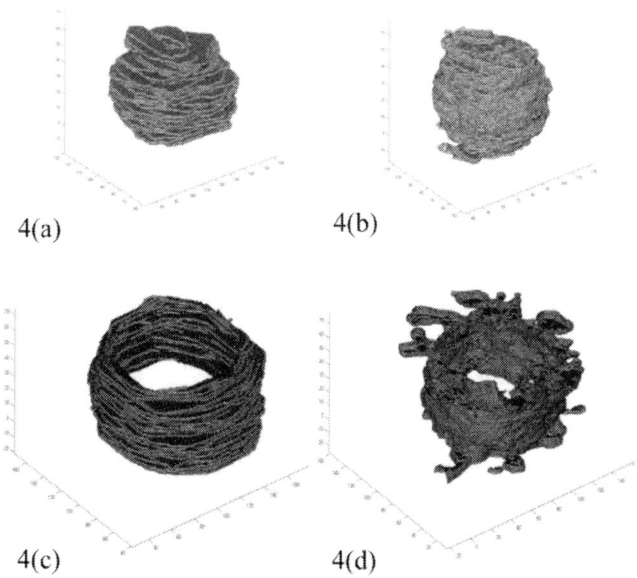

Figure 4: 3D core (a,b) and halo (c,d) segmentation. 4(a) and (c) done manually, (b) and (d) with our algorithm. Dimensions in pixels where pixel size = 3.8nm

3. GUI

The user interface is currently being tested and an example of the page for complete segmentation is shown in Figure 5

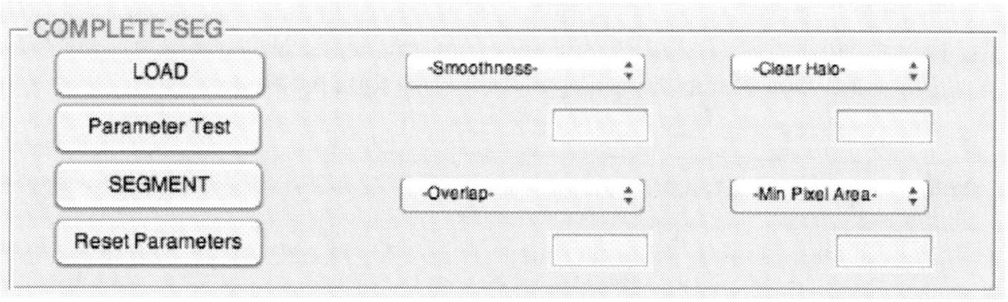

Figure 5: The Graphical User Interface

After loading an image, a ROI is selected and user selectable parameters for effective segmmentation can be input and evaluated. The values of the parameters depend on the character of the image and sample preparation method. "Smoothness" encourages granule circularity, for example. "Clear halo" removes contrast features between the granule and the membrane and "overlap" accounts for the minimum overlap of the segmented (possible) membrane and the initial segmentation. Guidance on appropriate values is given in a Help function. A test segmentation on the ROI can be performed and evaluated before committing to the complete segmentation. The results are output to Excel spreadsheets. A batch programme is also available to handle many images for say, overnight analysis.

4. Experimental Results

4.1 Validation of algorithms

The algorithm was initially applied to seven 3x3 tiled images each showing a complete beta cell prepared either by CF or HPF methods. The manual measurements made on the same images were considered to be "ground truth". The number of cores detected was within 96% of the number of cores counted manually. We also compared average core and membrane areas, with those measured manually. The metric we used to record granule sizes is the radius of a circle of the same area. On average the granule core radii difference for all images is 13.35% and for membranes 6.08%.

The analysis of the z-stack in the tomograms is in its early stages. The visual inspection of post processed granules, as in Figure 4, looks promising. The process compares with manually segmenting and building a model to generate volume data as with IMOD software, which is extremely labour intensive. Thus to be able to extend volume measurement to thousands of granules would only be realistic with automated software such as ours.

4.2 Comparison of Specimen preparation protocols

10 (4k x4k) EM images were taken from each of sixteen differently prepared samples and these were analysed using our software. The core and membrane equivalent radii for each granule were taken as the dependent variables and the fixation method as the independent variable. A MANOVA indicated that the fixation method does indeed affect both the core and membrane areas. An ANOVA test then confirmed that there is a significant difference in mean areas for both cores and membranes between the different sample preparation procedures. Tukey's HDS test was used to do multiple comparisons. We found that for our standard CF method (fixed with 2.5% glutaraldehyde in cacodylate buffer), there is a significant difference in membrane areas when compared with all other methods. For our standard HPF method (freeze substitution time 5hours (-90°C to 0°C), with 1% osmium + 0.1% uranyl acetate in acetone)[3], there was a significant difference in membrane areas when compared to the CF methods, but no significant difference in membrane areas when compared to most of the other freeze substitution methods. Detailed evaluation of the differences between the 16 methods is currently being done.

5. Conclusions

We have developed automated software using active contours to segment insulin granules more effectively than other published methods. A GUI has been designed to allow non experts to use the programs to analyse images. The algorithms have been extended to analyse 3D tomographic z-stack data. Ultimately a measure of the volume of insulin per cell or differences between healthy and diabetic animals will provide most useful physiological information. Furthermore the results are allowing a meaningful comparison of published EM sample preparation protocols.

References

[1] Nam D, Mantell J, Bull D, Verkade P and Achim A 2012 *Conf ProcIEEE Eng Med Biol Soc 2012* 5339-5342
[2] Fava E, Dehghany J, Ouwendijk J, Müller A, Niederlein A, Verkade P, Meyer-Hermann, M and Solimena, M (2012) *Diabetologia* **55** 1013-1023
[3] Verkade P 2008 *J.Microscopy* **230** 317
[4] Nam DC, Mantell J, Bull DR, Verkade P and Achim A 2012 *ConfProcEUSIPCO 20* 2228
[5] Kremer J.R, Mastronarde D.N and McIntosh J.R 1996 *J. Struct. Biol.* **116** 71-76.
[6] Nam D, Mantell J, Bull D, Verkade P and Achim A *BMVC 2013* (Accepted paper) Insulin Granule Segmentation in 3-D TEM Beta Cell Tomograms

Towards sub-10 meV energy resolution STEM-EELS

Ondrej L Krivanek[1], Tracy C Lovejoy[1], Matthew F Murfitt[1], Gwyn Skone[1], Philip E. Batson[2] and Niklas Dellby[1]

[1]Nion Co., 1102 8th St., Kirkland, WA 98033, USA
[2]IAMDN, Rutgers University, Piscataway, NJ 08854, USA

E-mail: krivanek@nion.com

Abstract. A monochromator we have introduced is improving the attainable energy resolution of electron energy loss spectroscopy (EELS) in a scanning transmission electron microscope (STEM) by more than 2x relative to what has been available until recently. Here we briefly review the design and the performance attained so far. We then investigate the ultimate resolution limits of our system and show that it should be able to reach an energy resolution of <10 meV.

1. Introduction

The high energy resolution monochromated EELS-STEM (HERMES™) system we have recently developed has introduced several new design concepts [1-3]. It has been able to reach 12 meV energy resolution (full-width at half maximum (FWHM) of the zero loss peak (ZLP)) in a short-exposure spectrum [2] and <20 meV in longer-exposure ones (e.g. Fig. 1). It can analyze samples at atomic or near-atomic resolution. When monochromating to about 100 meV, it has slightly *improved* the spatial resolution of the STEM it has been built into [2].

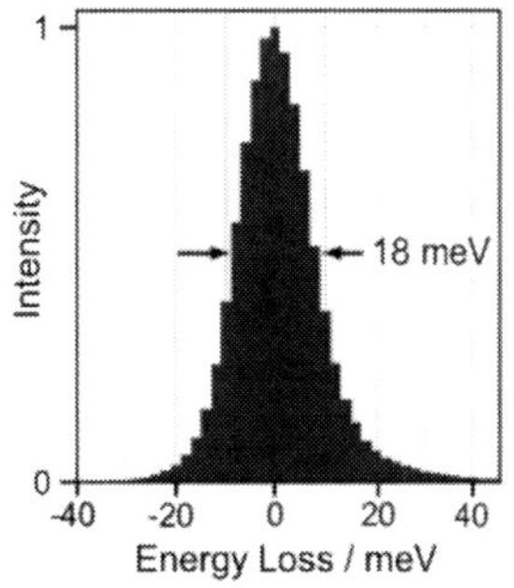

Fig. 1. Zero loss EELS peak obtained with Nion HERMES™, using a Gatan Enfinium EELS [4] equipped with ultra-high stability multipole power supplies. 100 keV, 1 s total acquisition.

In this paper, we review the reasons for the excellent energy resolution reached already, and we discuss the ultimate limits of the system.

2. Brief description of HERMES™

The system's monochromator uses an α-type electron trajectory, in which the beam traverses three magnetic prisms, the first one being traversed twice. The monochromator is located in the main column of the microscope, outside the electron gun. It energy-disperses the electron beam incident on a slit located in the mid-plane of the monochromator, and then undisperses it such that the beam re-inserted into the column has no energy dispersion in position or in angle.

Quadrupoles in the pre-slit half of the monochromator magnify the dispersion of the energy spectrum produced by the first prism, by an adjustable amount up to several hundred μm/eV (at 100 keV), and project the spectrum onto the energy-selecting slit. A post-slit set of quadrupoles, arranged symmetrically to the pre-slit set, works together with the system's prisms to undo the dispersion as the beam is steered back into the column. The electron trajectories fulfil precise symmetry requirements at the energy-selecting slit [1], without which complete un-dispersion would not be possible. Sextupoles and octupoles are provided so that important second and third order aberrations can be cancelled both at the slit and in the beam re-entering the microscope column.

3. Improved energy stability

The principal reason for the improved energy resolution of our system is that unlike most other designs, our monochromator produces an electron beam whose energy does not change when the high tension (HT) of the microscope changes. HT is difficult to stabilize to better than about 0.3 parts per million, i.e. about 30 mV r.m.s. for a 100 kV power supply. If the monochromator is located in the gun, upstream of the electron accelerator, and the spectrometer sits at ground potential, as is done in most monochromator designs, changes in the HT shift the EEL spectrum. In our design, both the monochromator and the spectrometer sit at ground potential. A change in the HT merely shifts the dispersed beam on the monochromator's energy-selecting slit, and the energy of the beam admitted into the microscope column and the position of the spectrum on the EELS CCD do not change.

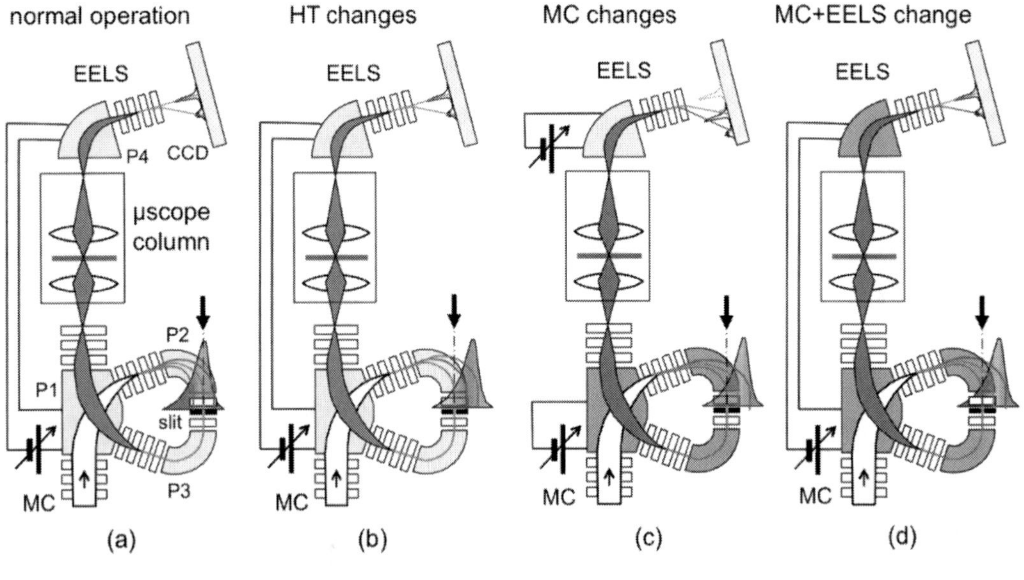

Fig. 2. Schematic diagram illustrating how various instabilities affect the HERMES™ energy resolution. MC = monochromator, P1 to P4 = prisms of the system. The microscope column is represented only by the condenser-objective lens, with a thin sample in its middle. Downward-pointing arrows mark the energy admitted by the energy-selecting slit.

Fig. 2(a) illustrates the normal operation of the monochromator: the energy-dispersed beam is centred on the slit, and only the central range of electron energies (green online) is admitted into the microscope column. The three magnetic prisms of the monochromator (P1 to P3) are connected in series. The prism of the electron energy loss spectrometer (P4) is also connected in series with the other prisms, and the "green" electrons arrive at the normal ZLP position on the EELS CCD. Electrons that have lost energy in the sample arrive on the CCD at a different position, forming a dispersed spectrum.

If the HT value changes, say by +200 mV, while the prism current remains the same in our system (Fig. 2(b)), the energy-dispersed beam at the slit shifts in position by an amount corresponding to an "energy gain" of 200 meV, as shown in the figure. The slit keeps on selecting the same energy, and the rest of the microscope column operates on an electron beam of exactly the same energy as before. Because the energy distribution of the cold field emission gun used by our system is rather narrow, changes in HT of as little as 20 mV can result in an intensity change of the beam admitted through the slit. We minimize such changes by sensing the current incident on the two slit halves, and using the difference signal in a feedback scheme controlling the fast loop of the HT generator so that the beam remains centred on the slit.

Fig 2(c) and (d) illustrate what happens when the monochromator prism current changes. Low-power current supplies are straightforward to stabilize to about 0.05 ppm r.m.s. (or better), but even such a small change alters the energy of the selected electrons by about 10 meV at 100 keV. A change in the energy of the selected beam due to a change of the prism current is indicated in the figure by the beam becoming red (online). If the monochromator prisms and the EELS prism were powered by separate power supplies (Fig. 2 (c)), the EELS spectrum would shift on the detector and the energy resolution in a time-integrated spectrum would become worse. A change in a separate current in the EELS prism would have a similar effect. However, when the EELS prism is supplied by the current that runs through the monochromator prisms (Fig. 2(d)), then the magnetic field in the EELS prism also changes, and the position of the spectrum on the EELS detector remains unchanged.

In reality, of course, the four prisms of the system cannot be tied together rigidly if the ability to change the prism excitations independently by small amounts is to be retained. Prisms 2-4 are therefore provided with weak auxiliary windings, whose power supplies run outside of the stabilization scheme. The magnetic fields produced by these windings are only about 3% of the main fields, and changes in their contributions to the total fields of the prisms thus produce beam deflections that are some 30x smaller than deflections due to fully independent prisms.

In our design, only instabilities that enter the microscope column between the slit of the monochromator and the EELS detector can directly affect the EELS energy resolution. Instabilities that enter the column between the MC slit and the sample (i.e., instabilities in the exit half of the MC, condenser lenses and their deflectors, corrector, quadrupole lens module, scan coils and the objective lens) also shift the probe on the sample and thus cause a loss of spatial resolution. Because of this sensitivity, the pre-sample part of the column is protected by several stability-enhancing measures, such as a mechanically rigid construction, triple and quadruple magnetic shielding, and highly stabilized power supplies. The post-sample part of the column, on the other hand, offers no similarly independent stability diagnostic, and EELS spectrometers up to now have typically only dealt with incident beams 50 meV and more in energy width. As a result, the rigorous stability measures of the pre-sample column are not always followed: the magnetic shielding tends to be single-layer, the detector column and the EELS are typically less rigid mechanically than the main column, and the power supplies are not always designed for the few-parts-per-10^8 stability that pre-sample power supplies aim for.

Fig. 3 illustrates the improvement possible when the performance of an important part of the post-column optics is improved in a stabilized monochromator system. It shows the position of the zero loss peak on the CCD detector of the Gatan Enfinium EELS, measured at 100 keV for a series of spectra recorded at about 100 spectra a second, and plotted as a function of time. Trace (a) shows the ZLP positions when the Enfinium multipoles are powered by Gatan's regular current supplies, trace (b) shows the positions when Gatan's newly developed "ultra-high stability" supplies are used.

The peak-to-peak variation is about 50 meV in trace (a) and about 12 meV in trace (b). The four-fold improvement in the stability meant that spectra acquired in 0.1 s with the improved power supplies gave ZLPs 17 meV wide (and 18 meV wide for 1 s acquisition), whereas spectra acquired in 0.1 s with the regular power supplies gave ZLPs 34 meV wide. The larger broadening would clearly not be of concern if the HT of the microscope caused say a 50 meV instability, or if no monochromator was being used. In the present case, however, the regular EELS power supplies were a major limit on the performance of the total system, and even the improved power supplies may not be sufficiently stable if the energy resolution is to be improved to <10 meV.

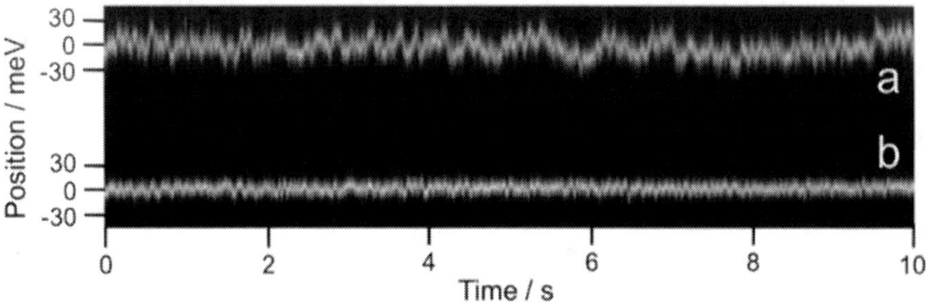

Fig. 3. Variation in the position of the zero loss peak on the EELS CCD as a function of time. a) regular Gatan Enfinium multipole power supplies, b) ultra-stable power supplies.

4. Fundamental limits on HERMES™ performance

The energy resolution of the monochromator and of the electron spectrometer is ultimately determined by the ratio of their energy dispersion to the size of the monochromatic beam crossover (measured at the monochromator slit and at the EELS detector, respectively).

As shown schematically in Fig. 2, the electron beam in our monochromated STEM-EELS system goes though a number of crossovers. The crossovers are simply successive images of the cold field emission source: highly magnified and energy-dispersed at the monochromator's slit and the EELS detector, highly demagnified and undispersed at the sample. Three of the crossovers play a fundamental role in determining the performance of the system. The crossover at the sample determines the spatial resolution, and the crossovers at the monochromator slit and at the EELS detector together determine the energy resolution. For each one of these crossovers, the range of the incident angles is set by an aperture whose size is chosen such that the crossover is not broadened by aberrations. The crossover size is then determined by diffraction due to the chosen aperture and by the chosen beam current, which is proportional to the projected area (magnification^2) of the source.

Many other factors can increase the crossover size: magnetic, electric and mechanical instabilities, charging of apertures or of the energy-selecting slit, stochastic Coulomb repulsion, and magnetic Johnson-Nyquist noise arising due to thermal motion of electrons in the drift tube, which has been pointed out recently (5,6). Provided that these effects are suitably minimized, the size of each crossover is given by:

$$d = 0.61\ (\lambda/\alpha)\ (1 + I_b/I_c)^{0.5} \quad (1)$$

where λ is the electron wavelength, α the semi-angle of the electron beam converging on the crossover, I_b the beam current at the crossover and I_c the coherent current of the electron source.

The electron-optical limit on the attainable energy resolution of the monochromator and the spectrometer is given by

$$\delta > d/\zeta \quad (2)$$

where δ is the energy resolution and ζ the energy dispersion of the spectrum in each device. The energy resolution is of course also broadened by the finite width of the slit in the monochromator and the finite spatial resolution of the EELS detector. However, the influence of these factors can be lessened by increasing the dispersion at the slit and at the detector, and they do not need to be discussed when evaluating the fundamental limits.

δ is determined most simply by examining the spectrum produced by the principal energy-dispersing element: prism 1 of the monochromator and prism 4 of the spectrometer. This is because subsequent magnification of the first spectrum by the system's quadrupoles magnifies the crossover and the energy dispersion by the same amount, and therefore does not change the energy resolution. The energy dispersions of the MC (Nion) and the EELS (Gatan Enfinium) prisms are 0.9 μm/eV for the main monochromator prism, and 0.8 μm/eV for the EELS one (at 100 kV). The preferred radius of the beam entering the MC in our present set-up is ~100 μm, which gives, for a beam current equal to 3 I_c, a crossover 7 nm in size formed in the (virtual) spectrum just after prism 1, and an energy resolution limit of 8 meV. For the EELS, our preferred entrance beam radius is 500 μm, and the beam current, which is limited to less than the monochromator value by a virtual objective aperture (VOA) positioned after the MC, is roughly equal to I_c (when the MC slit is open wide). This gives a theoretical EELS prism spectrum crossover size of 1.6 nm and an energy resolution limit of 2 meV. In other words, both the systems are capable of giving an energy resolution of 10 meV and better. We have not yet progressed to this level, most likely due to the individual limits all adding up (in quadrature) and due to remaining instabilities.

The EEL spectrometer comes off better in the above comparison, for two simple reasons: we admit a wider diameter beam into it than into the MC, and we restrict the beam current entering it more than the current entering the MC. The large entrance beam places strict requirements on the quality of the aberration correction performed in the spectrometer, but these can be readily met by an aberration-corrected spectrometer, as shown for instance in Fig. 7 of [2]. Fortunately, because the electrons stop at the EELS detector, a poor EELS set-up affects only the EELS energy resolution and no other aspects of the microscope performance. The EELS set-up is also simplified by the fact that only aberrations in the dispersion direction affect the quality of the EEL spectra.

On the minus side for the EELS, the phase space occupied by the electron beam can be greatly increased by scattering at the sample, particularly when the fast electrons are scattered to large angles (as they are when exciting phonons) *and* the sample is illuminated by a broadly defocused probe (to minimize radiation damage). In this case, the whole area illuminated on the sample becomes "a new source" that is imaged onto the EELS detector, and excellent energy resolution is likely to become impossible.

For the monochromator, aberrations in both the dispersion direction and perpendicular to it need to be controlled, because the crossover imaged at the MC slit is re-imaged onto the sample. Monochromator aberrations being able to spoil the probe-forming performance of the microscope is why we do not allow the beam entering the monochromator to be as wide as the beam entering the EELS. If we are to improve the energy resolution of our whole system to ~5 meV, we will need to work with an electron beam that is more than 300 μm wide (in the dispersion direction) in the monochromator's main prism (at 100 keV). Such a beam will be rather sensitive to stray magnetic fields and instabilities due to charging, but it should nevertheless be manageable with greater experience in monochromator tuning.

The path to energy resolution better than 10 meV in our system is thus rather clear: we have to clean up all remaining instabilities, and we have to improve the quality of the tuning of the monochromator and the spectrometer so that even the wide beams needed to minimize the diffraction limit do not give rise to appreciable aberrations. It is also worth noting that because the dispersion of magnetic prisms is proportional (for non-relativistic electrons) to $1/E_0$ (where E_0 is the primary energy) whereas the diffraction-limited size of a crossover only grows as

$1/\sqrt{E_0}$, the energy resolution of our system is expected to be proportional to $\sqrt{E_0}$. In other words, the energy resolution at 30 keV should be nearly 2x better than at 100 keV.

Another limit we need to be concerned about comes from the brightness of the electron source, as discussed in [2]. With an energy width of the electron source of around 300 meV (as is typical for W CFEG) and an energy interval of say 10 meV selected by the slit, 97% of the beam electrons are intercepted by the MC slit even when the slit and the focusing of the spectrum on it are absolutely perfect. A beam current of 300 pA entering the monochromator then gives a probe current of <10 pA on the sample. To maintain a bigger current the probe then has to become larger, i.e., we must give up spatial resolution in order to have a more substantial probe current at a very good energy resolution. A brighter / more monoenergetic electron source than tungsten CFEG would decrease this difficulty.

One more limit to which we are paying close attention is the extent of the "skirt" of the zero loss peak (also called "tail") that can arise due to effects such as mistuned aberrations, aperture charging, and the sideways spreading of the signal in the detector. Our goal is to reduce the skirt to $<10^{-5}$ of the ZLP's maximum intensity just 100 meV away from the maximum, and we are pursuing several measures to achieve this.

5. Conclusion

When the monochromator and the electron spectrometer overcome instabilities well enough to become diffraction-limited, *better energy resolution* will require that the beam entering either device *is made larger*. We are now approaching this limit in the Nion monochromator, and we are likely to reach it soon in the energy loss spectrometer too. We will then have entered the realm of "*diffraction-limited EELS*", in which the whole STEM-EELS system will be limited by the quality of its electron optics rather than by the energy spread of the electron source or the instabilities of the system. This type of performance limit is rather familiar to us: it is similar to the STEM spatial resolution limit due to spherical aberration, which we were able to improve substantially by developing a STEM aberration corrector [7,8].

A STEM-EELS system operating at this advanced level is likely to be able to attain <10 meV energy resolution with an atom-sized (<2 Å) electron probe, and thereby to open up a new field for experimental study: phonon spectroscopy with atomic spatial resolution. It is not every day that a new type of physical interaction becomes available in the electron microscope. It promises to make our efforts to improve the energy resolution further very worthwhile, every step of the way.

Acknowledgement

We are grateful to Profs R.W. Carpenter and P. Rez for stimulating discussions, and to Dr. C. Trevor of Gatan Inc. for loaning us ultra-high stability power supplies and for providing a DigitalMicrograph™ script that analyzes the ZLP positions in a sequence of EEL spectra. Financial support was provided by NSF grant DMR MRI-R2 #959905, DOE grant DE-SC0007694, and by Rutgers University and Nion Co.

References

[1] Krivanek OL et al 2009 Phil. Trans. R. Soc. **A367** 3683
[2] Krivanek OL et al 2013 Microscopy **62** 3
[3] Krivanek OL and Dellby N 2013 US patent #8373137 B2
[4] Gubbens AJ et al 2010 Ultramicroscopy **110** 962
[5] Uhlemann S et al 2013 Phys. Rev. Lett. **111** 046101
[6] Howie A 2013 these proceedings
[7] Krivanek OL Dellby N and Lupini AR 1999 Ultramicroscopy **78** 1
[8] Batson PE Dellby N and Krivanek OL 2002 Nature **418** 617

Journal of Physics: Conference Series **522** (2014) 012024 doi:10.1088/1742-6596/522/1/012024

Electron microscopy reveals unique microfossil preservation in 1 billion-year-old lakes

M Saunders[1], C Kong[2], S Menon[3] and D Wacey[4,5]

[1] Centre for Microscopy Characterisation and Analysis, The University of Western Australia, Crawley, WA 6009, Australia
[2] Electron Microscopy Unit, The University of New South Wales, Kingsford, NSW 2052, Australia
[3] Department of Mechanical and Aerospace Engineering, Naval Postgraduate School, Monterey, CA 93943, USA
[4] Department of Earth Science & Centre for Geobiology, University of Bergen, Bergen, N-5007, Norway
[5] Australian Research Council Centre of Excellence for Core to Crust Fluid Systems, Centre for Microscopy Characterisation and Analysis & Centre for Exploration Targeting, The University of Western Australia, Crawley, WA 6009, Australia

Email: Martin.Saunders@uwa.edu.au

Abstract. Electron microscopy was applied to the study of 1 billion-year-old microfossils from northwest Scotland in order to investigate their 3D morphology and mode of fossilization. 3D-FIB-SEM revealed high quality preservation of organic cell walls with only minor amounts of post-mortem decomposition, followed by variable degrees of morphological alteration (folding and compression of cell walls) during sediment compaction. EFTEM mapping plus SAED revealed a diverse fossilizing mineral assemblage including K-rich clay, Fe-Mg-rich clay and calcium phosphate, with each mineral occupying specific microenvironments in proximity to carbonaceous microfossil cell walls.

1. Introduction

One billion-year-old rocks from the Torridon Group of northwest Scotland contain abundant microfossils that are a rare example of a Precambrian (>540 million years ago) non-marine biological community [1,2]. Traditional palaeontological methods, such as light microscopy lack the spatial resolution to reveal the true morphology of these microfossils in three dimensions, and accurately decode the chemical environment of fossilization. The combination of high-spatial-resolution, *in situ* techniques applied here, namely focused ion beam (FIB) sample preparation combined with scanning electron microscopy and transmission electron microscopy (FIB-SEM and FIB-TEM) allows the chemical composition of the microfossils and associated fossilizing minerals, plus cellular morphology, to be studied at the micrometer to nanometer scale.

2. Experimental methods

Study samples were standard uncovered polished geological thin sections, about 30 μm in thickness. A *FEI Helios NanoLab* dual beam FIB system was used to prepare TEM wafers from the polished thin sections. Electron beam imaging was used to identify microfossils of interest in the thin sections allowing site-specific TEM samples to be prepared. The TEM sections were prepared using a series of steps involving different beam energies and currents [3], resulting in ultrathin wafers of ~100 nm thickness. The wafers were extracted using an ex-situ micromanipulator and deposited on continuous-carbon copper TEM grids.

TEM data were obtained using a JEOL 2100 LaB_6 TEM operating at 200 kV equipped with a Gatan Orius CCD camera and Tridiem energy filter. Energy-filtered (EFTEM) elemental maps were obtained using the conventional three-window technique [4], with energy windows selected to provide optimum signal-to-noise. Selected area electron diffraction (SAED) was performed using an aperture that selected a 200 nm diameter area of the sample.

Sequential FIB milling and SEM imaging was performed on *Zeiss Auriga Crossbeam* and *Zeiss Neon 40* dual-beam instruments, following the protocol outlined in [3]. Milling and imaging variables were optimized to suit the sample material (i.e. organic carbon in phosphate and clay), with ion beam currents of 2-3 nA for sequential slice milling, electron beam voltage of ~5 kV for imaging and an image capture time of ~30 seconds per frame.

FIB-SEM images were stacked, aligned and cropped using *SPIERSalign* [5]. The resultant stacks were imported into *SPIERSedit* [5] where a number of masks were added to segment individual components (e.g. mineral grains, cell walls) of the microfossil assemblage. The resulting files were exported and loaded into *SPIERSview* [5] to generate 3D volume renderings.

3. Results

3.1. Microfossil morphology

Over 500 serial slices through a cluster of small cells were milled and imaged using FIB-SEM to decode the true three-dimensional morphology of these microfossils (figure 1). The step size between images in z was 75 nm, the width of the milled area (x) was ~68 μm and the depth (y) was ~25 μm.

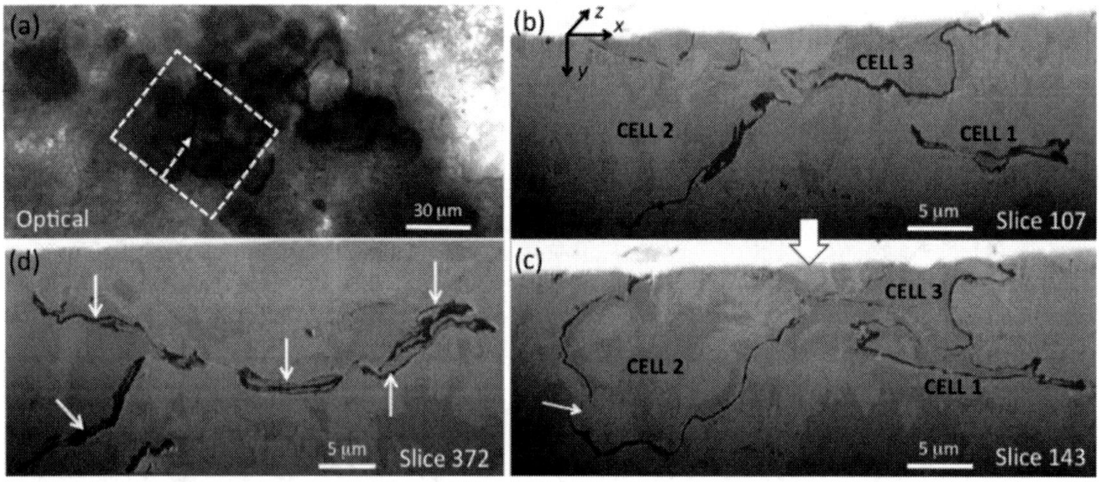

Figure 1. Morphology of microfossils from the 1 billion-year-old Torridon Group investigated in three dimensions. a) Optical microscopy image of a cluster of microfossils of unknown morphology. Dashed box shows region of interest (ROI) milled for 3D-FIB-SEM, where slicing proceeded in the direction of the arrow. b-c) FIB-SEM slices showing how the morphology of three cells changes over a distance of ~3 μm in the z direction. The increment between two adjacent slices is 75 nm, hence slice 107 is ~8 μm into the ROI, and slice 143 is ~11 μm into the ROI. Compression and folding of the microfossils is common. d) FIB-SEM slice 372 (~28 μm into the ROI) showing numerous examples (arrows) of more heavily folded and compressed cells.

Individual FIB-SEM slices (figure 1b-d) show the detailed morphology of several of these cells below the surface of the thin section. The cell walls (dark grey to black) have remained largely intact with only occasional minor degradation (e.g., arrow on slice 143), presumably allowing both the cell contents to escape and the fossilizing fluid to then permeate the interior of the cells. Mineral grains are rarely seen *within* the microfossil walls indicating a largely non-mineralized, purely carbonaceous cell wall composition.

All cell walls are folded, indicating that there was some degradation, dehydration and contraction of the walls prior to fossilization. Many microfossils are heavily compressed. However, some of these compressed microfossils occur in close proximity to those that have experienced little compaction (compare for example 'cell 1' with 'cell 2' in slice 143). This suggests that either the fossilizing fluids penetrated the interior of some of the microfossils earlier and to a greater extent than others affording them greater protection from later compaction, or that the fossilizing fluids crystallized at different rates within different fossil microenvironments.

The morphology of the cells can change rapidly over just a few micrometers. For example, in figure 1c slice 143 three cells are clearly visible and few would argue with an interpretation as microfossils. In contrast, figure 1d slice 372 shows at least five microfossils (arrows) but they are so compressed that they resemble simple carbonaceous laminations. It is only the high-spatial-resolution nature of these images that allow distinct microfossils to be visualized. In lower resolution microscopy, including standard light microscopy, the very narrow interiors (only ~100-300 nm wide) of these cells would not be visible. Instead, the microfossils would likely be mistaken for interwoven carbonaceous laminae and subsequent demonstration of their biological nature may be problematical.

3D volume renderings of the raw data from multiple FIB-SEM slices (figure 2) emphasize the highly compressed and folded nature of some of these small cells, which would have been spheroidal to ellipsoidal before death. They also show the extent to which the wall structure has been broken down during fossilization; in this case, 'cell 2' is significantly more degraded (higher volume of gaps within the red wall structure) than 'cell 1' despite the two cells being only a few micrometers apart.

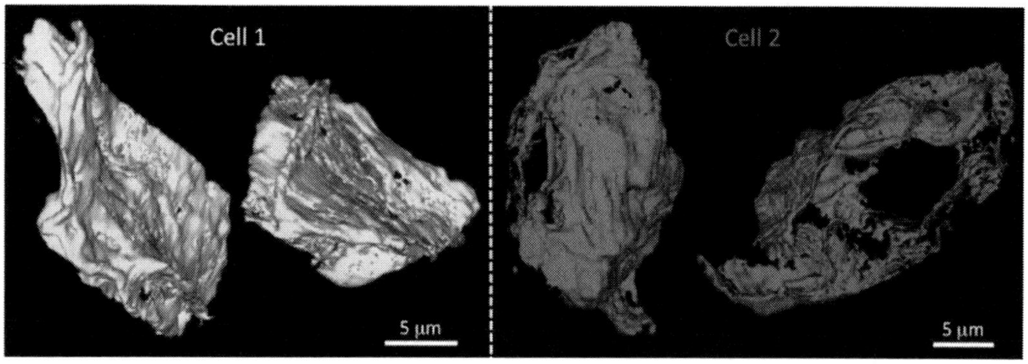

Figure 2. 3D reconstructions of the organic walls of cells 1 and 2 from figure 1, each viewed from two different perspectives. Cell 1 is very compressed but has a high percentage of its cell wall still intact. Cell 2 retains much of its original ellipsoidal shape but part of the cell wall is highly degraded.

3.2. Microfossil chemistry

FIB-SEM images suggest that the cells have been fossilized by more than one mineral phase (i.e., variable shades of grey within the cells in figure 1b-d likely correspond to different minerals). This was further investigated by analysis of FIB-milled ultrathin TEM wafers through several microfossils. EFTEM mapping (figure 3) confirms the carbonaceous nature of the cell walls and confirms that a variety of mineral phases have precipitated around the microfossil walls, probably subsequent to bursting of the cell and escape of the cell contents but prior to any further decay of the cell walls. In the example shown in figure 3, the cells have been fossilized by a mixture of alumino-silicates and silica. Calcium phosphate is also present in the vicinity of most cells. SAED and HRTEM show that the alumino-silicates are layered clay minerals, close in structure and composition to illite (K-rich varieties) and berthierine/chamosite (Fe-Mg-rich varieties). Different minerals are restricted to specific microenvironments with mineral composition often changing sharply across cell boundaries. Current and future work aims to relate these mineral phases to original cellular chemistry and/or sediment porewater chemistry.

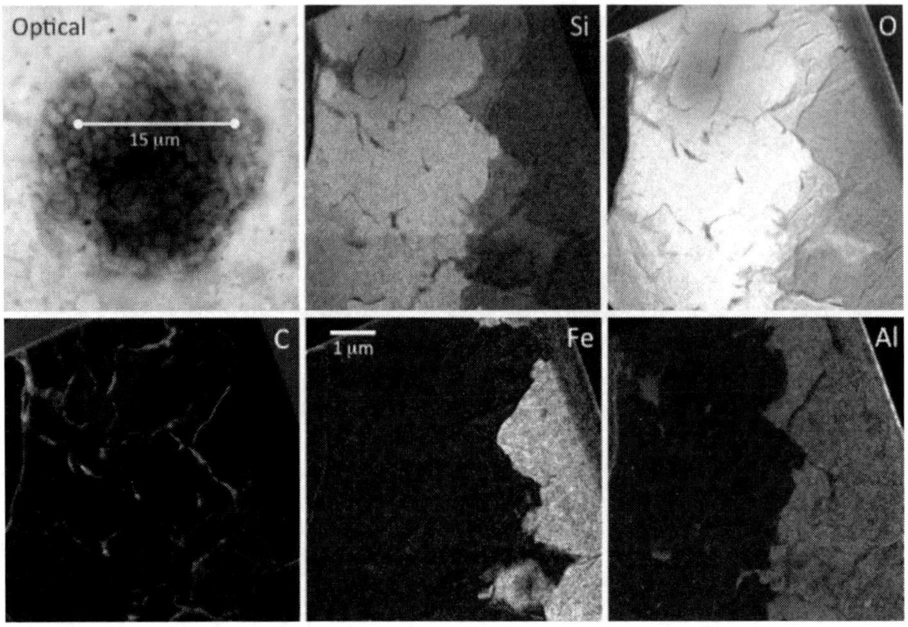

Figure 3. EFTEM elemental maps of a complex multicellular microfossil from the Torridon Group. The FIB wafer was extracted from the area indicated in the optical image. Cell walls appear white/light grey in the carbon map. Three distinct zones of minerals are seen and these change sharply across cell wall boundaries. The right hand zone is an iron-rich alumino-silicate approximating chamosite in composition, the narrow central zone is a potassium-rich alumino-silicate approximating illite in composition, and the left hand zone is silica.

4. Conclusion

The combination of high-resolution, *in situ* techniques used here allows the morphology and chemistry of Precambrian microfossils to be studied without the necessity to chemically extract the fossils from their rock matrix. Changes in cell wall structure, degree of wall degradation and deformation, together with mineralogy can be detected over a range of spatial scales from nanometers to millimeters within a single thin section. These data enhance our understanding of the composition of the preserved biological community and the environmental conditions that led to its fossilization.

References

[1] Strother P K, Battison L, Brasier M D and Wellman C H 2011 *Nature* **473** 505–509
[2] Battison L and Brasier M D 2012 *Precambrian Research* **196–197** 204–217
[3] Wacey D, Menon S, Green L, Gerstmann D, Kong C, McLoughlin N, Saunders M and Brasier M 2012 *Precambrian Research* **220–221** 234–250
[4] Brydson R 2001 *Electron Energy Loss Spectroscopy* (New York: Springer)
[5] Sutton M D, Garwood R J, Siveter D J and Siveter D J 2012 *Palaeontologia Electronica* **15(2)** 5T

Understanding the limitations of the Super-X energy dispersive x-ray spectrometer as a function of specimen tilt angle for tomographic data acquisition in the S/TEM

T J A Slater[1], P H C Camargo[2], M G Burke[1], N J Zaluzec[3] and S J Haigh[1*]

[1] School of Materials, University of Manchester, Manchester, M13 9PL, UK
[2] Departamento de Química Fundamental, Instituto de Química, Universidade de São Paulo, São Paulo, Brazil
[3] Argonne National Laboratory, Argonne, IL 60439 USA

*Sarah.haigh@manchester.ac.uk

Abstract. We have investigated the use of x-ray energy dispersive spectroscopy during tomographic hyperspectral imaging experiments in the scanning transmission electron microscope. In this work, we have found that for an analytical system employing a commercial high-tilt tomography holder the measured x-ray signal is limited by shadowing caused by the penumbra of the holder relative to the x-ray detector system. This limits the ability to perform quantitative, elemental tomographic analysis.

1. Introduction

Electron tomography is one of the more powerful techniques for retrieving nanoscale three-dimensional information, deduced from a series of two-dimensional projections. Tomographic reconstruction requires that the input data satisfies the projection requirement, which is that the signal intensity is a monotonic function of the quantity to be reconstructed. High angle annular dark field (HAADF) scanning transmission electron microscope (STEM) imaging satisfies the projection requirement and has thus far been one of the preferred imaging modes for crystalline materials specimens [1]. In cases where regions of the sample are sufficiently different in atomic number (Z), HAADF STEM tomography is readily able to distinguish between features via Z-contrast mechanisms. Elemental imaging is a useful augmentation to tomographic reconstruction and involves the use of complementary spectroscopic characterization using either electron energy loss (EEL) [2] or energy dispersive x-ray (EDX) [3] spectroscopic imaging. In the latter case, it is well known that detector/specimen holder geometry can limit successful data acquisition and analysis. This is due to the fact that, in some instrument configurations, x-ray signals may only be confidently acquired for specific ranges of specimen holder orientations. Newer generation analytical microscopes, such as the FEI Titan G2 80-200 S/TEM with ChemiSTEM™ technology have been designed to minimize such limitations by incorporating multiple silicon drift detectors (SDDs), positioned symmetrically about the tomographic axis [4]. While successful qualitative tomographic reconstructions using EDX data sets have recently been reported [5,6], some such studies have ignored the effects of x-ray detector shadowing introduced by some designs of tomographic stages. This shadowing, when it is present, compromises the ability of the EDX signal to provide quantitative three-dimensional information.

In this paper, we report on the magnitude of the shadowing observed when using a Fischione high-tilt tomography specimen holder for two different x-ray detector geometries. The spectrometers positions are compared in Figure 1d (SuperX EDX spectrometer on an FEI Titan G2 S/TEM microscope) and Figure 1e (Oxford Instruments X-Max 80 EDX spectrometer on an FEI Tecnai F30 S/TEM).

2. Experimental Method

The specimen chosen for this study was a single Au/Ag nanoparticle (NP) with a diameter of ~30 nm and a hollow geometry (Figure 1a) which was supported on a holey carbon +200 mesh Cu support film. The sample was synthesized as reported in reference [6] and has recently shown high efficiency as a catalyst in the conversion of 4-nitrophenol to 4-aminophenol. The nanoparticle selected for measurement was carefully chosen to be near the center of any support grid, so as to minimize the effects of grid bars. The same specimen and stage was used for all measurements reported in this work, but a different nanoparticle had to be used for each tilt series as relocating the same nanoparticle was problematic.

We have compared EDX detectors measurements as a function of tilt angle from two microscopes with different EDX detector geometries. Microscope 1 is the FEI Titan G2 80-200 which is fitted with an X-FEG high brightness source, and is probe-side aberration corrected to 3rd order. The instrument was operated at an accelerating voltage of 200 kV and with a beam current of 0.3 nA. STEM HAADF data were acquired with a convergence angle of 26.2 mrad and a HAADF inner angle of 52 mrad. Microscope 1 has the ChemiSTEM™ Super-X EDX detector whose geometry consists of two pairs of silicon drift EDX detectors (SDD) either side of the tomography holder primary tilt axis, as illustrated in Figure 1d. These pairs of detectors (1+2) and (3+4) are rotated about the optic axis symmetrically by ±45° in the eucentric plane tilt axis. Data from these detectors were recorded as pairs of spectrum images so as to allow assessment of any shadowing created by the penumbra of the tomography stage as the holder is tilted. Microscope 2 is the FEI Tecnai F30 S/TEM equipped with an Oxford Instruments X-Max 80 SDD EDX system. It was operated at 300kV using a beam current of ~1 nA. Microscope 2 has the traditional EDX system geometry in which a single x-ray detector is located perpendicular to the tomographic tilt axis (Figure 1e).

The EDX spectrum image data sets (512 x 512 pixels) were acquired every 10° between ±70° using a Fischione 2020 single tilt tomography holder. Drift correction was applied to ensure that the nanoparticle remained central within the field of view. The specimen pixel size used was the same for both microscopes and the live acquisition time per spectral image was chosen such that the total beam current was also the same (300 s for Microscope 1 and 108 s for Microscope 2). Spectrum images were exported from either Bruker Esprit or Oxford Aztec software and processed in Gatan DigitalMicrograph™. Total x-ray counts from the whole spectrum image were summed without background subtraction using a 400eV energy window centred on the energy of interest.

3. Results and Discussion

Theoretically, because the nanoparticle is small and of constant volume, one would expect negligible variation in the detected x-ray intensity (counts) as a function of holder tilt. The crystalline structure of the nanoparticle means that strong electron channelling at specific crystallographic orientations could perturb results at a single tilt angle but will not affect the general trend of behaviour observed for the whole tilt range [8]. Figure 1c compares the experimentally observed Au Lα x-ray counts for Microscope 1 using the full Super-X detector (4 SDDs) with that obtained by reading out two pairs of SDDs separately. Reference spectra acquired on a nearby region of carbon film contained no gold signal so that these counts are considered to result entirely from the specimen.

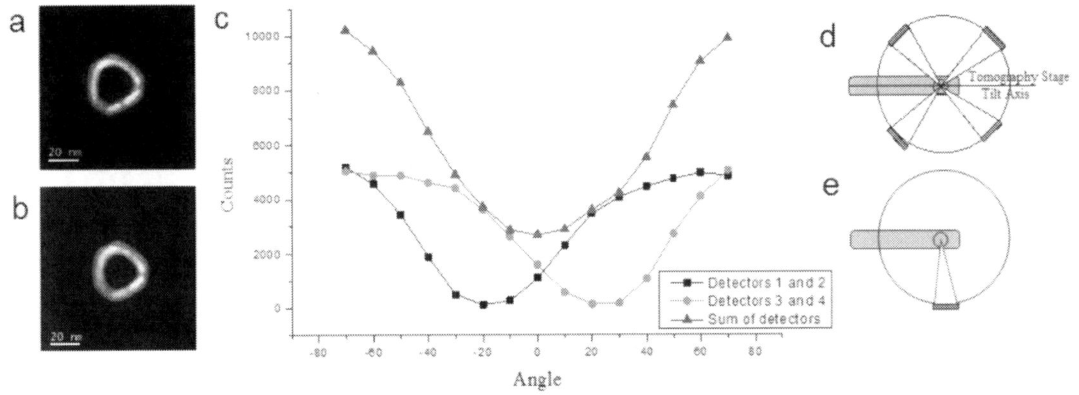

Figure 1. HAADF STEM images from Microscope 1 for a single NP **(a)** before and **(b)** after tilt series acquisition. **(c)** Au Lα x-ray counts from the NP shown in (a) as a function of holder tilt for Microscope 1 (Titan ChemiSTEM™/Super-X). SDDs to the right (1 and 2) and left (3 and 4) of the holder tilt axis are compared with the sum found for all four detectors the geometry of which is illustrated in **(d). (e)** Traditional EDX detector geometry of Microscope 2 (Tecnai/Xmax).

The variation of the x-ray counts when the signal from all of the Super-X detectors is summed (Figure 1c) demonstrates qualitatively that whilst signal counts are maintained above zero for the full angular range, there is a significant variation in the measured intensity with tilt angle. This intensity variation can be directly attributed to absorption of the NP x-rays by any intervening material along the line of sight path between the NP region of interest and the active area of the x-ray detector being used. The principal culprit in this regard is the tomographic stage side walls and support structures, with the maximum absorption (minimum signal) occurring when the tomography holder is oriented perpendicular to a given detector pair. This is most succinctly illustrated by studying the holder tilt-dependent signals for the two pairs of detectors, 1+2 and 3+4. These mirror each other, reflecting the symmetry of the tomographic stage and x-ray detector geometry on either side of the holder tilt axis. The minima in these latter two curves (at approximately ± 20°) is indicative of the elevation angle of the Super-X detectors.

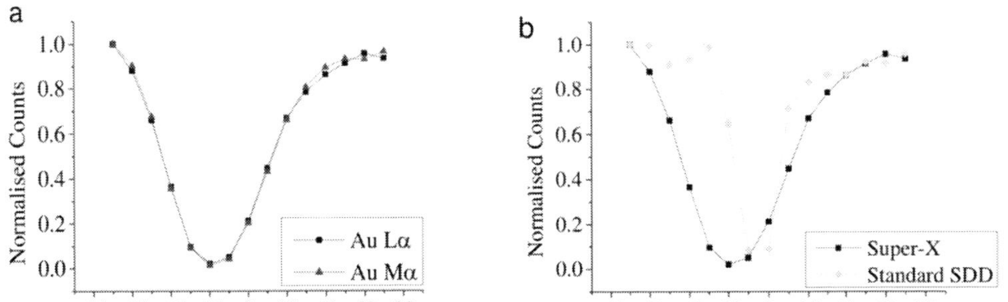

Figure 2. (a) Signal variation with tilt angle for Microscope 1 (Titan/Super-X) using a pair of detectors (1+2). Normalized x-ray counts are compared for Au Lα (9.7 keV) and Au Mα (2.1 keV) x-rays. **(b)** Comparison of the variation with tilt angle for the Au Lα x-ray signal for a pair of detectors (1+2) from Microscope 1 with the single SDD detector of Microscope 2.

To ascertain whether shadowing effects varied as a function of x-ray energy we have also compared the normalized signal counts for one pair (1+2) of the Super-X detectors signal counts as a function of tilt angle for different energy x-rays from the same element (Figure 2a). The tilt dependent variation of the Au Lα (9.7keV) signal and Au Mα (2.1keV) signal is identical in the absence of background subtraction suggesting that shadowing is independent of x-ray energy for the tomographic holder used. This is indicative of massive absorption effects as the energies of these two lines are significantly different.

Figure 2b compares the normalized Au Lα x-ray counts for one pair of detectors from Microscope 1, with similar measurements made using the single detector geometry of Microscope 2 (Tecnai/X-max). The minimum signal for the pair of detectors on Microscope 1 occurs at a larger tilt angle than that the single X-max detector of Microscope 2 from which one can directly conclude that the elevation angle of the Super-X SDD detectors is greater than that of the single X-max80 SDD detector. The detector pair of Microscope 1(Titan/Super-X) is also shadowed for a wider angular range than the conventional geometry of Microscope 2 (Tecnai/Xmax), due to its larger solid collection angle [9].

4. Conclusions

In this study we have shown that for a nanoparticle sample the measured signal varies significantly with tilt angle when using a Fischione 2020 high tilt tomography holder. This variation is sufficient that it is likely to compromise the reconstruction of quantitative tomographic x-ray spectroscopy data.

There are several approaches that could be used to improve the ability of the EDX signal to satisfy the projection requirement. The first is *a posteriori* compensation of shadowing for EDX tilt data obtained with a uniform acquisition time by applying a weighting factor. The risk of this method is that it could increase noise levels in the EDX maps and thereby lead to false data in the reconstruction. An alternative approach is to redesign the tomographic holder geometry so as to minimize shadowing of the detector during EDX measurements, which is work in progress. It is our expectation that different designs of tomographic holder will perform differently, as their construction varies. This work emphasizes the necessity of testing tomographic holders as to their suitability for tomographic hyperspectral imaging using x-ray spectroscopy.

References

[1] Midgley P A and Weyland M 2003 *Ultramicroscopy* **96** 413-31
Midgley P A and Dunin-Borkowski R E 2009 *Nature Materials* **8** 271-80
[2] Möbus G, Doole R C and Inkson B J 2003 *Ultramicroscopy* **96** 433-51
Jarausch K, Thomas P, Leonard D N, Twesten R and Booth C R 2009 *Ultramicroscopy* **109** 326-37
[3] Möbus G, Doole R C and Inkson B J 2003 *Ultramicroscopy* **96** 433-51
[4] von Harrach H S 2009 *Microsc Microanal* **15 S2** 208-9
[5] Genc A, Kovarik L, Gu M, Cheng H, Plachinda P, Pullan L, Freitag B and Wang C 2013 *Ultramicroscopy* **131** 24-32
[6] Lepinay K, Lorut F, Pantel R and Epicier T 2013 *Micron* **47** 43-9
[7] Petri M V, Ando R A and Camargo P H C 2012 *Chemical Physics Letters* **531** 188-92
[8] Zaluzec N J *Microsc. Microanal.* 18 (Suppl 2), 2012, 678-679
[9] Zaluzec N J *Microsc Microanal*, 15, 93-98, 2009

Journal of Physics: Conference Series **522** (2014) 012026 doi:10.1088/1742-6596/522/1/012026

Analytical transmission electron microscopy in the third dimension

Daniela Sudfeld[1], Oleg Lourie[1] and Stephan Kujawa[1]

[1]FEI Electron Optics B. V., Achtseweg Noord 5, 5600 KA Eindhoven, The Netherlands
E-mail: Daniela.Sudfeld@fei.com

Abstract

For researchers in metallurgy, energy storage or functional nanomaterials, where elemental distribution is crucial to performance, full characterization is more difficult by decreasing feature sizes and increasingly complex architectures. Two-dimensional imaging and analytical techniques often cannot characterize nanostructures completely. A new three-dimensional (3D) tomography technique for STEM XEDS, namely ChemiSTEM™ Technology [1], reveals the true microstructure of materials offering high X-ray collection efficiency via its windowless detector design, and fast XEDS mapping with tilt response at all angles. ChemiSTEM features the X-FEG and Super X, containing four symmetric detectors for 3D chemical imaging at 0 degree tilt, with significant EDX signal collected under all tilt conditions. The EDS signal of STEM XEDS image tilt series can be processed like normal z-contrast images, because of the monotonic relationship between the EDS signal and the thickness of the TEM foil. Four symmetrically arranged silicon drift detectors (SDD) provide extreme high output count rates (300 kcps) required for acquisition of EDX maps of thick samples, especially on 3 mm disk samples for 3D tomography studies. Examples of 3D chemical mapping using XEDS are given on (InGa)N Nanopyramid LEDs, Ni based super alloy material for turbine blades, dielectric transistor, and catalytic particles. In lithium ion batteries, the elemental segregation of manganese and nickel, which is responsible for the aging of electrode materials made from lithium-nickel-manganese oxide layered nanoparticles was analyzed in 3D XEDS mapping [2].

1. Introduction

As the feature sizes in material science continues to decrease to the nanometer regime, 3D chemical information is required to improve the function of many materials. STEM tomography is a well established technique in material science to provide 3D structural information, but is failing to provide the necessary chemical information [3-7]. A combination of STEM tomography with XEDS mapping solves the problem. Atomic level chemical measurements [8, 9] benefit from the high brightness of the X-FEG and the reduced probe size, of course supported by the Cs corrected optics in the STEM. And Super X is the only solution that can deliver 3D, nanoscale structural and compositional information on regular 3 mm disk samples.
It is well known that the performance of lithium ion batteries degrades over time. Understanding this behavior requires an understanding of the chemical composition changes that take place during the battery's use. One application example is presented which is showing the advantage of 3D chemical data analyses over the 2D data generation resulting in a better interpretation due to faster acquisition and higher quality data.
ChemiSTEM analysis is used for superalloy material studies enabling the design and development of a jet turbine fan blade by helping to characterize the physical and chemical properties in a S/TEM. To our knowledge, researchers developing new alloys are focused on reducing the weight, and improving oxidation and corrosion resistance while maintaining the strength of the alloy. Furthermore with the increasing demand for the use of superalloys in turbine blades for power generation, another focus of alloy design is to reduce their cost. CMSX-4 superalloy [13], a heat resistant material, was especially designed for gas turbine blades in jet aircraft engines and stationary turbines generating electricity and investigated for the elemental distribution of the different components.

2. Experimental data

Chemical analysis in TEM has been initiated by extracting a site specific thin lamella from a superalloy material using FIB lift out technique on the FEI Helios SDB [10]. Available on FEI's Tecnai Osiris™, Talos™ and Titan™ G2 S/TEM and Titan™ Themis™ S/TEM, ChemiSTEM features Super X, a symmetric four detector design that allows 3D chemical imaging at 0 degree tilt, with significant EDX signal collected under all tilt conditions. A signal under all tilt conditions is mandatory to acquire 3D EDS maps. An example is the sub-set of Ni elemental maps acquired from a Li ion battery cathode material in the tilt range of ∓70 degrees, see Figure 1. Fast mapping electronics are capable of 100,000 spectra/second in EDX spectrum imaging. On Titan Themis, atomic chemical EDS maps can be measured with 500 spc/ sec. High speed processing provides the extreme high output count rates (300 kcps) required for acquisition of EDX maps of thick samples, especially on regular 3 mm disk samples for 3D studies. The experimental parameters for the studies of the Li battery specimen are described elsewhere [2]. The EDS tomography experiments of the Ni-based superalloy were acquired on the microscope at 200 kV acceleration voltage with a probe current of ~1.05nA in STEM mode at a nominal magnification of 56000 x. The total pixels 512 x 512 pix^2 with a pixel size of 3.60 nm result in a total image size of 1.84 um with an exposure time of 15 seconds. The dwell time for the EDS mapping is 50 usec. The reconstruction was performed with the software Xplore3D, mode: simultenous iterative reconstruction technique (SIRT) and visualized with the software Avizo Fire 7.0.

3. Results

FEI's ChemiSTEM Technology provides 3D chemical mapping with directly interpretable 3D structural and chemical compositional information, all crucial for materials scientists to better understand the properties of complex 3D nanostructures. Nowadays, sensitive detectors as the XEDS Super X enable the required signal strength for faster time to get high quality data. The time which was needed in the past for measuring static, two dimensional information can be the same but now providing an additional dimension in 3D. The advantage of rapid data acquisition is to expose the specimen area of interest minimal and beam sensitive samples can withstand the damage under the electron beam. The quality of data is improved by the X-FEG due to its stable long term (important for tomography) and short term (vital for ultrafast imaging and spectroscopy) emission current within a ≤ 1 % range.

The EDS signal of STEM XEDS image tilt series can be processed like normal z-contrast images, because of the monotonic relationship between the increasing EDS signal as a function of the thickness of the TEM foil. Total EDS signal behaves like Z-contrast, if you compare the intensities

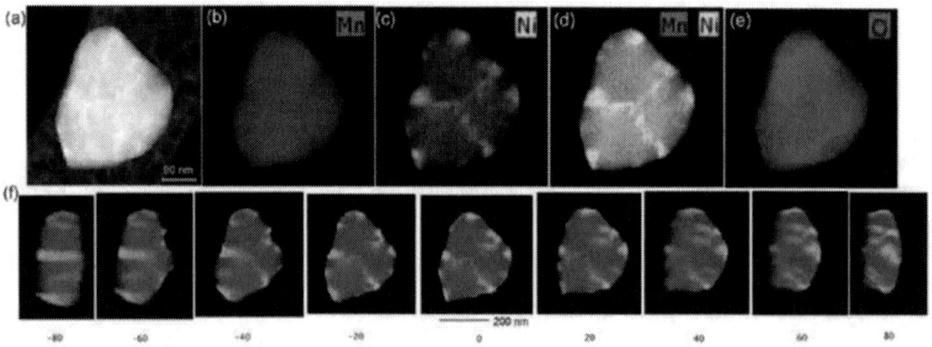

Figure 1 The STEM tomogram and corresponding elemental chemical maps, namely for Mn, Ni, Mn and Ni and O, were acquired from a Li ion battery cathode material [2]. The tilt series of +/- 70 degrees is presented.

and contrast of simultaneously acquired images corresponding to the HAADF signal and the total x-ray signal. The benefits of XEDS approach are that the majority of the elements is accessible via the XEDS signal (above Beryllium). On the one hand the signal increases monotonously with mass thickness while on the other hand the Peak-to-Background (P/B) ratio is independent and not varying with the sample thickness nor requiring a special sample geometry for investigation.

In the first application example, FEI's ChemiSTEM Technology was used to collect 2D and 3D chemical maps, see Figure 1, of a lithium ion battery [2, 11]. 3D EDX was applied to different parts in the battery to reveal the nanoscale chemical changes occurring during battery use. This technique has made it possible for the first time to observe the aging of electrode materials made from lithium-nickel-manganese oxide layered nanoparticles. What is now seen is that nickel segregation occurs, blocking the channels through which lithium normally travels during the charge-discharge cycle of the device.

The ability to characterize superalloy material and their design with 2D and 3D chemical analysis using FEI's ChemiSTEM Technology was critical to understand and successfully optimize micro-structures of gas turbines. Partition of alloying elements within the γ/γ' microstructure of the Ni-based superalloy is extremely important and defines the mechanical properties of the superalloy at high temperatures [12]. The 3D chemical information revealed necessary information to improve their performance, life time and efficiency.
The first single crystal CMSX-4 nickel-base superalloy (Figure 2) features a very complex chemical composition containing Ni – 8.4 Co – 6.4 Cr – 6.5 Ta – 6.4 W – 5.68 Al – 2.8 Re – 1.04 Ti – 0.58 (in wt %). The microstructure of the single crystal superalloys consists of a high volume fraction of cuboidal γ' precipitates surrounded by γ matrix channels with nanometric width and different chemical composition. The γ' precipitates are an ordered Ni_3Al-base intermetallic phase (Pm m) with the lattice parameter $a_{\gamma'}$ = 0.35718 nm which is only slightly smaller than that of the Ni base 3 solid solution (Fm m), $a_{\gamma'}$ = 0.3524 nm. EDX microanalysis of the γ and γ' phases allowed identification of the chemical variation with nanometer resolution and enabled detection at the highest intensities even from elements with low concentrations.

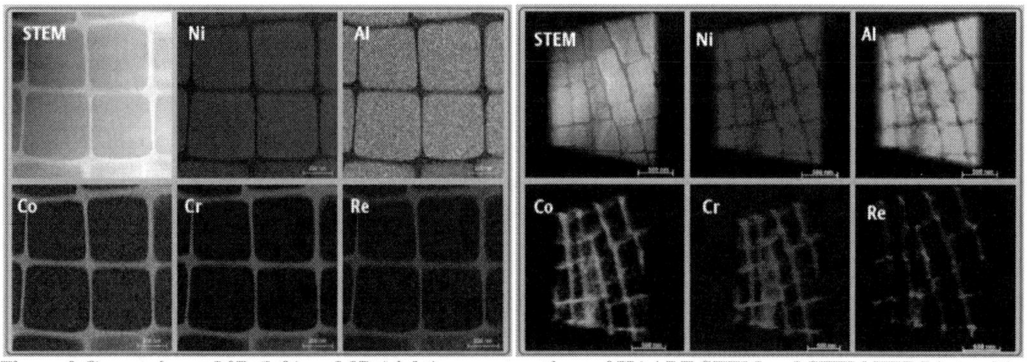

Figure 2 Comparison of 2D (left) and 3D (right) reconstructions of HAADF STEM and STEM XEDS intensities of Ni, Al, Co, Cr and Re. The 3D chemical mapping simply resolves the true structure of the material and reveals the 3D distribution of the individual elements through the the thickness of the TEM foil.

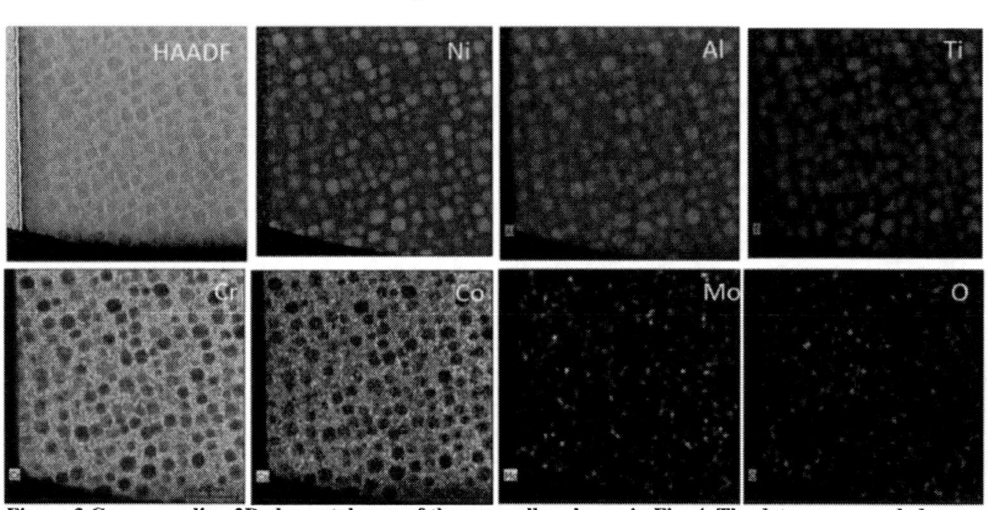

Figure 3 Corresponding 2D elemental map of the superalloy shown in Fig. 4. The data was recorded on a S/TEM with ChemiSTEM™.

The chemical composition of the other superalloy specimen is also very complex and contains Ni, Al, Ti, Cr, Co and Mo. The microstructure of single crystal superalloys consists of a higher volume fraction of cuboidal γ' precipitates, an ordered Ni_3Al-base intermetallic phase, surrounded by matrix channels with nanometric width [12]. XEDS microanalysis of the γ and γ' phases chemical composition is acquired with a few nanometer resolution on a superalloy heat treated to breaking point at 1150 °C. The results of STEM-EDX tomography (Figure 3) are compared to the 2D elemental maps (Figure 4) showing the advantage of this new technology in the accurate analysis of three dimensional elements distribution in a super alloy in both γ and γ' phases. There is no Oxygen observed on the surface indicating that no oxidation occurred. However, further investigations are necessary to accompaign the result quantitatively.

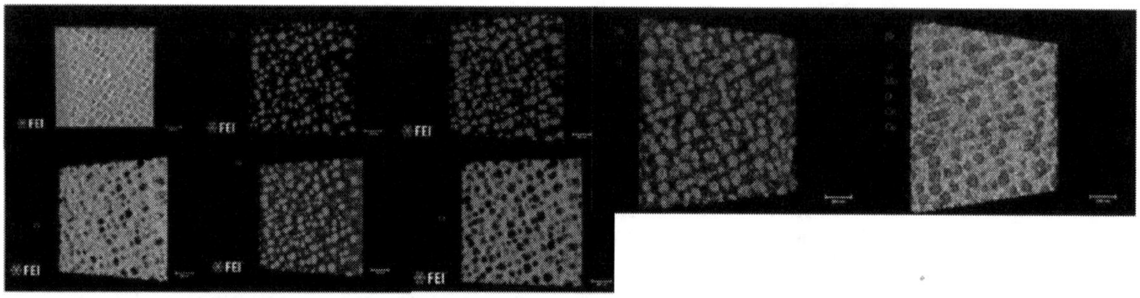

Figure 4 Corresponding 3D chemical tomograms to Figure 3 of the superalloy containing Ni, Al, Ti, Cr, Co and Mo. This comparison also depicts the superior chemical sensitivity of STEM XEDS tomography over conventional HAADF STEM tomography in a multi-component system resulting in the finding that there is no oxidation on the surface.

Acknowledgement

The authors acknowledge the experimental work performed by Dr. K. Song, Dr. C. Bouchet-Marquis, Y. Liu, R. Passey, Dr. A. Genc and Dr. L. Pullan from FEI Company, 5350 NE Dawson Creek Drive, Hillsboro, OR 97124, USA. The CMSX-4 single crystal nickel-base superalloy: Sample courtesy of Dr. Beata Dubiel, Prof. Aleksandra Czyrska-Filemonowicz, AGH University of Science and Technology, International Centre of Electron Microscopy for Materials Science IC-EM, Kraków, Poland. The Li ion battery cathode material: Courtesy of Dr. Chongmin Wang and Dr. Meng Gu, Pacific Northwest National Laboratory (PNNL), Richland, WA, USA.

References

[1] P. Schlossmacher et al., *Microscopy Today* **18** (4) (2010) 14-20.
[2] Meng Gu et al., *Nano Lett.* 2012, **12**, 5186–5191.
[3] K. Lepinary et al., *Micron Volume* **47**, April 2013, Pages 43–49.
[4] A. Genc et al., *Ultramicroscopy* **131** (2013) 24–32.
[5] K. Jarausch *et al., Ultramicroscopy* **109** (2009) 326-337.
[6] T. Yaguchi *et al., Ultramicroscopy* **108** (2008) 1603-1615.
[7] D. Huber et al., Late Breaking Poster, *Microscopy and Microanalysis* (2010)
[8] L. J. Allen et al., *MRS BULLETIN* VOLUME **37** JANUARY 2012.
[9] A. R. Akbashev et al., *CrystEngComm*, 2012, **14**, 5373-5376.
[10] [10] Poster, Gordon Research Conference on Physical Metallurgy, ME, 2013
[11] *DGE Journal "Elektronenmikroskopie"* **35/12**, December 2012.
[12] L. Kovarik *et al., Progress in Materials Science*, **54** (2009) 839–873
[13] A. Genc, H. Cheng, J.P. Winterstein, L. Pullan and B. Freitag, *M and M proceedings* 2012.

Journal of Physics: Conference Series **522** (2014) 012027 doi:10.1088/1742-6596/522/1/012027

Auger Electron Spectroscopy in high vacuum: Nanocharacterisation in the Scanning Electron Microscope

Xiaoping Zha[+], Christopher GH Walker, Mohamed El-Gomati

Department of Electronics, University of York, Heslington, York, YO10 5DD.
+ York Probe Sources Ltd, York, UK, www.YPS-Ltd.com

Mohamed.elgomati@york.ac.uk

Abstract. It is generally regarded as impossible to carry out Auger Electron Spectroscopy (AES) analysis in a scanning electron microscope (SEM) due to the high ambient gas pressure in an SEM. This is because standard electron energy analysers such as the Concentric Hemispherical Analyser (CHA) or Cylindrical Mirror Analyser (CMA) are devices that acquire data in a serial manner that can last up to few minutes. This is considered too slow for high vacuum (10^{-6}mbar) and a previously cleaned surface would be re-contaminated before a spectrum could be completed. This has led to AES being traditionally carried out under ultra-high vacuum (UHV) environment. We report on two devices for fast acquisition of AES data characterising nanoscale objects by the use of AES in an SEM.

1. Introduction

The characterisation of samples at the nanoscale is an important aspect of nanotechnology. Characterisation at the micron scale has been carried out for several decades using X-ray microanalysis – usually with the aid of Energy Dispersive Spectroscopy (EDS) as an add-on technique in scanning electron microscopy (SEM). In order to approach nanoscale characterization using EDS, the primary electron energy needs to be reduced. This means that the K shells of many elements can no longer be excited and the analysis needs to rely on L and M shell excitations. The peaks of these levels are often close together and quantification is difficult. An important recent advance in EDS is the use of Silicon Drift Detectors (SDDs) which have higher energy resolution and allow the L and M peaks to be separated. However, even at lower primary energies, the excitation volume is defined by the primary beam energy and is larger than the primary electron beam spot on the sample. In order to characterize samples at such a scale, it is necessary to resort to Auger electron spectroscopy (AES). In addition, AES could provide complementary information to EDS such as the nature of the chemical bonding, as well as electronic information provided by the plasmon losses. However, a major obstacle to the use of AES in standard SEMs is that AES normally requires Ultra High Vacuum (UHV) conditions. This is because at poorer vacuum conditions, the surface would be covered with a layer of mostly carbon and oxygen compounds in just a few seconds. Since AES normally requires at least a few minutes to acquire a spectrum, then this renders AES impractical in SEMs at their usual operating pressure. This paper discusses two approaches to the fast acquisition of Auger electron spectra – the Hyperbolic Field Analyser (HFA) [1] and the Magnetic Electron Energy Spectrometer (MEES) [2] which overcome these restrictions.

The first device for parallel AES data acquisition is the Hyperbolic Field Analyser (HFA) which has been previously described [1]. In this paper we will show AES spectra that have been acquired using the HFA on an SEM with a pressure of ~10^{-6} mbar (JEOL 6400F). The surfaces of the used samples are in-situ cleaned with energetic Ar ions. The rate of recontamination is determined by studying the rates of growth of Carbon and Oxygen Auger electron peak heights.

Another device for fast acquisition of AES data is the Magnetic Electron Energy Spectrometer (MEES). This device would be better suited to SEMs with higher magnetic field strengths in the sample region and can acquire electrons over a wide range of electron take-off angles. Theoretical background to the MEES will be presented as well as AES spectra obtained using a custom built UHV system operating at ~10^{-7} mbar and using a novel Active Pixel Sensor as electron detector.

2. The Hyperbolic Field Analyser (HFA) in the Scanning Electron Microscope

The HFA was mounted in the specimen chamber of the SEM. The sample surface was cleaned using an Ar ion gun (IQE 12/38 by SPECS). A number of materials were studied, but Silicon is the example presented here. The sample is rotated between the position for Ar ion cleaning and that for AES. The sample was cleaned using 5keV Ar ions (specimen current 1-2μA) for 10-20 minutes prior to analysis. Figure 1 shows the Auger spectrum of the Si just after cleaning. Acquisition started within 10s of finishing the Ar ion cleaning. Fifty spectra were taken with each spectrum taking 1s. These spectra were then averaged. Several such averaged spectra were taken over subsequent minutes.

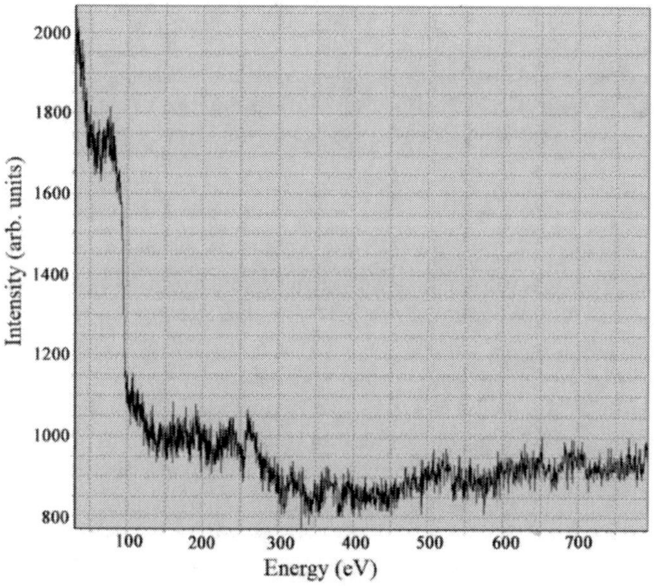

Figure 1. An Auger spectrum of Silicon taken using the HFA showing the Si LVV and C KLL Auger spectra.

The depth of the layer C_i is related to the atomic concentration of an overlayer of element i by

$$X_i = 1 - \exp\left(\frac{-C_i}{\lambda \cos\theta}\right)$$

where λ is the inelastic electron mean free path, θ is the angle between the surface normal and the detector [3].

The atomic concentration X_i is given by

$$X_i = \frac{\alpha_i I_i}{\sum_j \alpha_j I_j}$$

where α_i is the inverse relative sensitivity factor, I_i is the intensity of the Auger peak of element I [4,5].

The rate of growth of Carbon on the Silicon could be determined by measuring the depth of the layer Si and C in relation to elapsed time.

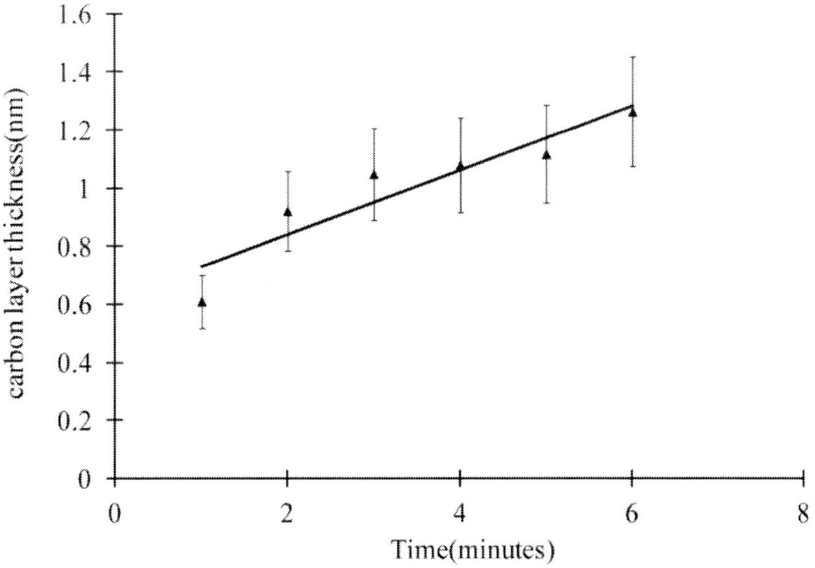

Figure 2. Rate of growth of Carbon with time as estimated using Si and C Auger signals.

3. The Magnetic Electron Energy Spectrometer (MEES)

The MEES uses a magnetic field to disperse the electrons. The magnetic field direction is arranged to be parallel to the primary beam. Electrons emitted from the surface pass through a small slit and travel in a helical fashion until they strike a 2D detector. The detector used in this experiment was an Active Pixel Sensor [6]. Electrons of the same energy but varying take off angles form a curve as shown in Figure 3. In this figure the intense line is caused by elastic peak electrons being emitted at a range of take off angles. A spectrum of the output can be determined by integrating in a line from top left (= low energy) to bottom right (=high energy) and the result can be seen in Figure 4.

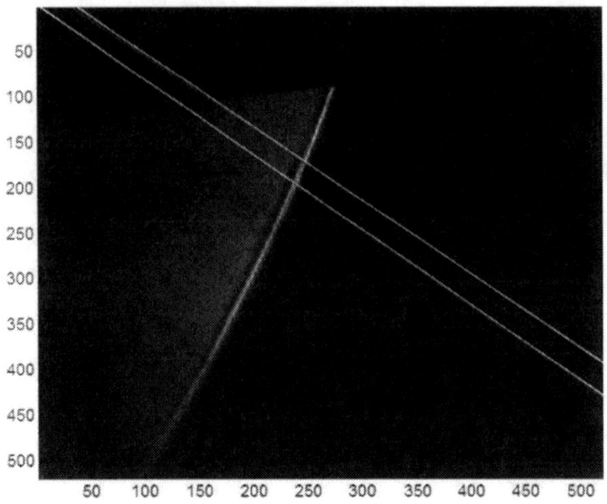

Figure 3. Image of an elastic peak at ~900eV as acquired on the Active Pixel Sensor. The two straight lines define the solid angle in which the data will be integrated to form a spectrum.

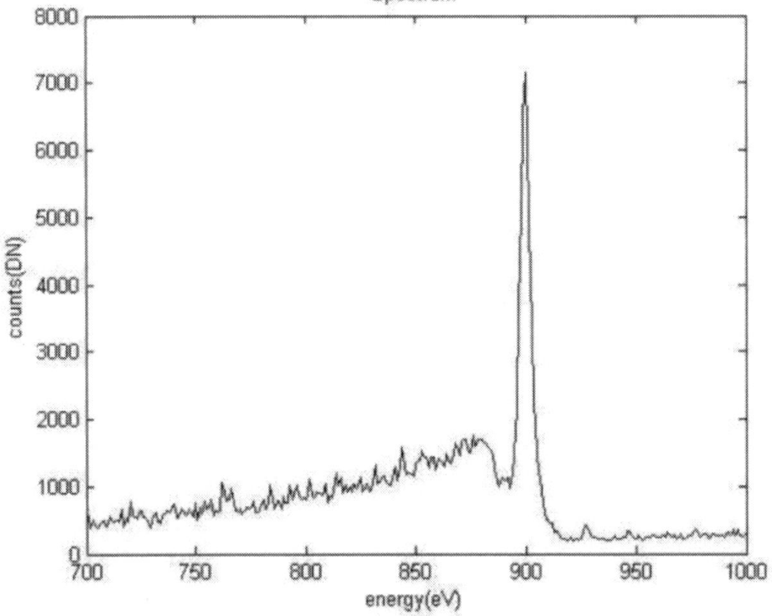

Figure 4. A spectrum of the elastic peak determined from figure 3 collected with an electron beam of ~900eV, after image processing [7].

4. Conclusions

Two different approaches to rapidly acquiring Auger spectra using two entirely different electron energy analysers have been presented. These class of analysers are >x50 faster than conventional AES analysers. The ability to acquire Auger spectra in high vacuum has been shown to be possible and that Auger spectroscopy can be carried out in a standard SEM. This is most relevant in applications requiring elemental and chemical analysis at high spatial resolution that conventional X-ray spectroscopy (EDS) cannot provide. In addition, the magnetic spectrometer allows the possibility to acquire Auger spectra even in the presence of relatively strong magnetic fields and in confined spaces.

References

[1] Jacka, M; Kirk, M; El Gomati, M M; Prutton, M A fast, parallel acquisition, electron energy analyzer: The hyperbolic field analyzer., 1999, *Rev. Sci. Instr.* Vol. 70, p. 2282-87.
[2] Zha, X; Walker, Christopher G; El Gomati, Mohamed M A magnetic electron energy analyser for fast data acquisition. 2011, *Nucl. Instr. Meth. Phys. Res. A* Vol. 645, p. 257-259.
[3] Chang C C, General formalism for quantitative Auger analysis. 1975, *Surface Science* 48, p. 9-21.
[4] Mroczkowski S, Calculated Auger yields and sensitivity factors for KLL-NOO transitions with 1-10 kV primary beams. 1985, *Journal of Vacuum Science and Technology - A, Vacuum, Surfaces, and Film 3*, p. 1860-65.
[5] Seah M P, Gilmore I S, Quantitative AES. VIII: Analysis of Auger Electron Intensities from Elemental Data in a Digital Auger Database. 1998, *Surface and Interface Analysis* 26, p. 908-929.
[6] Zha X, El-Gomati M M, Chen L, Walker C, Clark A T and Turchetta R, Direct Detection of Low-Energy Electrons With a Novel CMOS APS Sensor, 2012, *Electron Devices, IEEE Trans.* Vol. 59, p3594-3600
[7] Zha X, Magnetic electron energy spectrometer Ph.d. University of York, (2009).

Comparison of atom probe tomography and transmission electron microscopy analysis of oxide dispersion strengthened steels

AJ London[1], S Lozano-Perez[1], S Santra[2], S Amirthapandian[2], BK Panigrahi[2], CS Sundar[2] and CRM Grovenor[1]

[1] Department of Materials, University of Oxford, Parks Road, Oxford, OX1 3PH, UK
[2] Indira Gandhi Centre for Atomic Research, Kalpakkam - 603 102, TN, India

E-mail: andrew.london@materials.ox.ac.uk

Abstract. Oxide dispersion strengthened steels owe part of their high temperature stability to the nano-scale oxides they contain. These yttrium-titanium oxides are notoriously difficult to characterise since they are embedded in a magnetic-ferritic matrix and often <10 nm across. This study uses correlated transmission electron microscopy and atom probe tomography on the same material to explore the kind of information that can be gained on the character of the oxide particles.

The influence of chromium in these alloys is of interest, therefore two model ODS steels Fe-(14Cr)-0.2Ti-0.3Y_2O_3 are compared. TEM is shown to accurately measure the size of the oxide particles and atom probe tomography is necessary to observe the smallest sub-1.5 nm particles. Larger $Y_2Ti_2O_7$ and Y_2TiO_5 structured particles were identified by high-resolution transmission electron microscopy, but the smallest oxides remain difficult to index. Chemical data from energy-filtered TEM agreed qualitatively with the atom probe findings. It was found that the majority of the oxide particles exhibit an unoxidised chromium shell which may be responsible for reducing the ultimate size of the oxide particles.

1. Introduction & Background

Ferritic oxide dispersion strengthened (ODS) steels are being considered as structural materials for fission and fusion reactors due to their low activation, high temperature stability and radiation resistance [1]. The 1–10 nm yttrium-titanium rich oxides that hinder dislocation motion at high temperature also harmlessly recombine radiation defects and trap inert gases away from grain boundaries [2]. Understanding the particle character is critical because the size distribution and dispersion influence the mechanical properties [3] and their chemical nature determines their thermal and radiation stability [4].

Many methods have been used to study these nano-scale oxides [5, 6, 7] but only transmission electron microscopy (TEM) and atom probe tomography (APT) provide sufficiently spatially resolved chemical and structural information to distinguish the smaller oxides from larger oxides located at the surface, grain boundaries and in the bulk.

The methodology of this study includes both TEM and APT of the same specimens to give a more convincing and complete characterisation of the same volume of material. The focus of the TEM examination was the spatial distribution, size and the structure of the oxides. With atomistic sensitivity and sub-nanometer resolution, APT was used to study the chemistry of individual oxide particles and their chromium content.

Part of the uncertainty relating to the identity of the oxides undoubtedly comes from the many different alloy compositions and processing routes used to produce different ODS alloys. In this study, the alloys have identical processing history and a systematic variation in chemistry to determine the influence of chromium on the oxide particles. Chromium is a key component of structural steels, providing enhanced strength and corrosion resistance. The model ODS steels are nominally Fe-0.2Ti-0.3Y_2O_3 and Fe-14Cr-0.2Ti-0.3Y_2O_3 (all wt%) and are produced by the Indira Gandhi Centre for Atomic Research (India).

Titanium has been attributed to having a strong refining influence on the oxides [8] but the effect of chromium has been less fully investigated. Chromium rich shells around the Y-Ti-oxides have been observed by TEM [9] and APT [10]. By comparing model alloys with and without a chromium addition the effect of chromium can be evaluated.

2. Materials & Methods

A pre-alloyed powder containing Fe-(14Cr)-0.2 Ti was mechanically alloyed with 0.3 Y_2O_3 powder (all in wt%) using a horizontal high-energy Simoloyer attritor mill. Milling under an argon atmosphere took 4 hrs using a ball-to-powder ratio of 10:1. Consolidation was achieved using hot extrusion at 1050 °C and 1150 °C for the Cr-free and Cr containing alloys respectively.

Bulk material were cut to 0.5x0.5x20 mm and electropolished into needles using perchloric acid by standard methods [11]. TEM samples were also prepared using a FIB lift-out technique.

200 keV TEM observations were made using a Philips CM20 and a JEOL 2200MCO FEG TEM fitted with a monochromator, two spherical aberration (Cs) correctors and an Ω-filter. The largest (120 μm) condenser aperture was used and the beam was focused to give an an image with a diameter slightly bigger than the CCD to maximise intensity. The convergence and objective aperture half-angles were 6 mrad and 20 mrad, respectively. Low-loss energy-filtered series collected from the 10–80 eV energy-loss range included the $Y_{N2,3}$ (26 eV), $Cr_{M2,3}$ (42 eV) and $Fe_{M2,3}$ (54 eV) edges. Steps of 1 eV, using a 2 eV slit and 6 s acquisition time per image were used and images were hardware binned by 4 to 1024×1024 pixels. The EFTEM series were processed after acquisition using multivariate statistical analysis (MSA) in order to reduce statistical noise, as described elsewhere [9].

Atom probe analyses were performed using a LEAP-3000X HR microscope. During analysis, samples were maintained at 50 K and a 532 nm wavelength laser with a pulse width of 10 ps, repetition rate of 200 kHz, pulse energy of 0.3 nJ and spot size of less than 10 μm were used to promote field evaporation. Oxide clusters in the data were identified using the method of maximum separation of Y, YO, TiO and O ions. A maximum separation of core ions of 0.7–1.1 nm was used with a N_{min} between 8–15. The concentration line profiles were calculated in MATLAB for every cluster in the data scaled to each cluster's size. In this way profiles of large and small clusters can be combined on the same plot. Previously, representative clusters had to be chosen for analysis but using this method all the clusters can be investigated.

3. Results & Discussion

The particle radii were measured using diffraction contrast (Fig. 1a), Fresnel imaging and EFTEM (Fig. 1b) techniques. All of these gave consistent results and the size distributions are shown in figure 2b. The TEM imaging of electropolished atom probe needles leads to a bias in counting medium sized particles as they are more obvious in the thicker shank of the tip (leftmost Fig.1a), but very large particles are physically too large to exist in the end of the tip. EFTEM provides a wider field of view but is limited to thin regions (< 100 nm). The APT data is in the form of a point cloud and the particles exist as more diffuse clusters. It is common to report the radius of gyration (R_g) of these clusters to compare their size. Trajectory aberrations can influence the lateral dimensions of the particles and oxygen spreading can increase their size perpendicular to the analysis direction, Z. But the size along Z is relatively unchanged in-spite of these aberrations [12], therefore only the component of R_g along Z is reported. The R_{gz} is much smaller than the radius measured by TEM as shown in the size distributions in figure 2b. Because

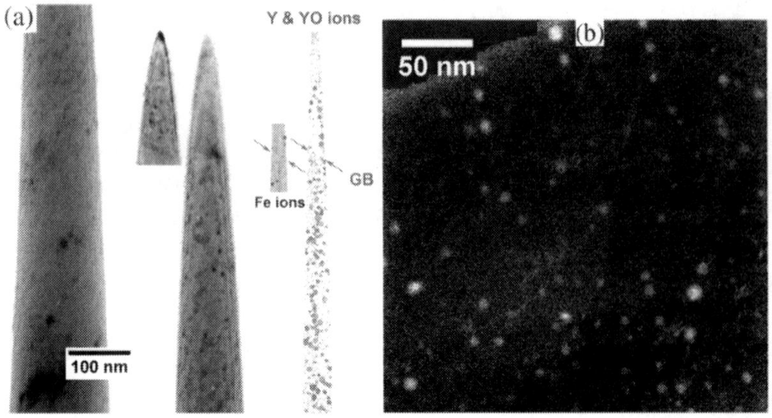

Figure 1. (a) TEM image of a needle with accompanying APT data showing the Y and YO ions with yttrium iso-concentration surfaces to highlight the oxides. Two tilt angles are used to show clearly the oxides in the top and bottom grains. The grain boundary (GB) is visible in TEM and APT, 200 nm from the tip. All images are shown at the same scale.
(b) EFTEM image of the $Y_{N2,3}$ edge showing yttrium rich oxides in a Fe-14Cr ODS alloy. With a 2 eV energy slit and 6 s acquisition using MSA, oxides as small as 1.5 nm in diameter were observed.

only 37% of the atoms are detected in the LEAP some reduction of size is expected, especially for the smaller clusters. The extent of the clusters (in Z) was measured to be 3 times R_{gz} and may therefore give a more representative size measurement. In the Fe-14Cr alloy, the number density of particles as measured by TEM was 3.0–4.4x10^{22} per m^3 but ten times greater than this when measured by APT (Fig.2b), suggesting that the APT is indeed more sensitive to the numerous smaller oxide particles.

Faceted particle edges parallel to $\{110\}_{matrix}$ were observed when viewing down $[002]_{matrix}$. No structural determination was possible for the <2 nm particles, but larger particles were structurally consistent with the bulk structures of $Y_2Ti_2O_7$ and Y_2TiO_5. The composition of the oxides from APT in both alloys shows a range of Y:Ti ratios consistent with either bulk stoichiometry but the oxygen content is only 50–60 at%. Fe is present in the clusters because of trajectory aberrations [12]. Fe & Cr are not expected in the oxide core by energetic considerations [13] and Fe has not been found in extracted particles [5].

In the Fe-14Cr alloy, a Cr-rich shell was apparent using EFTEM in >10 nm particles and in most clusters observed using APT. As shown in figure 2a the Cr ions have a shell-like profile (11.5 at% in the bulk) but the CrO ions take a profile similar to the O and FeO ions. We do not believe that there is either oxidised Fe or Cr in the particles, but that oxygen atoms originating from the particles combine with Fe or Cr from the matrix and evaporate as FeO or CrO molecular species. Similarly partially-oxidised Cr-rich shells have been reported in similar alloys [10].

The addition of Cr to the alloy reduces the average size and distribution of sizes of the Y-Ti oxides (Fig. 2b). The addition of Cr could increase the rate of nucleation of the Y-Ti oxides or reduce the rate of coarsening in order that the average size is reduced even with the increased consolidation temperature. The smallest clusters contain fewer than 100 atoms making the evaluation of shell size uncertain. For the larger clusters, the shell size increases with cluster radius. Quantification of the size of the Cr-shell will help to clarify the mechanism of its formation. This shell should not be ignored as it reduces the oxide-metal surface energy, could increase thermal stability and has implications for the response of the oxide particles to radiation.

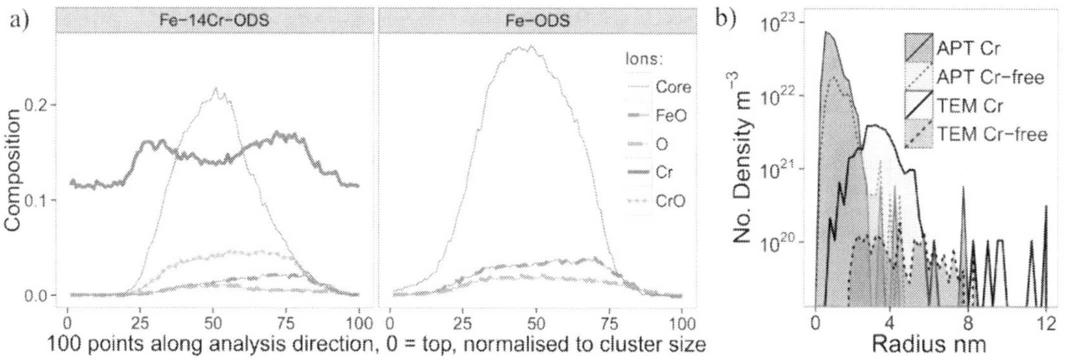

Figure 2. (a) Average line profiles along the analysis direction from atom probe (Cr-free = 200 and Cr-alloy = 500 clusters total). Each line profile was normalised to the size of the oxide clusters so that different sized clusters could be compared on one plot. Core ions=YO+Y+TiO.
(b) Size distributions of oxide clusters in the same alloys measured by TEM and APT (R_{gz}).

4. Conclusions

TEM and APT have been used to characterise model ODS steels and investigate the oxides present in detail. For larger particles diffraction contrast was suitable to determine their size and EFTEM imaging of the $Y_{N2,3}$ edge showed yttrium rich oxides as small as 1.5 nm in diameter. The smallest most numerous (< 1.5 nm) oxides were only visible using atom probe tomography but the size measurement by APT requires careful calibration of the reconstructed data and analysis to reduce the influence of trajectory aberrations. Low-loss EFTEM images showed Y-rich particles with a Cr-rich shell and the bulk oxides $Y_2Ti_2O_7$ and Y_2TiO_5 were identified by structural measurements. Y-Ti-oxide clusters in the APT data had compositions consistent with these two bulk oxides but were oxygen deficient. From the profile of the CrO ions collected by APT, it is more likely the Cr shell is metallic or only partially oxidised. The influence of the Cr on the oxides is to reduce their average size and increase their number density.

Acknowledgments

Dr. Tom Black for funding part of this project, Prof. A.Kirkland and & Dr. Judy Kim for the support on the 2200 microscope, Paul Bagot & Oxford FIM group (University of Oxford).

References

[1] Baluc N, Schäublin R, Spätig P and Victoria M 2004 *Nuclear fusion* **44** 56
[2] Odette G, MJ A and BD W 2008 *Annual Review of Materials Research* **38** 471–503
[3] Ukai S, Miztua S, Fujiwara M, Okuda T and Kobayshi T 2002 *J. Nucl. Sci. and Tech.* **39** 778–788
[4] Williams C, Smith G and Marquis E 2012 *Scripta Materialia* **67** 108–111
[5] Wu Y, Haney E, Cunningham N and Odette G 2012 *Acta Materialia* **60** 3456–3468
[6] Mathon M, de Carlan Y, Averty X, Alamo A and de Novion C 2006 *ASTM Spec. tech. pub.* **1475** 120
[7] Ortega Y, de Castro V, Monge M, Munoz A, Leguey T and Pareja R 2008 *J. Nuc. Mat.* **376** 222–228
[8] Ratti M, Leuvrey D, Mathon M and De Carlan Y 2009 *J. Nuc. Mat.* **386** 540–543
[9] Lozano-Perez S, de Castro Bernal V and Nicholls R 2009 *Ultramicroscopy* **109** 1217–1228
[10] De Castro V, Marquis E, Lozano-Perez S, Pareja R and Jenkins M 2011 *Acta Mater.* **59** 3927–3936
[11] Miller M 2000 *Atom Probe Tomography: Analysis at the Atomic Level* (Plenum Pub Corp)
[12] Oberdorfer C and Schmitz G 2010 *Microscopy and Microanalysis* **17** 15
[13] Murali D, Panigrahi B, Valsakumar M, Chandra S, Sundar C and Raj B 2010 *J. of Nuc. Mat.* **403** 113–116

Orientation dependence of shock induced dislocations in tantalum single crystals.

B Pang[1], I Jones[1], Y Chiu[1], J Millett[2], G Whiteman[2], N Bourne[2]

[1]*School of Metallurgy and Materials, the University of Birmingham, Birmingham B15 2TT, UK,*

[2]*AWE, Aldermaston, Reading, Berkshire RG7 4PR, United Kingdom*

E-mail: BXP171@bham.ac.uk

Abstract: A planar shock wave with a peak pressure of 6.2 GPa and duration of 1.7 μs followed by a lateral release wave generates profuse dislocations in single crystalline tantalum. Three orientations [100], [110], [111] were tested to examine the orientation dependence of the dislocation generation. The dislocations were characterised by transmission electron microscopy. The Burgers vectors and morphology of the primary dislocations in the specimens with different orientations showed a distinct orientation dependence and will be discussed in light of the model of slip behaviour in one-dimensional strain of C.S. Smith [1].

1. Introduction

There is a need to understand the deformation mechanism of materials in high rate shock loading-unloading phenomena. The deformation behaviour of material under one dimensional shock loading has been extensively studied over the past few decades. The tested materials were generally shock loaded by a plane wave and unloaded in the same direction with a planar release wave. The experimental data is usually taken from the shock front to investigate the material behaviour at high pressure, with less attention paid to the effect of the release wave. However, plate impact experiments by Stevens show that in some cases the plastic strain made by radial release is even bigger than that of the shock wave [2]. In a real shock loading condition like foreign object damage on aircraft, the loading cannot always be one dimensional. Therefore, for strengthening materials, we should consider the lateral release as as important as the shock wave front.

Experimental results on laser compressed tantalum single crystals have revealed that dislocation walls formed in samples with a loading axis of [001] when the local pressure is higher than 21GPa [3]. In explosive shock experiments on polycrystalline tantalum, it was found that dislocations formed a cell structure under a pressure of 45GPa but not under 15GPa [4]. Long screw dislocations, dislocation tangles and loops were detected in polycrystalline tantalum shock loaded using a gas gun recovery technique[5]. Most previous research focused on dislocation density and morphology, without much attention to the slip direction of the dislocations. As predicted by the model of Smith [1], the slip systems activated in a 1-D shock wave should tend to resolve the uniaxial principal strain into the other two directions, which means that dislocations should slip on a direction close to 45° to the loading direction[1]. In this paper, tantalum was chosen as a model material representing the body centred cubic structure. Single crystals with principal orientations [111], [011] and [001] were shocked with a planar wave and released radially. The defects generated by the shock and release wave were characterised by TEM to see if the deformation mechanisms of tantalum match with the predictions of Smith's model.

2. Experimental procedure

Single crystal tantalum with 12mm diameter and 4mm thickness were subjected to a plate impact experiment along three directions, [111], [011] and [001] to a peak pressure of 6GPa. Instead of using a recovery assembly with momentum rings, the specimen plates were mounted in epoxy to provide a lateral release wave.

Foils for TEM examination were obtained from the as-received materials and from the front surface impacted by the projectile of the shock recovered specimens, as shown in Figure 1. The front surface of the samples are sliced by EDM spark machine and 3~4 discs were punched from each specimen. They were prepared by electropolishing at about ~25V and ~ -15°C in an electrolyte containing 86 vol. % methanol, 10 vol. % sulphuric acid and 4 vol. % hydrofluoric acid. The deformation substructures were then examined using a JEOL 2100 TEM.

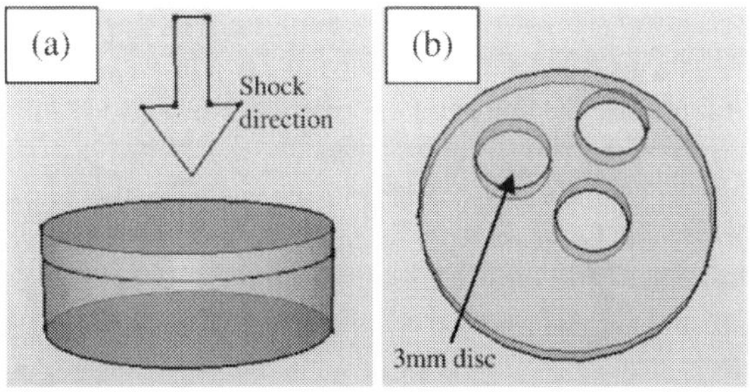

Figure 1 Sketch shows shock direction and sample position.

3. Results and Discussion

3.1. Dislocation morphology and density

TEM bright field images of the as-received and the recovered materials are presented in Figure 2. The TEM foils taken from same sample reveal the same characteristics. The as-received tantalum single crystals have a dislocation density of about $6 \times 10^8 cm^{-2}$.

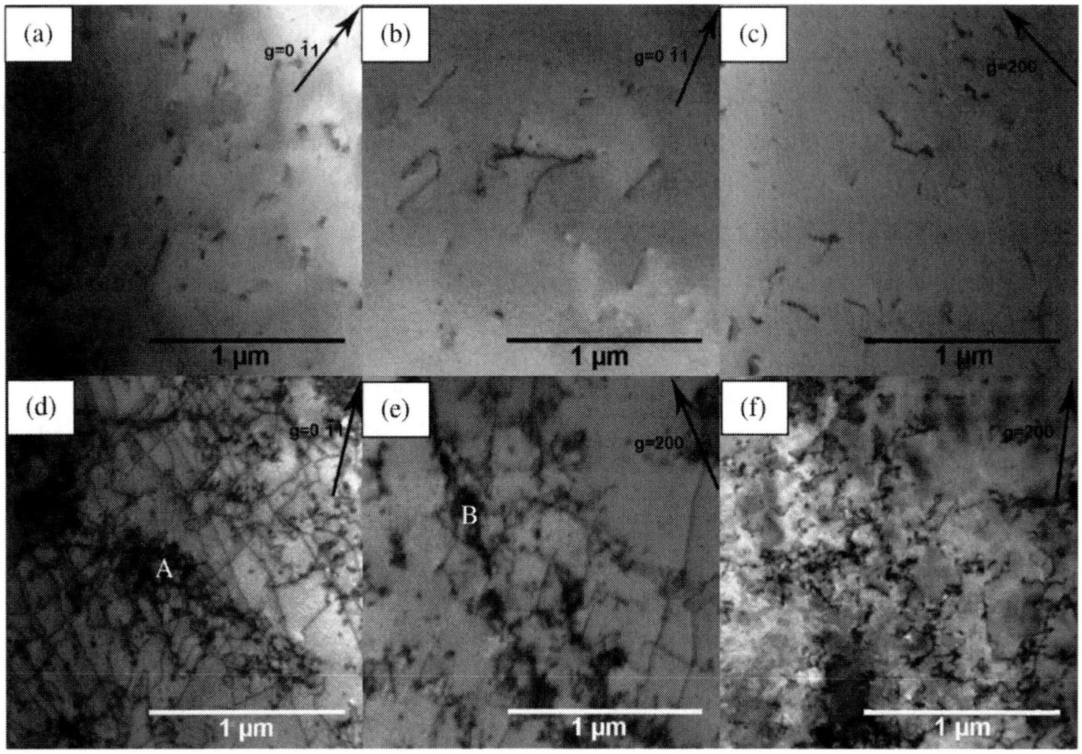

Figure 2 Dislocation substructure of (a) the as-received [111] (b) as-received [011] (c) as-received [001] (d) shocked [111] (e) shocked [011] (f) shocked [001]

After shock recovery, the substructures in the [111] and [011] samples are similar in that both contain dislocation walls constructed with tangled dislocations, aligned with a separation of around 1 μm (A and B in Figure 2) and a low density of long screw dislocations and high density of small dislocation loops in between the walls. In the [001] loaded sample, no dislocation walls were detected. The dislocation densities calculated from the TEM images are $6 \times 10^{10} cm^{-2}$, $4 \times 10^{10} cm^{-2}$and $2 \times 10^{10} cm^{-2}$ for samples loaded along [111], [011] and [001], respectively, which agree with the expected dislocation densities from [4] and [5].

The driving force for dislocation wall formation is recognised as arising from a reduction in the total elastic energy of the dislocations because of clustering [6]. It means the material needs a very high dislocation density with a high energy state for the dislocation wall to form. The dislocation density in the [001] sample is considered to be too low for it to decompose into dislocation walls. As shown in the TEM image it is lower than in the other two orientations [3]. The morphology of the dislocation walls is similar to those observed in fatigued specimens. A probable reason for their formation is the compression-tension cycle made by the shock and then lateral release waves. These last in the current experiment, are much stronger than the laser and explosive induced lateral release waves. The compression-tension cycle is similar to fatigue, but with fewer cycles and a higher loading stress[7].

3.2. Burgers vector analysis

The Burgers vectors of the dislocations were analysed, except the ones tangled in the dislocation walls. Figure 3 shows an example of images obtained using different reflections. The results of the analysis are shown in table 1. From the TEM images the densities of the activated slip systems are uniform in each sample. It seems that in the sample with [111] and [001] loading directions, the activated slip systems are those which resolve the strain from 1-D to 3-D as predicted by the model of Smith [1], but the dislocations in the sample loaded along [011] have their Burgers vector perpendicular to the loading direction. These dislocations are not expected to be activated in uniaxial loading. The reason for the anomalous slip in the [011] loaded sample is still unclear.

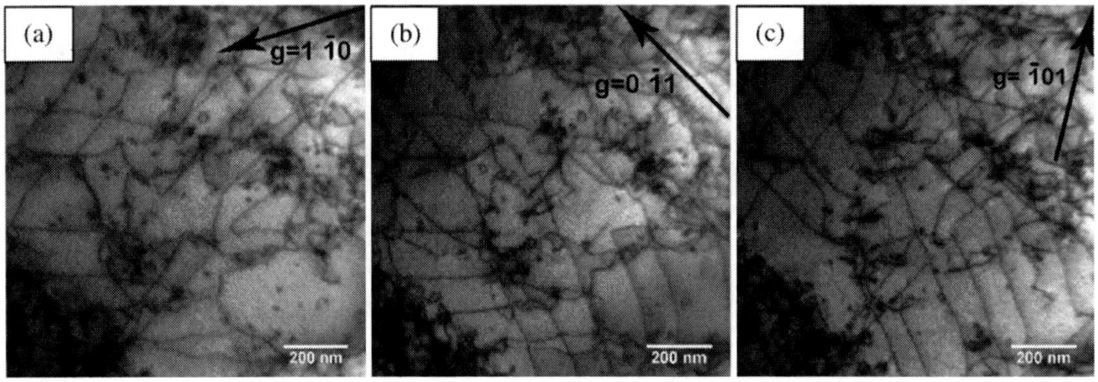

Figure 3 Dislocation substructure obtained from [111] sample with B=111: (a) g=$1\bar{1}0$ (b) g=$0\bar{1}1$ (c) g= $\bar{1}01$

Table 1 The Burgers vectors of dislocations in the shocked specimens

Sample	**111**	**011**	**001**
Burgers vector	$[1\bar{1}\bar{1}]$, $[11\bar{1}]$, $[1\bar{1}1]$	$[11\bar{1}]$, $[1\bar{1}1]$	$[\bar{1}\bar{1}1]$, $[\bar{1}11]$, $[111]$, $[1\bar{1}1]$

3.3. Twins

The twin area fractions are about 1%, 2% and 5% in samples loaded along [111], [011] and [001]. Figure 4 shows examples of twins in the three specimens. It has been reported that the twinning threshold pressure in planar shock experiments is higher than 7GPa [8].

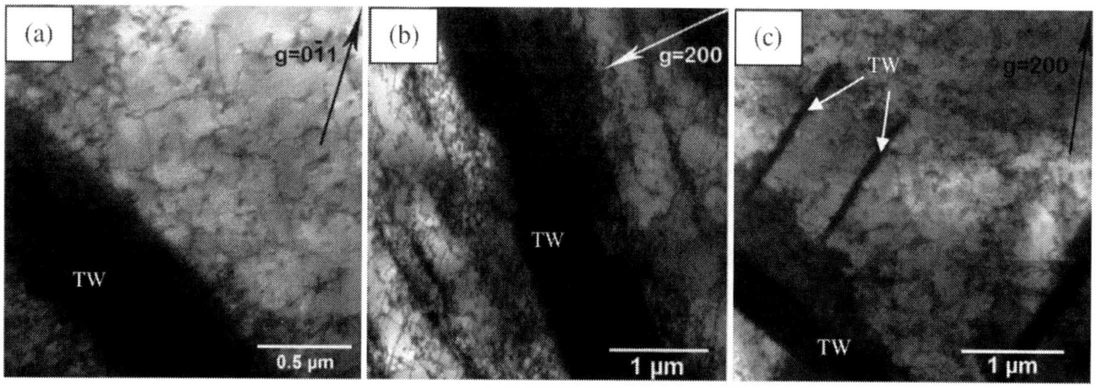

Figure 4 TEM images showing twins in sample shock loaded along (a) [111] (b) [011] (c) [001]

Since the peak pressure employed in the current study (6GPa) is lower than the twinning threshold pressure reported, it is possible that the twins observed in this work may originate from the lateral release wave. This is different from the initial shock for various probable reasons: (i) the dislocations already present may not be the appropriate ones to accomplish the reverse strain via a lateral release wave, (ii) the dislocations present will almost certainly be tangled, raising the stress necessary for glide, (iii) the tensile strain may be higher than the compressive strain if release waves from (laterally) opposite directions arrive at the centre of the specimen together.

4. Conclusions

1. In the [111] and [011] specimens, dislocations are tangled into dislocation walls, similar to those after fatigue. This might result from the compression-tension cycle made by the shock and lateral release waves.
2. Dislocations in the samples with [111] and [001] loading follow Smith's model. However in the [011] specimen unexpected slip in the direction perpendicular to the loading direction occurs.
3. Lateral release induced twins are found at a pressure lower than the published threshold.

5. Reference

[1]. Smith C *Transactions of the American Institute of Mining and Metallurgical Engineers*, 212 (1958) 574-589.
[2]. Stevens A and O Jones *Journal of Applied Mechanics*, 39 (1972) 359-366.
[3]. Lu C, et al. *Acta Materialia*, 60 (2012) 6601-6620.
[4]. Hsiung L and D Lassila *Acta Materialia*, 48 (2000) 4851-4865.
[5]. Gray G and K Vecchio *Metallurgical and Materials Transactions a-Physical Metallurgy and Materials Science*, 26 (1995) 2555-2563.
[6]. Holt D *Journal of Applied Physics*, 41 (1970) 3197-3201.
[7]. Meyers M, *Dynamic Behavior of Materials*. 1994: Wiley.
[8]. McNaney J, et al. *Shock Compression of Condensed Matter - 2009, Pts 1 and 2*, 1195 (2009) 1127-1130.

Structural investigation of precipitates with Cu and Zn atomic columns in Al-Mg-Si alloys by aberration-corrected HAADF-STEM

Takeshi Saito[1], Calin D. Marioara[2], Sigmund J. Andersen[2], Williams Lefebvre[3] and Randi Holmestad[1]

[1]Department of Physics, Norwegian University of Science and Technology (NTNU), 7491 Trondheim, Norway
[2]SINTEF Materials and Chemistry, 7465 Trondheim, Norway
[3]GPM, Université de Rouen, 76801 Saint Etienne du Rouvray, France

Email: takeshi.saito@ntnu.no

Abstract. Precipitates in Al-Mg-Si alloys with Cu addition (~0.1 wt%) and Zn addition (~1 wt%) were investigated by aberration corrected high angle annular dark field scanning transmission electron microscopy (HAADF-STEM). Most precipitates had no overall unit cell but contained ordered network of Si atomic columns for both the Cu and the Zn containing precipitates. It was found that both Cu and Zn atomic columns are located at specific sites and producing characteristic local configurations on the Si atomic columns.

1. Introduction

Al-Mg-Si alloys (i.e. 6xxx series) are important structural materials characterized by formation of high number density of nano-sized metastable precipitates. The strength in these alloys is achieved by precipitates setting up a surrounding strain fields, which prevent dislocation movements. There are several types of metastable precipitates in the alloy depending on the composition and thermo-mechanical processing. In the Al-Mg-Si alloy system, all precipitates are needle/rod/lath-like with main growth direction along <100> Al. In recent years, their crystal structures have been investigated in details by transmission electron microscopy (TEM) [1-8]. All metastable precipitates in Al-Mg-Si alloys are structurally connected via a network of Si atomic columns (Si-network) with hexagonal projected symmetry along needle growth direction [7, 9].

Influences of additional alloying elements on the Al-Mg-Si alloys are of interest as they affect the precipitate structure and consequently the mechanical properties. For instance, an addition of ~0.4 wt% Cu drastically alters the precipitation sequence [9-12]. Although Cu often makes the precipitates disordered, it does not break up the Si-network. The Cu atomic columns resides in specific local atomic configurations connected with the Si-network [13]. Zn is one of the main alloying elements in a different family of aluminum alloy (i.e. 7xxx series) and forms different types of metastable precipitates together with Mg. Our interest here concerns how Cu and Zn atoms interact with Mg and/or Si in forming precipitates, especially what atomic sites they take in the Al-Mg-Si alloys. In order to investigate the atomic structure of the precipitates, aberration corrected TEM plays an important role. High angle annular dark field scanning TEM (HAADF-STEM) has distinctive advantages since it provides atomic number (Z) contrast intensity, namely it is possible to distinguish

Cu (Z_{Cu} = 29) and Zn (Z_{Zn} = 30) atomic columns from Al (Z_{Al} = 13), Mg (Z_{Mg} = 12) and Si (Z_{Si} = 14) columns in the precipitates. In the present study, the structures of Cu and Zn atomic columns in precipitates in the Al-Mg-Si alloy system have been investigated by aberration-corrected HAADF-STEM.

2. Experimental procedure

Two Al-0.47Mg-0.42Si (wt%) alloys were prepared by extrusion from cast billets; one added 0.1 wt% Cu and the other added 1 wt% Zn. The Cu-added alloy was solution heat treated at 545°C for 5 minutes, kept 30 minutes at room temperature and isothermally heat treated at 190°C for 300 minutes. The Zn-added alloy was solution heat treated at 540°C for 60 minutes, kept 240 minutes at room temperature and isothermally heat treated at 185°C for 720 minutes. The final microstructures correspond to peak hardness condition. STEM specimens were prepared by electro polishing using a Tenupol 5 machine (Struers) on transversal slices along the extrusion direction. The electrolyte consisted of 1/3 vol% HNO_3 in methanol, and the solution was kept at a temperature between -20°C and -35°C.

The HAADF-STEM images were taken by a spherical aberration probe corrected JEOL ARM200F equipped with a Schottky field emitter operated at 200kV. The probe diameter was 0.1 nm and the collection angle of the HAADF detector was in the range of 45-150 mrad. All specimens were ion-milled by precision ion polishing system (PIPS Gatan) and plasma cleaned by plasma cleaner (SOLARUS Gatan) before the HAADF-STEM observation in order to minimise contamination on the specimens during the observation.

3. Results and discussion

All presented HAADF-STEM images have <001>Al orientation, namely along the precipitates needle growing direction, thus corresponding to the cross section of the precipitates. Projected positions of all atomic columns could be resolved in the structure owing to the high resolution of aberration-corrected HAADF-STEM. Based on Z contrast, Cu and Zn atomic columns could be identified even in the unprocessed HAADF-STEM images. Fast Fourier transform (FFT) filtering was applied to reduce noise using a circular band pass mask that removed all period shorter than 0.15 nm. After filtering the images, low contrast Si atomic columns could also be identified, as distinguished from those Al and Mg columns. Local symmetries with known structures have also been employed to identify the different atomic columns. The atomic overlays presented in figure 1 were readily determined by the method described. Figure 1 shows one of the Cu-containing precipitates consisting of perfect β" and a disordered part in the same precipitate needle (β"/disordered precipitate in [13]). In this image, the periodicity of the perfect β" part is easily identified (left side of the precipitate) while there is no clear periodicity in the disordered part. Bright spots, representing Cu atomic columns, are distinguished only in the disordered part and interface, but not in perfect β" part. However, the disordered part consists of ordered Si-network [13]. Although Cu atomic columns at the interface appear in a less systematic manner (as some take Al fcc positions), in the bulk structure they are consistently were identified with two specific configurations on the Si-network; these are *replacing the Si-network columns* and *in-between the Si-network columns*, see figure 1 (b) and (c). Note that the first case resembled the local configuration of the Cu atomic columns in β'_{Ag} phase [14] and the second case in the Q' phase [15]. In the first case Cu atomic columns are located precisely in Ag positions. Apart from this the local surrounding column symmetry is similar. It is worth mentioning that both atomic configurations have a 3-fold rotational symmetry around the Cu atomic column in this projection. In the Zn-containing precipitates, shown in figure 2, Zn atomic columns appear evenly in the bulk structure and interface. It is interesting to note that the two specific configurations as observed in Cu-containing precipitates, were also observed in a Zn-containing precipitate, see figure 2. In addition, the Zn atomic columns reside also in a disordered part. This indicates that Zn may also have an effect on nucleation and precipitation in the Al-Mg-Si alloys in a similar manner as Cu. Further analysis and investigations will be published elsewhere.

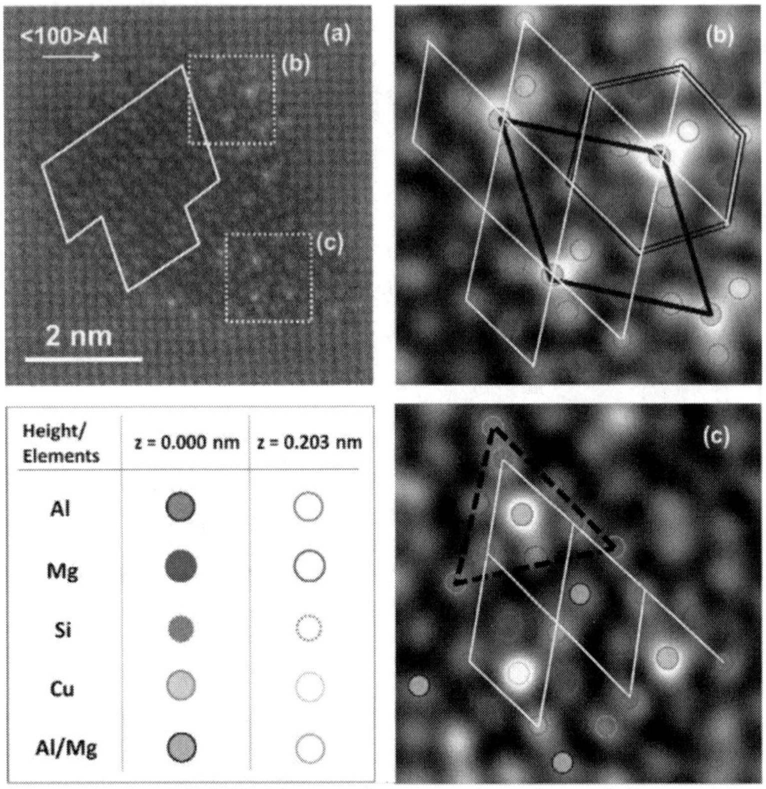

Figure 1. HAADF-STEM image of cross-section of one of Cu-containing precipitates (β"/disordered precipitate) taken along <001>Al. (a) unprocessed image with ordered β" part (white solid line) and two disordered part (b, c) (dotted line squares). (b, c) magnified and inverse FFT filtered disordered parts with hexagonal Si-network indicated (thin white lines). Legend is shown on lower left side. Different Cu atomic columns are shown with Si-network columns; these are (b) *replacing the Si-network columns* marked with black double lines and (c) *in-between the Si-network columns* marked with black thick dashed line. In (b), one unit cell of β'_{Ag} phase having Cu in the Ag site, is indicated with black solid line.

Figure 2. Inverse FFT filtered HAADF-STEM image of a Zn-containing precipitate cross-section taken along <001>Al. Local atomic configurations as in (b) and (c) in figure 1, but with Zn instead of Cu are indicated (hexagon and triangle). An area with disordered Zn distribution is marked by black solid line circle.

4. Conclusions

The Cu and Zn atomic columns in Al-Mg-Si alloys with respective Cu and Zn additions were investigated by aberration corrected HAADF-STEM. Results revealed that although most precipitates lack an overall periodicity, Cu and Zn columns form similar and specific local symmetries connected with the Si-network. Cu and Zn atomic columns were observed also on Al fcc positions in the precipitate/matrix interface. The results suggest that Cu and Zn atoms can have a considerable influence on precipitation in Al-Mg-Si alloys.

Acknowledgement

The authors would like to thank Dr. Olaf Engler, Hydro Bonn Germany, for the composition measurement by inductively coupled plasma optical microscopy. This research is supported by Hydro Aluminum and the Research Council of Norway through the bilateral KMB project: 193619 "The Norwegian-Japanese Al-Mg-Si Alloy Precipitation Project".

References

[1] Edwards G A, Stiller K, Dunlop G L and Couper M J 1998 *Acta Mater.* **46** 3893

[2] Andersen S J, Zandbergen H W, Jansen J, Træholt C, Tundal U and Reiso O 1998 *Acta Mater.* **46** 3283

[3] Hasting H S, Frøseth A G, Andersen S J, Vissers R, Walmsley J C, Marioara C D, Danoix F, Lefebvre W and Holmestad R 2009 *J. Appl. Phys.* **106** 123527-1

[4] Marioara C D, Andersen S J, Zandbergen H W and Holmestad R 2005 *Metall. Mater. Trans. A* **36** 691

[5] Marioara C D, Nordmark H, Andersen S J, Zandbergen H W and Holmestad R 2006 *J. Mater. Sci.* **46** 471

[6] Vissers R, van Huis M A, Jansen J, Zandbergen H W, Marioara C D and Andersen S J 2007 *Acta Mater.* **55** 3815

[7] Andersen S J, Marioara C D, Vissers R, Frøseth A and Zandbergen H W 2007 *Mater. Sci. Eng. A* **444** 157

[8] Matsuda K, Sakaguchi Y, Miyata Y, Uetani Y, Sato T, Kamio A and Ikeno S 2000 *J. Mater.Sci.* **35** 179

[9] Marioara C D, Andersen S J, Stene T N, Hasting H, Walmsley J, Van Helvoort A T J and Holmestad R 2007 *Philos. Mag.* **87** 3385

[10] Chakrabarti D J and Laughlin D E 2004 *Prog. Mater. Sci.* **49** 389

[11] Cayron C, Sagalowicz L, Beffort O and Buffat P A 1999 *Philos. Mag. A* **79** 2833

[12] Matsuda K, Uetani Y, Sato T and Ikeno S 2001 *Metall. Mater. Trans. A* **32** 1293

[13] Saito T, Muraishi S, Marioara C D, Andersen S J, Røyset J and Holmestad R, 2013 *Metall. Mater. Trans. A* **44** 4124

[14] Marioara C D, Nakamura J, Matsuda K, Andersen S J, Holmestad R, Sato T, Kawabata T and Ikeno S 2012 *Philos. Mag.* **92** 1149

[15] Torsæter M, Vissers R, Marioara C D, Andersen S J and Holmestad R 2008 *Proceedings of ICAA11* vol.2 p 1338

Nanocharacterisation of precipitates in austenite high manganese steels with advanced techniques: HRSTEM and DualEELS mapping

J. Bobynko[1], A.J. Craven[1], D. McGrouther[1], I. MacLaren[1], and G. Paul[2]
1 Kelvin Nanocharacterisation Centre, SUPA School of Physics and Astronomy, University of Glasgow, Glasgow G12 8QQ, UK.

2 ThyssenKrupp Steel Europe AG, Research and Development, FuE-E Modelling and Simulation, Kaiser-Wilhelm-Straße 100, 47166 Duisburg, Germany

E-mail: Ian.MacLaren@glasgow.ac.uk

Abstract. To achieve optimal mechanical properties in high manganese steels, the precipitation of nanoprecipitates of vanadium and niobium carbides is under investigation. It is shown that under controlled heat treatments between 850°C and 950°C following hot deformation, few-nanometre precipitates of either carbide can be produced in test steels with suitable contents of vanadium or niobium. The structure and chemistry of these precipitates are examined in detail with a spatial resolution down to better than 1 nm using a newly commissioned scanning transmission electron microscope. In particular, it is shown that the nucleation of vanadium carbide precipitates often occurs at pre-existing titanium carbide precipitates which formed from titanium impurities in the bulk steel. This work will also highlight the links between the nanocharacterisation and changes in the bulk properties on annealing.

1. Introduction

The growing demand for increased fuel-efficiency in automotive applications is driving research on the development of higher strength ductile steels for body applications in order to reduce weight without compromising safety. One alloy system of interest for this application is a high-manganese (HiMn) steel microalloyed with either Nb or V to provide high densities of nanoscale carbide precipitates. Manganese stabilises the austenite phase and thereby increases ductility. The main characteristic of HiMn steels is their extraordinary combination of high tensile strength and ductility due to the work hardening effect [1-3], introducing additional dispersion hardening through the precipitation of nanosized carbides provides an effective way of further increasing strength. To effectively investigate the size, morphology and chemistry of such nanoscale precipitates, we are using scanning transmission electron microscopy, especially in combination with Electron Energy Loss Spectroscopy (EELS) and using the new DualEELS technique [4] to allow absolute quantification of the chemistry of nanosized precipitates. Clear differences are noted between precipitation behaviour in V and Nb microalloyed steels. In Nb steels, the precipitates are predominantly NbC_x and are noticeably faceted, and rather fine (typically 5-10 nm). In V steels, the precipitates are less strongly faceted, slightly larger (often > 10 nm) and usually contain significant Ti impurities, possibly suggesting that they nucleate on pre-existing TiC precipitates.

2. Experimental

2.1. Materials and preparation
The base chemical composition of the alloy investigated is 20 wt. % Mn, 0.6 wt. % C, 1.5 wt. % Al, and balance Fe. We study two compositions microalloyed with either 0.1 wt. % Nb or 0.3 wt. % V. Transmission Electron Microscope (TEM) and Scanning Transmission Electron Microscope (STEM) specimen were prepared by Focused Ion Beam technique (FIB) using a FEI Nova 2000 DualBeam FIB (FEI, Eindhoven, NL).

2.2. Microscopy investigation
Precipitates were studied by STEM using a JEOL ARM200F equipped with a cold Field Emission Gun (FEG) source and operated at 200kV. EELS-spectrum imaging was performed using Gatan GIF Quantum 965 spectrometer using the DualEELS acquisition mode. Spectra were recorded using a 29mrad convergence semiangle and collection semiangles of either 36mrad or 72mrad. Atomic resolution HAADF imaging was also used to investigate the atomic structure and the faceting of the precipitates.

2.3. Post-processing procedure
The DualEELS technique makes it easy to use Fourier Deconvolution on an entire spectrum image, which makes it possible to remove the effects of plural scattering from core-loss edges, thereby enhancing the signal to background ratio and the reliability of quantification. We used a multistep procedure to process the data and to analyse the precipitates apart from the matrix. This starts by splicing the low loss and core loss spectra, followed by the reduction of noise using the weighted Principal Component Analysis (PCA) technique. It is then possible to deconvolute to remove plural scattering from each spectrum in the spectrum image. Finally, following [5], we calculated a matrix only spectrum image and subtracted this from the deconvoluted spectrum image to leave just a spectrum image of the precipitate. This final precipitate spectrum image could then be analysed using existing quantification routines to reveal precipitate chemistry and thickness without interference from the matrix.

3. Results and discussion
Figure 1a-d shows an example of the elemental maps of a typical carbide precipitate in a vanadium steel. One notable feature is the significant concentration of Ti throughout the particle, which is not an intentional part of the alloy. We had possibly expected to see the core shell structure, as this is commonly observed and published in the literature [6], which would arise if V nucleates on pre-existing TiN or TiC resulting from the interaction of Ti impurities from the foundry with C or N in the alloy. However in our case a range of thickness profiles for the different elements and the total inelastic scattering thickness are compared using the line shown on the inelastic thickness map in Figure 1 and these tell a different story. The Ti is evenly distributed through the particle, and the shape of its profile (blue line) has the same shape as the V profile (green). The thickness measured using the calibration of the mean free path with the Iakoubovskii formula [7] fits with expectations and shows a thickness close to the lateral dimensions. By comparing the thickness profiles for the different elements to this, it is clear that there are some systematic errors arising from incorrect cross-section estimation for the core-loss edges. The V profile follows the total thickness closely, when in reality it should be the V+ Ti thickness that matches the total thickness. This would indicate a slight underestimate of the V-$L_{2,3}$ and Ti-$L_{2,3}$ cross sections. However, considering that the quantification of both V and Ti is based on $L_{2,3}$ edges, we can probably trust ratios of Ti/V, and for different precipitates this has been observed to vary from 1:2.5 up to 1:4. These differences would probably lead to slight variations in lattice parameter from 4.16 Å up to 4.2 Å. The C profile follows the form of the thickness profile but is much higher, suggesting a significant underestimate of the C-K cross section

(especially as a metal carbide, $MC_{1-\delta}$, of this sort is likely to be substoichiometric with a noticeable C deficiency). It may be noted that there were some problems with background subtraction around the carbon edge which lead to negative counts outside the precipitate.

Figure 2 shows the results of a similar procedure when applied to a carbide precipitate in a Nb steel. As before, the total inelastic scattering thickness is consistent with the lateral dimensions of the particle and is probably reliable. In the precipitate, Nb and Ti is found and the form of their thickness traces matches well to the total thickness profile. As before, Ti substation occurs throughout the particle. Nevertheless, the absolute value of the thickness calculated for the Nb is very low, suggesting that there are systematic errors in the procedure. One likely source of error is an overestimate of the cross section for the Nb $M_{4,5}$ edge in Digital Micrograph. We also noted, however, that background subtraction for this edge is difficult due to the shape of the background before the edge not being well-behaved, possibly due to interference from the extended fine structure of lower energy edges. This can result in negative counts in some places, and these are particularly noticeable in the upper left of these maps. As before, the carbon thickness is significantly overestimated, suggesting again that the cross section for the C-K edge is significantly underestimated – this result has been consistently found in every precipitate we have mapped. Also, as before, carbon maps have significantly negative counts outside the precipitate due to background subtraction issues.

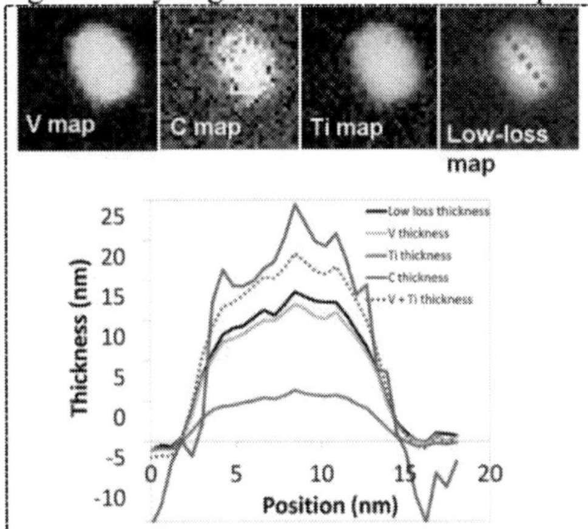

Figure 1. Maps with calculated thickness: a) V-$L_{2,3}$, b) C-K, c)Ti-$L_{2,3}$, d) calibrated inelastic scattering thickness, e) Profile of each thickness along the line drawn on d).

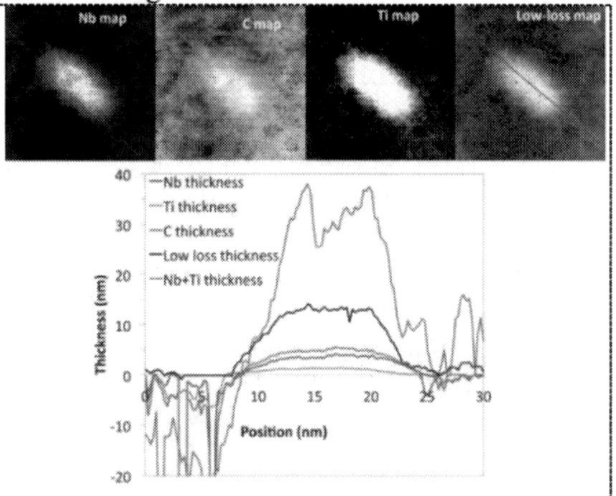

Figure 2. Maps with calculated thickness: a) Nb-$M_{4,5}$, b) C-K, c)Ti-$L_{2,3}$, d) calibrated inelastic scattering thickness, e) Profile of each thickness along the line drawn on d).

Figure 3. HRSTEM of Nb-particle, [011] direction.

An HRSTEM image of a NbC precipitate is presented in figure 3. HRSTEM can be used for evaluation of lattice parameter in the precipitate since there is a very clear lattice mismatch between precipitate and the matrix. Using the matrix lattice parameter of 3,668Å we can calculate the precipitate to have a lattice parameter of 4,41Å, which fits well with expectations for NbC [8]. The Nb precipitates have a tendency for strong faceting and the interface between matrix and precipitate is sharp on the following planes – (111), (200) and (220). The largest facets are on {111} planes, which are the close-packed planes in the face-centred cubic austenite, suggesting that strain can be minimised by forming semi-coherent interfaces on close-packed planes.

In the case of Vanadium steels, the imaging of precipitates is more difficult due to the lower Z of Vanadium, resulting in $(V,Ti)C_x$ precipitates appearing as darker than the austenite matrix. This makes HRSTEM extremely difficult in all but the thinnest areas, and consequently no detailed results can yet be presented on the interfaces of precipitates to the matrix in vanadium steels.

4. Conclusion

DualEELS spectroscopy allows absolute quantification of nanoscale precipitates in steels in terms of equivalent thicknesses of different elements, as well as total thickness from the total inelastic scattering. This also reveals that the cross sections used for quantifying elements in EELS are in need of correction. Future work will concentrate on accurate determination of partial cross sections for the elements found in our steels under the conditions used in this work.

The work reveals that nanoscale precipitates in V and Nb steels contain some fraction of Ti impurities in both cases, and it is likely that the Ti plays some role in the nucleation of the precipitation. In contrast to some studies, we found little evidence for core-shell structures involving Ti cores in precipitates. Atomic resolution STEM imaging also allowed us to investigate the faceting of NbC precipitates in some detail, showing preferential faceting to {111} planes, and also allowed us to measure the lattice parameter of these precipitates as 4,41 Å.

Acknowledgements

We are grateful to the European Commission for the funding of this work as part of a Research Fund for Coal and Steel Project (Precipitation in High Manganese Steels, RFSR-CT-2010-00018). This work was only possible because of the generous provision of the MagTEM facility by SUPA and the University of Glasgow.

References

[1] Bouaziz O *et al* 2011 High manganese austenitic twinning induced plasticity steels: A review of the microstructure properties relationships. Curr Opin Solid St M **15(4)** 141-168.

[2] Hamada, AS and Karjalainen LP 2011 Hot ductility behaviour of high-Mn TWIP steels. Mater Sci Eng A, **528(3)** 1819-1827.

[3] Sil'man GI 2006 Alloys of the Fe–C–Mn system part 4 special features of structure formation in manganese and high-manganese steels Met Sci heat treat+ **48(1-2)**.

[4] Scott J et al 2008 Near-simultaneous dual energy range EELS spectrum imaging. Ultramicroscopy **108 (12)** 1586-1594

[5] Craven AJ *et al* 2008 Spectrum imaging and three-dimensional atom probe studies of fine particles in a vanadium micro-alloyed steel Mat Sci Tech Ser **24(6)** 641-650.

[6] Show BK *et al* 2010 Effect of vanadium and titanium modification on the microstructure and mechanical properties of a microalloyed HSLA steel Mat Sci Eng A **527(6)** 1595-1604

[7] Iakoubovskii K *et al*.2008 Thickness measurements with electron energy loss spectroscopy Micros Res Techniq **71**(8) 626-31

[8] Tirumalasetty G K *et al* 2011 Characterisation of NbC and (Nb,Ti)N nanoprecipitates in TRIP assisted multiphase steels Acta Mater **59** 19 7406-7415

Journal of Physics: Conference Series **522** (2014) 012032 doi:10.1088/1742-6596/522/1/012032

In-situ Fe^+ Ion Irradiation of an Oxide Dispersion Strengthened Steel

V. de Castro[1], M. Briceno[2, 4], M. L. Jenkins[2], M. Kirk[3], S. Lozano-Perez[2], S. G. Roberts[2]
[1] Departamento de Física, Universidad Carlos III de Madrid, 28911 Leganés, Spain
[2] Department of Materials, University of Oxford, Parks Road, OX1 3PH, Oxford, UK
[3] Materials Science Division, Argonne National Laboratory, Argonne, IL 60439, USA
[4] Johnson Matthey Technology Centre, Blount's Court Rd, Sonning Common, RG4 9NH, UK

E-mail: vanessa.decastro@uc3m.es

Abstract. Oxide Dispersion Strengthened (ODS) reduced activation ferritic steels are promising candidate materials for structural components of both nuclear fission and fusion reactors. However, when irradiated with energetic particles, they may suffer changes on their microstructures that degrade their mechanical performance. In-situ transmission electron microscopy studies on ion-irradiated ODS steels can give remarkable insights into fundamental aspects of radiation damage allowing dynamic observations of defect formation, mobilities, and interactions during irradiation. In this investigation, a commercially available PM2000 ODS steel was in-situ irradiated with 150KeV Fe^+ at room temperature and 700ºC. These experiments showed that the oxide nanoparticles in these steels remain stable up to the higher irradiation dose (~ 1.5 dpa), and that these particles seem to be effective sinks for irradiation induced defects.

1. Introduction

Oxide Dispersion Strengthened (ODS) ferritic steels are one of the best considered materials for structural components in advanced fission power plants and future fusion reactors [1]. In these steels, nanometer sized Y-rich oxide particles homogeneously dispersed in the matrix strengthen the alloy providing better mechanical and creep resistances which would allow higher operating temperatures to be used, resulting in enhanced efficiencies [2]. ODS ferritic and ferritic-martensitic steels also show a high resistance to void swelling and better thermal properties than austenitic steels [3]. It is well known that irradiation induces undesirable microstructural changes that may lead to severe degradation of mechanical properties. It is expected that ODS ferritic steels will present a lower damage accumulation rate than the base steels, due to the oxide particles acting as trapping sites for irradiation induced defects and transmutation gasses [4]. However, there has not been much systematic work to determine the effect of irradiation on ODS steels.
This investigation reports the microstructural characteristics of a commercially available PM2000 ODS ferritic steel in-situ irradiated with Fe^+ ions at room temperature (RT) and 700ºC up to calculated doses of ~ 1.5 dpa. These Fe^+ ions simulate the damage produced by energetic neutrons.

2. Experimental methods

The PM2000 ODS steel used in this study was produced by Plansee Industries in form of bars. The nominal chemical composition given in the Plansee datasheets is Fe-19Cr-5.5Al-0.5Ti-0.5Y_2O_3 (wt%). Before irradiation, disks of diameter 3 mm were cut perpendicular to the bar axis, thinned to a thickness of ~ 60 μm to minimise magnetic effects within the transmission electron microscope (TEM) and finally JET electro-polished to get electron transparency. Electro-polishing was carried out in a TENUPOL 5 polishing unit using 5 % $HClO_4$ + 95 %CH_3OH as electrolyte, voltages of ~ 20 V and temperatures below -30°C. The disks were in-situ irradiated with 150KeV Fe^+ ions at RT and 700°C at the intermediate voltage electron microscope (IVEM) facility available at Argonne National Laboratory (ANL) [5]. One of the advantages of using this facility compared to other in-situ irradiation facilities is that it allows the angle between the sample axis and the ion beam to be as low as 15°, also permitting continuous observation and data recording during irradiation. The ion energy (150KeV) was chosen to mimic high-energy fusion cascades and for comparison with previous work conducted by Jenkins and co-workers [6-8]. The maximum irradiation dose was $4x10^4$ ions/cm^2. This fluence corresponds to a damage of ~ 1.5 dpa in the region of the maximum damage (~ 30 nm from the irradiated surface) as calculated with the SRIM program [9].

During the experiment, TEM characterization was carried out on a Hitachi 9000 NAR microscope operated at 300 kV. TEM studies before and after irradiation were carried out by bright-field (BF), dark-field weak beam (DF-WB) and energy dispersive spectroscopy (EDS) with a CM20 TEM operated at 200 kV. Energy filtered TEM (EFTEM) maps were acquired using a JEOL 3000F operated at 297 kV.

3. Results and discussion

Figures 1(a) to (d) show BF-TEM images of an area of the PM2000 steel irradiated to different damage levels at RT. This area was recorded during irradiation. Irradiation induced defects start to be visible at doses of ~ 0.4 dpa, see figure 1 (b). They are revealed as small black dots which appear to be associated to ODS nanoparticles. No damage is observed in the matrix below ~ 0.8 dpa (figure 1(c)). At this irradiation dose, the few defects present in the matrix have sizes < 5 nm. The defect density increases as the dose does, up to the maximum irradiation dose of ~ 1.5 dpa (figure 1(d)).

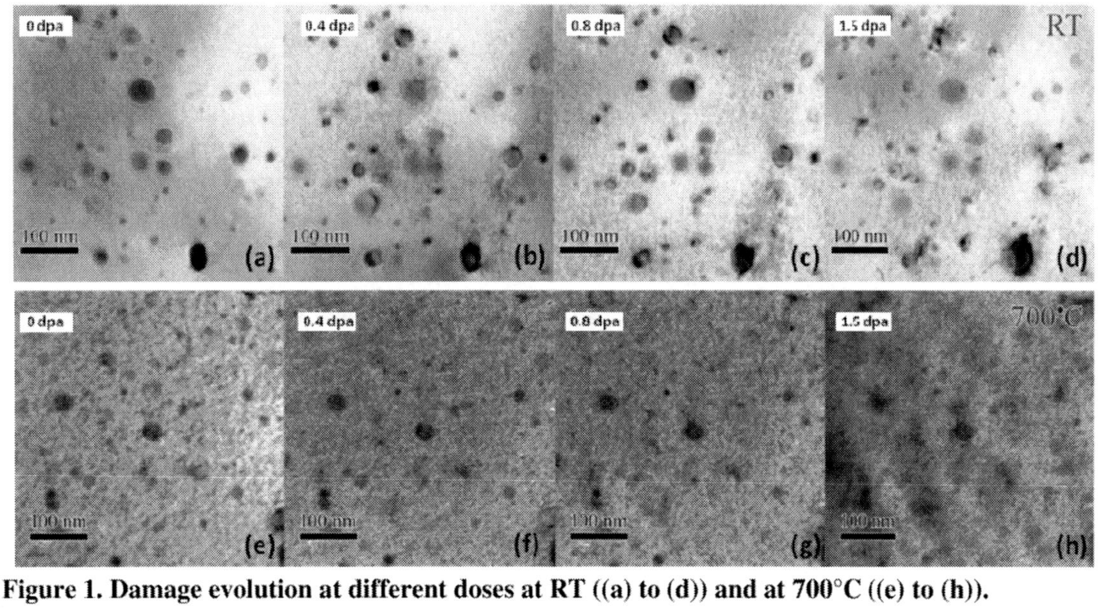

Figure 1. Damage evolution at different doses at RT ((a) to (d)) and at 700°C ((e) to (h)).

Figures 1 (e) to (h) show the evolution of the PM2000 steel under irradiation at 700ºC. Irradiation induced defects could not be studied as surface degradation due to high temperature oxidation prevented their visualization at this temperature.

BF and DF-WB micrographs of two other areas irradiated to the maximum dose are presented in Figure 2. The area shown in figures 2 (a) and (b) contains a higher particle density while the particle density is lower in the area depicted in figures (c) and (d). In the high particle density region there are few defects present in the matrix; most defects are observed at particles. The number of defects present in the matrix is higher in the area containing a lower particle density. These results reveal that ODS particles act as sinks for irradiation induced defects, absorbing the vacancies or interstitials formed during the irradiation and subsequently expelling them as dislocations giving rise to the debris detected especially in the area containing a high particle density. This sequence of events was also observed during irradiation.

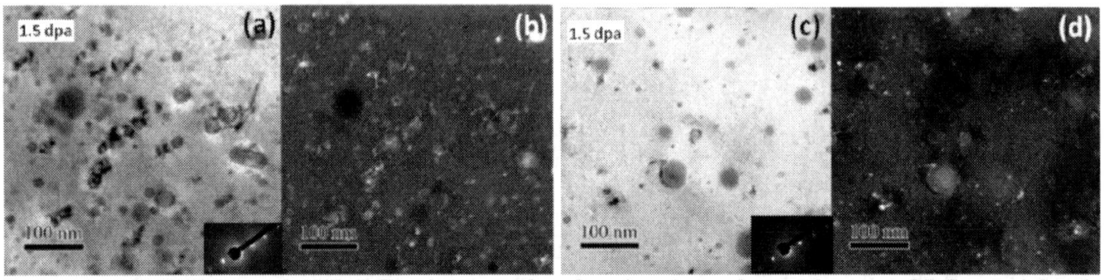

Figure 2. BF-TEM and WB-DF micrographs showing loop formation on ODS particles in PM2000 irradiated at RT. The grains are oriented along g (110).

A previous investigation of as received PM2000 steel sheets produced by the same distributor revealed that the Y_2O_3 dispersoids added to the starting powder are not retained in the final product. They react with Al during high temperature pressing and form Y–Al–O ternary oxides with sizes < 40 nm [10]. In the present work, the ODS particle dispersion was analysed before and after irradiation using EFTEM to detect possible irradiation induced changes. The particle distribution before and after the RT irradiation is depicted in the elastic images, Y $N_{2,3}$ and O K EFTEM elemental maps presented in figure 3.

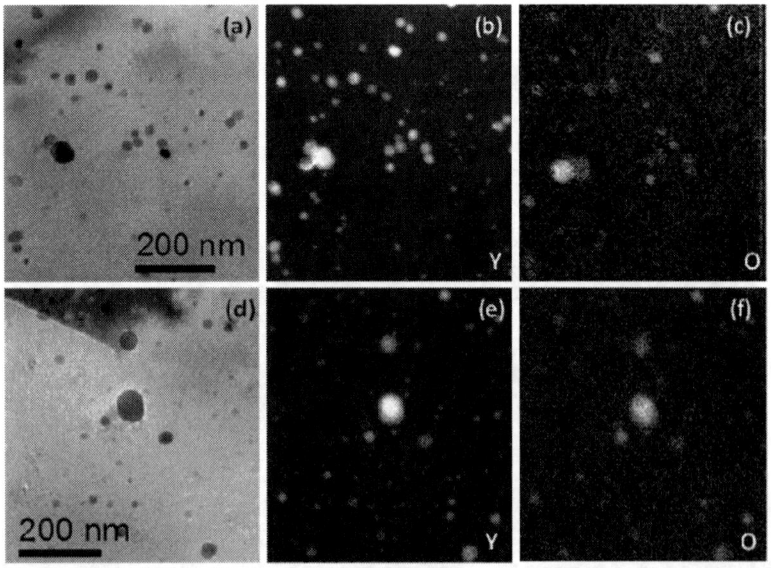

Figure 3. Elastic images, and Y $N_{2,3}$ and O K EFTEM maps of PM2000 showing the distribution of ODS particles before ((a) to (c)) and RT after irradiation ((d) to (f)) at 1.5 dpa.

The Al L elemental map was too noisy and is not presented here, but point EDS analyses reveal that the particles contain Al. EFTEM analyses did not reveal any differences in the particle morphology, sizes or chemical composition due to irradiation. Particles are nearly spherical and have (Y, Al, O) – rich compositions. The size distributions are similar before and after irradiation at both RT and 700°C. Average particle sizes are 20 ± 10 nm and 22 ± 11 nm for the PM2000 steel before and after RT irradiation and 21 ± 10 nm after irradiation at 700°C.

4. Conclusions

This work analyses the microstructure of a PM2000 steel in-situ irradiated at RT and 700°C up to doses of ~ 1.5 dpa. The experiments showed that the ODS nanoparticles present in these steels are effective sinks for irradiation induced defects. The number of defects present in the matrix was lower in regions containing a higher particle density. Ion irradiation up to the higher irradiation dose had no measureable effect on the sizes of the ODS particles, caused no measureable change in particle composition, and moreover the particles kept their spherical shapes. ODS particles are therefore very stable in this material under these experimental conditions.

5. References

[1] Boutard J L, Alamo A, Lindau R, Rieth M 2008 *CR Physique* **9** 287.
[2] Allen T, Burlet H, Nanstad R K, Samaras M, Ukai S 2009 *MRS Bulletin* **34** 20.
[3] Klueh R L, Harries D R 2001 *High Chromium Ferritic and Martensitic Steels for Nuclear Applications* Ed ASTM (West Conshohocken, PA, USA) chapter 2 pp 5–27.
[4] Odette G R, Alinger M J, Wirth B D 2009 *Annu. Rev. Mater. Res.* **38** 471.
[5] Kirk M A, *et al* 2009 Microsc. Res. Tech. 72-3 182.
[6] Jenkins M L, Yao Z, Hernandez-Mayoral M, Kirk M A 2009 *J. Nucl. Mater.* **389** 197.
[7] Hernandez-Mayoral M, Yao Z, Jenkins M L, Kirk M A 2008 *Phil. Mag.* **88- 21** 2851.
[8] Hernandez-Mayoral M, Yao Z, Jenkins M L, Kirk M A 2008 *Phil. Mag.* **88- 21** 2881.
[9] Ziegler J F, Biersack J P, *Stopping and Range of Ions into Matter (SRIM) software.*
[10] Klimiankou M, Lindau R, Möslang A, Schröder J 2005 *Powder Metall.* **48-3** 277.

Acknowledgments

This research has been funded by EPSRC under the Program Grant "Materials for Fusion & Fission Power" EP/H018921, the Spanish Ministry of Science and Innovation (project ENE2010-17462), the EU FP7 under Grant Agreement 312483 - ESTEEM2 (Integrated Infrastructure Initiative-I3) and the Royal Society International Exchanges Scheme 2011/R1 (ref. IE110136). The IVEM-Tandem Facility at ANL is supported by the US DOE Office of Science and operated under contract no.DE-AC02-06CH11357. The authors thank Dr. Michael Klimenkov for providing the PM2000 steel used for this investigation and Dr. Peter Baldo for providing scientific and technological advice.

Microstructure of as-cast Mg-4.2Zn-0.8Y (at.%) alloys containing Gd

J. Wu[1], Y.L. Chiu, I.P. Jones

School of Metallurgy and Materials, University of Birmingham, Edgbaston, Birmingham, UK, B15 2TT

JXW256@bham.ac.uk

Abstract. Mg-Zn-Y is a high performance magnesium ternary alloy with high specific strength. A second rare earth element Gd with different content was added into Mg-4.2Zn-0.8Y (at.%) alloys to investigate the effect of Gd addition on the microstructure and mechanical properties. In the as cast alloys, the following phases have been identified: in the Gd-free alloy, an H phase (hcp, a=0.918nm, c=0.95nm), W phase (fcc, a=0.682nm) and icosahedral quasicrystalline phases (i-phase). With 0.5at.% Gd and 1 at.% Gd, W phase, long-period stacking ordered phase (LPSO phase) and i-phase were observed. The coexistence of i-phase and LPSO phase was observed after Gd addition. As the content of Gd increased, the grain size decreased, the volume fractions of second phases increased, and the hardness of the as-cast alloys increased.

1. Introduction

In recent years, Mg-Zn-Y alloys have been a focus of research interest due to the good mechanical properties due to the presence of an icosahedral quasicrystalline phase (i-phase) or long period stacking ordered phase (LPSO phase), which provides great resistance to dislocation movement [1-3]. I-phase possesses a quasiperiodic lattice structure and exhibits high hardness, good thermal reliability, and low interface energy [1]. The LPSO phase was reported to be the main strengthening phase in $Mg_{97}Zn_1Y_2$ (at. %) with the high yield strength of 610 MPa [2]. Many researchers have reported that the main phases in Mg-Zn-Y alloys are always closely connected to the atomic ratio Zn/Y and follow a formation trend with increasing Zn/Y: β ($Mg_{24}Y_5$) → LPSO ($Mg_{12}ZnY$) → W ($Mg_3Zn_3Y_2$) → I-phase (Mg_3Zn_6Y) [1, 4, 5]. However, the phase evolution was obtained when only the rare earth element Y is added, no effect of second rare earth element, e.g Gd, Dy, Tb, Ho or Er has been studied in Mg-Zn-Y alloys. Gd has high solubility in magnesium and is popularly investigated for possible development of magnesium alloys [6, 7]. The present paper reports the results of a detailed microstructure examination of three as-cast Mg-Zn-Y(-Gd) alloys, including precipitates, grain size and hardness as a function of Gd addition.

2. Experiments

Ingots with nominal composition Mg-4.2Zn-0.8Y-xGd (at.%, x=0, 0.5, 1) were melted using pure Mg (99.9%), pure Zn (99.99%), Mg-30Y (wt.%) and Mg-30Gd (wt.%) master alloys in an induction furnace, and then cast into a steel mould (67 × 100 × 34 mm). The as-cast specimens were examined using a PHILIPS X' Pert X-ray diffractometer (XRD) with the scanning angle ranging from 20° to 100° and a scanning speed of 2°/min. The microstructures were characterized by optical microscopy (OM, ZEISS Axioskop 2), scanning electron microscopy (PHILIPS XL-30), and transmission electron microscopy (JEOL 2100 with Oxford Inca EDS). The OM samples were etched using a solution of 4 mL nitric acid and 96 mL ethanol. The TEM foils were mechanically grinded to 150 μm first, and then twin-jet polished using a solution containing 40 mL perchloric acid and 960 mL ethanol at -10°C combined with ion-beam thinning. The Vickers hardness was measured using an automated Struers DuraScan-50 micro-hardness machine with 0.1 kg load, and each measurement was repeated 20 times.

3. Results and discussion

Figure 1 shows XRD patterns from the as-cast Mg-4.2Zn-0.8Y (at.%, alloy I), Mg-4.2Zn-0.8Y-0.5Gd (at.%, alloy II) and Mg-4.2Zn-0.8Y-1Gd (at.%, alloy III) alloys. The matrix in all three alloys is magnesium (hcp, a = 0.3209 nm, c = 0.5211 nm). Besides the strong peaks from the magnesium, a few weak peaks were observed in alloy I and indexed as H phase ($MgZn_3Y$, hcp, a = 0.918 nm, c = 0.95 nm) and W phase ($Mg_3Zn_3Y_2$, fcc, a = 0.683 nm). After adding 0.5 at.% Gd, the intensity of the peaks belonging to W phase increased in alloy II and H phase was not observed. When the fraction of Gd increased to 1 at.%, the diffraction pattern of alloy III is similar to that of alloy II, i.e. W phase is the main precipitate.

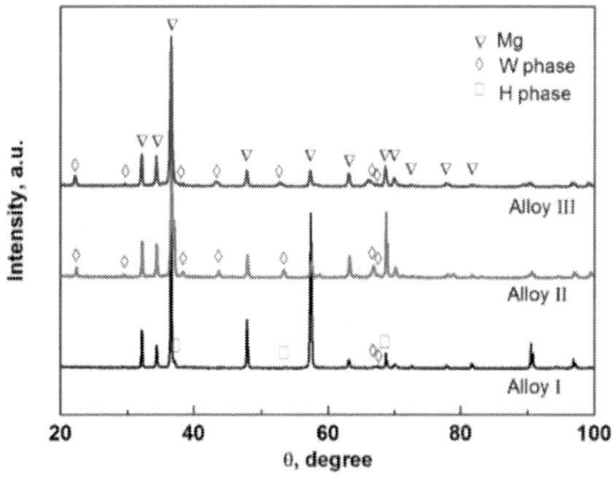

Figure 1 XRD spectra obtained from the as-cast alloys.

Figure 2 shows the microstructure of the as-cast alloys. The grain size of alloy I (Figure 2a) is about 400 μm. After adding 0.5 at.% Gd and 1 at.% Gd, the grain sizes of alloy II and alloy III decreased to 200 μm and 115 μm respectively. In alloy I, spot-like precipitates were dispersed inside the grains, while long second phase precipitates were scattered along the grain boundary. The area fraction of the precipitates based on micrographs obtained using OM is about 4 %. With Gd addition, the morphology and distribution of the precipitates changed. There are some grey phases with streaks inside the grain as shown in Figure 2b, which is confirmed by TEM as lamellar LPSO phase (see Figure 4a). The microstructure of alloy III in Figure 2c is similar to that of alloy II, but the area fractions of the second phases in alloy III increased. The area fractions of the dark phase in alloy II and alloy III are 12 % and 18 % respectively, while that of the grey phase in alloys II and III are 16 % and 19 % respectively. Both area fractions increased with the addition of Gd.

The significant changes in microstructure of the three alloys results in changes in microhardness. The Vickers hardness values of alloy I, alloy II and alloy III are 64±3.5, 76±5.5 and 85±4.8 respectively, higher than that of pure magnesium which is about 40 HV. The hardness increased as the result of Gd addition, due to the decrease in grain size and increasing volume fraction of precipitates.

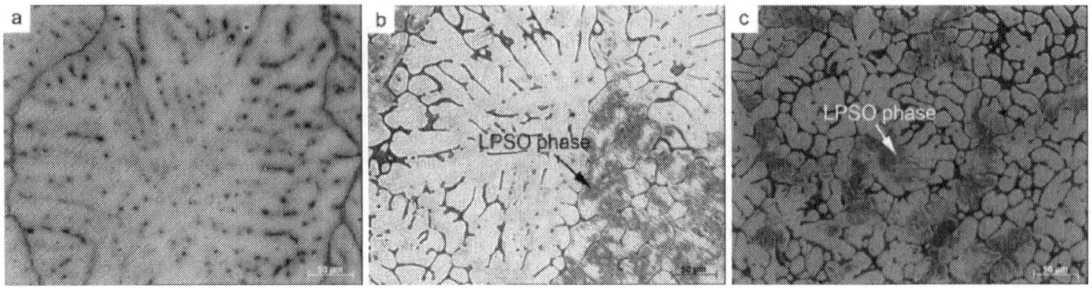

Figure 2 OM images of alloy I (a), alloy II (b), and alloy III (c).

I-phase was observed in all three alloys. Figure 3a shows a typical morphology of i-phase in alloy II. The diffraction patterns along the 5-fold, 2-fold and 3-fold rotational axes of i-phase are displayed in Figure 3b. The tilting angles between <5-fold, 2-fold>, <5-fold, 3-fold>, and the <2-fold, 3-fold> axis are 31.66°, 37.38° and 21.36° respectively, as measured using TEM. The chemical composition of the i-phase measured using EDS in the current work is $Mg_{26\pm0.9}Zn_{62.6\pm0.6}Y_{5.8\pm0.2}Gd_{5.3\pm0.2}$ at.%, when the content of Y and Gd in i-phase are about equal. This is similar to the $Mg_3Zn_6Y_1$ reported in Mg-Zn-Y alloys in the literature [1]. Tsai et al. [8] also reported that the i-phase in Zn-Mg-RE alloys close to $Zn_{50}Mg_{42}RE_8$ (RE≡Y, Gd, Dy, Ho or Er). There are no obvious peaks of i-phase observed in the XRD patterns presumably due to its low volume fraction.

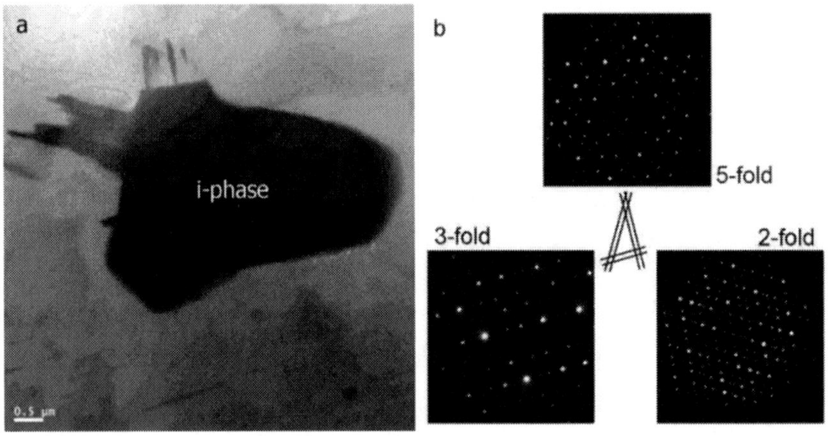

Figure 3 (a) Bright field image and (b) corresponding SAD patterns obtained from the i-phase in the as-cast alloy II.

Figure 4a shows TEM of the LPSO phases in alloy III. Two lamellar shaped LPSO phases b and c are identified as 18R and 14H type using diffraction patterns, as shown in Figure 4b and 4c respectively. The [010] pattern of 18R type in Figure 4b clearly shows there are 6 spots at regular intervals between the transmitted beam and the spot corresponding to 0002 of the pure magnesium. This agrees with the structure reported by Zhu et. [9]: an ordered base-centred monoclinic structure (a = 1.112 nm, b= 1.926 nm, c = 4.689 nm, β = 85.25°). Figure 4c shows there are 14 spots at equal intervals in the [010] pattern of 14H type between the transmitted beam and 0002 of the pure magnesium. It has been confirmed that this is a hexagonal structure with lattice parameters a = 0.321 nm and c = 3.694 nm [10]. Both 18R and 14H LPSO phases are observed in the same grain. The

$[010]_{18R}$ and $[010]_{14H}$ zone axis of the LPSO phase are parallel to $[1\bar{2}10]_{Mg}$, the $(001)_{18R}$ and $(001)_{14H}$ planes are parallel to $(0001)_{Mg}$.

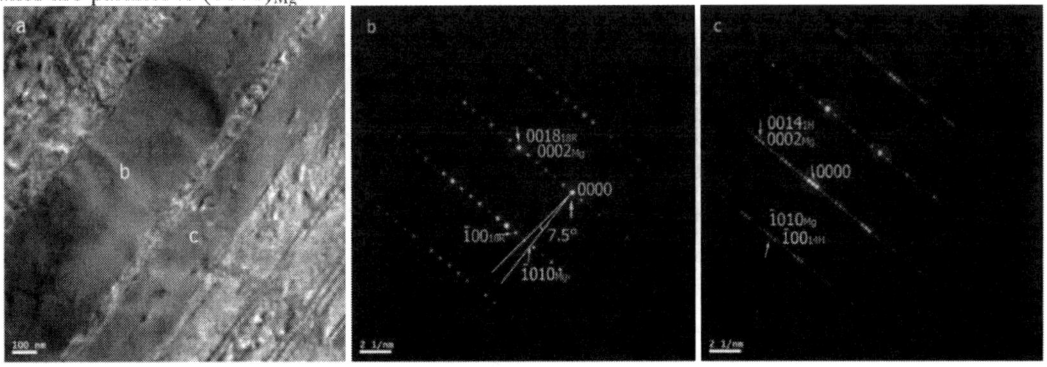

Fig. 4 Bright field image (a), SAED patterns of 18R (b) and 14H type (c) LPSO phases. The beam direction is $[1\bar{2}10]_{Mg}$.

W phase and H phase were main precipitates in alloy I, and only a little i-phase was observed, which is different from the common opinion that mainly i-phase formed in as-cast Mg-4.2Zn-0.8Y alloy [1]. According to Mg-Zn-Y ternary phase diagram [11], W phase can directly form during solidified from the melt and is in equilibrium with i-phase at certain temperature. So W phase formed in alloy I possibly because of the low cooling rate during the solidification. Normally, LPSO phase precipitates when the ratio of Zn/Y is about 0.5 [12]. In our work, the coexistence of i-phase and LPSO phase was observed in alloys II and III when the Zn/(Y+Gd) ratios are 3.2 and 2.3 respectively. So the Gd addition encourages the formation of LPSO phase in Mg-4.2Zn-0.8Y systems.

4. Conclusion

The effects of Gd content on the microstructure and mechanical properties of Mg-4.2Zn-0.8Y alloys have been investigated. The addition of Gd significantly decreases the grain size, increases the volume fraction of precipitates and increases the hardness of the as-cast alloys. LPSO formed after adding Gd, and a small amount of i-phase is observed at the same time.

Acknowledgments
The authors would like to thank Yu Lu for the sample preparation. The first author thanks for the Li Siguang Scholarship found by University of Birmingham and China Scholarship Council.

References

[1] Bae D H, Kim S H, Kim D H and Kim W T 2002 *Acta Mater.* **50** 2343-2356
[2] Kawamura Y, Hayashi K, Inoue A and Masumoto T 2001 *Mater. Trans.* **42** 1172-1176
[3] Mora E, Garces G, Onorbe E, Perez P and Adeva P 2009 *Scripta Mater.* **60** 776-779
[4] Liu Y, Yuan G, Zhang S, Zhang X, Lu C and Ding W 2008 *Mater. Trans.* **49** 941-944
[5] Lee J Y, Kim D H, Lim H K and Kim D H 2005 *Mater. Lett.* **59** 3801-3805
[6] Nie J F, Oh-ishi K, Gao X and Hono K 2008 *Acta Materialia* **56** 6061-6076
[7] Peng Q, Hou X, Wang L, Wu Y, Cao Z and Wang L 2009 *Mater. Des.* **30** 292-296
[8] Tsai A P, Niikura A, Inoue A, Masumoto T, Nishida Y, Tsuda K and Tanaka M 1994 *Philos. Mag. Lett.* **70** 169-175
[9] Matsuda M, Ii S, Kawamura Y, Ikuhara Y and Nishida M 2005 *Mater. Sci. Eng. A* **393** 269-274
[10] Itoi T, Seimiya T, Kawamura Y and Hirohashi M 2004 *Scripta Mater.* **51** 107-111
[11] Tsai A-P, Murakami Y and Niikura A 2000 *Philos. Mag. A* **80** 1043-1054
[12] Zhu Y M, Morton A J and Nie J F 2010 *Acta Mater.* **58** 2936-2947

Microstructural characterisation and microanalysis of creep resistant steels

M Wang[1], Y Chiu[1], I Jones[1], N Rowlands[2], J Holland[2], Z Zhang[3] and D Flahaut[4]
[1]School of Metallurgy and Materials, University of Birmingham, Birmingham, B15 2TT, UK
[2]Oxford Instruments NanoAnalysis, Halifax Road, High Wycombe, HP12 3SE, UK
[3]Doncasters Technical Centre, George Baylis Road, Worcestershire, WR9 9RB, UK
[4]Doncasters Paralloy Ltd, Paralloy House, Nuffield Road, Cleveland, TS23 4DA, UK
E-mail: MXW158@bham.ac.uk

Abstract: Steels for high temperature applications require good creep resistance which is controlled by the chemistry and microstructure of the materials. This paper focuses on the microstructural characterisation of a creep resistant steel using electron microscopy. The existence of various primary carbides, e.g. NbC, M_7C_3 and $M_{23}C_6$ was confirmed by electron diffraction. The primary chromium carbides transformed from M_7C_3 to $M_{23}C_6$ during creep while the niobium carbides were nearly unaltered. In addition, secondary precipitates ($M_{23}C_6$) were observed within the matrix after creep. The size and distribution of the secondary carbides were analysed by a 80 mm^2 windowless X-Max^N SDD at 3 kV on an SEM. Scanning transmission electron microscopy (STEM) observations showed the appearance of fine NbC, G phase ($Ni_{16}Nb_6Si_7$) and (Nb, Ti)(C, N) particles.

1. Introduction

As the service conditions in steam reforming involve a high temperature and a high pressure, the materials for this application require superior creep resistance [1]. One of the basic ways in which creep resistant steels can be strengthened is by precipitation. The precipitates, such as M_7C_3, $M_{23}C_6$ and MX, play an important role in the achievement of good creep properties. Traditional Si(Li) detectors, however, have a relatively poor performance in the identifications of carbides or nitrides due to limited spatial resolution. Silicon drift detectors (SDD), which were invented by Gatti and Rehak in 1984 [2], have become very powerful instrumentation tools for industrial and scientific application in X-ray spectroscopy due to their good energy resolution, high count rate capability and convenient experimental conditions [3]. Furthermore, a new generation of windowless SDD enhances the collection efficiency for low energy X-rays to improve any analyses which need to be based on low energy lines, such as Nb M lines. The present work focuses on microstructure characterisation and microanalysis of creep resistant steels by using windowless SDD on both SEM and TEM.

2. Experimental Methods

A creep resistant steel --- *Alloy A* was supplied by Doncasters Group Ltd. The alloy was crept at 1000°C and 40 MPa for 12 hours. Back-scattered electron (BSE) imaging and energy dispersive X-ray spectrometry (EDS) were used to investigate the element distribution. The size and distribution of the secondary chromium carbides were analysed with an 80 mm^2 windowless X-Max^N SDD at 3 kV on an FEG-SEM. Scanning transmission electron microscopy (STEM) was carried out on a Tecnai F20 TEM operating at 200 kV with an 80 mm^2 windowless X-Max^N TLE SDD.

3. Results and discussion

3.1 As-cast microstructure

Figure 1 (a) shows the microstructure of as-cast *Alloy A*. The grains are surrounded by Nb-rich carbides and Cr-rich carbides which form a fragmented network. Nb is confined to the primary carbide network [4], whereas Si and Fe remain in solid solution in the Ni-rich matrix. The EDS analyses show that the concentration of carbon in the chromium carbide is around 30 at.%, which indicates that it should be the M_7C_3 type. As shown in Table 1, Ni and Fe are found to substitute partially for Cr [5]. No fine precipitation has been observed in the matrix.

Table 1 EDS analyses obtained from the points shown in Figure 1

Atomic %	C	Si	Cr	Fe	Ni	Nb
1—Cr-carbides (as-cast)	31.85±0.72	---	59.63±1.13	7.30±0.68	1.21±0.22	---
2—Nb-carbides	41.73±0.81	---	9.56±1.33	3.46±0.52	2.67±0.14	42.36±0.33
3—Matrix	6.99±0.67	2.12±0.54	22.41±1.04	35.77±0.43	31.94±0.20	---
4—Cr-carbides (crept)	21.67±1.21	---	66.26±0.84	8.35±0.39	3.02±0.47	---

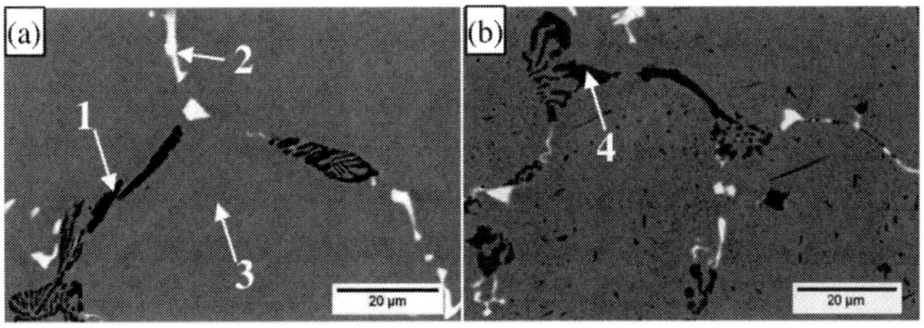

Figure 1 BSE micrograph obtained from (a) as-cast *Alloy A* and (b) crept *Alloy A*

3.2 Microstructure evolution

The microstructure of the crept *Alloy A* is illustrated in Figure 1 (b). The content of carbon in primary chromium carbides decreased from 30% to 21% (c.f. Table 1), indicating transformation from M_7C_3 to $M_{23}C_6$. The NbC morphologies are nearly unaltered during the creep, except for slight coarsening. Secondary precipitation can be observed within the matrix resulting from the supersaturated solid solution of carbon. Figures 2 (a-d) shows the results, using an SEM operating at 3 kV, of background and overlap corrected X-ray mapping (AZtec TruMap). The X-ray lines used in the SEM were low energy Cr L and Nb M, while the lines used in the TEM were Cr K and Nb K. Although the spatial resolution is not as good as that using TEM (c.f. Figures 2 (e-h)), they are in agreement that secondary chromium carbides are 100~250 nm. The secondary chromium carbides of such a size can act as barriers to dislocation movement contributing to good creep resistance.

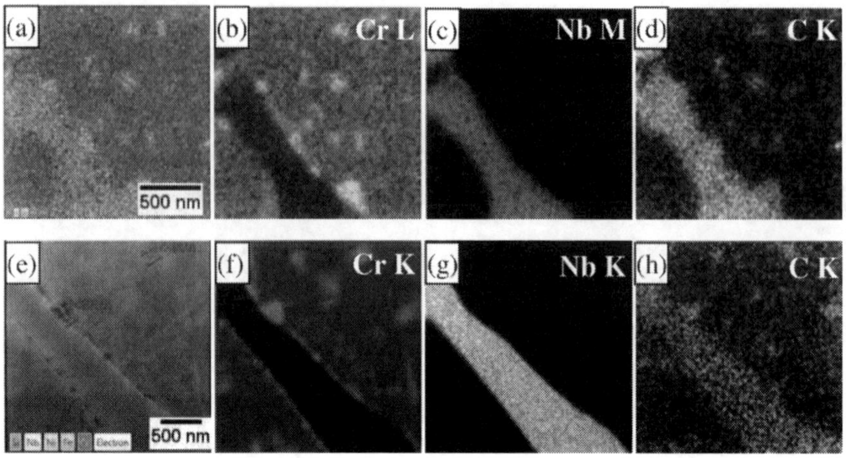

Figure 2 Overlap corrected TruMaps collected on (a-d) SEM at 3kV with an 80 mm^2 windowless X-Max^N SDD (e-h) TEM at 200 kV with an 80 mm^2 windowless X-Max^N TLE SDD

3.3 TEM observations

As shown in Figure 3, the STEM analysis was carried on a selected area of crept *Alloy A* containing a primary chromium carbide. Nb carbides were mainly observed at the interface between the primary chromium carbide and the matrix. A Nb-Ni-Si particle was found associated with the primary chromium carbide; it has a fcc structure with a lattice parameter of 1.13 nm identifying it as G-phase ($Ni_{16}Nb_6Si_7$). Although G phase does not appear to be intrinsically detrimental to the mechanical properties, the transformation from NbC (a=0.44 nm) into G phase (a=1.12 nm) induces a volume expansion and leads to cracking along the grain boundary.

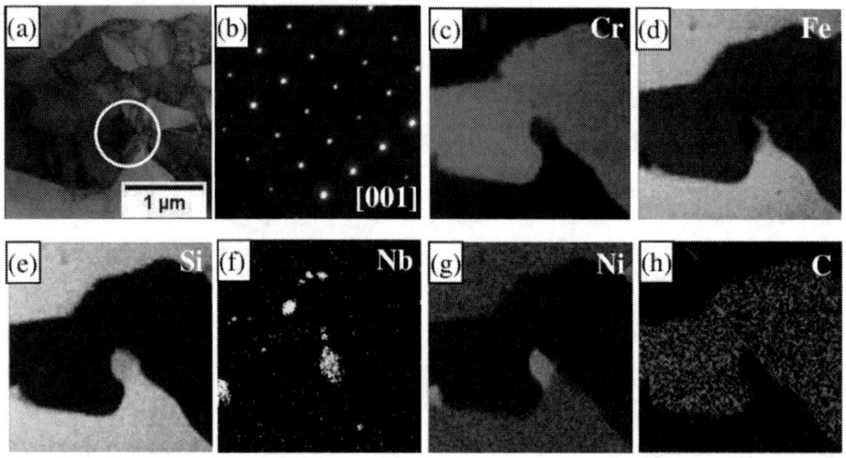

Figure 3 Cr carbide (a) bright field image; (b) selected area diffraction and (c-h) overlap corrected TruMaps

The EDS spectrum shows that the fine particle highlighted in Figure 4 (a) is rich in chromium, niobium, carbon and nitrogen. However, the contents of Nb, Cr and N are different from that of a typical nitride such as Z-phase (NbCrN) in austenitic stainless steel [6]. The diffraction patterns obtained from the particle were indexed as fcc with a lattice parameter of 0.429 nm close to NbC with its lattice parameter of 0.44 nm. EDS mapping shown in Figure 5 leads to a similar result --- observation of a (Nb, Ti)(C, N) particle combined with a chromium carbide.

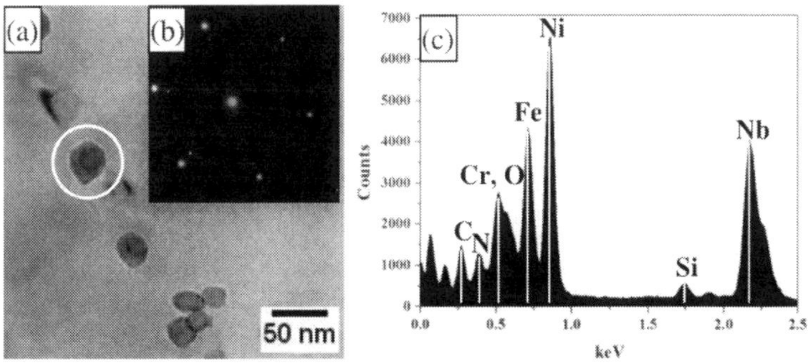

Figure 4 Selected particles (a) bright field image; (b) selected area diffraction and (c) EDS spectrum

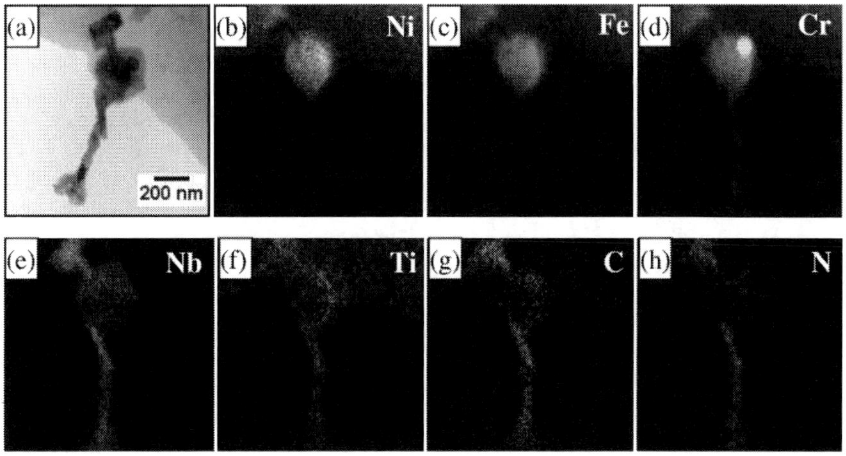

Figure 5 (Nb, Ti)(C, N) particle (a) bright field image and (b-h) overlap corrected TruMaps

4. Conclusion

- The existence of NbC, M_7C_3 and $M_{23}C_6$ was confirmed;
- The primary chromium carbides transformed from M_7C_3 to $M_{23}C_6$ during creep;
- Secondary chromium carbides $M_{23}C_6$ (100~250 nm) were observed within the matrix;
- Scanning transmission electron microscopy (STEM) showed the appearance of fine NbC, G phase ($Ni_{16}Nb_6Si_7$) and (Nb, Ti)(C, N) particles.

Reference

[1] Yan J B *et al Mat Sci Eng A-Struct.* **529** (2011) 361-369.
[2] Gatti E *et al Nucl Instrum Meth A.* **226** (1984) 129-141.
[3] Gatti E and Rehak P *Nucl Instrum Meth A.* **541** (2005) 47-60.
[4] Wang W Z *et al Mater Design.* **32** (2011) 4010-4016.
[5] Sourmail T *Mater Sci Tech Ser.* **17** (2001) 1-14.
[6] Danielsen H K and Hald J *Mat Sci Eng A-Struct.* **505** (2009) 169-177.

Microstructure characterisation of an X70 grade pipeline steel and its dislocation development during Bauschinger testing

T S Pereira, Y L Chiu and I P Jones

School of Metallurgy and Materials, University of Birmingham, Edgbaston, B15 2TT, UK.

E-mail: TXS890@bham.ac.uk

Abstract. The microstructure of an X70 alloy and its dislocation development during one cycle of compression/tension strain has been analysed. The microstructure of the as received material and the dislocation arrangements before, during and after the Bauschinger test have been characterised. Results for this particular alloy show no evidence of grain boundary dislocation pile-ups or of precipitates pinning dislocations. A comparison between the dislocation arrangement in the as received, compressed and compressed-tensioned samples shows the formation of small dislocation cells, implying that the conception of the Bauschinger effect comes from dislocation-dislocation interactions.

1. Introduction

Due to their versatility, high strength low alloy steels have been widely used for the past few decades in structural, automotive and pipeline applications. Their high strength, combined with toughness and corrosion resistance give these alloys a great advantage over plain carbon steels without an extensive cost rise. Extensive research has been carried out in the past few decades to keep up with these industries requirements and this research field will continue to evolve as industries develop.

Pipeline steels are directly linked to the Bauschinger effect due to the UOE pipe making process. During this process, the wall of the pipe goes through successive steps of compression and tension plastic deformation but, depending on its position, the wall receives a different number of steps, resulting in a strain gradient in its transverse direction. As a consequence, after the UOE process, one side of the pipe will have work hardened and the opposite side will be affected by the Bauschinger effect, thus, presenting a lower yield stress.

The mechanisms that explain the Bauschinger effect are likely to be a combination of the Orowan mechanism and the long range internal stress [1]. According to Sleeswyk et al. [2], during work-hardening, gliding dislocations interact with defects/obstacles and, as they are subjected to stress in the opposite direction, these dislocations overcome the same defects/obstacles more easily. In addition, Mughrabi [3] suggested that the long range internal stress in heterogeneous dislocation arrangements contributed to the Bauschinger effect considering that dislocation walls and cell interiors yield at different stresses, so that, when unloaded, one stresses the other, creating a stress field individual to each dislocation cell, even though the average stress is null.

The aim of this research was to explore the dislocation evolution of a commercial API X70 pipeline steel during Bauschinger testing. The as received material was analysed using TEM and compared to the compressed and compressed-tensioned samples.

2. Experimental methods

2.1. Material

The as-received samples were extracted from a rolled API X70 plate. Table 1 shows the chemical composition of the steel plate.

Table 1. Chemical composition of the as-received API X70 alloy (wt%).

C	Si	Mn	P	S	Cr	Mo	Ni	Nb	V	Ti	Cu	B	Al
0.042	0.23	1.53	0.0088	0.0025	0.021	0.14	0.19	0.062	0.0039	0.014	0.16	0.0002	0.03

2.2. SEM and TEM analysis

Samples for SEM analysis were ground and polished before being etched with 3% nitric acid in ethanol for 10 seconds in order to reveal its microstructure. These samples were analysed using a JEOL 7000 operating at 20kV. EDM was used to cut 300 µm slices, from which 3 mm TEM discs were extracted using the same technique. The discs were then thinned to 50 µm before being electropolished using a Tenupol machine at 25 V, in a solution containing 95% ethanol and 5% perchloric acid at -30 °C. A JEOL 2100 operating at 200kV was used to analyse TEM samples.

2.3. Mechanical tests

An ESH Servo Hydraulic Universal Testing machine with a 200 kN load cell was used to perform the compression and Bauschinger tests at a strain rate of $1.6x10^{-4} s^{-1}$. Rectangular test samples of 4x4x8 mm were compressed to 2% plastic strain to simulate the "U" step of the UOE process.

Bauschinger tests consisted of a 2 % plastic strain under compression, followed by a 2 % plastic strain under tension. Cylindrical samples of 3.5 mm diameter and 13 mm gauge length were used for the Bauschinger tests.

The parameters measured to quantify the Bauschinger effect were the stress parameters at 0.2 and 0.5 % strain ($\beta_{\sigma r0.2}$ and $\beta_{\sigma r0.5}$ respectively), explained by equation 1 and 2, where σp is the value for maximum forward-stress, $\sigma r_{0.2}$ and $\sigma r_{0.5}$ are the values for stress in the opposite direction at 0.2 and 0.5 % strain.

$$\beta_{\sigma r0.2} = (\sigma p - \sigma r_{0.2})(\sigma p)^{-1} \quad (1)$$

$$\beta_{\sigma r0.5} = (\sigma p - \sigma r_{0.5})(\sigma p)^{-1} \quad (2)$$

3. Results and discussion

3.1. Microstructure

The API X70 alloy presented a predominantly ferritic structure with the average grain size of 2.8 µm, and 1.7 and 2.3 % area fractions of perlite and austenite phases respectively. SEM images showed a small amount (0.0873 % area fraction) of relatively large precipitates, with EDX data revealing them to be Ti(C, N). The diameter of these precipitates ranges from 50 to 100 µm.

Titanium-type particles start to precipitate at high temperatures, near the liquid-austenite transformation temperature. Therefore, they act mainly as ferrite grain size controllers, refining the microstructure of the plates, but not playing an important role in precipitate strengthening [4].

EDX data from TEM revealed a second type of precipitate, Nb(C, N), of size 25 µm, with a much smaller area fraction. No evidence of vanadium type precipitates was found. Because niobium and vanadium particles precipitate at lower temperatures, they tend to be smaller than Ti(C,N)and are able to act as barriers to the movement of dislocations. However, with a small area fraction of the niobium particles and the absence of vanadium particles, very little or no precipitate strengthening is expected

Journal of Physics: Conference Series **522** (2014) 012035 doi:10.1088/1742-6596/522/1/012035

for this particular alloy and, therefore, none of the Bauschinger effect can be attributed to dislocation-particle interactions.

3.2. Bauschinger test

The results of the Bauschinger test curve are shown in figure 1.

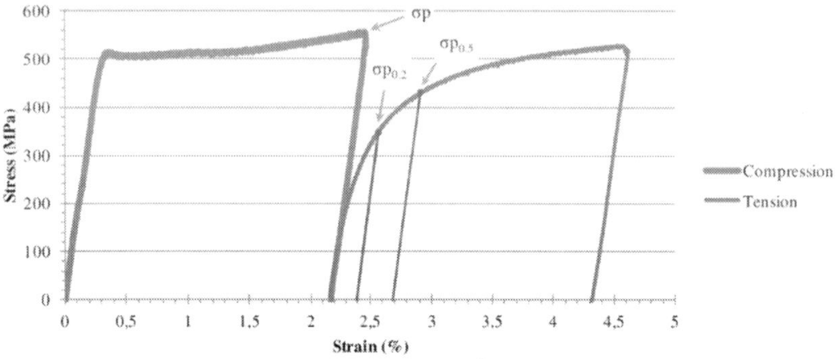

Figure 1. Bauschinger test curve, showing the Bauschinger parameters.

The Bauschinger parameters $\beta_{\sigma r0.2}$ and $\beta_{\sigma r0.5}$ were calculated using equations 1 and 2 from the parameters σp, $\sigma r_{0.2}$ and $\sigma r_{0.5}$ from the Bauschinger test curve, which values are shown in table 2.

Table 2. The Bauschinger parameters.

σp	$\sigma r_{0.2}$	$\sigma r_{0.5}$	$\beta_{\sigma r0.2}$	$\beta_{\sigma r0.5}$
561 MPa	405 MPa	480 MPa	0.278	0.144

From the tension curve of the Bauschinger test, it is noticeable that plastic deformation starts as early as half of the compression yield stress value and the flow stress changes drastically. However, after 2% tension strain, the stress value has not returned to its maximum forward stress level.

3.3. Dislocation analysis

The typical dislocation microstructure of the as-received, 2 % compressed and Bauschinger tested samples are shown in figure 2. The dislocation density analysis shows that the as-received specimens have a dislocation density of 4.7×10^{8} cm^{-2}, the compressed specimens 7.8×10^{9} cm^{-2} and the Bauschinger tested specimens 6.2×10^{9} cm^{-2}.

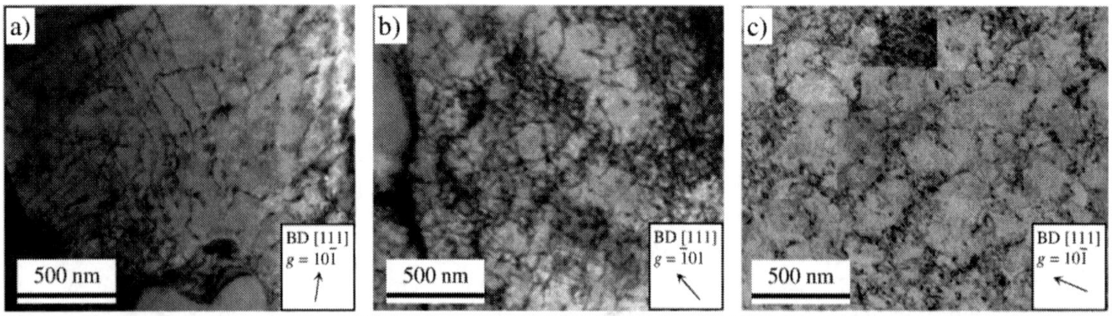

Figure 2. Typical two beam TEM micrographs showing the dislocation microstructure in a) as-received, b) compressed and c) Bauschinger tested specimens.

The dislocation structure of the as-received samples showed a predominantly straight dislocation lines arrangement, with homogeneous spacing in between lines. After 2% compression, the dislocation density increased by over one order of magnitude, starting to present high dislocation density walls but not well defined cells. The Bauschinger tested samples showed a similar dislocation density to the compressed samples, but the cell structure was much better defined, with the interior of the sub-grains almost dislocation free.

Values for the Bauschinger parameter $\beta_{\sigma r 0.2}$ after 2% plastic strain have been found of up to 0.756 due to the presence of precipitates [5], which, compared to 0.278 of the present X70 alloy, is much more significant. However, as previously mentioned, there was no evidence of dislocation pile-ups at grain boundaries or precipitates pinning dislocations, thus, the Bauschinger effect generated by dislocation-dislocation interactions only may not be neglected.

4. Conclusions

The main findings of this study are:

- The microstructure of the as-received X70 alloy is predominantly ferritic with a small fraction of perlite and austenite grains. The particles identified are the titanium and niobium-type precipitates, although they are either too large or occupy a considerable small area fraction in order to produce any pronounce precipitate strengthening. Vanadium-type precipitates are not present;
- The Bauschinger parameters $\beta_{\sigma r 0.2}$ and $\beta_{\sigma r 0.5}$ are calculated to be 0.278 and 0.144 respectively. After 2% reverse strain, the stress value did not return to the maximum value of forward stress;
- The dislocation density increases after compression but there isn't any significant change from the compressed state to the Bauschinger tested state. However, there is a significant change in the dislocation structure of the specimens when comparing the as-received, compressed and Bauschinger tested samples. High dislocation density walls are formed after 2% compression, which are transformed into well defined dislocation cells after the Bauschinger test;

Acknowledgments

The financial support from the Roberto Rocca Fellowship Programme along with the support from the School of Metallurgy and Materials of the University of Birmingham have been greatly appreciated.

References

[1] Kassner M E, Geantil P, Levine L E and Larson B C 2009 Backstress, the Bauschinger effect and cyclic deformation *Mat. Sci. Forum* **604-605** 39-51

[2] Sleeswyk A W, James M R, Plantinga D H and Maathuis W S T 1978 Reversible strain in cyclic plastic deformation *Acta Metall.* **26** 1265-71

[3] Mughrabi H 1983 Dislocation wall and cell structures and long range internal stresses in deformed metal crystals *Acta Metall.* **31** 1367-79

[4] Gladman T 1997 *The Physical Metallurgy of Microalloyed Steels* (Cambridge, Institute of Materials)

[5] Aran A, Demirkol M and Karabulut A 1987 Bauschinger effect in precipitation-strengthened aluminium alloy 2024 *Mater Sci Eng* **89** 35-39

A STEM study of twin defects in $Fe_3O_4(111)$/YZO(111)

D Gilks[1], L Lari[1,2], K Matsuzaki[3], R Evans[1], K McKenna[1], T Susaki[3], V K Lazarov[1].

[1] Department of Physics, University of York, Heslington, York, YO10 5DD, UK
[2] York-JEOL Nanocentre, University of York, Heslington, York, YO10 5BR, UK
[3] Secure Materials Centre, Materials and Structures Laboratory, Tokyo Institute of Technology, 4259 Nagatsuta, Midori-ku, Yokohama 226-8503, Japan

E-mail: dg522@york.ac.uk

Abstract. We observe twin defects on both the (111) and (11-2) planes of thin film Fe_3O_4 using atomically resolved high angle annular dark field (HAADF) STEM imaging. These defects are significant for understanding the previously reported anomalous properties of Fe_3O_4 films as they generate non-bulk bonding configurations leading to non-bulk superexchange interactions in these regions.

1. Introduction

Bulk Fe_3O_4 (Figure 1) is a predicted half metal with a 100% spin polarisation at the Fermi level. [1, 2]. This bulk property makes thin film Fe_3O_4 of interest for future spintronics device applications [3-6]. However, structural defects including anti-phase boundaries (APB) in thin film Fe_3O_4 prevent the fabrication of films with bulk Fe_3O_4 properties. Thin film Fe_3O_4 exhibits negative magnetoresistance and does not magnetically saturate, e.g. bulk Fe_3O_4 shows no magneto-resistance and magnetically saturates at low applied fields. This makes a thorough understanding of defect formation in thin film Fe_3O_4 essential. The presence of APB defects is well documented in the literature [7-9]. In this paper we show that twin defects are also present in epitaxially grown single crystal films and their presence should be taken into account to fully understand the functional properties of Fe_3O_4 thin films. We observe twin defects in Fe_3O_4 films grown on Yttrium stabilized Zirconium Oxide (YZO). Currently twin defects in thin film Fe_3O_4 have

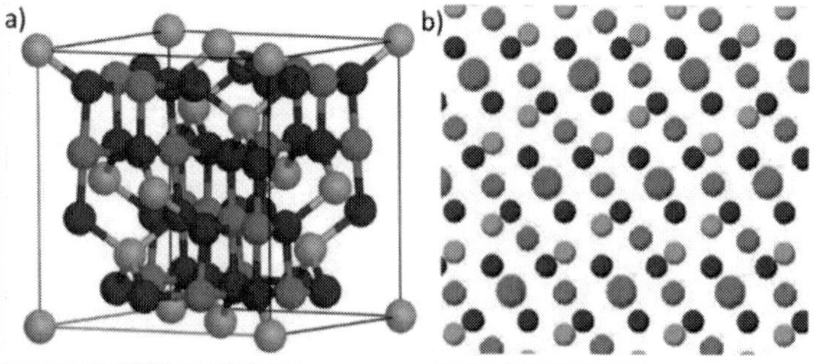

Figure 1. a) The unit cell of Fe_3O_4 consisting of 32 O atoms, 16 Fe in octahedral (B) positions and 8 Fe in tetrahedral positions (A). b) Fe_3O_4 viewed along the [1-10] direction showing all the atomic columns occupied by single atom types.

not received significant research attention, consequently the impact of twin defects on the overall magnetic and transport properties of thin film Fe_3O_4 are not well known at present.

2. Methods

High angle annular dark field scanning transmission electron microscopy (HAADF-STEM) data has been collected using a JEOL 2200FS microscope operating at 200kV and equipped with CEOS 3rd order imaging and probe aberration correctors. Diffraction patterns have been acquired using a JEOL 2011. Image simulations have been calculated using parallelised QSTEM image simulation software [10] with 30 thermal diffuse scattering iterations performed for each simulation. Aberration coefficients values used for the simulations were C_s(1.1 μm), C_5 (1.76 mm) and C_c (1.6 mm) respectively, as measured by the CEOS corrector software before HAADF image acquisition. Fe_3O_4 films have been grown using PLD methods with a KrF excimer laser incident on a Fe_3O_4 target, films were grown at 300 °C in an oxygen partial pressure of 2×10^{-4} Pa followed by high temperature post-annealing at 1100°C in a CO/CO_2 atmosphere as discussed elsewhere [11]. TEM samples have been produced by mechanical thinning and low angle argon ion milling to electron transparency following the protocol described in [12].

3. Results and Discussion

The annealed films show atomically sharp film/substrate interfaces and uniform film thickness across the samples as shown in Figure 2. We find two types of twin defects in these films. The first type of twin defect propagates in the film growth direction from the film/substrate interface to the film surface (Figure 3). These defects are not seen to nucleate or discontinue within the film, nor are they seen to deviate from the single (11-2) planes on which they are located. The second type of twin defect is perpendicular to the growth direction on a (111) twin-plane, Figure 4; as the first type of defects they are also atomically sharp.

In the (11-2) twins, we observe defects originating from bunched step edges in the YZO substrates indicating that surface steps could be potential twin nucleation sites. We have also observed that the twins are highly mobile whilst being annealed at 1100°C, as their density drops considerably as a result of the annealing process. From this it is unclear whether the twin defects nucleate as a result of the substrate step edges, or migration of twin defects during the annealing procedure pins them to the step edges. By studying a series of films with varied annealing times (or temperatures) and the resultant defect density and morphology it should be possible to identify a relationship between initial deposition characteristics in the films and the evolution of these defects during the annealing procedure.

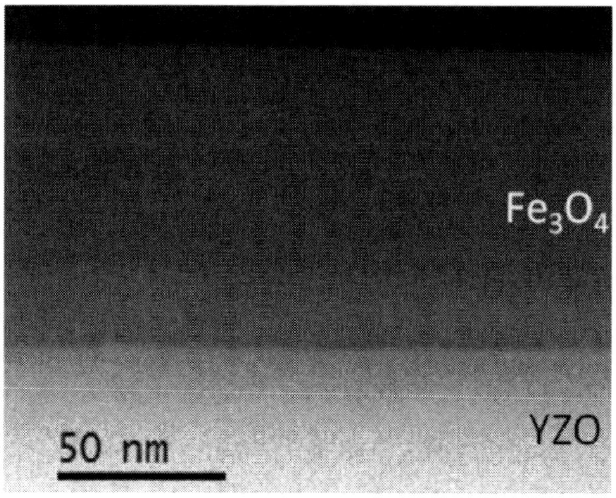

Figure 2. HAADF-STEM image of a Fe_3O_4 film grown on YZO substrate.

Several scenarios are possible to create twin defects in Fe_3O_4 grown on YZO. The underlying stacking sequence of the YZO substrate should 'direct' the stacking of the Fe_3O_4 film and give a preferred stacking sequence in the deposited film, however, in these films it seems that the stacking sequence through the interface is often not maintained. This suggests that stacking faults at the Fe_3O_4/YZO interface are either of low formation energy or different surface termination of YZO can provide growth templates for different atomic films-substrate interfaces which ultimately result in twin-related stacking sequence that propagate as the films is grown.

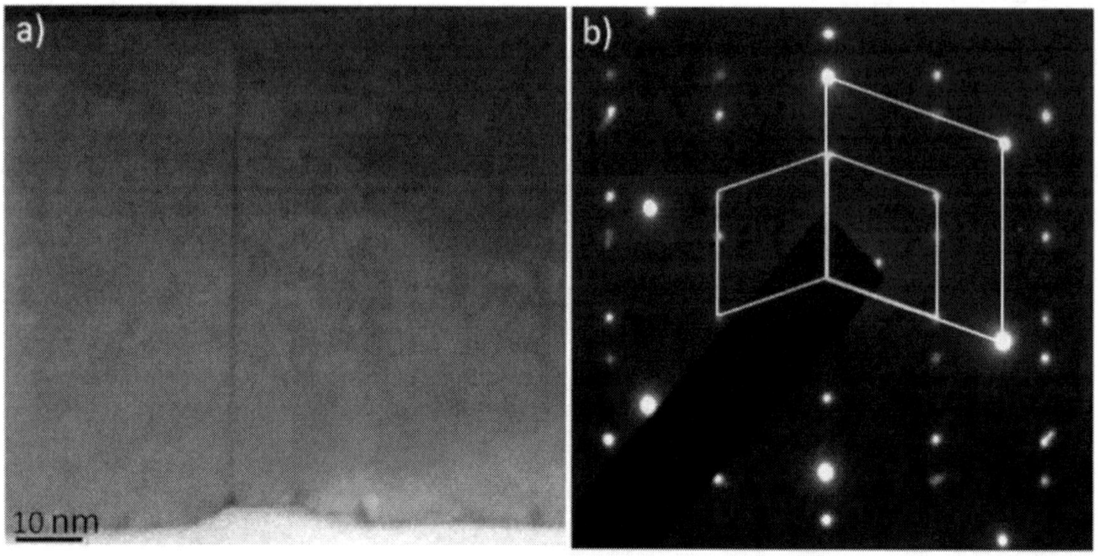

Figure 3. HAADF-STEM and diffraction patterns of a twin defect in Fe_3O_4 imaged in the [1-10] zone axis. a) Shows a twin running vertically from a stepped region of the film/substrate interface to the film surface. b) Shows the overlaid diffraction patterns of the YZO substrate (white/largest rhombus) and two Fe_3O_4 diffraction patterns (two yellow/smaller rhombuses) the twinned structure can be identified through the mirror symmetry of the Fe_3O_4 diffraction spots from left to right in the pattern.

The twin structures we observe running perpendicular to the growth direction demonstrate that Fe_3O_4 can form twins even when the underlying stacking sequence of well ordered Fe_3O_4 is already present in the film. This indicates the formation energy of twins defects on the (111) plane of Fe_3O_4 is low. In the twin defects we have observed the octahedrally co-ordinated Fe_B plane appears to form the twin with structural symmetry about this plane.

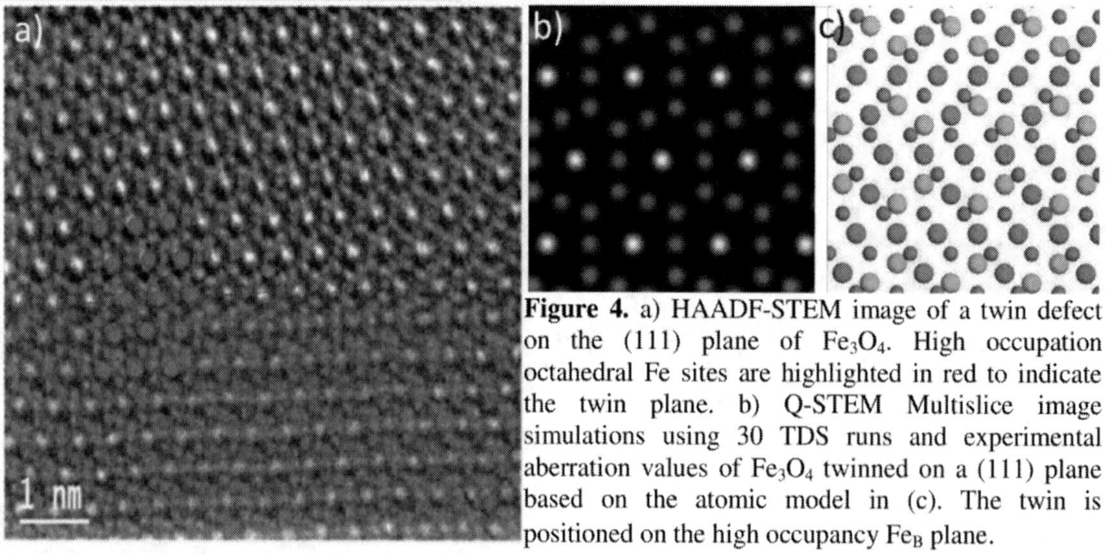

Figure 4. a) HAADF-STEM image of a twin defect on the (111) plane of Fe_3O_4. High occupation octahedral Fe sites are highlighted in red to indicate the twin plane. b) Q-STEM Multislice image simulations using 30 TDS runs and experimental aberration values of Fe_3O_4 twinned on a (111) plane based on the atomic model in (c). The twin is positioned on the high occupancy Fe_B plane.

Unlike in the (11-2) plane defect, this (111) twin does not lead to a heavily relaxed interfacial structure, rather the Fe_3O_4 structure simply continues but with an inversion of the stacking sequence across the twin boundary. From the Figure 4 it can be observed that the octahedral Fe-O-Fe bond angles are significantly increased compared to 90° bonds in bulk Fe_3O_4. This is of importance as the bond angles define the superexchange interactions between Fe-O-Fe
sites. As bond angles between Fe sites increase the superexchange interaction become strongly anti-ferromagnetic.

4. Conclusions

We show direct evidence of two distinct styles of twin boundaries in thin film Fe_3O_4. While the origin of the vertical (11-2) twin defects could be influenced by several mechanisms, the formation of perpendicular (111) twins is most likely driving by their low formation energy. The effect of twin structural defects on the magnetic and magneto-transport properties of magnetite thin films is still to be explored. These properties will be studied using atomistic magnetic modelling and *ab. Initio* structural refinement and electronic structure calculations. This will allow a comparison between simulated magnetisation properties and structure with experimental in-plane magnetisation and transport measurements.

[1] Yanase A and Siratori K 1984 Band-Structure in the High-Temperature Phase of Fe_3O_4 *J. Phys. Soc. Jpn* **53** 312-7

[2] Zhang Z and Satpathy S 1991 Electron states, magnetism, and the Verwey transition in magnetite *Phys. Rev. B* **44** 13319-31

[3] Bataille A M, Moussy J B, Paumier F, Gota S, Guittet M J, Gautier-Soyer M, Warin P, Bayle-Guillemaud P, Seneor P, Bouzehouane K and Petroff F 2005 Crystalline gamma-Al_2O_3 barrier for magnetite-based magnetic tunnel junctions *Appl. Phys. Lett.* **86** 012509-3

[4] Kado T, Saito H and Ando K 2007 Room-temperature magnetoresistance in magnetic tunnel junctions with Fe_3O_4 electrode *J. Appl Phys.* **101** 09J511-3

[5] Alexe M, Ziese M, Hesse D, Esquinazi P, Yamauchi K, Fukushima T, Picozzi S and Gösele U 2009 Ferroelectric Switching in Multiferroic Magnetite (Fe_3O_4) Thin Films *Adv. Mater.* **21** 4452-5

[6] Hu G and Suzuki Y 2002 Negative Spin Polarization of Fe_3O_4 in Magnetite/Manganite-Based Junctions *Phys. Rev. Lett.* **89** 276601

[7] Margulies D T, Parker F T, Rudee M L, Spada F E, Chapman J N, Aitchison P R and Berkowitz A E 1997 Origin of the Anomalous Magnetic Behavior in Single Crystal Fe_3O_4 Films *Phys. Rev. Lett.* **79** 5162-5

[8] Celotto S, Eerenstein W and Hibma T 2003 Characterization of anti-phase boundaries in epitaxial magnetite films *Eur. Phys. J. B* **36** 271-9

[9] Ziese M and Blythe H J 2000 Magnetoresistance of magnetite *J. Phys.: Cond. Matt.* **12** 13

[10] Koch C T 2002 Determination of Core Structure Periodicity and Point Defect Density along dislocations, *PhD Thesis*, Arizona State University

[11] Kosuke M, Vlado K L, Leonardo L, Hideo H and Tomofumi S 2013 $Fe_3O_4(111)$ thin films with bulk-like properties: growth and atomic characterization *J. Phys. D: Appl. Phys.* **46** 022001

[12] Lari L, Lea S, Feeser C, Wessels B W and Lazarov V K 2012 Ferromagnetic InMnSb multi-phase films study by aberration-corrected (scanning) transmission electron microscopy *J. Appl. Phys.* **111** 07C311-3

A dissociation mechanism for the [a+c] dislocation in GaN

P.D. Nellist[1], P.B. Hirsch[1], S. Rhode[2], M.K. Horton[3], J.G. Lozano[1] A. Yasuhara[4], E. Okunishi[4], S. Zhang[2], S.-L. Sahonta, M.J. Kappers[2], C.J. Humphreys[2], M.A. Moram[3]

[1]Department of Materials, University of Oxford, Parks Road, Oxford, OX1 3PH, UK.
[2]Department of Materials Science and Metallurgy, University of Cambridge, Pembroke Street, Cambridge , CB2 3QZ , UK.
[3]Department of Materials, Imperial College London, Exhibition Road, London, SW7 2AZ, UK.
[4]JEOL Ltd., EM Application Group , 1-2 Musashino 3-Chome, Akishima, 196-8558 , Tokyo , Japan.

E-mail: peter.nellist@materials.ox.ac.uk

Abstract. Mixed-type [**a**+**c**] dislocations can be identified in atomic-resolution high-angle annular dark-field scanning transmission electron microscope images of GaN viewed along [0001] by use of a Burgers loop analysis and by observation of the depth-dependent displacements associated with the Eshelby twist. These dislocations are found to be able to dissociate resulting in a fault that lies perpendicular to the dislocation glide plane. Consideration of the bonding that occurs in such a fault allows the dissociation reaction to be proposed, and the proposed fault agrees with the experimental images when kinks are incorporated into the model.

1. Introduction

Despite the relatively high density of threading dislocations (with densities typically ranging from 10^7–10^9 cm^{-2}) that occur in GaN and its alloys because of growth on poor lattice-matched substrates, high-efficiencies can be achieved in blue emitting InGaN/GaN-based optoelectronic devices. The dislocations can, however, have a deleterious effects on lifetimes [1] and leakage currents [2,3] within these devices. It is known that the dislocations introduce electronic states into the band gap and act as non-radiative recombination centres [4,5]. Transmission electron microscopy (TEM) and scanning TEM (STEM) studies have been carried out to determine the core structure of these dislocations, with a recent publication noting the effect on core structures of Mg doping [6].

The general direction of these dislocations in (0001) oriented films is along [0001]. Three types of dislocations are generally found: edge-type dislocations with Burgers vector **b** = [**a**], where **a** is the unit cell lattice vector in the basal (0001) plane, screw dislocations with **b** = [**c**], the unit cell translation along the [0001] axis, and mixed dislocations with **b** = [**a** + **c**]. This paper presents high-angle annular dark-field (HAADF) STEM images of the [**a** + **c**] dislocation that show that a dissociation reaction can occur with this dislocation, and the proposed dissociation reaction and resulting fault are described.

2. Experimental details

Specimens of GaN (0001) layers grown on sapphire substrates, using the metalorganic chemical vapour deposition technique (MOCVD), were examined in an aberration corrected JEOL ARM 200F operating at 200 kV with a cold field emission gun. The 4.2 μm thick GaN layer was undoped (not intentionally doped) and had a relatively low threading dislocation density of $4\text{-}5\times10^{8}$ cm^{-2}. Plan view TEM samples were prepared using mechanical and ion polishing techniques.

3. Results and discussion

Figure 1 shows an image of a dislocation. It can be seen that the dislocation core appears to have a structure that is elongated with a fault parallel to the **a** plane (ie perpendicular to the **a** direction). The larger Burgers circuit shown in Fig 1(b) shows that the dislocation burgers vector has a component in the plane of the image of **a** and therefore it has an edge component. The atomic positions in the image do not, however, allow us to determine whether this dislocation has a screw component. A smaller circuit that encloses only one end of the fault shows a loop closure error of **a**/2, suggesting that the fault has been caused by a dissociation of the dislocation into partials. The Burgers loop analysis suggests that the edge component of the dislocation has dissociated by the following reaction

$$[\mathbf{a}] \rightarrow [\mathbf{a}/2] + [\mathbf{a}/2].$$

The glide plane (which must contain the **a** component of the Burgers vector and the dislocation line) is also marked on Fig. 1(b). It is clear that the dissociation has occurred in a direction perpendicular to the glide plane, and that therefore the edge component of the dislocation has dissociated in the climb direction.

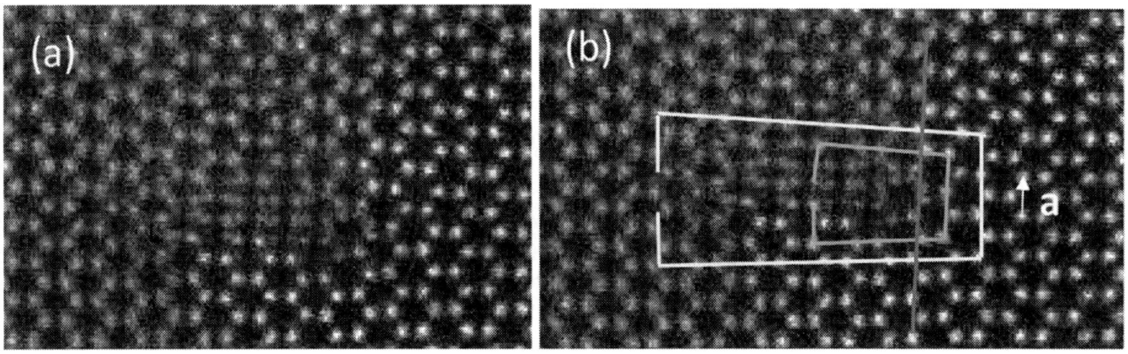

Figure 1. (a) HAADF STEM images of a dislocation viewed end on along [0001] in GaN; (b) Burgers loops showing the edge component of the Burgers vector associated with the complete dislocation (yellow) and that associated with the defect at the end of the fault (green). The red line indicates the glide plane. The lattice vector **a** has a length of 0.319 nm.

We now turn to the possible structure of the planar fault caused by the dissociation. Figure 2(a) shows a model of a single layer of Ga and N atoms in one (0002) plane viewed along the [0001] direction. After removal of a line of atoms to create the **a**/2 displacement observed in the image, we find that the resultant fault consists of a large number of wrong bonds (Ga bonded to Ga and N bonded to N) that is likely to result in a high-energy fault and making dissociation unlikely (Fig 2b).

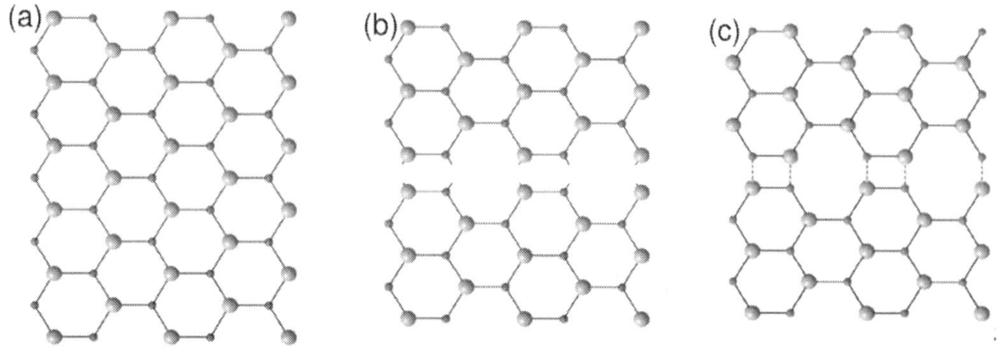

Figure 2. (a) a single (0002) plane of GaN; (b) removal of a line of atoms to give a shift by **a**/2 resulting in wrong bonds; (c) a shift of the top part by **c**/2 gives 4 and 8 membered rings with no wrong bonds.

The screw component offers a route to overcome this difficulty. Optical sectioning of the lattice surrounding this fault [7] shows depth-dependent displacements that are consistent with those expected to arise from the Eshelby twist. The Eshelby twist displacements [8] arise when a dislocation with a screw component intersects the free surfaces of the thin TEM sample, and their observation shows that the dislocation does have a screw component, which is likely be have a value of **c**, and the complete dislocation is therefore of the **a**+**c** type. If we make the assumption that the screw component of the dislocation dissociates through the following reaction

$$[\mathbf{c}] \rightarrow [\mathbf{c}/2] + [\mathbf{c}/2].$$

then the lattice on one side of the fault will be shifted by half a unit cell along the **c**-axis relative to the opposite side. Figure 2(c) shows that such a shift removes the wrong bonds, and results in a fault with alternating 8 and 4 membered rings but with no wrong bonds. Such a fault is likely to be relatively low in energy, as confirmed by numerical calculations and by considering the repulsion of the dislocations balancing the fault energy [9].

The projected structure of the fault (Fig. 2c) does not entirely correspond to the high resolution image of the fault seen in Fig. 3a. The 8 and 4 membered rings are marked in Fig. 3b, but additional atomic columns may also be seen. Optical sectioning focal series data of the dislocation [9] show that the dislocation is inclined at approximately 12° to the **c**-axis in a plane inclined to both the glide and climb planes. The dislocation must therefore contain both kinks and jogs along its length. Figure 3c shows the projected structure of the fault if we assume that there are two elementary kinks within the projected thickness of the sample. This shows better agreement with the experimental data. A more sophisticated analysis allows a more detailed model to be proposed [9].

Given that many more than two elementary kinks must exist along the dislocation within the thickness of the sample given the 12° inclination, the question then arises of why a reasonable agreement with the experimental image is achieved with only two kinks. The explanation is likely to be that the reduced depth of focus of the microscope only allows a thickness of sample to be imaged in high resolution in which approximately two kinks are contained given the 12° inclination, and indeed that it is the optical sectioning effect of aberration-corrected HAADF STEM that has allowed clear images of the fault to be achieved.

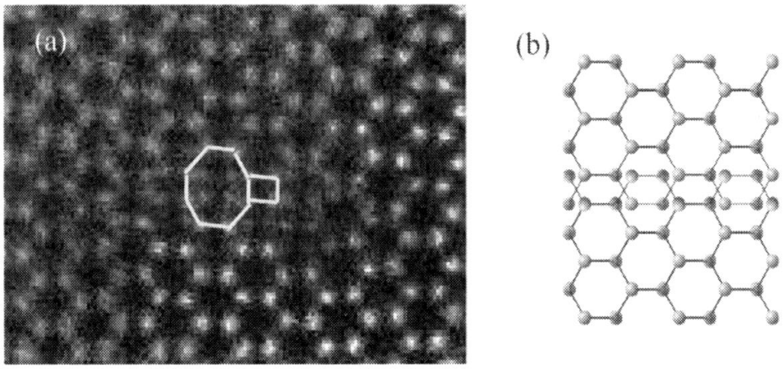

Figure 3. (a) The image from Fig. 1(a) reproduced with examples of the 8 and 4 membered rings highlighted. (b) A model containing two elementary kinks within the projected thickness gives better agreement with the image.

4. Conclusions

Mixed type [**a**+**c**] dislocations can be identified in GaN viewed along [0001] by Burgers circuit analysis and by the use of optical sectioning to identify the presence of the screw dislocation. It is found that such mixed dislocation can dissociate, and the following dissociation reaction is proposed

$$[\mathbf{a}+\mathbf{c}] \rightarrow [\mathbf{a}/2+\mathbf{c/2}] + [\mathbf{a}/2+\mathbf{c}/2].$$

The resulting fault has the same structure as that observed by Drum [9] in AlN, and contains no wrong bonds. Agreement with the experimental images is made when the kinks associated with the inclination of the dislocation are included in the model.

5. References

[1] M.F. Schubert, J. Xu, J.K. Kim, E.F. Schubert, M.H. Kim, S. Yoon, S.M. Lee, C Sone, T. Sakong, and Y. Park, Appl. Phys. Lett. **93**, 041102 (2008).
[2] M. S. Ferdous, X. Wang, M. N. Fairchild, and S. D. Hersee, Appl. Phys. Lett. **91**, 231107 (2007).
[3] J. W. P. Hsu, M. J. Manfra, R. J. Molnar, B. Heying, and J. S. Speck, Appl. Phys. Lett. **81**, 79 (2002).
[4] D. Cherns, S. J. Henley, and F. A. Ponce, Appl. Phys. Lett. **78**, 2691 (2001).
[5] S. E. Bennett, Mat. Sci. Tech. **26**, 1017 (2010).
[6] S. K. Rhode, M. K. Horton,2 M. J. Kappers, S. Zhang, C. J. Humphreys, R. O. Dusane, S. -L. Sahonta and M. A. Moram, Phys. Rev. Lett. **111**, 025502 (2013).
[7] J G Lozano, M P Guerrero-Lebrero, A Yasuhara, E Okinishi, S Zhang, C J Humphreys, P L Galindo, P B Hirsch and P D Nellist, this volume.
[8] J.D. Eshelby and A.N. Stroh, Phil. Mag. **42**, 1401 (1951).
[9] P. B. Hirsch, J. G. Lozano, S. Rhode, M. Horton, M.A. Moram, S. Zhang, M. J. Kappers, C. J. Humphreys, A. Yasuhara, E. Okunishi and P. D. Nellist, Phil. Mag. **in press (available online) DOI:10.1080/14786435.2013.797617 (2013).**
[10] C.M. Drum, Phil. Mag. **11**, 313 (1965).

6. Acknowledgments

PDN and JGL gratefully acknowledge financial support from the EPSRC (EP/F048009/1). SR from the Cambridge Commonwealth Trust, MJK from the EPSRC (EP/H049533/1).

Characterisation of Defects at Non-Polar GaN/InGaN Junctions in Novel Materials for Application in Light Emitting Diodes

J Severs[1], J G Lozano[1], S Hooper [2] and P D Nellist[1]

[1]University of Oxford, Department of Materials, Parks Road, Oxford, OX1 3PH, United Kingdom
[2] Sharp Laboratories of Europe Limited, Edmund Halley Road, Oxford Science Park, Oxford, OX4 4GB, United Kingdom

Email: jack.severs@materials.ox.ac.uk

Abstract. Despite the significant progress in the area, III-nitride LEDs still suffer from reduced efficiency due to a high dislocation density associated with a lack of suitable growth substrates and strong polarisation fields along the c-axis, the predominant growth direction, resulting in the quantum confined stark effect (QCSE). The novel materials studied here have been developed with the aim of reducing both of these effects and consist of a series of GaN and InGaN layers deposited by chemical vapour deposition on the non-polar, *m*-plane facets of core structures prepared from *c*-axis GaN thin films. A dual-beam FIB has been used to prepare a and *c*-plane sections through these novel structures in order to characterise the defects in the core and non-polar junction using, both diffraction contrast and high resolution, TEM and STEM. The dislocations in the *c*-axis GaN were generally observed to be confined to this region but the non-polar layers had a high density of defects. The latter include unusual flag type defects which are shown to be comprised of pyramidal dislocations and basal-plane stacking faults, with the Burgers vectors determined via **g·b** analysis.

1. Introduction

The III-nitride family of semiconductors are firmly established as the primary material system of interest in the development of short wavelength LEDs for solid state lighting. The favourable properties include the possibility of tuning the bandgap over a wide range of energies from infrared to deep UV via alloying and the strong chemical stability giving a reduced rate of degradation in comparison to other candidate materials. There are however, well-documented difficulties including the high dislocation densities associated with a lack of suitable growth substrates and the strong polarisation fields along the *c*-axis, the predominant growth direction, both of which are believed to reduce the efficiency of devices based upon these materials. The former is believed to lead to non-radiative recombination although the exact role the defects play in the luminescence from these materials is still under debate. The effect of the piezoelectric fields across the junctions is to spatially separate the electron and hole wavefunctions, limiting the efficiency of radiative recombination, and to introduce a current-dependent red-shift in wavelength referred to as the quantum confined stark effect (QCSE).

With the motivation of overcoming the above issues, there has been an increasing research effort in the area of growth of nitrides in the nonpolar *a*- and *m*- directions [1] with demonstration of quantum wells free of the spontaneous and piezoelectric fields which limit the performance of *c*-axis devices. Thus far, non-polar material grown on foreign substrates has had a high density of defects whereas the price of native non-polar GaN substrates reduces their viability as a commercial possibility. The following work involves the characterisation of defects in *m*-axis GaN/InGaN junctions deposited by chemical vapour deposition on the exposed *m*-plane facets of *c*-axis grown GaN core structures. A detailed understanding of the defects present in non-polar nitride junctions is of particular interest due to the impact of the microstructure on the optical properties of the material [2] and a complete characterisation is required in order to provide a basis for managing the density of defects in future devices. This task is more complex in the case of *a*- or *m*-plane films in comparison with conventional *c*-plane GaN due to the reduced symmetry.

2. Experimental details

The materials studied here are n-GaN/InGaN/p-GaN junctions deposited by MOCVD on the *m*-plane facets of *c*-axis grown core structures and illustrated in figure 1(a) below. The precise fabrication details of the latter are the subject of an ongoing patent application and cannot be fully disclosed presently. The following will focus on the details of the defect characterisation in the non-polar layers. The layers are ~50nm thick with the indium composition in the InGaN, measured to be 1-2%. This was lower than the intended value, which may be attributed to known issues with indium incorporation in non-polar InGaN [3]. A Zeiss NVision dual-beam FIB has been used to prepare and lift out *a*- and *c*-plane cross sections in order to study the structural defects in the core and shell regions via diffraction contrast TEM using a JEOL 2000FX and annular dark field STEM using a JEOL 3000F with high resolution data collected with a JEOL 2200MCO.

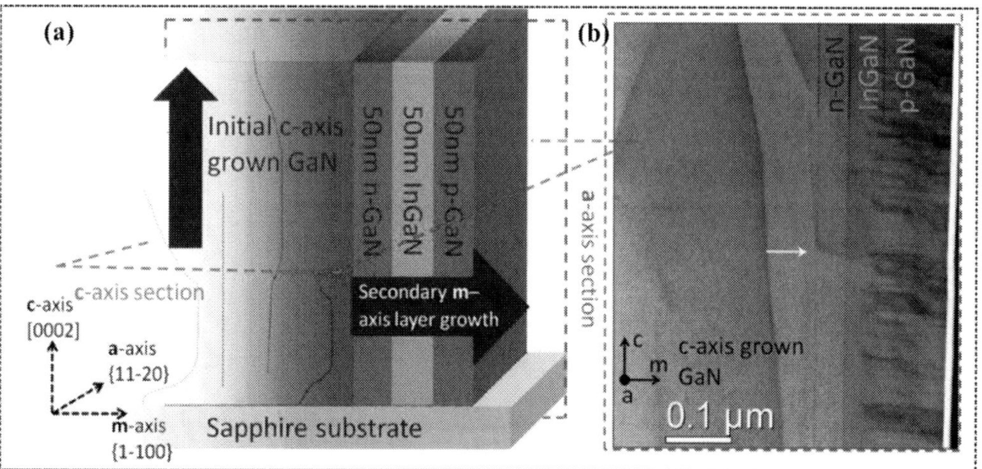

Figure 1 (a) A schematic illustration of the n-GaN/InGaN/p-GaN junctions deposited on exposed *m*-facets of conventional *c*-axis grown GaN with the geometry of the FIB-prepared cross sections indicated. (b) ADF STEM image of an *a*-plane section.

3. Results

An ADF STEM image of part of an *a*-plane section is given in figure 1(b) above. The threading dislocations in the *c*-axis GaN core region are clearly visible and were identified via a **g·b** analysis – not included here - as mostly pure edge (**b** = **a** for this orientation) and mixed character (**b** = **a** + **c**) which are commonly observed in *c*-axis GaN films. The majority were confined to the core structure, which is encouraging for the potential of using *c*-axis GaN as a substrate in this way, although a few instances were observed in which the dislocations reached the interface appearing to nucleate further defects. An example of this is indicated with the white arrow in figure 1(b).

In addition to large numbers of basal plane stacking faults (BSFs), as widely reported for non-polar nitride material, there was also a high density of unusual 'flag-shaped' defects in the *m*-plane layers comprised of a basal plane section and inclined section. These usually started in the InGaN layer nearer the p-GaN interface and reached the free surface - at the far right in the image - and also frequently appeared to bound regions of differing contrast in ADF STEM images. The inclined side of the defects always pointed in the same *c*-axis direction and is likely to be related to the polarity of the material as has been confirmed the case for pyramidal defects in p-GaN [4]. A detailed **g·b** analysis was carried out with a large number of **g** vectors over several zone axes on two such flag defects. Figure 2(a) includes three example images, taken along] under the diffraction conditions indicated, along with a schematic diagram of the two flag defects: the upper flag, referred to as defect 1, comprising basal plane segment 1a and inclined segment 1b, and lower, defect 2 comprising segments 2a and 2b. The results of the contrast analysis suggested the basal plane defect segments, 1a and 2a, were basal plane stacking faults which was shown conclusively in the case of the latter via high resolution ADF STEM. Figure 2(b) illustrates the change in stacking sequence across 2a comprising ABAB<u>ABC</u>BCBC, corresponding to an I_1 type BSF with fault vector R=1/6[20 3].

The inclined sections of the defects, 1b and 2b, most closely match those of a perfect dislocation with burgers vectors of type **b**=1/3[1 10] and a partial with **b**=1/3[20 3] respectively, both of which refer to a mixed character for dislocations in this orientation. By comparing the geometry of these dislocations in weak beam images along different zone axes it was possible to determine that they lay in different pyramidal planes. Figure 2c illustrates this approach in the case of 1b and includes two weak beam images taken along z=[10 0] and [110] in which the diffraction conditions have been chosen so as to set the excitation error to satisfy the weak beam criterion, $\mathbf{s}_g$>0.2nm^{-1}, wherever possible. Under these conditions the observed peak should refer fairly closely to the position of the dislocation core and by measuring the geometry it was possible to determine that, in this case, the dislocation lay in the (02 3) pyramidal plane.

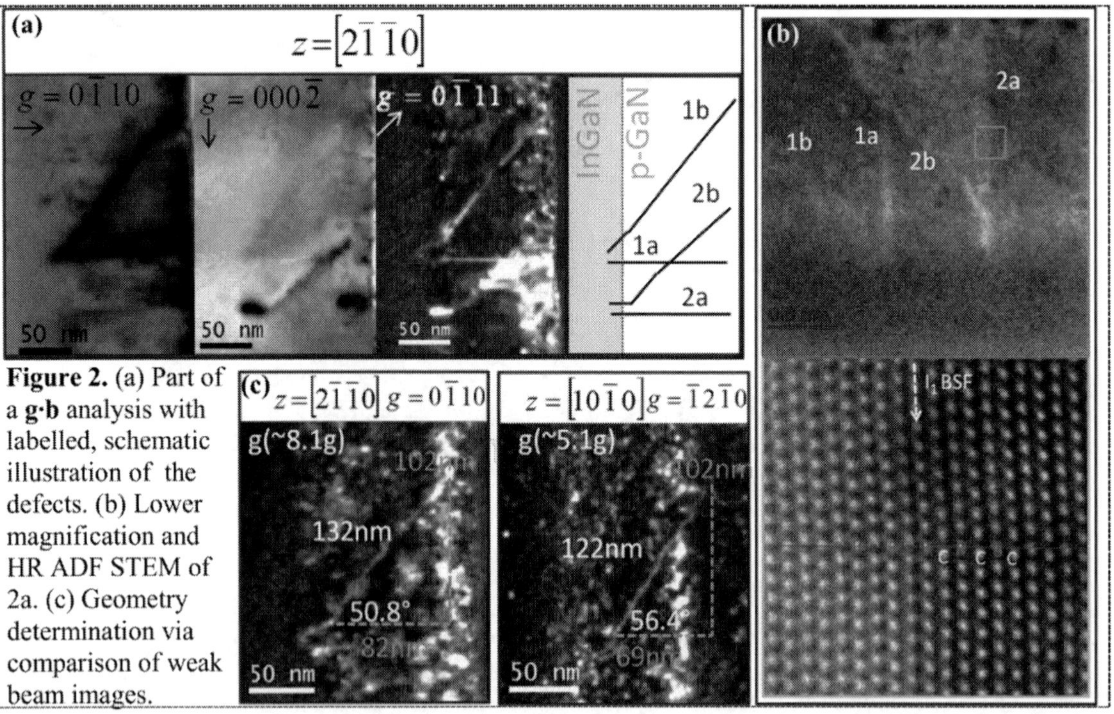

Figure 2. (a) Part of a **g·b** analysis with labelled, schematic illustration of the defects. (b) Lower magnification and HR ADF STEM of 2a. (c) Geometry determination via comparison of weak beam images.

Journal of Physics: Conference Series **522** (2014) 012038 doi:10.1088/1742-6596/522/1/012038

Considering defect 2, for which the high resolution data was acquired, the contrast in the top ADF STEM image of figure 2(b) suggests that the change in stacking imparted by the I_1 stacking fault, labelled 2a, is reversed by 2b as the contrast changes back to that of the surrounding material. The burgers vector of a Frank-Shockley partial which was assigned to 2b via the **g·b** analysis is the correct value to achieve this change in stacking and it could therefore by hypothesised that defect 2 consists of a partial dislocation loop of this type on a pyramidal plane meeting an I_1 type basal plane stacking fault. Observations along the *c*-axis appear to show loops, marked by dashed arrows in the weak beam images of figure 3(a), which display contrast changes consistent with a burgers vector of this type, supporting this theory. Such a structure would not however, explain the presence of perfect dislocations with burgers vectors matching that found for dislocation 1b. Figure 3(b) shows further evidence of perfect dislocations, with two defects which are invisible under one of the three **g**={1 00} vectors, indicative of perfect dislocation burgers vectors of **b**=**a** or **b**=**a**+**c** and further work needs to conducted to determine how these are consistent with the observed flag defects.

Specific luminescence peaks have been directly related to basal plane stacking faults and the partial dislocations terminating them [5] so it is likely the density of these defects will need to be managed in order to precisely control the optical properties of these materials.

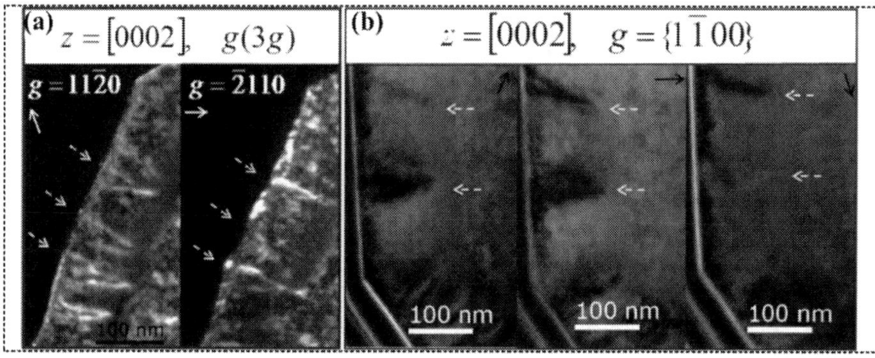

Figure 3. *C*-plane section images of different sets of defects (a) weak beam images using **g**=**a** (b) bright field images under two beam conditions with **g**=**m**.

4. Conclusions

A dual-beam FIB has been used to prepare sections through novel materials comprising non-polar n-GaN/InGaN/p-GaN junctions grown on exposed *m*-plane facets of *c*-axis grown GaN core structures. The dislocations in the *c*-axis GaN remain confined to the cores for the most part but a high density of basal plane stacking faults and unusual flag shaped defects were observed in the *m*-plane junction layers. The latter were analysed via a detailed **g·b** contrast analysis and HR STEM imaging and were found to be formed from pyramidal dislocations and basal plane stacking faults.

References

[1] U. T. Schwarz and M. Kneissl, "Nitride emitters go nonpolar," *physica status solidi (RRL) – Rapid Research Letters*, vol. 1, no. 3, pp. A44–A46, May 2007.

[2] M. A. Reshchikov and H. Morkoc, "Luminescence properties of defects in GaN," *Journal of Applied Physics*, vol. 97, no. 6, p. 61301, 2005.

[3] Y. Zhao, Q. Yan, C.-Y. Huang, S.-C. Huang, P. Shan Hsu, S. Tanaka, C.-C. Pan, Y. Kawaguchi, K. Fujito, C. G. Van de Walle, J. S. Speck, S. P. DenBaars, S. Nakamura, and D. Feezell, "Indium incorporation and emission properties of nonpolar and semipolar InGaN quantum wells," *Applied Physics Letters*, vol. 100, no. 20, p. 201108, 2012.

[4] , M. Benaissa, B. Beaumont, E. Feltin, P. De Mierry, S. Dalmasso, M. Leroux, and P. Gibart, "Pyramidal defects in metalorganic vapor phase epitaxial Mg doped GaN," *Applied Physics Letters*, vol. 77, no. 6, p. 880, 2000.

[5] R. Liu, A. Bell, F. A. Ponce, C. Q. Chen, J. W. Yang, and M. A. Khan, "Luminescence from stacking faults in gallium nitride," *Applied Physics Letters*, vol. 86, no. 2, p. 021908, 2005.

Aberration corrected STEM of iron rhodium nanoislands

M J McLaren[1], F S Hage[2], M Loving[3], Q M Ramasse[2], L H Lewis[3], C H Marrows[4], R M D Brydson[1]

[1]Institute of Materials Research, SPEME, University of Leeds, LS2 9JT, UK
[2]SuperSTEM Laboratory, SciTech Daresbury Campus, Daresbury WA4 4AD, UK
[3]Department of Chemical Engineering, Northeastern University, Boston, MA 02115, USA
[4]School of Physics and Astronomy, University of Leeds, LS2 9JT, UK

E-mail: py06mjm@leeds.ac.uk

Abstract. Iron-rhodium (FeRh) nanoislands of equiatomic composition have been analysed using scanning transmission electron microscopy (STEM) electron energy loss spectroscopy(EELS) and high angle annular dark field (HAADF) techniques. Previous magnetometry results have lead to a hypothesis that at room temperature the core of the islands are antiferromagnetic while the shell has a small ferromagnetic signal. The causes of this effect are most likely to be a difference in composition at the edges or a strain on the island that stretches the lattice and forces the ferromagnetic transition. The results find, at the film-substrate interface, an iron-rich layer $\sim$ 5 Å thick that could play a key role in affecting the magnetostructural transition around the interfacial region and account for the room temperature ferromagnetism.

1. Introduction

Iron-rhodium (FeRh) alloys exhibit an unusual magnetostructural transition, making them an interesting area of research. When in an approximately equiatomic state ($Fe_{48}Rh_{52}$ to $Fe_{56}Rh_{44}$), the alloy is in an ordered α' state and arranged in a CsCl structure[1]. At room temperature, FeRh is antiferromagnetic, but makes a first-order phase transition upon heating to a ferromagnetic (FM) state at ~350 K[2].

The transition also has a uniform 1% increase[3] in unit cell lattice volume when heating to the ferromagnetic phase, which correlates to a lattice parameter increase of 2.987 Å to 2.997 Å[1] (a change of 0.3%). Other characteristics associated with the phase transition include an increase in entropy[4] and drop in resistivity[5] when changing to the FM phase as well as a ~10 K temperature hysteresis[6] in bulk. The properties of the phase change can be controlled by altering the structure of the FeRh *via* multiple methods, including ion beam irradiation[7] and chemical doping with various elements[8]. One other technique of altering the magnetic transition discussed here is growing FeRh epitaxially onto a substrate with a small lattice mismatch. The epitaxial growth causes a compressive or tensile strain on the FeRh lattice (depending on whether the lattice parameter of the substrate is larger or smaller than that of FeRh), thereby causing an increase in the transition temperature with a compressed lattice and vice versa[9].

Due to the magnetic phase change and its ability to be tuned to specific temperatures, FeRh has potential uses for spin valves in magnetic data storage or microelectromechanical systems[10].

There is an interest in using electron microscopy to fully characterise the FeRh system when epitaxially grown onto a substrate. Much work has been done on the magnetic properties of the transition but little has been done to relate these properties to the nanostructure of the thin films. In particular, there is much interest in the regions surrounding the interfaces because of the potential changes in structure and magnetism located here. In this paper we discuss FeRh samples that have grown in an island-like structure. These are found to have a ferromagnetic signal in the nominally AFM phase with close contact between the two regions as determined by SQUID magnetometry[11]. It is hypothesised that each nanoisland has an AFM core with a FM shell that surrounds it. The reason for this phase coexistence is unknown but could either be as a result of strain, a change in composition around the shell or a combination of the two.

2. Experimental

The FeRh film was grown by sputtering onto a magnesium oxide (MgO) substrate, deposited from a single target. The samples were annealed at 700 °C then capped in-situ with a 20-30 Å layer of Al using methods that have been described in detail elsewhere[11][12].

For observation in the TEM, the sample was prepared into a cross section by focused ion beam (FIB) using a FEI Nova 200 NanoLab dual beam FIB with a Ga ion source. A layer of platinum was deposited over the area of interest to protect from damage. A 30 kV beam at 20 nA excavated both sides of the region to a depth of 5 μm. The sample was then thinned further using a beam current of 100 pA. The sample was lifted out in-situ by welding to a micromanipulator and moved to a TEM sample holder.

The sample was characterised in the Nion UltraSTEM100 operated at 100 kV using high angle annular dark field (HAADF) scanning TEM (STEM) and STEM-electron energy loss spectroscopy (EELS) techniques with the EELS data denoised in DigitalMicrograph using principle component analysi[13].

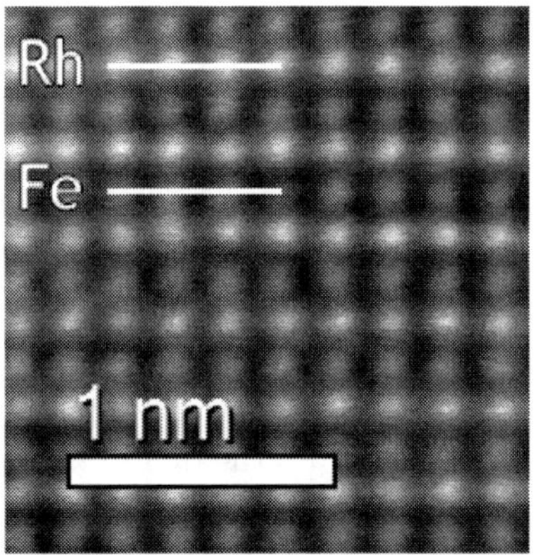

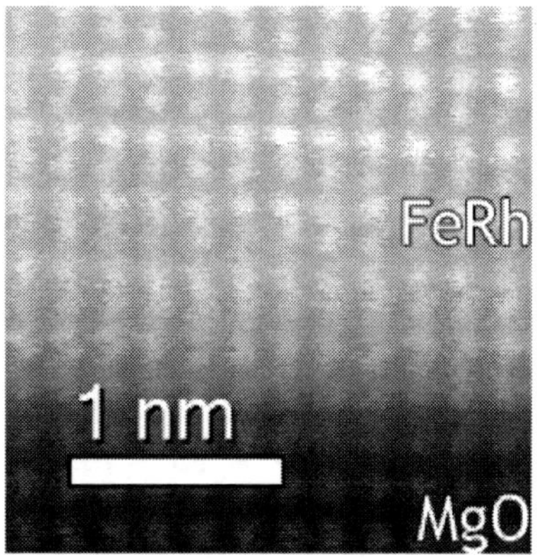

Figure 1. HAADF-STEM images showing the structure at the centre of the island averaged from an image stack (left) and at the interface with MgO (right). With both images, the expected CsCl structure can be seen viewing along the (110) zone axis, with alternating rows of Fe and Rh (dark and bright respectively, due to the differences in their atomic number), however any minor changes in FeRh composition are not observable through HAADF alone.

3. Results and Discussion

Figure 1 shows HAADF-STEM images comparing the structure in the middle of an island (a) and at the FeRh-MgO interface (b). The difference in contrast between Fe (dark) and Rh (bright) is visible due to the large difference in atomic number. Viewing the structure along the

(110) axis, the CsCl structure can be seen. X-ray diffraction techniques are generally applied to determine a chemical order factor, S, to determine the chemical order of the sample[12]. This value is typically ~ 0.8 for FeRh films grown using the described method, meaning 10% of Fe will be located on the Rh sites and vice versa. Directly detecting this chemical disorder in the HAADF of this scale is difficult in these images so it is unlikely that, unless the displacement of Fe and Rh is significant, a contrast change will be noticeable.

Figure 2a shows EELS maps for Fe $L_{2,3}$, Rh $M_{4,5}$ and O K edges at the FeRh interface with MgO. The intensities of Fe and Rh alternate with each row with a terminating Fe layer at the interface with MgO. This initial observation fits well with the expected crystal structure of FeRh. The oxygen signal drops significantly across the interface, showing no discernible diffusion into the FeRh. For the first two rows of Fe, the signal is overwhelmingly strong relative to the other rows. The first Rh row is extremely weak in comparison, suggesting that the Fe content in this area is higher than Rh. It is known that rhodium oxides do not readily occur due to their small free energies of formation[14]. In comparison, iron oxides have a greater tendency to form[15]. The fine structure data in figure 2b shows an O-K pre-peak indicative of a transition metal oxide bond[16].

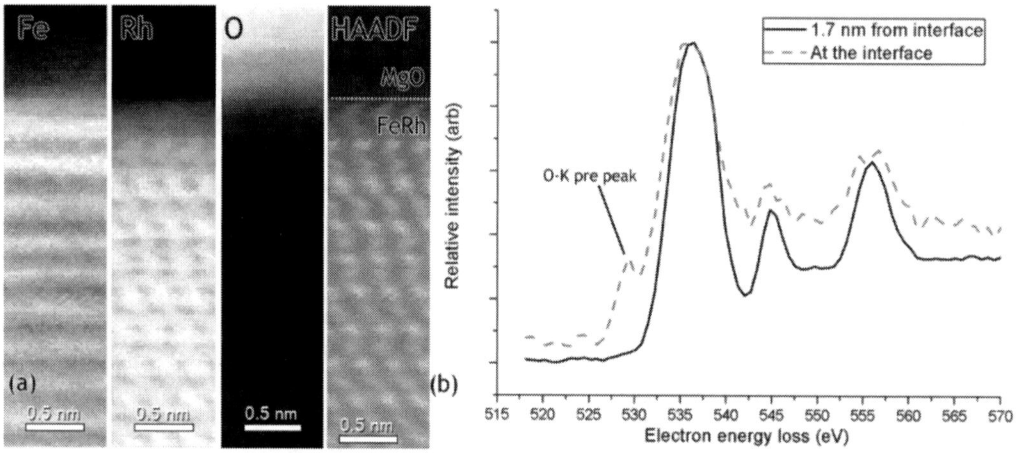

Figure 2. (a) STEM-EELS maps acquired at the FeRh-MgO interface, with maps for iron $L_{2,3}$, rhodium $M_{4,5}$ and oxygen K edges. A HAADF image of the same region is provided for comparison, with a dotted line indicating the FeRh-MgO interface. There is a variation in intensity for the first two Fe rows, which appear brighter than the other Fe layers. The distortion in the FeRh lattice is believed to be as a result of interaction between the beam and the magnetic sample. (b) Typical O-K spectra along and 1.7 nm from the interface into the MgO, showing the difference in fine structure. There is an obvious pre-peak on the interfacial spectrum indicating a transition metal oxide bond which, combined with the EELS maps, confirms an Fe terminating layer.

Taking 20 pixel integrated line profiles of the EELS maps obtained in figure 2a, the interfacial changes can be seen much more clearly. Figure 3 plots the relative intensities of the profiles with distance, with a comparison to the HAADF of the same area. The change in intensity for the first three layers of FeRh atoms is very clear, with the first two Fe layers appearing stronger and the first Rh layer showing a diminished peak. This result shows clearly that there is a relative change in composition at the interface that results in a narrow (~ 5 Å) Fe-rich band relative to the surrounding area. As these islands are expected to be equiatomic, there should be diffusion of Rh from this Fe-rich region to elsewhere in the island. However, the small compositional variations expected (e.g. $Fe_{48}Rh_{52}$) were not immediately detectable with the equipment used.

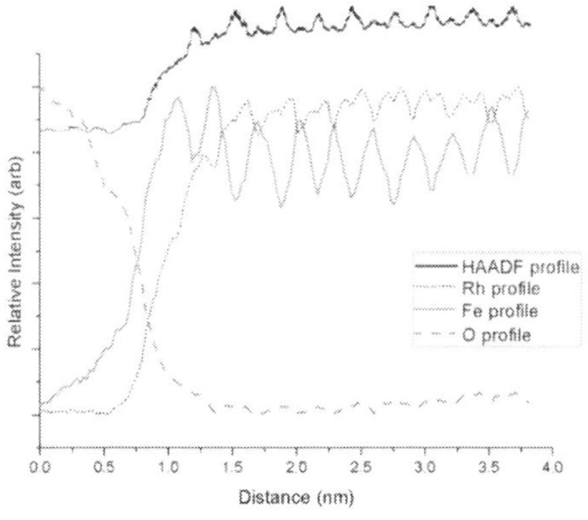

Figure 3. Line profiles taken from the above EELS maps, showing the relative change of Fe, Rh and O across the FeRh-MgO interface. The first two rows of Fe atoms have a noticeably stronger signal relative to the rest of the atoms and the first Rh layer (in between the two Fe rows) is diminished in comparison to the remaining layers. The HAADF profile for the same area is offset above. The intensity change at the interface cannot be seen on the HAADF.

4. Conclusions

The interfacial EELS maps show that, in general, the FeRh crystal is layered as expected with alternating Fe and Rh layers with a strong Fe signal detected for the first two rows of Fe atoms. Closer examination using line profiles shows very clearly that the Rh concentration is significantly lower in the first nominally Rh layer and the increased Fe signal is the result of an Fe-rich area in the first ~ 5 Å of the FeRh film at the interface with the substrate. An Fe-rich band along the interface would likely be ferromagnetic as expected from the FeRh phase diagram[1] and provide an explanation for ferromagnetic signal in the nominally antiferromagnetic phase[17]. A Rh-rich region, while not seen, is still quite likely as the FeRh is expected to be equiatomic and diffusion of Fe into the interface would result in displaced Rh. Further investigation with more quantitative analysis could confirm the diffusion of Rh to nearby areas and further investigate the role of composition and interdiffusion on the magnetic transition.

Acknowledgments

This grant has been supported by EPSRC grants EP/G065640/1, EP/F056311/1 and the National Science Foundation grant DMR-0908767.

References

[1] Swartzendruber L J, 1984 Bulletin of Alloy Phase Diagrams 5 45662
[2] Kouvel J S and Hartelius C C, 1962 *J. Appl. Phys.* **33** 1343
[3] Khan M J, 1979 *Phys. F: Metal Phys.* **9** 457
[4] Annaorazov M P, Nikitin S A, Tyurin A L, Asatryan K A, and Dovletov A K,h 1996 *J. App. Phys.* **79** 3
[5] Kouvel J S 1966 *J. App. Phys.* **37** 1257
[6] Ibarra M and Algarabel P, 1994 Phys. Rev. B 50 4196
[7] Aikoh K, Kosugi S, Matsui T and Iwase A, 2011 *J. App. Phys.* **109** 07E311
[8] Schinkel C, Hartog R, and Hochstenbach F, 1974 *J. Phys. F: Metal Phys.* **4** 1412
[9] Maat S, Thiele J and Fullerton E, 2005 *Phys. Rev. B* **72** 21
[10] Thiele J ,Maat S and Fullerton E, 2003 *Appl. Phys. Let.* **82** 17
[11] Loving M et al, 2013 *J. Phys. D* **46** 16
[12] Le Graet C et al, 2013 *J. Vis. Exp.* **in press**
[13] M. Watanabe et al, 2007 *Microsc. Microanal.* **13** 1264. MSA plugin is available from www.hremresearch.com
[14] Jacob K T, Prusty D 2010 *J. Alloys Compd.* **507**
[15] Ragone D 1994 *Thermodynamics of Materials* (Wiley)
[16] Cakvert C C, Brown A, Brydson R 2005 *J. Electron. Spectrosc. Relat. Phenom.* **143**
[17] Fan R et al, 2010 *Phys. Rev. B* **82** 2-6 184418

Characterization of nanoembedded alloyed thermoelectrics

L. Lari[1,2,3], C.J. Nuttall[4], M.P. Copley[4], R.J. Potter[4] J. Simon[5], N. Mingo[5], and D.Ozkaya[4]

[2]Department of Physics, University of York, Heslington, YO10 5DD, York, UK.
[3]The York JEOL Nanocentre, York Science Park, Heslington, YO10 5BR, York, UK
[4]Johnson Matthey Technology Centre, Sonning Common, Reading, RG4 9NH, UK
[5] CEA–LITEN, 17 rue des Martyrs, BP166, 38042 Grenoble, France

E-mail: leonardo.lari@york.ac.uk

Abstract. Nano-structured alloyed materials have been demonstrated with a large improvement of their thermoelectric properties. These improvements have been obtained by tailoring a reduction in the material thermal conductivity. However, such improvements have not yet been demonstrated for bulk materials. Recently, a new approach based on phonon scattering theory has been successfully applied to bulk materials. In this approach, nanoparticles are embedded in a hosting matrix and act as multi-scale phonon scattering centres providing a reduction in the material thermal conductivity. In this work, {$CoSi_2$ (nanoparticle): SiGe (host)} nanocomposite thermoelectric materials and their precursors for the synthesis were investigated. The purpose of the investigation is to understand the inherent interactions of the nanoparticles with the support material.

1. Introduction

The performance of a thermoelectric device, in terms of energy conversion, depends on its thermoelectric *figure of merit* (*ZT*) which is defined as $ZT=S^2\sigma T/k$, where k, T, σ, and S are respectively the thermal conductivity, the absolute temperature, the electrical conductivity, and the Seebeck coefficient [1]. While it has been a challenge for decades to increase *ZT* beyond unity, the so-called "alloy limit", in the last decade $ZT>1$ has been demonstrated by nanostructuring thermoelectric materials [2,3]. These improvements have been obtained by tailoring a reduction in the material thermal conductivity. However, such improvements have not yet been demonstrated for bulk materials. Recently, a new approach based on phonon scattering theory has been successfully applied to bulk materials demonstrating a $ZT\sim1.7$ [4]. In this approach, nanoparticles (NPs) are embedded in a hosting matrix and act as multi-scale phonon scattering centres providing a reduction in the material thermal conductivity [5]. In this work, {$CoSi_2$ (NPs): SiGe (host)} nanocomposite thermoelectric materials and their precursors for the synthesis were investigated. $CoSi_2$ was chosen here because of a close cell parameters match (within 2%) with the hosting SiGe matrix. Due to similar contrast for precursors and product NPs in the images, as well as a close lattice match between matrix grains and product NPs, analytical and high resolution electron microscopy is needed to verify the transformation

[1] To whom any correspondence should be addressed.

occurrence during the material synthesis. Here we apply these techniques to understand the inherent interactions of the nanoparticles with the support material.

2. Methods

Two samples, labelled here *A* and *B* were provided for electron microscopy investigations. Sample *A* consisted of Co_3O_4 nanoparticles (NPs) in a hosting $Si_{0.8}Ge_{0.2}$ matrix. A homogenised powder was obtained from the sample using standard industrial mixing-milling equipment. Sample *B* was obtained from Sample *A* after proprietary thermal treatment of the homogenised powder under controlled catalytic environment.

TEM samples were prepared by using the following method: firstly holey carbon films supported by standard Cu TEM grids were heat treated by contact with a hot surface (lamp bulb) for 30 second on each side with temperature ramping from about 80 °C to 130°C; powders were then finely crushed between two glass slides and the grids were finally dusted with the crushed powders.

Electron microscopy analyses were performed using a CEOS double aberration corrected JEOL 2200FS (scanning) transmission electron microscope operating at 200keV equipped with an in-column Omega filter and a Thermo Scientific Noran 7 EDX detector system. Samples were "beam showered" [6] for 20min at the beginning of the investigations to limit hydrocarbon contamination.

3. Result and discussion

Figure 1a) shows a characteristic overview bright field (BF) STEM image of the homogenised powder from Sample *A* dispersed on holey carbon film (scale bar 1 micron). Nanoparticles are combined with the hosting matrix in clusters up to few microns in size. Figure 1b) shows a high resolution (HR) BF-STEM image of a NP with lattice resolution. Fast Fourier Transform (FFT) analysis was applied to the area in the red box inset in Figure 1b). Figure 1c) is a Bragg filtered image obtained by selecting the major Bragg spots (see inset) in the FFT image. Lattice spacing as shown in the FFT (figure 2d) were measured and compared with simulated diffractograms obtained using JEMS software [7].

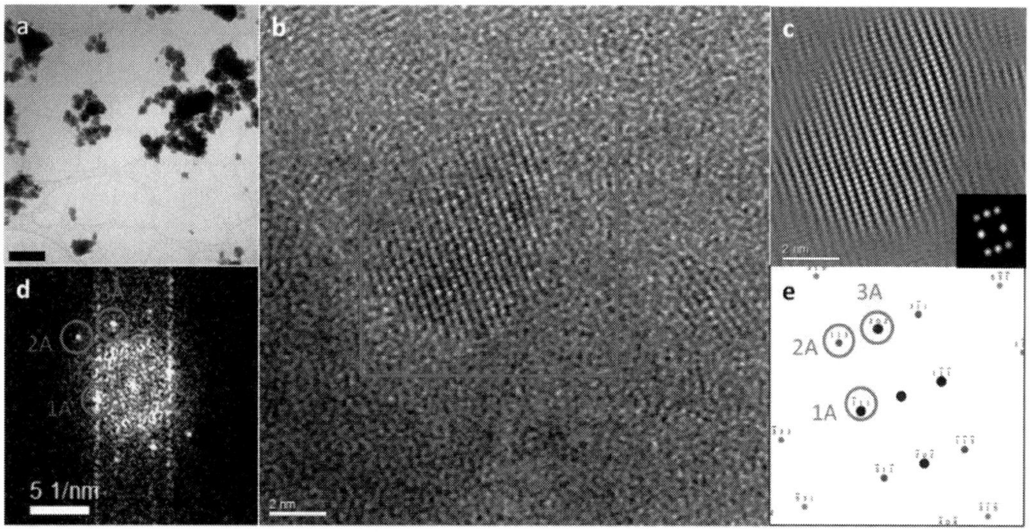

Figure 1. a) Overview of the sample (scale bar 1micron); b) BF-STEM image of a NP on the matrix; c) FFT filtered image using Bragg spots as in the inset; d) FFT from the area in the red box in b), spot labels refer to Table 1; e) simulated diffraction pattern for $Si_{0.8}Ge_{0.2}$ along the [-1-21] zone axis

Lattice parameters were found to be consistent within 1-2% with $Si_{0.8}Ge_{0.2}$ (a=0.54722 nm, space group Fd-3m) [8] imaged along the [-1-21] zone axis as in Figure 1e) (see Table 1 for the details). The closest diffractogram configuration for Co_3O_4 (a=0.8084 nm, space group Fd-3m) [9] was found to be along [11-6] zone axis direction with a discrepancy of about 10% on the expected value. Other crystallographic directions for the two crystals were excluded being inconsistent with the measured angle and distances. This result shows that the mix-milling process is able to finely grind the matrix down to nanoparticles with 5-10 nm in diameter.

The presence of the precursor oxidised Co nanoparticles (namely Co_3O_4) onto the matrix of sample *A* was confirmed by EDX analysis. The HAADF image in Figure 2a) shows a large grain decorated by NPs with sizes up to 20nm. EDX point analyses were applied to the NPs as well as the matrix. Co:Si and O:Si ratios are larger when the spectrum is acquired on the particle (point 2 Figure2) respect to when on the matrix (point 1 figure2) indicating a Co-O compound for the particle.

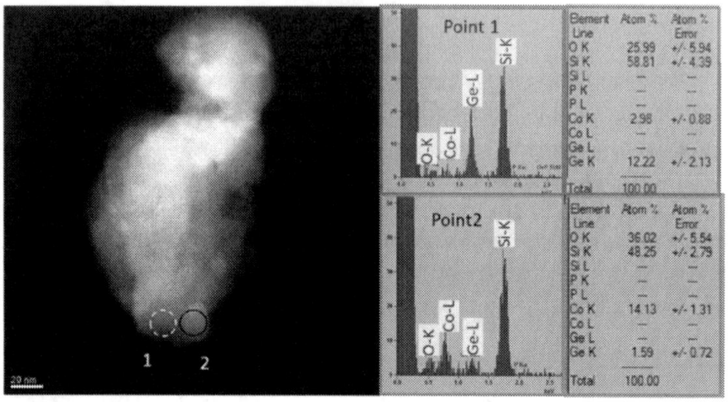

Figure 2. Sample *A*; EDX point analysis from the hosting matrix (point 1) and a Co_3O_4 particle on top of the matrix (point 2).

After thermal treatment in catalytic environment the material (Sample *B*) was investigated to look for traces of the transformation of the precursors Co_3O_4 nanoparticles into $CoSi_2$. FFT analysis of HR-STEM images was applied to NPs found on the surface of the matrix (see Figure 3a). The lattice spacing visible along the zone axis imaged were measured in the Fourier transform and compared against the possible combination of crystals available within the set of elements present in the sample. These include: Co_3O_4[9], CoGe (high pressure crystal, a= 0.4637nm, space group P 21 3)[10], Co_2Ge (space group P n m a) [11], $Co_{0.9}Ge_{0.1}$ (a=0.35515 nm, space group F m -3 m) [12], $Co(SiO_3)$ (space group P b c a) [13], polymorph crystals of cobalt orthosilicate $Co_2(SiO_4)$ (I m m a, P b n m [14]; F d -3 m S[15]), and CoSi (a= 0.4438 nm, space group P 21 3)[16]. Analysis of the lattice spacings and their relative angles show values are consistent within 1% experimental error with $CoSi_2$ imaged along [001] zone axis as shown in Figure 3d) and Table 1. However, also SiGe grains along the same zone axis show close distances within 2-3%. This is expected since this combination of materials is chosen to minimise the lattice mismatch between hosting matrix and nanoparticles. Figure 3e) shows the simulated diffractogram for SiGe with additional crossed spots corresponding to Bragg spots with half lattice distance visible in the FFT which would be a forbidden reflection on a real electron diffraction pattern.

Table 1. FFT analysis of particles as from Figure 1 and 3.

Spot #	*lattice planes*	*Measured lattice (nm)*	*Expected values (nm)*	*Measured angle to Spot 1*	*Expected angle to Spot 1*
Results from image 1A respect to Si0.8Ge0.2					
1A	(-111)	0.320	0.3159(1.3%)	0.00	0.00
2A	(113)	0.168	0.1650(1.8%)	60.2	58.52
3A	(202)	0.194	0.1935(0.2%)	91.5	90.00
Results from image 3a respect to CoSi2					
1Ba	(-220)	0.188	0.1897 (+0.7%)	0.00	0.00
2Ba	(-200)	0.268	0.2682(-0.1%)	44.40	45.0
3Ba	(0-20)	0.267	0.2682(+0.4%)	135.46	135.0
Results from image 3a respect to Si0.8Ge0.2					
1Bb	(2-20)	0.188	0.1935 (-3%)	0.00	0.00
2Bb	(200)	0.268	0.2737 (-2%)	44.40	45.0
3Bb	(020)	0.267	0.2737 (-2%)	135.46	135.0

To ascertain the presence of Co so the composition of the particle shown in Figure 3 can be attributed to $CoSi_2$, EDX spectra were acquired from the particle+matrix (point *pt1*) and from the area adjacent the particle (matrix only) (dashed box in figure3a, point *pt2*). Spectra are shown overlapped in Figure 3F with normalisation done on the area of the Ge-Lα1 peak. The spectrum from the particle+matrix shows larger Co and Si peaks. Quantification with calculated K factors, under thin film approximation, confirms that with the same amount of Ge on the particle+matrix and on the matrix, the changes in Co are due to the particle being $CoSi_2$ and not $Si_{0.8}Ge_{0.2}$.

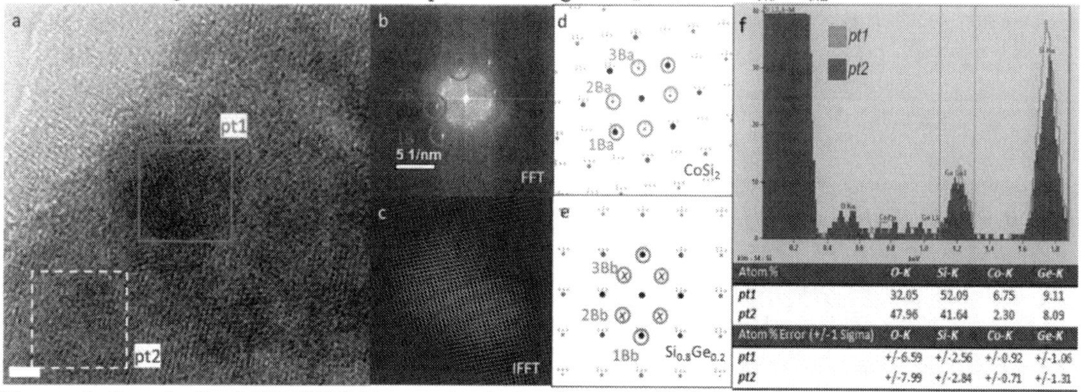

Figure 3. Sample *B*; a) BF-STEM image (scale bar 2nm); b) FFT from red box in a); c) Inverse FFT selecting 1B,2B,3B spot pairs; d) simulated DP for $CoSi_2$ along [100]; e) simulated DP for $Si_{0.8}Ge_{0.2}$ along [100]; f) EDX analyses from red and dashed boxes in Figure 3a).

4. Conclusions

In this paper a thermoelectric nano-composite material and its precursors were investigated by high resolution and analytical STEM. Structural analysis shows that the mix-milling preparation of the precursor materials produced a mixture of Co_3O_4 NPs, large grain and NPs from the SiGe matrix. Cobalt from the dispersed Co_3O_4 nanoparticles interacts with the SiGe host and forms $CoSi_2$ as a result of the proprietary thermal treatment. This is a desired transformation and should result in improved performance.

Acknowledgements

Financial support from EU Seventh Framework Programme FP7 NEAT "Nanoparticle Embedded in Alloys Thermoelectrics" is gratefully acknowledged.

References

[1] Majumdar A., 2004 Science **303**
[2] Hsu K. F. *et al.*, 2004 Science **303** 818.
[3] Harman T. C. *et al.*, 2002 Science **297** 2229
[4] Kim W *et al.*,2006 Physical Review Letters 045901
[5] Mingo N. *et al.*, 2009 Nanoletters **9** 711
[6] Egerton RF, et al. 2004 Micron **35** 399
[7] P. Stadelmann software http://cimewww.epfl.ch/people/stadelmann/jemswebsite/jems.html
[8] Dismukes J.P. *et al.* 1964 *J. Phys. Chem.***68** 3021-7
[9] Smith, W.L. *et al.* 1982 Acta Crystallographica B **24** 1968
[10] Larchev, V.I. *et al.* 1982 Journal of the Less-Common Metals **87** 53-57
[11] Panday, P.K. *et al.* 1969 Journal of the Less-Common Metals **18** 175-202
[12] Lecocq, P. *et al.* 1963 Annales de Chimie (Paris) **1963** 85-116
[13] Sasaki, S. *et al.* 1982 Zeitschrift fuer Kristallographie **158** 279-297
[14] Morimoto, N. *et al.* 1974 American Mineralogist **59** 475-485
[15] Marumo, F . *et al.* 1977 Acta Crystallographica B **33** 713-716
[16] P.A.Doyle and P.S.Turner 1968 Acta Cryst. **A24** 390.

Electron microscopy of multi-layered polymer-nanocomposite based dielectrics

Amoghavarsha Mahadevegowda, Neil P Young and Patrick S Grant
Department of Materials, University of Oxford, Parks Road, Oxford, OX1 3PH, UK
E-mail: amogh.gowda@materials.ox.ac.uk

Abstract. Vacuum deposited polymer-nanocomposites (PNCs) comprising alternate layers of metal (Al/Ag) (filler) and polymer (nylon-6) (matrix) have been investigated using chemical, impedance spectroscopy and microstructural characterisation techniques. Electron microscopy investigations revealed the morphology, nanostructure and phases of nano-scale core (metal)-shell (oxide) particles and metallic nano-islands in Al and Ag based PNCs respectively. Evaporation of Al yielded islands of angular core-shell nanoparticles in an Al-oxide/nylon-6 matrix whereas Ag yielded rounded, discrete nanoparticles in nylon-6 matrix. The high particle surface area and an affinity for oxygen formed oxide shells in Al nanoparticles and was critical to charge accumulation and enhanced dielectric behaviour; in contrast, Ag showed little oxidation and less charge accumulation. With an increase in the thickness of the deposited metal layer, Al formed a continuous film of particles whereas Ag condensed to form interconnected nano-islands. This microstructural study is useful in conceptualising better dielectrics based on PNCs.

1. Introduction

Polymer-nanocomposites (PNCs) consisting of a highly polarisable minority phase (filler) in an easily processable polymer (matrix) are promising dielectrics for capacitive applications. Recently, core-shell particles have been used as fillers to yield PNCs with a high dielectric constant k of up to ~60 [1]. Experimental insights [2] and computational studies [3] have suggested that the shell thickness of the core-shell particles influences their polarisability, along with filler shape, size, and size distribution [4–6]. Therefore, knowledge of the chemical and structural arrangement of phases at fine-scale is crucial to understand capacitative mechanisms in PNCs and to guide the design and fabrication of polymer-based capacitors with enhanced properties.

In this paper, we use a simple vacuum based deposition technique to fabricate Al/Ag and nylon-6 based PNCs. The formation and growth of Al/Ag nanoparticles on the nylon-6 surface and their encapsulation is studied by employing various electron microscopy techniques and microstructure is correlated with the PNC dielectric response. Based on the importance of the formation of *in-situ* interfacial layers on reactive metals for charge accumulation, different nano-arrangements of Al and Ag are proposed, and the resulting dielectric behaviour is described and compared.

2. Experimental

Alternately-deposited nanocomposite (ANC) films were fabricated by tandem deposition of five individual layers of metal and nylon-6 in an Edwards Auto 306 Cryo evaporator, where the evaporator-chamber was evacuated to a pressure of $5x10^{-4}$ Pa before evaporation. Aluminium wire (99.999% pure, Alfa Aesar, 0.5 mm diameter) and nylon-6 (Aldrich) were thermally evaporated alternatively using two different tungsten boats onto a glass slide at a deposition rate of 0.5 nm/s and 0.1 nm/s respectively. The top and bottom (1st and 5th) layers of Al (40 nm each) formed the electrodes of a simple parallel-plate capacitor. The growth of Al (3rd layer), sandwiched in between the nylon-6 layers (2nd and 4th - 20 nm each), was studied at various thicknesses (5, 10, and 15 nm). Silver based ANCs were fabricated by replacing Al with Ag (99.99% pure, Advent, 0.125 mm diameter) in the W-boat with a similar procedure. The resulting capacitors were dried at 40^0C for 20 hours to negate any water vapour effects, and then the dielectric properties were studied using a Solartron SI 1260 impedance analyzer operating at 1 kHz. The thickness of the ANC films was measured using a DekTak 6M profilometer (Veeco Instruments, Inc) and allowed estimation of k from the impedance data.

JEOL transmission electron microscopes, JEM-2000FX and JEM-3000F, operating at 200 kV and 300 kV respectively were used for ANC characterisation on identical samples produced in parallel but with no top/bottom Al-electrodes. Energy dispersive X-ray spectroscopy (EDS) and scanning transmission electron microscopy (STEM) were carried out via the 3000F.

3. Results and discussion

Figure 1 shows the variation of dielectric constant of Al (■) and Ag (•) of the ANCs with the thickness t of deposited metal layer at 1 kHz and 40^0C. For reference, k of "pure" deposited nylon-6 was ~4.3. At t=15 nm, k of Al-ANCs was approximately twice that of Ag-ANCs; the k of both ANCs also tended to increase with increasing t.

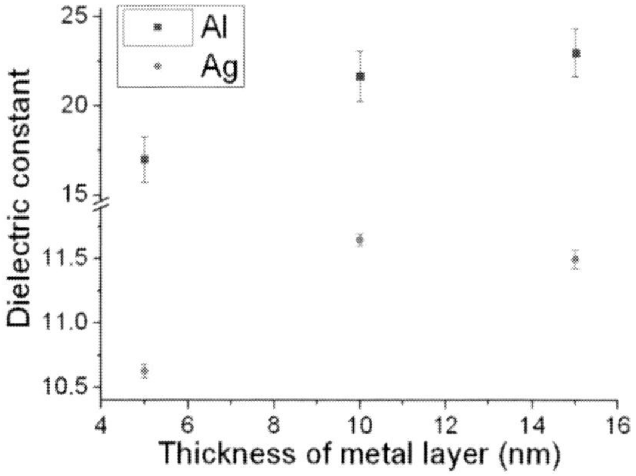

Figure 1: Variation of dielectric constant k of Al (■) and Ag (•) of the ANCs with the thickness t of metal layer at 1 kHz and 40^0C.

Bright field (BF) TEM images of Al-ANCs at t=5 nm and t=15 nm are shown in Figs. 2(a) and (b) respectively. The images show dark nanoparticles of 5-60 nm diameter dispersed in a relatively bright matrix. The presence of any Al-oxide shell was not resolvable from the nylon-6 matrix in the BF-TEM images and both were expected to be substantially or wholly amorphous. Nonetheless, previous work involving the evaporation of Al [7] and the pervious nature of nylon-6 strongly suggested the oxidation of Al. Figure 2 suggested that the increase in k from 17±1 (at t=5 nm) to 23±1.5 (at t=15 nm) was due to an increase in the size of the Al nanoparticles. Figure 2(c) shows a high resolution transmission electron microscopy (HRTEM) image of the Al-ANC. The d-spacings measured from the HRTEM image and the spots from the fast Fourier transform (FFT) of the image [Fig. 2(d)] corresponded to Al [d~0.23 nm (111) and d~0.2 nm (002)].

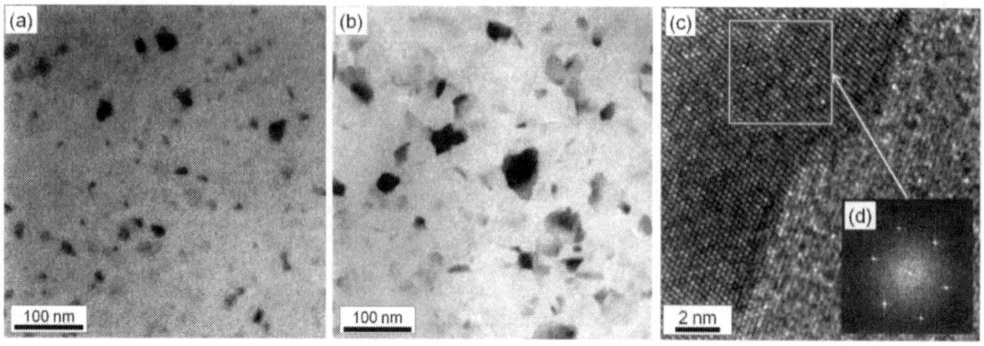

Figure 2: Bright field (BF) TEM images of Al-ANCs at (a) t= 5 nm and (b) t=15 nm with a bright matrix containing relatively dark nanoparticles; (c) a HRTEM image of the Al-ANC and (d) corresponding FFT from highlighted region.

A further BF-TEM image of Al-ANC at t=5 nm is shown in Fig. 3(a) with the marked region containing nanoparticles. This region was darker than the nylon-6 "matrix", which appeared as a light contrast, branched dendritic continuous phase separating dark regions. The contrast was due to differences in atomic number Z of Al and lighter elements in nylon-6 which is shown more clearly in the annular dark field (ADF)-STEM image in Fig. 3(b). The crystalline nanoparticles appeared very bright due to high diffraction contrast. The corresponding EDS map in Fig. 3(c) revealed that Al was also distributed throughout the film, not only in the nanoparticles, and suggested that the evaporated Al not only formed discrete, crystalline nanoparticles but was also relatively uniformly deposited as aluminium oxide-based islands delineated by the branched nylon-6 "matrix". A cross-sectional schematic representation of this arrangement is shown in Fig. 3(d).

Figures 4(a) and (c) show BF-TEM images of Ag-ANCs at t=5 and 15 nm respectively. Spherical nano-islands of Ag formed at t=5 nm, and these spheres grew and formed inter-connected islands at t=15 nm due to Volmer-Weber growth [8]. The corresponding ADF-STEM image in Fig. 4(b) confirmed discrete Ag-islands (bright) in a nylon-6 matrix (dark). A cross-sectional schematic representation of Ag-ANC based capacitor at t=5 nm is shown in Fig. 4(d).

The enhancement in k for Al was primarily due to the differences in the oxygen reactivity of Al ($\Delta G_{Al_2O_3}$=-1582.3 kJ/mol at 25^{0}C) and Ag (ΔG_{Ag_2O}=-11.21 kJ/mol at 25^{0}C). Although the ANCs were processed in vacuum, it is well-known that Al will nonetheless undergo some

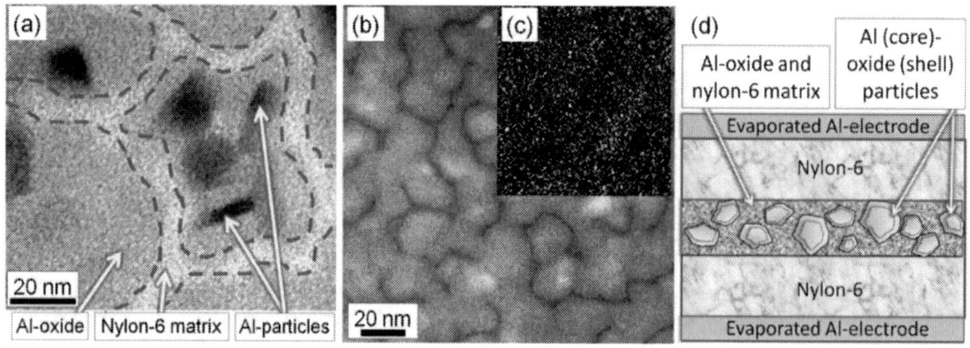

Figure 3: (a) A BF-TEM image of Al-ANC at t=5 nm, (b) corresponding ADF-STEM image, (c) corresponding EDS map and (d) a cross-sectional schematic of the Al-ANC microstructure.

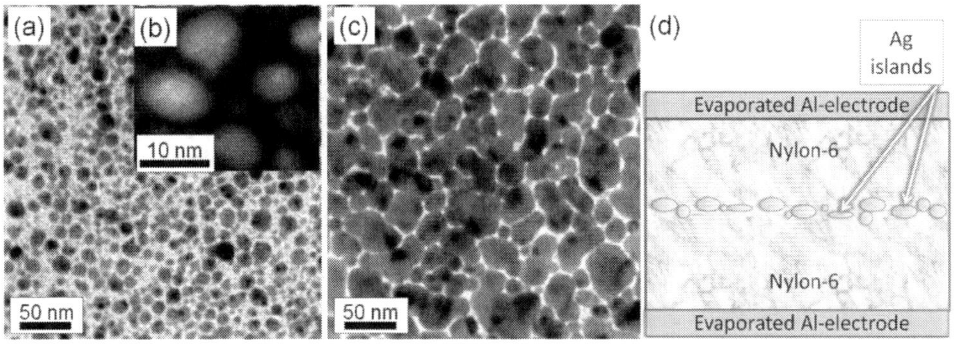

Figure 4: BF-TEM images of Ag-ANCs at (a) t=5 and (c) 15 nm. The formation of discrete nano-islands of Ag on nylon-6 at t=5 nm is seen in the ADF-STEM image, (b). A cross-sectional schematic representation of Ag-ANC based capacitor at t=5 nm is shown in (d).

oxidation either *in-situ* [7] or on removal from vacuum. In this case, deposited Al particles formed Al (core)-aluminium oxide (shell) particles. The presence of Al (core)-aluminium oxide (shell) particles leads to an enhancement in interfacial polarisability resulting in high-k. Interfacial polarisation takes place due to the piling up of charges at the heterogeneous interfaces of the core-shell particles and polymer. Although there were interfacial polarisation effects in the Ag-ANCs, the increase in k was less pronounced, since oxide formation was much reduced or absent.

4. Conclusions

Dielectric materials consisting of Al or Ag in nylon-6 films were fabricated by an alternate vacuum deposition technique. Electron microscopy studies revealed that the deposition of Al led to the formation of aluminium (core)-aluminium oxide (shell) nanoparticles, a branched nylon-6 matrix with fine-scale aluminium oxide "islands". On the other hand, because Ag was much less-reactive with ambient gases (especially oxygen), it formed well-separated, spherical Ag-nanoparticles. Enhanced interfacial polarisation in Al-ANCs resulted in k that was two-times higher than Ag-ANCs, attributed to the formation of a comparatively polarisable oxide containing shell around an aluminium nanoparticle core.

Acknowledgments

The authors would like to thank the UK Engineering and Physical Sciences Research Council grant EP/I034548/1 and a Felix Scholarship for financial support.

References

[1] Zhang Y, Wang Y, Deng Y, Li M and Bai J 2012 *ACS Appl. Mater. Interfaces* **4** 65
[2] Dang Z M, You S S, Zha J W, Song H T and Li S T 2010 *Phys. Status Solidi A* **207** 739
[3] Wang Y U, Tan D Q and Krahn J 2011 *J. Appl. Phys.* **110** 044103
[4] Xu P and Li Z Y 2004 *Physica B* **348** 101
[5] Tang H, Lin Y, Andrews C and Sodano H A 2011 *Nanotechnology* **22** 015702
[6] Rasul A, Zhang J, Gamota D, Singh M and Takoudis C 2010 *Thin Solid Films* **518** 7024
[7] Silvain J F and Ehrhardt J J 1993 *Thin Solid Films* **236** 230
[8] Venables J A, Spiller G D T and Hanbucken M 1984 *Rep. Prog. Phys.* **47** 399

Quantitative Electron Tomography of Rubber Composites

Lech Staniewicz[1], Thomas Vaudey[2], Christophe Degrandcourt[2], Marc Couty[2], Fabien Gaboriaud[2] and Paul Midgley[1]

[1] Department of Materials Science and Metallurgy, University of Cambridge, Pembroke Street, Cambridge, CB2 3QZ, UK

[2] Manufacture Française des Pneumatiques Michelin, 23 Place des Carmes Déchaux, 63040 Clermont Ferrand Cedex 9, France

E-mail: ltls2@cam.ac.uk

Abstract.

Rubber composite materials have many applications, one example being tyre manufacture. The presence of a filler material in the composite (such as carbon black or silica) causes its mechanical properties to differ in several ways when compared to pure rubber such as viscoelastic behaviour (the Payne effect), increased tensile strength and improved wear resistance. To fully understand these properties, it is necessary to characterise how the filler material is organised on the nanoscale. Using composite materials representative of those found in tyres, this work illustrates the use of electron tomography and machine learning methods as tools to describe the percolation behaviour of the filler; in this case, we focus on the largest proportion of particles absorbed into one single object as a function of particle spacing.

1. Introduction

The change in the mechanical properties of a rubber-filler composite material with the concentration of filler particles was first characterised by Payne in 1962 [1]. He demonstrated that the bulk modulus increased with both the quantity of filler particles in the composite and the effective surface area of the filler particles themselves. Filler particles will form a percolating network when the quantity present in the material exceeds a certain threshold, as shown by a sudden rise in electrical conductivity [2] at the threshold filler concentration. The mechanical properties of a rubber composite show a dependency on the filler concentration [3, 4] with a similar percolation transition - implying that there is a critical distance between filler particles before an electrical or mechanical network can be formed.

The structural properties of rubber-filler composites are key issues in designing new materials. As these composites are heterogeneous, bulk measurements of the volume fraction alone are not sufficient and local, nanoscale information is necessary to fully understand these materials and their properties. Electron tomography enables a complete 3D volume of a nanoscale object to be reconstructed and can provide quantitative structural information from this volume.

Previous work [2] has emphasised the percolation structure of filler particles by skeletonising networks of particles. In contrast, this work is intended to determine what proportion of the filler contributes to the percolation network by measuring the distribution of particle spacings.

Journal of Physics: Conference Series **522** (2014) 012042 doi:10.1088/1742-6596/522/1/012042

2. Experimental Methods

5 μm long, 100–150 nm diameter cylindrical pillars were fabricated from rubber composite samples using an FEI Helios FIB/SEM dual-beam instrument and mounted on Omniprobe TEM grids. Pillars were imaged in an FEI Tecnai F20 instrument at 200 kV under HAADF STEM conditions (detector collection angle of 23.7–118 mrad) using a pixel size of 0.56 nm, a dwell time of 11.7 μs and an image size of 1024x1024 pixels (field of view: 573x573 nm). A tilt series was recorded using the FEI Xplore3D software from $-76°$ to $+76°$ with automatic focus and manually-assisted tracking. Tilt series were aligned using cross-correlation and reconstructed using 20 iterations of the SIRT algorithm [5] with the FEI Inspect3D software.

3. Data Processing and Results

Because the exact shape of each filler particle is intimately linked to the inter-particle spacings, it is important that the reconstruction is not biased in any post-processing operation. Although image filters such as anisotropic diffusion may improve the contrast and make it easier to delineate the boundaries of filler particles, these operations may change the position of the particle boundaries and introduce inaccuracies into any measurements made.

Unfortunately, SIRT reconstructions are not 3-level images. The voxel intensity values for each component (vacuum, rubber, filler) overlap and so a simple thresholding algorithm will not provide an accurate distinction between particle and matrix. Figure 1 illustrates the effect of a 3-level Otsu thresholding operation carried out on one "Z-slice" of a tomographic reconstruction, along with one example of mis-classification.

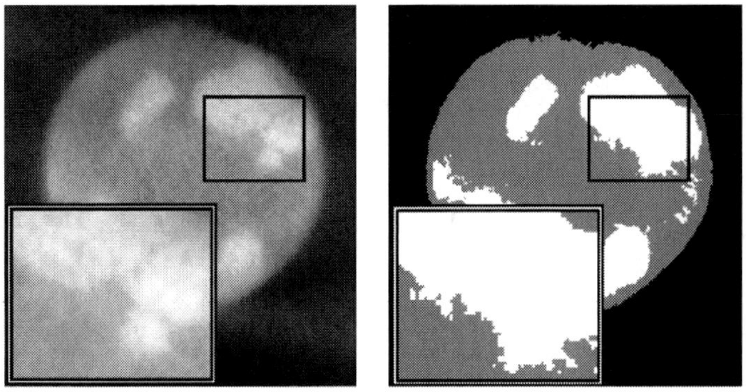

(a) (b)

Figure 1: The results of running a 3-level Otsu thresholding algorithm (b) on an unprocessed reconstruction (a). Inset regions are 2x magnified - the raw data is visibly more concave than the Otsu-thresholded area. Pillar diameter = 110 nm.

A more accurate option is to have a trained operator go over each image and manually delineate parts of the image as belonging to a specific phase. While a human operator can easily work around uneven particle intensities (such as in Figure 1), manual classification is time-consuming, susceptible to operator bias and will not give identical results when repeatedly applied to the same data due to human error.

Machine learning tools can be used to find a compromise between these two extremes. They function by having the user mark portions of the unchanged source image as belonging to one specific class or phase, just as in manual image classification. The software is then taught to recognise those phases by analysing the marked regions on "feature" images which are generated by running image filters (such as gradient operations and smoothing) on the source image. Some form of decision algorithm is then used to identify the un-marked parts of the image (and other images, if operating on an image stack) as belonging to one of the defined classes and a map is created from this information. Any mis-classifications can be corrected and the procedure repeated until the operator is satisfied with the result. Because the original image is used as the reference for the human operator to compare and correct mistakes against and because multiple filtered images are taken together to obtain a result, any applied filtering operations should not alter the quantitative nature of the data.

This method offers a combination of the potential accuracy of a trained human operator with the repeatability of an automated routine (the classification data can be saved for later use). Training the software, although much faster than manually marking out every particle in a reconstruction image stack, can be time consuming. Classifier training on this reconstruction took approximately 16 hours of operator time, including processing time for the software.

This work uses a machine learning package called Trainable Weka Segmentation, implemented in the freely-available Fiji software [6]. It offers a choice from multiple decision algorithms and feature image - we used the default random forest algorithm [7] with the following features: Gaussian Blur and Difference of Gaussians (feature size detection); Sobel filter, Laplacian filter, Derivatives, Gabor filter, Structure Tensor, Membrane Projections and Hessian Matrix (edge detection); and Variance, Anisotropic Diffusion, Bilateral filter, Kuwahara filter, Entropy and Neighbours (pixel intensity detection).

Since the entire feature stack is held in memory while training, using such a large number of features requires a great deal of RAM. Therefore, training is carried out on a substack consisting of every 15th image in the reconstruction, taken along the pillar axis. The pillar reconstruction used in this work is 250x250x950 voxels in size and the training procedure on a 1-in-15 substack consumed approximately 10 GB of RAM. Applying the final classifier to the full reconstruction took approximately 6 hours of processing time on an Intel Core i7-2600 CPU and used substantially less RAM, since each image slice is calculated separately.

Figure 2 shows the results of a trained classifier on the same data as in figure 1. The classified image appears to match the raw data much more accurately than with global thresholding. Since the source image has remained unchanged, there is no ambiguity as to whether any filtering process has altered the data and so the results can be considered accurate enough for pixel-level quantification.

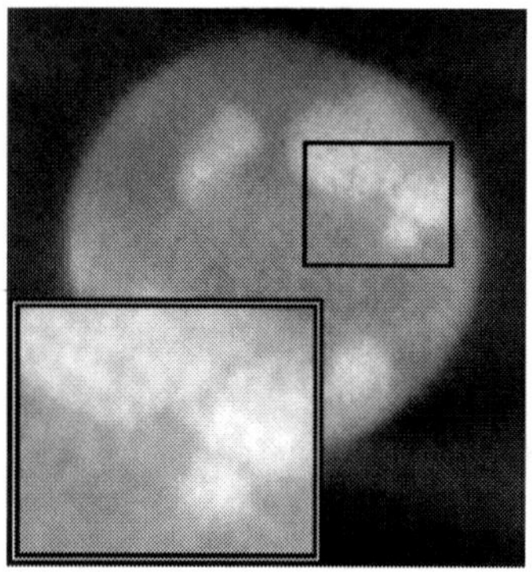
(a)

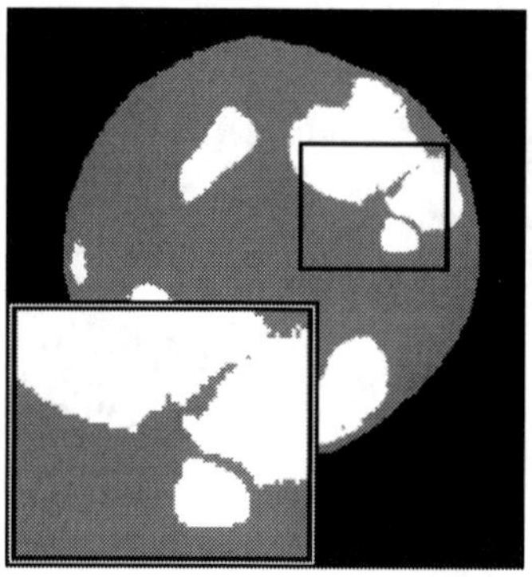
(b)

Figure 2: The same reconstruction as in Figure 1, but processed with the Trainable Weka Segmentation software. Inset region is the same as in Figure 1. Pillar diameter = 110 nm.

The classified data is then segmented and percolated. Percolation is quantified by the closest approach distance between two particles and the degree of percolation by how many of them are included within one network for any given "added radius" or particle spacing.

Numerical values for percolation are obtained using a custom C++ program. The classified volume is segmented into discrete objects, which are then dilated by one voxel and segmented again. If two or more particles merge, the new one is given a list of which original particles it contains. This process repeats until there is only a single object remaining. The final data is presented as the largest fraction of the original particles assimilated into any one single object (the percolation fraction) against the particle separation (2 x number of dilate steps x voxel size). Figure 3 is the result of running this procedure on the machine learning-classified volume.

From this, we can see that 25% of the particles percolate at a spacing of 1.3 nm, 50% of the particles percolate at 4.8 nm and 75% of the particles percolate at 16.1 nm. Because these spacings are very close to the voxel size (0.56 nm), it is easy to recognise the importance of accurately classifying the reconstruction.

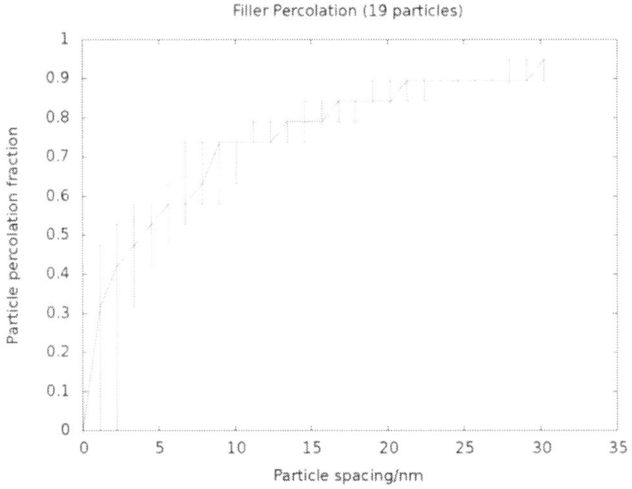

Figure 3: Percolation data from the reconstruction in Figure 2. The volume contained 19 separate particles.

4. Conclusions

We have demonstrated that machine learning methods offer a reliable means of identifying regions of a tomographic reconstruction without using filters to alter the source data or by using time-consuming and potentially inaccurate manual segmentation. Data processed in this manner retains fine details on the order of one or two voxels, meaning that quantitative information can usefully be obtained on particle spacings even when particles approach one another very closely. The next step would be to combine this information with mechanical testing in order to relate the detailed mechanical properties of these rubber composites to the distribution of particle spacings (as opposed to the bulk volume fraction used in current work). Understanding this relationship will allow rubber composite manufacturers to actively design their filler particles in order to more accurately specify and create new materials.

5. Acknowledgements

Funding and rubber composite samples were provided by Michelin. P.A.M. thanks the ERC grant number 291522 '3DIMAGE' for funding.

References

[1] Payne A R 1962 *J. Appl. Polym. Sci.* **6** 57–63 ISSN 1097-4628 URL `http://dx.doi.org/10.1002/app.1962.070061906`

[2] Kohjiya S, Katoh A, Suda T, Shimanuki J and Ikeda Y 2006 *Polymer* **47** 3298 – 3301 ISSN 0032-3861 URL `http://www.sciencedirect.com/science/article/pii/S0032386106002916`

[3] Poompradub S, Tosaka M, Kohjiya S, Ikeda Y, Toki S, Sics I and Hsiao B S 2005 *J. Appl. Phys.* **97** 103529 (pages 9) URL `http://link.aip.org/link/?JAP/97/103529/1`

[4] Wang Z, Liu J, Wu S, Wang W and Zhang L 2010 *Phys. Chem. Chem. Phys.* **12**(12) 3014–3030 URL `http://dx.doi.org/10.1039/B919789C`

[5] Gilbert P 1972 *Journal of Theoretical Biology* **36** 105 – 117 ISSN 0022-5193 URL `http://www.sciencedirect.com/science/article/pii/0022519372901804`

[6] Schindelin J, Arganda-Carreras I, Frise E, Kaynig V, Longair M, Pietzsch T, Preibisch S, Rueden C, Saalfeld S, Schmid B, Tinevez J Y, White D J, Hartenstein V, Eliceiri K, Tomancak P and Cardona A 2012 *Nat Meth* **9** 676–682 ISSN 1548-7091 URL `http://dx.doi.org/10.1038/nmeth.2019`

[7] Breiman L 2001 *Machine Learning* **45** 5–32 ISSN 0885-6125

Investigating the medium range order in amorphous Ta_2O_5 coatings

Riccardo Bassiri[1,a], Martin Hart[2,b], Robert L. Byer[1], Konstantin B. Borisenko[3], Keith Evans[4], Martin M. Fejer[1], Angie C. Lin[1], Ian MacLaren[2], Ashot S. Markosyan[1], Iain W. Martin[2], Roger K. Route[1], Sheila Rowan[2]

[1] E.L. Ginzton Laboratory, Stanford University, Stanford, California, USA
[2] SUPA School of Physics and Astronomy, University of Glasgow, Glasgow, UK
[3] Department of Materials, University of Oxford, Oxford, UK
[4] Department of Physics, University of Warwick, Warwick, UK
[a] rbassiri@stanford.edu
[b] m.hart.1@research.gla.ac.uk

Abstract. Ion-beam sputtered amorphous heavy metal oxides, such as Ta_2O_5, are widely used as the high refractive index layer of highly reflective dielectric coatings. Such coatings are used in the ground based Laser Interferometer Gravitational-wave Observatory (LIGO), in which mechanical loss, directly related to Brownian thermal noise, from the coatings forms an important limit to the sensitivity of the LIGO detector. It has previously been shown that heat-treatment and TiO_2 doping of amorphous Ta_2O_5 coatings causes significant changes to the levels of mechanical loss measured and is thought to result from changes in the atomic structure. This work aims to find ways to reduce the levels of mechanical loss in the coatings by understanding the atomic structure properties that are responsible for it, and thus helping to increase the LIGO detector sensitivity. Using a combination of Reduced Density Functions (RDFs) from electron diffraction and Fluctuation Electron Microscopy (FEM), we probe the medium range order (in the 2-3 nm range) of these amorphous coatings.

1. Introduction

Amorphous heavy metal oxides, such as Ta_2O_5, are widely used as the high refractive index layer of highly reflective dielectric coatings. Ion-beam sputtered Ta_2O_5 is used in the dielectric coatings of high performance applications such as the ground based Laser Interferometer Gravitational-wave Observatory (LIGO). Mechanical loss, which is directly related to Brownian thermal noise from the coatings, forms an important limit to the sensitivity of the LIGO detector [1].

It has previously been shown that heat-treatment and TiO_2 doping of the Ta_2O_5 coatings can cause noticeable changes to the levels of mechanical loss measured [2, 3]. This is thought to be the result of changes in the atomic structure, and recent observations have shown a direct correlation between the mechanical loss and local atomic structure [4]. Therefore, understanding the atomic structure and properties of these coatings is an area of intensive investigation. We ultimately aim to find ways to reduce the levels of mechanical loss in the coatings, and thus help to increase the LIGO detector sensitivity.

In this work, the atomic structure of heat-treated Ta_2O_5 coatings is studied and compared to previously measured values of mechanical loss, and the first results of the medium range order from a heat-treated Ta_2O_5 coating is presented.

2. Experimental details

The coatings studied in this work were 500 nm thick argon ion-beam sputtered Ta_2O_5 on silica (SiO_2) glass substrates, manufactured by the Commonwealth Scientific and Industrial Research Organisation (CSIRO, Materials Science and Engineering Division, West Lindfield, NSW, Australia). Heat-treatment of the samples was carried out at 300, 400 and 600°C for a period of 24 hours in air. TEM samples were prepared using a conventional cross section method, finished with argon ion polishing. Full details of the TEM sample preparation technique can be found in a detailed guide prepared by Muller *et al.* [5]. Investigations of the medium range order were carried out on a JEOL ARM 200FCS at 25 cm camera length, using a 2 nm FWHM probe and recording onto an 11 megapixel Gatan Orius SC1000A CCD at 4x binning. The high dynamic range imaging routine used to improve the short range order measurements was provided by Bernhard Schaffer of Gatan and scales and stitches together multiple diffraction patterns taken at a range of exposure times. The FEM data collection entailed formation of the desired probe size, positioning of the probe on a bright field image where a line of measurements could be taken from, thereafter in diffraction mode taking roughly seventy exposures, each two seconds long. The resultant diffraction patterns were then cropped around their common centre and stacked together. Each diffraction pattern was then radially averaged before the variance was computed through the stack.

3. Linking atomic structure and mechanical loss

To study the atomic structure of the Ta_2O_5 coatings, the Short Range Order (SRO) of the material is studied. The SRO is measured by computing the Reduced Density Function (RDF) from radially averaged electron diffraction data. RDFs give a statistical representation of where atoms sit with regards to a central atom [6].

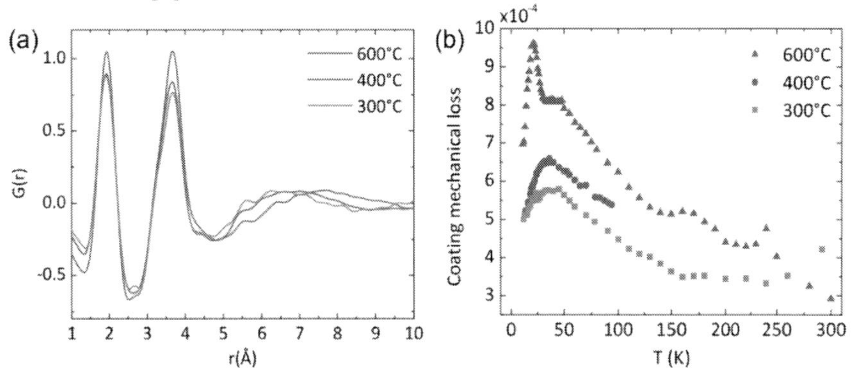

Figure 1. (a) Reduced density functions [7], (b) mechanical loss [2] of heat-treated Ta_2O_5 coatings.

Figure 1(a) contains the RDFs obtained from the diffraction patterns from coatings that have undergone different heat-treatments [7, 8]. The RDFs show subtle changes in SRO, the peaks become taller and narrower, indicating a slight increase in short-range ordering with increasing heat-treatment temperature. Figure 1(b) shows previously measured mechanical loss measurements by Martin *et al.* [2] and highlights the substantially different mechanical losses measured from these coatings, showing low temperature loss peaks which grow with increasing heat-treatment temperature. The substantial changes observed in the temperature dependent mechanical loss, induced by heat-treatment, is interesting in that the atomic structure shows only small and subtle changes in the SRO. It is therefore hypothesised that there may be larger scale changes in the atomic structure responsible for the changes observed in the mechanical loss, and that these are occurring at a longer range than the SRO is sensitive to, i.e., greater than 1 nm.

4. Improving the short range order

A recent improvement in RDF measurements has been developed which may help to probe any structural changes in more detail. High Dynamic Range (HDR) diffraction imaging is the process of taking multiple diffraction patterns at different exposure times, and stitching them together to create one image. This increases the SNR at the faintest area of the image (high-q), whilst maintaining an accurate representation of the centre of the diffraction pattern.

This is an important improvement as it allows the possibility to easily obtain a q-range of 25 $Å^{-1}$ with 0.0031 $Å^{-1}$ $pixel^{-1}$ resolution (highlighted in figure 3(a)), with planned experiments out to a 50 $Å^{-1}$ range. Assuming elastic scattering we define $q = 4\pi \sin\theta/\lambda$, with 2θ being the scattering angle and λ the wavelength of the radiation used. The RDFs in figure 1(a) were computed for data recorded out to 12 $Å^{-1}$. The new procedure with a doubled q-range allows much better background fitting which in turn allows better peak representation in the RDF, especially at low r, and thus improves the reliability of our RDFs. This will be implemented in all future RDF investigations.

5. Probing the medium range order

One way to measure changes in the atomic structure over a longer range is to use Fluctuation Electron Microscopy (FEM). FEM is a diffraction and/or imaging technique that probes the Medium Range Order (MRO), in the roughly 1 to 3 nm range. FEM measures the MRO by measuring intensity fluctuations from nano-volumes in the sample material. These fluctuations are quantified by computing the normalised variance of the diffracted intensity, which is maximally sensitive when the electron probe size is of comparable length scale to the MRO structural ordering being probed. Originally proposed by Gibson and Treacy [9], the technique was initially carried out using dark-field imaging in the TEM. We have used nanodiffraction FEM to collect data, in a similar way to that described by Voyles and Muller's [10].

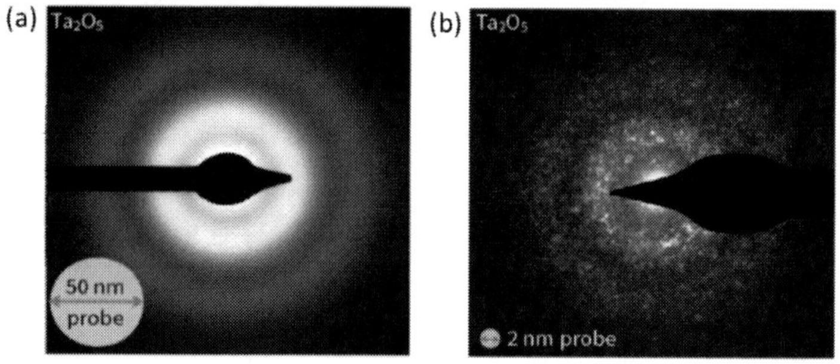

Figure 2. (a) Electron diffraction pattern with 50 nm probe, (b) FEM electron diffraction pattern with 2 nm probe from a 600°C heat-treated Ta_2O_5 coating.

Figure 2(a) shows the diffuse rings of the electron diffraction pattern obtained from an amorphous Ta_2O_5 coating using a 50 nm probe that is used routinely in RDF analysis. Here, any MRO structural detail is washed out by the inherent averaging over the relatively large volume probed. Figure 3(b) shows the FEM diffraction pattern collected from a 600°C heat-treated Ta_2O_5 coating, and highlights the presence of spatial fluctuations when using a 2 nm probe. Quantification of these spatial fluctuations is determined by examining the normalized variance of the diffracted intensity, $V(q)$.

The 600°C heat-treated Ta_2O_5 coating sample was investigated. Figure 3(a) shows the HDR intensity profile from a 50 nm probe (for use in RDF investigations), and figure 3(b) shows the corresponding normalized variance when a 2 nm probe is used, which confirms the presence of MRO.

Planned work in the future will use this method of FEM to study the MRO of Ta_2O_5 coatings as a function of heat-treatment temperature. Here it will be expected that heat-treatment causes larger changes in the MRO than are obvious in the SRO, and the aim will be to correlate this to the observed changes in mechanical loss.

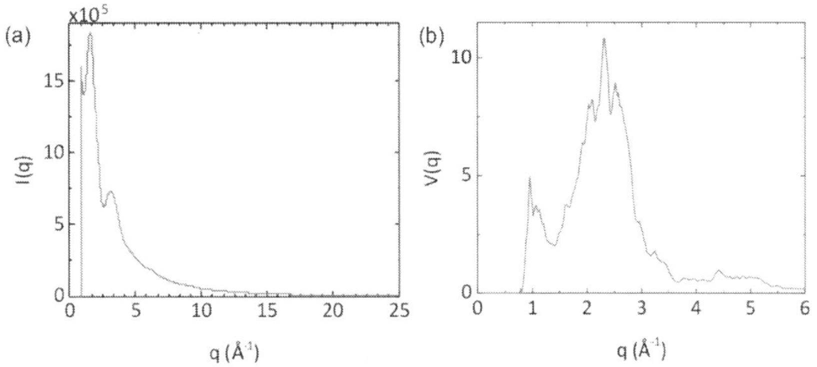

Figure 3. (a) HDR intensity profile and (b) normalised variance V(q) plot from a 600°C Ta_2O_5 coating.

6. Conclusions and future work

Heat-treatment causes subtle changes in the SRO atomic structure of amorphous Ta_2O_5 coatings. However, there are substantial changes observed in temperature dependant mechanical loss measurements. Larger changes in the atomic structure, which may more readily explain the observed changes in the mechanical loss, are thought to be occurring in the MRO, and work has begun to probe this using FEM.

Further work will be performed using FEM at varying probe sizes to pin down the length scale of the MRO as described by Voyles *et al.* [10]. We also plan on using FEM, in conjunction with improved high dynamic range RDF data, to create atomic models that include MRO as a key input into the reverse Monte Carlo and molecular dynamics based atomistic simulations.

Acknowledgements

We would like to thank the support of colleagues in the LIGO Scientific Collaboration, SUPA and the University of Glasgow for the provision of the MagTEM facility used in this work and Bernhard Schaffer of Gatan for the HDR imaging script. We also wish to acknowledge funding from the NSF (Grant No: PHY-1068596) and STFC (Grant No: ST/J000361/1).

References

[1] Harry G M, Armandula H, Black E, Crooks DRM, et al. 2010 Appl. Opt. **45** 1569
[2] Martin I W, Bassiri R, Nawrodt R, Fejer M M, et al. 2010 Class. Quantum Grav. **27** 225020
[3] Harry G M, Abernathy M R, Becerra-Toledo A E, Armandula H, et al. 2007 Class. Quantum Grav. **24** 405-415
[4] Bassiri R, Evans K, Borisenko K B, Fejer M M, et al. (2013) Acta Mat. **61** 1070–1077, 2013
[5] Muller E, Abou-Ras D, 2004 *Preparation of Cross-section Samples for Transmission Electron Microscopy* (Swiss Federal Institute of Technology Zurich)
[6] Cockayne D J H 2007 Annu. Rev. Mater. Res. 37 159-87
[7] Bassiri R 2011 *The Atomic Structure and Properties of Mirror Coatings for use in Gravitational Wave Detectors* (University of Glasgow) Ph.D. Thesis
[8] Bassiri R, Borisenko K B, Cockayne D J H, Hough J, et al. 2011 App. Phys. Lett. **98** 031904
[9] Treacy M M J and Gibson J M 1996 Acta Cryst. A **52** 212
[10] Voyles P M and Muller D A 2002 Ultramicroscopy **93** 147

Preparation of plan-view Co-doped FeSi thin film TEM specimens using FIB

M B Ward, N A Porter, P Sinha, R Brydson and C H Marrows

Institute for Materials Research, University of Leeds, Leeds, LS2 9JT
Department of Physics and Astronomy, University of Leeds, Leeds, LS2 9JT

m.b.ward@leeds.ac.uk

Abstract. When examining thin films using transmission electron microscopy (TEM), it is usually necessary to image a cross-section of the film (i.e. parallel to the film). However sometimes it is favourable to image thin films in plan-view (i.e. perpendicular to the film). This is the case for Co-doped FeSi thin films, which possess chiral symmetry along certain zone axes. In order to view these zone axes it is necessary to prepare the films in plan-view. There exist various ways to produce plan-view TEM specimens of thin films, such as back etching, ion milling and mechanical polishing. Here, a method using focused ion beam (FIB) is described in detail. Benefits of using FIB are that it is a quick process, there are no limitations in terms of substrate material, and samples can be produced from sections of substrate/film that may be too small to prepare any other way. The effectiveness of the preparation technique is also discussed here, with some preliminary energy dispersive X-ray (EDX) mapping data.

1. Introduction

Focused ion beam (FIB) is a well-established technique that has found applications in a number of areas due to its versatile fabrication capabilities and ability to image (either using secondary electrons generated by the ion beam or using an integrated scanning electron microscope (SEM)). One area that has become routine is in the preparation of samples for transmission electron microscopy (TEM) [1, 2].

The main benefits of using FIB to prepare samples for TEM are the ability to be site specific and the ability to apply the technique to a wide range of materials. The technique has uses in both the preparation of bulk materials and also thin film/multilayer materials. With respect to the latter, there is usually a requirement to view films in cross section using TEM. This allows for the determination of factors such as film thickness, epitaxy, de-bonding and elemental segregation. However in some cases it is favorable to view thin films in plan-view, and this is the case for the thin film Co-doped FeSi discussed in this paper.

Co-doped FeSi thin films grown in the 111 direction on a Si substrate are being investigated in order to establish a relationship between their structure and magnetic properties. Films composed of MnSi have been shown to form chiral domains, where reversals in the chiral symmetry in the [221] direction occur between regions of the order of a few microns [3]. In order to carry out similar investigations into Co-doped FeSi thin films, plan-view specimens need to be produced so that they can be analysed using TEM.

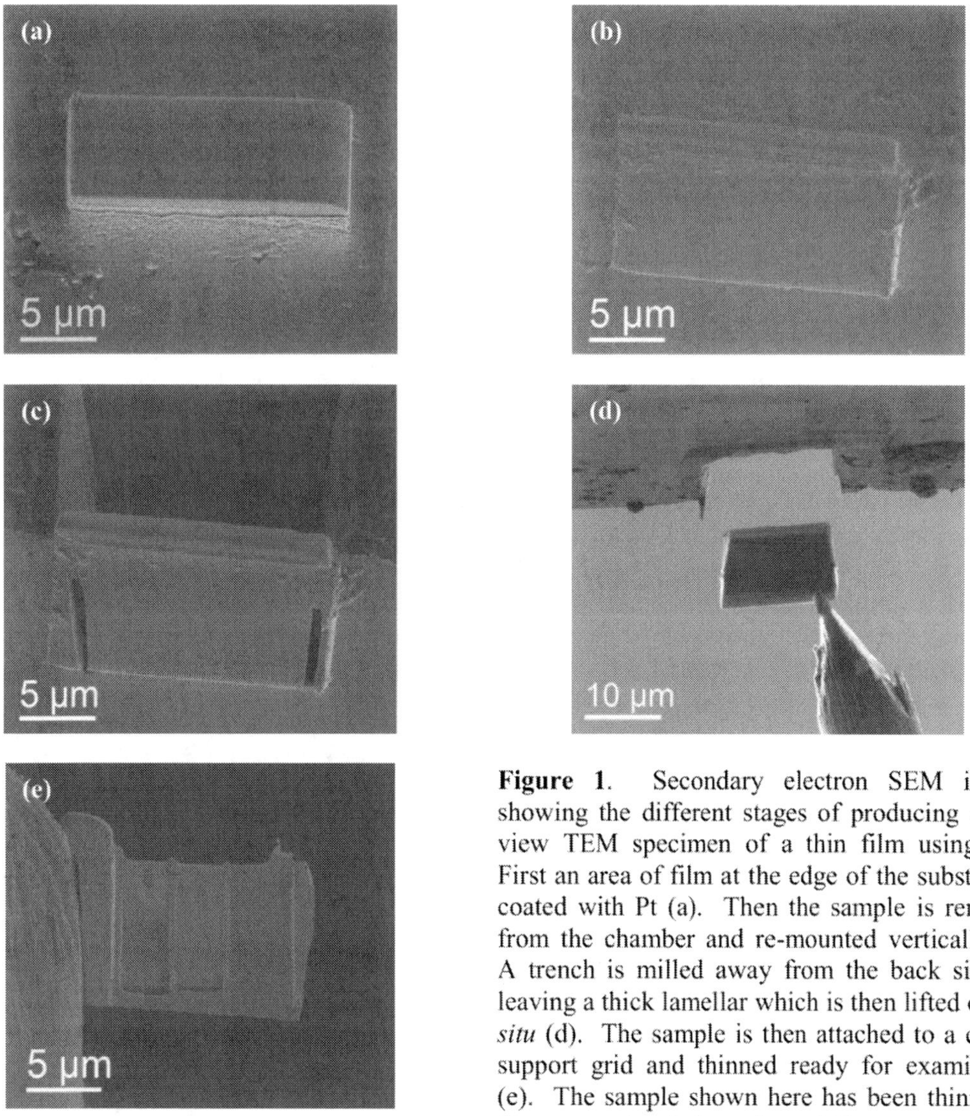

Figure 1. Secondary electron SEM images showing the different stages of producing a plan view TEM specimen of a thin film using FIB. First an area of film at the edge of the substrate is coated with Pt (a). Then the sample is removed from the chamber and re-mounted vertically (b). A trench is milled away from the back side (c) leaving a thick lamellar which is then lifted out *in-situ* (d). The sample is then attached to a copper support grid and thinned ready for examination (e). The sample shown here has been thinned in three separate windows for stability.

It has been demonstrated that plan-view thin film specimens can be produced via mechanical polishing [3], back etching [4] and ion milling [5]. It has also been shown that FIB can be used to produce plan-view TEM specimens [6, 7], but here a new FIB method for specifically producing plan-view TEM specimens of thin films is described. The reason for using FIB as an alternative method is that samples can be made from very small pieces of film on substrate (too small for fabrication by any other technique). This may be the case if samples can only be made in small dimensions and/or they need to be divided for characterisation by different techniques. Additionally it may be that FIB is the only technique that is available or the only that expertise exists for. Finally, if the films have been grown on a substrate material that is not compatible with any other sample preparation technique then FIB may be an alternative.

2. Methods

For this investigation a 50 nm thin film of Co-doped FeSi was grown via MBE on a <111> oriented Si substrate as described in [8]. TEM specimens were prepared using a FEI Nova200 Nanolab dual beam SEM FIB fitted with a Kleindiek micromanipulator for *in situ* lift-out. The ion column was operated at 30 kV for all steps except final cleaning of the specimen, where a voltage of 5 kV was used. Beam currents varied between 100 and 5000 pA. TEM was carried out using a Philips CM200 FEGTEM fitted with an Oxford Instruments X-Max 80mm^2 SDD EDX detector running Aztec software.

Figure 1 shows the main stages in the production of plan-view TEM specimens of thin films. The sample was first mounted horizontally in the chamber (that is with the thin film perpendicular to the axis of the SEM column) and a region at the edge of the thin film located. The reason for choosing an area at the edge of the film was so that the edge could serve as the top surface for the final prepared specimen and it could be easily lifted out. The electron beam was used to deposit a layer of Pt 200 nm thick over an area equal to the proposed specimen size (approximately 18 μm wide and 10 μm tall). The ion beam was then used to deposit a layer 1 μm thick over the same area (figure 1a). Next the sample was removed from the chamber and re-mounted vertically (with the thin film parallel to the axis of the SEM column) with the edge of the film that was identified in the previous step now topmost. The same area that was coated previously was located and an 18 μm × 2 μm layer of ion beam deposited Pt 1 μm thick was deposited on top (figure 1b). Preparation could now proceed in a similar fashion to standard TEM specimen preparation [1, 2] except material only had to be removed from one side of the specimen. This was done using a standard cross-section milling pattern, to reach a final depth of around 10 μm. The specimen was then partially cut-out (figure 1c). The sample was lifted out *in situ* (figure 1d) and then attached to a copper support grid. Thinning of the specimen could then proceed. This was first performed on the front face of the specimen that was covered in platinum during the first stage of preparation. The sample was oriented so that this front face was in-

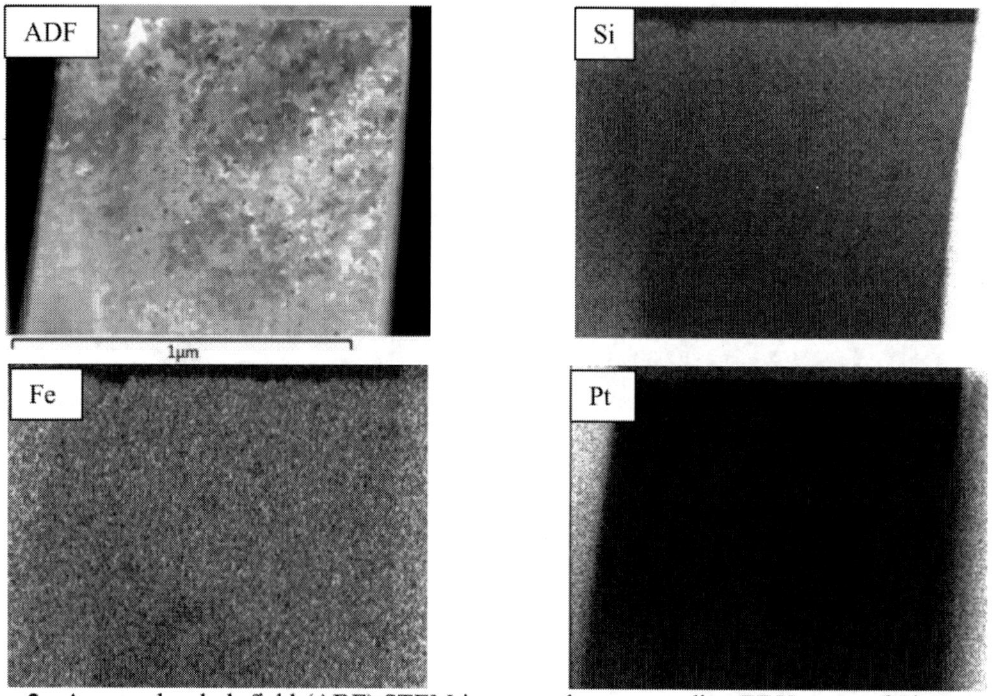

Figure 2. An annular dark-field (ADF) STEM image and corresponding EDX maps of Si, Fe and Pt of one of the thinned windows in the specimen. The Fe map confirms that the Co-doped FeSi film has largely remained intact over the entire window, and the absence of Pt signal at the very top of the window suggests that the protective Pt layer on the face of the specimen has mostly been removed by the thinning process.

line with the ion-column. This was checked by imaging the face with the ion column whilst varying the tilt of the specimen in increments of 1°. At the point where the face disappeared from view, it was assumed to be in-line with the column. Thinning then proceeded using a gradual cross-section cleaning milling pattern, with a final milling beam current of 100 pA. The tendency for TEM specimens to be left thicker at the base during the thinning process [1] meant that the Co-doped FeSi film was first revealed at the top of the specimen, and this could be easily observed through a change in contrast between the thinned region at the top and that below, with an abrupt interface between (this can be seen in the left-hand thinned window in figure 1e). At this point, thinning was halted so that the Co-doped FeSi was not completely milled away. Then thinning of the back-side was carried out in the same way with a counter-tilt of 1-2° to ensure the specimen had parallel sides [1]. Finally a 5 kV cleaning step was carried out on each side.

3. Results and discussion

Once the sample had been prepared, it was characterised via Scanning (S)TEM energy dispersive X-ray (EDX) mapping in order to check that the Co-doped FeSi film still remained and that the specimen was free from the protective platinum layer. STEM-EDX maps were collected for Si, Fe and Pt (figure 2). The Pt map shows that the protective layer has been mostly removed, which was important so that the Pt did not affect any subsequent TEM data (imaging of spectroscopic) collected from the specimen. The Fe map shows the coverage of the Co-doped FeSi map, which is largely complete across the entire thinned window. The Si map will contain signal from the Co-doped FeSi film, but also contributions from any substrate material that remains. Therefore, it is difficult to assess how effectively it has been removed by the thinning process. The presence of some Si substrate material should not affect the analysis in this case however, as any chiral domains (which are the features of interest in this sample) should still be visible via diffraction contrast.

4. Conclusions

A plan-view TEM specimen of a 50nm film of Co-doped FeSi on a Si substrate has been produced using a new FIB method. In principle the method can be used for any film/substrate materials. STEM-EDX mapping of the specimen has shown that the Co-doped FeSi film remains largely intact, and free from the deposited platinum used to protect the film during the process. The technique could certainly be used to produce specimens from thicker films (>50 nm), however great care would have to be taken when making specimens from thinner films so that the film was not lost during the thinning process.

Acknowledgements

This work was funded by EPSRC grant number EP/J007110/1.

References

[1] Giannuzzi L A and Stevie F A 1999. *Micron*. **30** 197.
[2] Schaffer M, Schaffer B and Ramasse Q 2012. *Ultramicroscopy*. **114** 62.
[3] Karhu E, Kahwaji S, Monchesky T L, Parsons C, Robertson M D and Maunders C 2010. *Phys. Rev. B*. **82** 184417.
[4] Yao B and Coffey K R 2008. *J. Electron Microsc.* **57** (2) 47.
[5] Eberg E, Monsen Å F, Tybell T, van Helvoort A T J and Holmestad R 2008. *J. Electron Microsc.* **57**(6) 175.
[6] Stevie F A, Irwin R B, Shofner T L, Brown S R, Drown J L and Giannuzzi L A 1998. *AIP Conf. Proc.* **449** 868.
[7] Langford R M, Huang Y Z, Lozano-Perez S, Titchmarsh J M and Petford-Long A K 2001. *J. Vac. Sci. Technol. B*. **19** 755.
[8] Porter N A, Creeth G L and Marrows C H 2012. *Phys. Rev. B*. **86** 064423.

Resistive switching in ZrO_2 films: physical mechanism for filament formation and dissolution

Pedro Parreira, Stephen McVitie and DA MacLaren

SUPA, School of Physics & Astronomy, University of Glasgow, G12 8QQ Glasgow

Email: p.raimundo-parreira.1@research.gla.ac.uk

Abstract. Resistive switching devices, also called memristors, have attracted much attention due to their potential memory, logic and even neuromorphic applications. Multiple physical mechanisms underpin the non-volatile switching process and are ultimately believed to give rise to the formation and dissolution of a discrete conductive filament within the active layer. However, a detailed nanoscopic analysis that fully explains all the contributory events remains to be presented. Here, we present aspects of the switching events that are correlated back to tunable details of the device fabrication process. Transmission electron microscopy and atomically resolved electron energy loss spectroscopy (EELS) studies of electrically stressed devices will then be presented, with a view to understanding the driving forces behind filament formation and dissolution.

1. Introduction

The need for enormous data density and non-volatile storage has been the motivation behind the development of new memory materials and design concepts [1]. Several competing technologies have been developed in the pursuit of a good candidate that matches the density of Si-based Flash but with increased performance, scalability and power requirements [2]. In recent years, resistive switching random access memory (ReRAM) devices have been extensively studied due to the fact that ReRAM possesses all the properties required for computing and/or data storage, including scalability, non-volatility and rapid switching times [3, 4]. Although much work has been done with resistive switching systems in general, a complete model that describes all known switching phenomena is still lacking and in many ways, recent progress has outpaced a detailed understanding [5]. This is particularly true for the case of memristors based on tantalum and hafnium oxides, which are now commercially available despite the fact that the underlying physics of resistive switching is still under debate [6-8]. It is known that ReRAM relies on electrically induced ion migration in the metal-oxide layers of TaO_x, TiO_x, ZrO_x and HfO_x systems [1-4]. In these systems a conductive filament (CF) is formed (after an initial electroforming step) by local chemical reduction of an insulating oxide layer located between two contacts [1]. It is the formation of this CF that dramatically lowers the device's resistance between the two electrodes. The application of an electric field with the opposite bias will suppress the CF due to its re-oxidation and restore the high resistance condition (often mediated by local Joule heating) [1-3].

Reproducible resistive switching (RS) is then obtained be repeated CF formation and dissolution, which has recently been observed by *in-situ* TEM measurements [9, 10]. It is accepted that the active electrode plays an important role in this reproducible RS process by acting as a source/sink of oxygen ions [1-5]. A major difference between basic ReRAM devices and these studied here was the introduction of a redox-active insertion layer in the middle of the active layer, which was included to promote the reproducible reduction of the active medium and hence facilitate CF formation [11, 12].

A detailed nanoscopic analysis that fully explains all the contributory events will be essential if issues of device scalability and failure are to be addressed. Given the redox mechanism proposed for CF formation and dissolution, electron transmission microscopy (TEM) and nano-resolved electron energy loss spectroscopy (EELS) are essential tools for unravelling all the mechanisms involved in resistive switching. In this work we present preliminary results of electrically stressed zirconium oxide (ZrO_2) based memristors deposited via pulsed laser deposition (PLD) in an effort to shed some light on some of the resistive switching phenomena.

2. Experimental

Pulsed laser deposition was used to deposit the zirconium oxide active layer using a chamber with a base pressure of 8×10^{-7} Pa and equipped with a load lock and a dry pumping system. The zirconia layer was obtained by ablating a ZrO_2 target with a pulsed 248 nm laser (KrF excimer) with a repetition rate of 18 Hz, a fluence of 3.0 J/cm^2 and an oxygen partial pressure of 0.5 Pa. The deposition temperature was set to 500 °C during deposition to crystallise the zirconia and activate the Ti insertion layer, substrates were held 8 cm away from the target and were rotating to ensure uniform film coverage. All the devices in this work make use of crossbar architecture with a bottom inert electrode (IE) and a top electrochemically active electrode (AE). The bottom inert electrodes, 90 nm thick, are made from platinum and were deposited using an e-beam thermal evaporation system whereas the active electrodes, also 90 nm thick, are made from titanium and were deposited using a sputtering system. The oxide layer includes a thin titanium insertion layer also deposited by sputtering. Electrical characterisation was made using a LabView controlled Keithley 2602 sourcemeter connected to a temperature controlled probe station. The TEM sample was prepared using a FEI NovaNanolab 200 where a focused ion beam (FIB) was used to isolate a cross section of a memristor and mill it to the required thickness. TEM was performed using a probe corrected JEOL ARM200 CFEG system operating at 200 kV while EELS was done in scanning mode with a Gatan Quantum 965 spectrometer.

3. Results and Discussion

Figure 1 illustrates typical electrical characterisation, which comprises (a) voltage sweeps and (b) endurance cycles where the voltage is pulsed to induce resistive switching. Devices without the Ti insertion layer showed bipolar resistive switching (BRS) but with significantly lower on/off ratios and endurance (Fig. 1, dark curve). These devices also required the current limit enforced by the sourcemeter so that several switching cycles could be performed. The introduction of the Ti insertion layer increases the voltage required for switching but greatly enhances both the on/off ratio and the endurance and reproducibility significantly (Fig.1, light curve). Fig. 2a shows a dark field scanning TEM (STEM) image of the complete device stack, moving left to right, from the bottom platinum IE to the titanium AE and platinum capping layer. Fig. 2b shows normalised elemental distribution measured by EELS in the area indicated. It is clear from elemental quantification analysis that the titanium insertion layer is at least partially oxidised after device cycling, indicating oxygen ion migration within the applied electric field. This insertion layer oxidation is likely to enhance the reduced zirconium CF formation and hence explain the enhanced BRS observed. Another interesting observation lies in the fact that the active layer/active electrode interface (which is approximately 15 nm thick) seems to be oxidised as well, since the oxygen signal drops significantly after the Zr signal and within the Ti electrode itself.

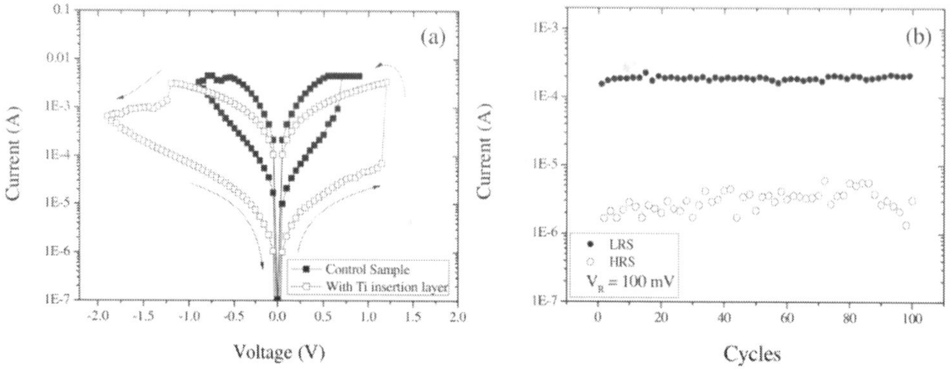

Figure 1 *– (a) iV curve illustrating different switching characteristics for devices with and without a Ti insertion layer, the later shows much better switching. Arrows indicated the sense of voltage sweep. (b) Resistance switching endurance for 100 write/erase cycles for devices with a Ti insertion layer only.*

The preferential oxidation of the titanium AE in comparison with the IE is also coherent with RS model, where switching from a HRS to a LRS is achieved by the formation of a CF and oxygen migration towards the AE. A further interesting and surprising feature is the apparent reduction of the zirconia active layer back to zirconium metal at the interface with the IE, presumably accounting for the oxygen within the Ti insertion layer. This metallic zirconium layer appears to be at least 15 nm thick, although interface roughening effects should be taken into account.

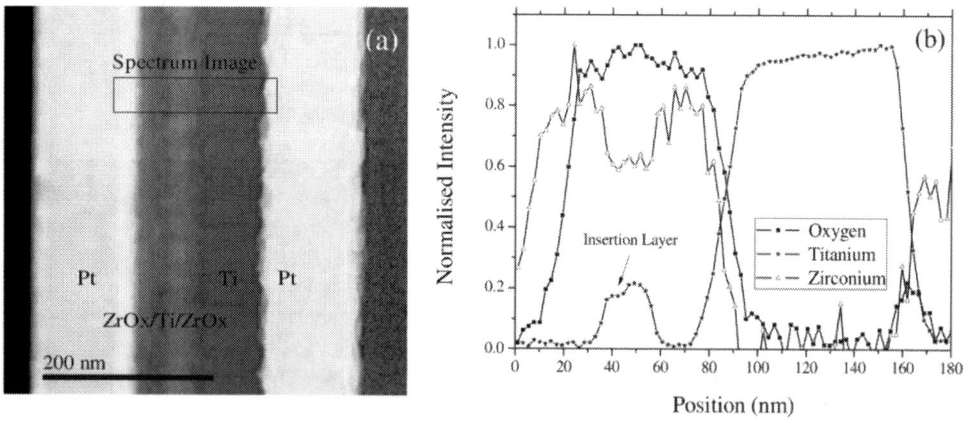

Figure 2 *– (a) Dark field image of a device in cross-section illustrating where EELS analysis was performed and (b) relative elemental quantification. Apparent Zr counts at 160 nm are due to the proximity of the Pt edge whereas the Zr negative counts are attributed to difficulties in background subtraction.*

This implies that it is likely that switching is governed mainly by ionic oxygen movement within the active layer and AE, in agreement with the generally accepted model for CF formation.

4. Conclusions

It was shown that the use of an insertion layer made from an easily oxidisable element enhances the RS performance of memristor devices. This can be explained by the fact that the titanium is oxidised at the expense of zirconia reduction. The localised reduction of the zirconia active layer promotes the conductive filament formation responsible for resistive switching. It was also clear that the use of tools like TEM and EELS is essential if the chemical changes that underpin resistive switching, are to be completely mapped and understood. Future work will focus on more detailed quantification of oxidation states so that in-situ TEM resistive switching can be performed and studied with unprecedented detail.

References

[1] Balatti, S., Larentis, S., Gilmer, D. C. & Ielmini, D. *Advanced materials* **25,** 1474 (2013).

[2] Waser, R., Dittmann, R., Staikov, G. & Szot, K. *Advanced Materials* **21,** 2632 (2009).

[3] Yu, S., Lee, B. & Wong, H. S. P. *Functional Metal Oxide Nanostructures.* **149,** 303 (2012).

[4] Yang, J. J., Strukov, D. B. & Stewart, D. R. *Nature Nanotechnology* **8,** 13 (2013).

[5] Lohn, A. J., Stevens, J. E., Mickel, P. R. & Marinella, M. J. *Applied Physics Letters* **103,** 063502 (2013).

[6] Lee, C. B. *et al. IEEE Electron Device Letters* **32,** 399 (2011).

[7] Ninomiya, T. *et al.* **48,** 2011 (2012).

[8] Lee, H. D. *et al. 2012 Symposium on VLSI Technology (VLSIT)* 151–152 (2012).

[9] Kwon, D.-H. *et al. Nature Nanotechnology* **5,** 148 (2010).

[10] Yang, Y. *et al. Nature Communications* **3,** 732 (2012).

[11] Ryu, S. W., Ahn, Y. B., Kim, H. J. & Nishi, Y. *Applied Physics Letters* **100,** 133502 (2012).

[12] Zhang, H. *et al. Applied Physics Letters* **98,** 042105 (2011).

Journal of Physics: Conference Series **522** (2014) 012046 doi:10.1088/1742-6596/522/1/012046

Direct Detectors for Electron Microscopy

R N Clough[1], **G Moldovan**[2], **A I Kirkland**[1]

[1] Department of Materials, University of Oxford, Parks Road, OX1 3PH, UK

[2] Oxford Instruments NanoAnalysis, Halifax Road, High Wycombe, HP12 3SE, UK

E-mail: robert.clough@materials.ox.ac.uk

Abstract.

There is interest in improving the detectors used to capture images in transmission electron microscopy. Detectors with an improved modulation transfer function at high spatial frequencies allow for higher resolution in images at lower magnification, which leads to an increased effective field of view. Detectors with improved detective quantum efficiency are important for low dose applications. One way in which these performance enhancements can be achieved is through direct detection, where primary electrons are converted directly into suitable electrical signals by the detector rather than relying on an indirect electron to photon conversion before detection. In this paper we present the characterisation of detector performance for a number of different direct detection technologies, and compare these technologies to traditional indirect detectors. Overall our results show that direct detection enables a significant improvement in all aspects of detector performance.

1. Introduction

The majority of detectors in use for transmission electron microscopy utilise indirect electron detection. Indirect detection requires that primary elections are converted to photons in the scintillator, which is coupled to the (CCD or CMOS) sensor through a lens or fibre optic coupling. Direct detection is an attractive alternative where signal in the CMOS or CCD sensor is generated directly by the primary electron beam. However CCD and CMOS sensors are prone to radiation damage which has a detrimental effect on detector performance and lifetime[1]. The use of enclosed-gate layout transistors (ELT) and thin gate oxide improves the radiation hardness of devices[2], which are now attractive as a technology for electron imaging.

There are a number of fundamental disadvantages to indirect detection. Scattering of the primary electrons in the scintillator leads to generation of photons in a volume larger than a single pixel. This implies that each electron generates signal in a cluster of pixels on the sensor. The effect of an electron being detected as signal in multiple pixels can be described quantitatively using the modulation transfer function (MTF). Direct detectors offer the possibility of significantly improving this MTF. Of particular interest are thin detectors where much of the inactive supporting substrate has been removed, leaving a thin active layer. Signal is generated in the active layer and then the electrons exit the active layer before significant lateral scattering has occurred. The removal of the supporting substrate prevents electrons scattering from the substrate back into the active layer, which considerably improves MTF[3].

The detective quantum efficiency (DQE) describes how the detector affects the signal to noise ratio in the image, as defined by Equation 1.

$$DQE(\omega) = \frac{(SNR_{out})^2}{(SNR_{in})^2} \quad (1)$$

The DQE is a performance metric of particular interest in low dose work, as an improved DQE implies that a lower dose is required to separate signal from noise. A perfect detector would have a DQE of 1 over all spatial frequencies. However in practice there are no perfect detectors and the DQE is affected by both the MTF and noise added to the image by the detector. The noise in an image is described by the noise power spectrum (NPS) and it can be shown that the DQE is related to the MTF through Equation 2.

$$DQE(\omega) = \frac{Dose \times MTF(\omega)^2}{NPS(\omega)} \quad (2)$$

The MTF and DQE are a function of spatial frequency from 0 to 0.5 pixel^{-1}. The upper limit is known as the Nyquist frequency or Nyquist limit. The Nyquist limit corresponds to a wavelength of 2 pixels and is the largest possible frequency obtainable by a pixelated image, spatial frequencies above the nyquist limit undergo aliasing.

2. Methodology

Two different direct electron detectors have been characterised together with one indirect detector as a basis for comparison. The commercial Gatan K2 Summit has a 5μm pixel size, with 3838 x 3710 pixels full frame and achievable sensor readout speeds of 400fps. We note that this detector also has a "Counting Super-resolution" mode which is not dealt with here, but can achieve up to an effective 7676 x 7420 pixels by splitting each physical pixel into 4 sub-pixels and assigning each electron a sub-pixel by analysis of the resulting cluster from an electron event[4]. This detector has been characterised in both integrating and counting mode[5] at 300kV with 2x2 binning giving a Nyquist frequency of 50 lp/mm. The DUOS detector is a non-commercial integrating direct detector with a 20μm pixel size, manufactured on 20μm thick EPI SOI wafers, mechanically thinned to 50μm with a 1024 x 1024 pixel full frame readout at 30fps. This was characterised at 200kV with no pixel binning giving a Nyquist frequency of 25 lp/mm. A Gatan UltraScan 4000 is an indirect detector with 4080 x 4080 15μm pixels and 0.2 fps full frame readout which was characterised at 200kV with 2x2 binning, giving a Nyquist frequency of 17 lp/mm. For all of the above detectors the MTF was measured using the edge method, where a suitably sharp edge is projected onto the detector, for the data reported here using a beam stop[6]. The NPS was calculated from an analysis of 32 uniformly illuminated images. The pixel values were converted to electron units by taking a series of images with the entire beam focused on the detector, the pixel values summed over the detector were compared to the total number of electrons in the image to yield a conversion factor (Equation 3).

$$C = \frac{Current \times ExposureTime}{PixelSum \times e} \quad (3)$$

The electron beam current was measured with a Faraday cup for the DUOS and UltraScan detectors, and for the Gatan K2 camera the current measurement from the microscope fluorescent screen was used. The DQE was acquired from the experimental measurements of MTF and NPS and the relationship given in Equation 2.

3. Results and Discussion

3.1. MTF results

Comparative MTFs are shown in figure 1. It is immediately obvious that the performance of the CCD detector is significantly worse than the direct detectors. Comparison of the Gatan K2

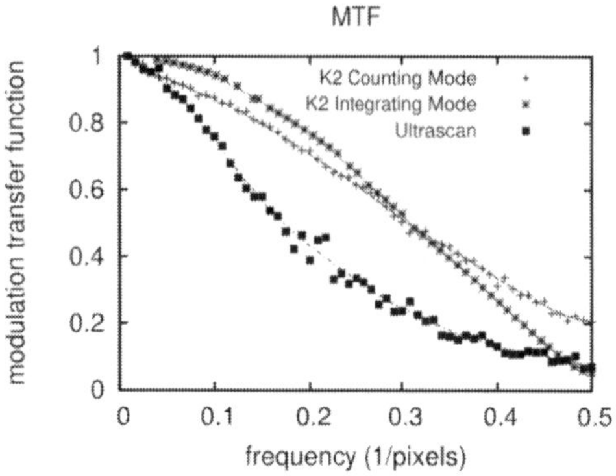

Figure 1. MTF for the Gatan K2 detector in counting and integrating mode and the indirect UltraScan 4000 camera up to the Nyquist limit.

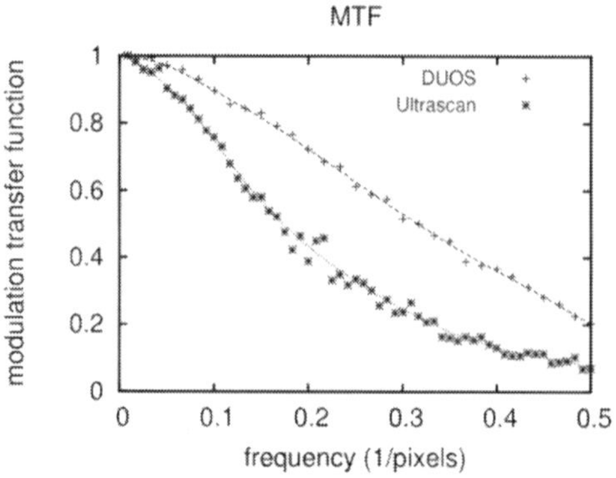

Figure 2. MTF for the DUOS and the indirect UltraScan 4000 cameras up to the Nyquist limit.

operating in counting mode with the same detector operating in integrating mode shows that the MTF at the Nyquist limit in counting mode is 4x higher than that in integrating mode. This is expected based on previous models of these two modes. However the integrating mode outperforms the counting mode at low spatial frequency. This indicates that in counting mode a larger proportion of the signal is recorded in pixels far away from the initial electron impact position on the sensor. We believe this can be attributed to signals that arise from electrons which undergo a large lateral scattering producing a secondary cluster far from the primary cluster. In integrating mode the signal from this secondary cluster would be smaller, as in counting mode any cluster that exceeds an energy threshold value is counted with equal weight. Figure 2 shows the DUOS detector performance is similar to that of the Gatan K2 detector operating in counting mode. This can be attributed to the larger pixel size of the DUOS detector which enables a larger proportion of signal to be deposited in the pixel corresponding to the electron impact.

The above data indicates that increasing the pixel size from $10\mu m$ to $20\mu m$ has a similar effect as switching to counting mode. There are advantages of both methods. Integrating mode is often desirable as the usable primary beam current is much higher, since there is no requirement that the detector needs to be able to distinguish individual events. Larger pixel sizes improve detector resolution but lead to slower sensor readout due to the physical distances that any signal needs to travel prior to readout. Conversely, smaller pixels in counting mode can achieve higher frame rates. Counting mode also makes sub pixel resolution possible by allocating each event to a pixel subarray.

3.2. DQE Results

The DQE data shown in Figure 3 show a similar pattern to the MTF, integrating mode performs better at low spatial frequencies, but drops off sharply at higher spatial frequency. The cross over point for the two modes is at a higher spatial frequency due to counting mode increasing high frequency noise[7]. The DQE results presented here for the Gatan K2 Camera are different to those published by Gatan Inc[8]. DQE results in figure 4 show a marked improvement for the DUOS sensor compared to the UltraScan camera. Considering the DUOS and Gatan K2 integrating modes we can see an improvement in DQE arising from the improvement in MTF. Comparing with the the data published by Gatan Inc for the counting mode of the detector at

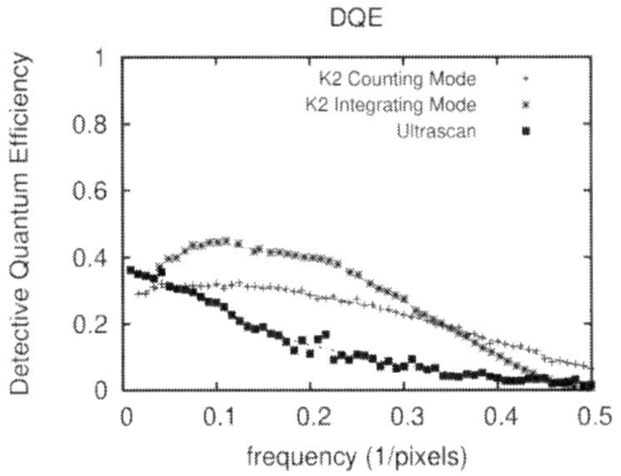

Figure 3. DQE for the Gatan K2 detector in counting and integrating mode and the indirect UltraScan 4000 camera up to the Nyquist limit.

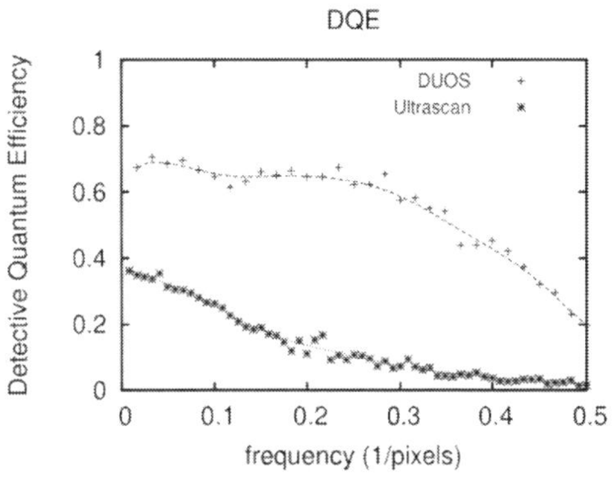

Figure 4. DQE for the prototype DUOS and the indirect UltraScan 4000 cameras up to the Nyquist limit.

200kV, at low spatial frequencies the counting detector performs better, as the counting mode records every event with equal weight low frequency noise is reduced. At the Nyquist frequency the DUOS detector has a DQE 50% higher due to lower high frequency noise.

4. Conclusions

Comparison of direct electron detection with indirect detection shows significantly improved performance in both MTF and DQE performance for the former. Within direct detection there are also variations in readout modes. Counting mode where each electron is given a binary assignment to a pixel significantly improves the MTF at high spatial frequency for detectors with small physical pixels. However increasing the physical pixel size, in an integrating mode yields a similar MTF as the interaction volume of the electron approaches the pixel size.

Acknowledgements

We are grateful to the CFN at Brookhaven National Lab for use of their equipment in the characterisation of the Gatan K2 Summit Detector, and Dr. Thalis Anaxagoras and Prof. Nigel Allinson from the University of Shefeld for provision of CMOS wafers and readout system.

References

[1] Faruqi A R, Henderson R, Pryddetch M, Allport P and Evans A 2006 *Nuclear Instruments & Methods in Physics Research Section A-Accelerators Spectrometers Detectors and Associated Equipment* **566** 770–770

[2] Battaglia M, Contarato D, Denes P, Doering D, Giubilato P, Kim T S, Mattiazzo S, Radmilovic V and Zalusky S 2009 *Nuclear Instruments & Methods in Physics Research Section A-Accelerators Spectrometers Detectors and Associated Equipment* **598** 642–649

[3] McMullan G, Faruqi A R, Henderson R, Guerrini N, Turchetta R, Jacobs A and van Hoften G 2009 *Ultramicroscopy* **109** 1144–1147

[4] GatanInc 2012 Direct detection - configurations and specifications `http://www.gatan.com/products/digital_imaging/products/K2/configuration.php`

[5] Mooney P, Contarato D, Denes P, Gubbens A, Lee B, Lent M and Agard D 2011 *Microscopy and Microanalysis* **17** 1004

[6] Meyer R R, Kirkland A I, Dunin-Borkowski R E and Hutchison J L 2000 *Ultramicroscopy* **85** 9–13

[7] McMullan G, Turchetta R and Faruqi A R 2011 *Journal of Instrumentation* **6** C04001 pT: J; TC: 0; UT: WOS:000294491300001

[8] GatanInc 2012 Direct detection - dqe graphs `http://www.gatan.com/products/digital_imaging/products/K2/dqe.php`

TEM imaging and application of thin-film magnetic rings for phase plates

C J Edgcombe[1] and J C Loudon[2]

[1]TFMM Group, Dept of Physics, University of Cambridge, CB3 0HE, UK
[2]Electron Microscopy Group, Dept of Materials Science, University of Cambridge, CB3 0FS, UK

email: cje1@cam.ac.uk

Abstract. The maximum out-of-plane field that some thin-film cobalt rings resisted while remaining in the vortex state was found experimentally to be lower than expected from simulation, possibly due to tilt. Analysis of a Zernike phase plate shows that the observed bright outlines around large objects result from the low-spatial-frequency limit of correction imposed by the beam hole. The maximum size of phase object that can be imaged accurately with a Zernike plate can be raised by increasing the effective focal length of the objective lens.

1. Introduction

The use of phase plates in transmission microscopy (TEM) offers the benefits of (1) increased contrast when the image is in focus, without contrast reversal over a wide range of spatial frequency, and (2) reduced electron dose to the specimen, of particular value for cryomicroscopy.

Of the various devices that have been built or proposed for phase plates, ferromagnetic rings are of interest because (1) the radial extent over which they intercept scattered electrons is expected to be smaller than for other plates; (2) all electrons scattered beyond the minimum effective angle experience the same phase shift. Previous work [1] demonstrated that the magnetization of thin-film rings of cobalt can be switched between two stable states, one of which has the continuous loop or vortex configuration of flux desirable for use as a phase plate in TEM. Further work on the fabrication of these rings is in progress. Here we report first some observations on the effect of axial (out-of-plane, OOP) field on rings in the vortex state, then some more general calculations on the imaging of extended objects by phase plates. These results show why bright outlines have been observed around some objects when phase plates are used, and suggest how to increase the size of objects that can be imaged without exciting these bright outlines.

2. Toleration of axial field in TEM

The most obvious location at which to place a phase plate is close to the back focal plane (BFP) of the objective lens. At this position, the axial field due to the objective lens may be as large as 2T in some TEMs, but appreciably smaller in others designed for biological applications. Simulations suggest that both circular and semicircular cobalt rings in the vortex state may tolerate an out-of-plane (OOP) field of ~1T. Thus it seemed possible that a ring of cobalt might be able to maintain the vortex state of magnetization in the presence of the axial field of the objective lens of a biological TEM.

Semicircular rings were deposited on SiN as described in [1]. The lift-off procedure succeeded in removing the centres of two of the rings, of which the narrower had width 30 nm. The rings were magnetized in the vortex state, inserted in the specimen position of a TEM and inspected by holography. Initially both were found to switch to the onion (or dipole) state when an OOP field as small as 0.1 T was applied. After remagnetisation in the vortex state, both were found to remain in that state when the OOP field was increased to 0.3T, but switched to the onion state when the field was raised to 0.5T (Fig. 1). This value of switching field was substantially lower than that expected from simulation. The reason for this low experimental switching field and its variability has not been identified but may be related to initial tilt of the specimen on insertion in the column.

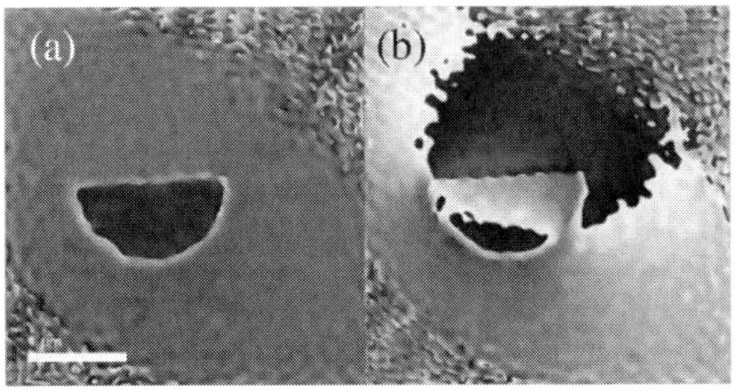

Fig. 1 Wrapped phase distributions (including jumps of 2π) near one Co ring after exposure to out-of-plane fields of (a) 0.3T and (b) 0.5T. Scale bar: 1 micrometre.

Even if the alignment of the magnetic ring in the region of the objective aperture is controlled sufficiently precisely for a ring to remain in the vortex state in an axial field of 1T, its sensitivity to OOP field is obviously inconvenient for use in practice. However, while a phase plate at the BFP is known to increase the contrast at focus, it has a further effect on imaging which suggests that modifying the TEM lens system may have advantages.

3. Effect of a phase plate on imaging

The type of Zernike phase plate that has been most widely reported (eg [2]) consists of a plain disc of material such as carbon, of controlled thickness, with a central hole to pass the direct beam. Images of biological specimens made with this type of plate are observed to show bright outlines or halos around certain features [3, 4]. Analysis of imaging has recently shown how these halos occur. It is necessary to consider the response not just to a single frequency but to the distribution of frequencies present in a typical object. In principle this can be done straightforwardly by Fourier transforming the object phase to find its spatial frequency distribution at the BFP, multiplying by the response of the phase plate and further transforming to find the resulting spatial distribution. In practice, for most objects it is difficult to do the second transform analytically for an arbitrary diameter of the central aperture of the phase plate. Some results of analysis are presented here for an object that produces a uniform phase shift within a circular disc; further details are given elsewhere [5].

An object wave with a small phase advance ($A\phi$) relative to free space is represented here approximately by

$$f \approx 1 + \mathrm{i}\, A\, \phi(r_o, \theta_o)$$

where r_o and θ_o are transverse coordinates at the object and the propagation factor $\exp \mathrm{i}(kz - \omega t)$ has been omitted. Here A is the amplitude of phase variation, assumed $\ll 1$. The plate is assumed to advance the phase of scattered wave components by α at spatial angular frequencies that are greater than a value q_0. This q_o is 2π times the quantity described elsewhere as the 'cut-on' frequency. Detailed analysis for an ideal lens shows [5] that for some typical forms of ϕ the image can be expressed as

$$g = 1 + \mathrm{i}\, A\, \phi(r_i, \theta_i) - \mathrm{i}\left(1 - \mathrm{e}^{\mathrm{i}\alpha}\right) A\, \eta(r_i, \theta_i, q_0)$$

where r_i and θ_i are transverse coordinates at the image. The intensity at the detector is then

$$|g|^2 = 1 - 2\, A\, \eta \sin\alpha + O(A^2)$$

showing that where (η sin α) is non-zero, the variation of image intensity from the mean is proportional to A. The ratio $\eta\,(r_i, \theta_I, q_0)\,/\,\phi\,(r_o, \theta_o)$ may vary over the image. At frequencies below

q_0, the phase contrast transfer function is much less than unity, so the plate behaves as a high-pass filter.

The uniform phase shift produced by a circular disc of radius b is defined here without dependence on θ_o as

$$\phi(r) = \begin{cases} 1, & 0 < r < b \\ 0, & b < r \end{cases}$$

and the result for η with unit magnification can be written with similar independence of θ_i as

$$\eta(r_i, q_0) = \phi(r_i) - B\int_0^{r_i/b} \frac{1}{(\rho^2 - 1)} [\rho\, J_0(\rho B) J_1(B) - J_0(B)\, J_1(\rho B)]\, d\rho$$

where $\rho = r_i / b$ and $B = q_0 b$. The resulting intensities are shown in figure 2 for object discs of unit amplitude and a range of values of B. The uniform object distributions are imaged with little overshoot when B is less than about 1. In Fig. 1(d) of [6] the image intensity was found to follow the object phase variation well when the cut-on periodicity was 50 nm, while there was appreciable overshoot for periodicity 25 nm. The corresponding values of B are estimated as 0.96 and 1.91.

As shown in [5], the magnitudes of steps at r_i = b are always imaged fully but as B increases beyond 1, the low-frequency components are progressively lost from the image and the *mean* value at a step falls to the background value ($\eta = 0$). Since the unit range of the step is maintained, η then has to be + 0.5 at radii just less than b and − 0.5 just outside the step. Consideration of $|g|^2$ with $0 < \alpha < \pi$ shows that as η becomes more negative, the magnitude of *g* and the image intensity increase. Thus a bright halo or outline is produced just outside the boundary r = b, for objects with B > ~1.

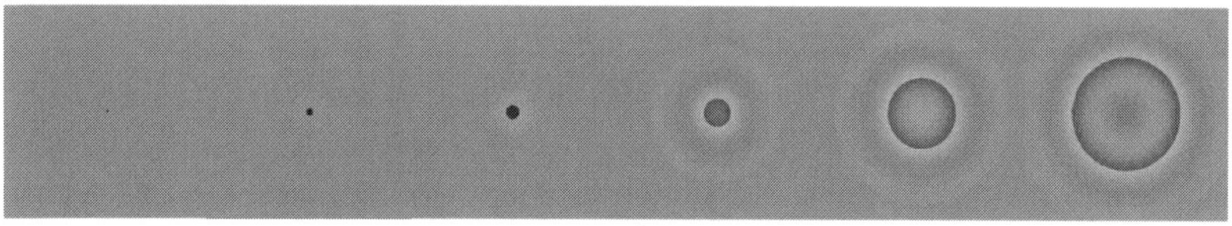

Figure 2. Image intensities produced by a $\pi/2$ phase plate with fixed q_0 for uniform disk objects with a range of diameters 2b. Responses are shown for values of $B = q_0 b$ (increasing from the left) of 0.2, 0.5, 1, 2, 5 and 8.

On the axis, $\eta\,(0,q_0)$ is given by the Bessel function $J_0(B)$ and so the intensity on axis (and elsewhere) oscillates as B varies. This effect can be seen in Fig. 1(b) of [4] where the distribution of intensity in the body of phage T4 appears to correspond to the distribution of η for B = 8 in figure 2.

The minimum radius s_2 at which a simple Zernike plate can be effective is that of the hole needed to pass the direct beam. The corresponding minimum scattering angle at which phase will be corrected, θ_{min}, is given approximately by geometry as

$$\theta_{min} \approx s_2 / f$$

The same angle equals the ratio of transverse and axial propagation constants q_0 (at cut-on) and k, where $k = 2\pi / \lambda$ and λ is the wavelength of electrons in the beam. Hence the maximum diameter corresponding to the maximum B for accurate imaging is

$$2b_{max} = 2B_{max} / q_0 \approx B_{max}\, \lambda\, f / \pi\, s_2$$

When f = 3 mm, s_2 = 0.5 μm and λ = 3.7 pm (100kV), then θ_{min} = 0.167 mrad, q_0= 0.283 rad $(nm)^{-1}$ and with B_{max} = 1, $2b_{max}$ ~ 7 nm. This b_{max} is inversely proportional to θ_{min}. To increase the size of object that can be imaged accurately, it will be necessary to reduce θ_{min} by reducing s_2 or increasing f, the focal length of the objective lens.

4. Location of phase plate in TEM column

The most obvious location for a phase plate is the objective aperture, nominally at the BFP of the objective lens. This location in the conventional TEM has some disadvantages. The space available for a traverse mechanism, heating arrangements or thermal insulation for cryogenic specimens is very limited. If the phase change is provided by a ferromagnetic ring, there is a question whether the ring can be aligned accurately enough to withstand the axial field of the lens. Also the maximum size of object that will be imaged accurately (as shown above) may be smaller than is desired for biological applications.

As deduced above, the maximum object size can be increased by increasing the effective focal length of the lens used as objective. Some instruments have been reported in which additional transfer optics are used. In many cases, however, the most practical alternative location for a phase plate is the selected-area (SA) position. Since this must still be the BFP of the lens that is to act as the objective, it suggests the use where available of a Lorentz lens, for which the BFP is at the SA height. Such lenses are liable to show large spherical aberration, but it has been reported [7] that this has been successfully reduced by using an existing aberration corrector.

5. Implications for biological microscopy

Published results (eg [2]) for Zernike phase plates show clearly that they provide the expected gain in contrast at focus. The theory outlined above suggests that accurate information transfer from the object phase to the image intensity is obtainable for object dimensions satisfying $q_0b < \sim 1$. For a conventional TEM with f = 3mm and beam energy 100kV, the maximum object diameter for accurate transfer is thus about 7 nm. Broader variations within the object will be attenuated, and sharp phase changes within such broad features are likely to produce overshoots which show up in the image as bright lines. As described above, if the effective focal length of the objective lens can be increased by use of a Lorentz lens or otherwise, this should allow proportionately larger object features to be imaged accurately.

6. Summary

The white outlines or halos seen around some phase objects when a Zernike plate is used have been shown [5] to be due to the high-pass response of the phase plate, defined by its central hole.

The disadvantages of placing a Zernike phase plate at the BFP of the objective lens can be reduced by placing it at the SA location and modifying the objective lens to have greater effective focal length, for instance by using a Lorentz lens where this is available. Doing so will increase the largest diameter of object features whose phase will be imaged accurately.

Acknowledgment

The authors thank A. M. Blackburn for ebeam lithography of the semicircular rings and R. M. Glaeser for stimulating discussions. Fig. 2 was produced by use of MATHEMATICA.

References

[1] Edgcombe C J and Loudon J C 2012 *J. Phys.: Conf. Ser.* 371 ed G Moebus *et al* (Bristol: IoP) p 012001
[2] Danev R and Nagayama K 2001 *Ultramicroscopy* **88** 243-52
[3] Fukuda Y, Fukazawa Y, Danev R, Shigemoto R and Nagayama K 2009 *J Struct Biol* **168** 476-84
[4] Danev R and Nagayama K 2011 *Ultramicroscopy* **111** 1305-15
[5] Edgcombe C J 2014 *Ultramicroscopy* **136** 154-9, http://dx.doi.org/10.1016/j.ultramic.2013.09.004
[6] Hall R J, Nogales E and Glaeser R M 2011 *J Struct Biol* **174** 468-75
[7] Freitag B, Bischoff M, Mueller H, Hartel P and von Harrach H S 2009 *Microscopy and Microanalysis*, **15** (Suppl. 2) pp 184-5

Observation of depth-dependent atomic displacements related to dislocations in GaN by optical sectioning in the STEM

J G Lozano[1], M P Guerrero-Lebrero[2], A Yasuhara[3], E Okinishi[3], S Zhang[4], C J Humphreys[4], P L Galindo[2], P B Hirsch[1] and P D Nellist[1]

[1] Department of Materials, University of Oxford, OX1 3PH, Oxford, United Kingdom
[2] Departamento de Ingenieria Informatica, CASEM, Universidad de Cadiz, Pol. Rio San Pedro s/n, 11510 Puerto Real (Cadiz), Spain
[3] JEOL Ltd., 1-2 Musashino 3-Chome, Akishima, 196-8558 Tokyo (Japan)
[4] Department of Materials Science and Metallurgy, University of Cambridge, Pembroke Street, Cambridge CB2 3QZ (UK)
E-mail: juan.lozano@materials.ox.ac.uk

Abstract. We demonstrate that it is possible to observe depth-dependent atomic displacements in a GaN crystal due to the sufficiently small depth of field achievable in the aberration-corrected scanning transmission electron microscope. The depth-dependent displacements associated with the Eshelby twist of screw dislocations in GaN viewed end on are directly imaged, and makes possible the determination of the sign of the Burgers vector of the dislocation. The experimental results are in good agreement with theoretical images.

1. Introduction

Dislocations in GaN, especially those with a pure screw nature, are known to have a strong detrimental effect on the optoelectronic properties of this material, as they are believed to behave as strongly active non-radiative recombination centers [1]. For this reason, a significant effort to determine the atomic structure of pure edge (***a***), mixed (***a+c***) and pure screw (***c***) dislocations in GaN has been made in the last decades [2-4]. These investigations, however, have so far ignored the surface relaxation effects, known at the Eshelby twist, that occurs when a dislocation with a screw component intersects the free surface of the sample [5]. In materials where the screw component of the Burgers vector is relatively large, as is the case for GaN, the Eshelby twist can produce displacements that are a significant fraction of the projected interatomic spacing of the material, and can therefore have a significant impact on the structure observed in (scanning) transmission electron microscopy (S)TEM images [6,7].

Additionally, in previous work [8] it was suggested that strain contrast in high angular annular dark field (HAADF)-STEM images of screw dislocations viewed end-on in GaN arose because of the large displacements parallel to the dislocation line. Such displacements parallel to the electron beam direction can only affect the higher-order Laue zone diffraction, which is known to make a negligible contribution to HAADF imaging [9], and only perpendicular displacements have a significant effect

[5] To whom any correspondence should be addressed.

on the contrast, so no image intensity variations are expected from displacements parallel to the screw dislocation. We will show that this contrast variation around the core of dislocations with a screw component is likely to be due to the Eshelby twist.

The increased numerical aperture allowed by the development of aberration correction has also led to a reduction in depth of field, which may be just a few nanometers and therefore smaller than typical sample thicknesses of tens of nm. This creates an opportunity to "optically section" a sample by simply imaging with the focal plane set to a specific depth within the sample. Indeed, a full three-dimensional image can be obtained by recording a series of images over a range of focus settings. For a general sample, the full depth resolution can only be obtained in a confocal geometry [10-12], but at atomic resolution optical sectioning at a nanoscale resolution is possible using a conventional STEM [13]. This approach can be applied to atomic-resolution imaging of dislocations in GaN. In this paper we show it is possible that the depth-dependent displacement associated with the Eshelby twist in c dislocations can be directly imaged, immediately allowing the determination of the sign of theirs Burgers vector.

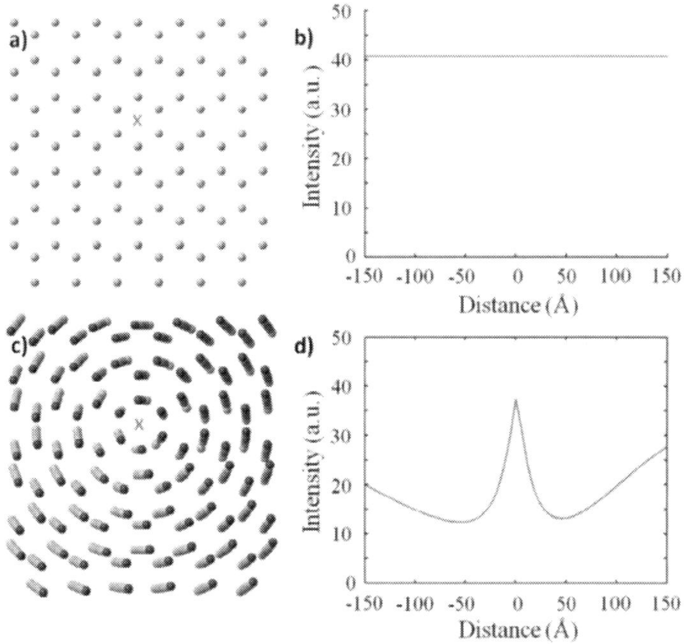

Figure 1: (a) Atomic model of a GaN crystal projected along the [0001] direction including a right handed pure screw dislocation (centre marked with a red cross). (b) HAADF STEM intensity profile across the core of the dislocation for a simulated 10 nm thick GaN crystal containing a right handed pure screw dislocation. The zero position corresponds to the core of the dislocation. (c) Atomic model of a GaN crystal along the [0001] direction including a right handed screw dislocation (marked with a red cross), and also including the displacements associated with the Eshelby twist. The colour of the top layer of atoms has been changed for a better visualization of the sense of the rotation. (d) HAADF STEM intensity profile across the core of the dislocation for the the model shown in (c).

2. Results and discussion

Figure 1(a) shows an atomic model consisting of a GaN crystal projected along [0001]. The displacements associated with a right-handed singular screw dislocation, assuming isotropic elasticity, have been introduced. These displacements are parallel to the [0001] direction and do not show in this projection..To explore possible strain contrast, Figure 1(b) shows an interpolated plot of HAADF image peak intensities across the area containing the dislocation core for a 10 nm thick GaN crystal. It can be clearly seen that no strain-induced contrast variations around the dislocation core exist if we consider only the displacements parallel [0001]. In contrast to this, Figures 1(c) and (d) show the same model and an interpolated HAADF intensity line profile across the dislocation core, where the displacements associated with the Eshelby twist have also been included. The assumption is made that the displacements are those for an isotropic solid as given by Eshelby and Stroh [5]. The assumption of isotropic elasticity stems from the fact that GaN is isotropic on the basal plane [14]. In this case, an area of reduced HAADF intensity can be seen surrounding the core of the dislocation induced by disruptions to the electron channelling conditions caused by the twist. These results suggest that the Eshelby twist provides the likely origin of the reduced intensity seen experimentally in Fig 3 of Ref. [15], and not the displacements parallel to the beam that the authors of that paper suggest. It is apparent that any quantitative interpretation of HAADF STEM imaging of dislocations with a screw component normal to the foil must therefore include the impact of the Eshelby twist.

A focal series of high angle annular dark field (HAADF) images were calculated using the model shown in Figure 1c and at defocus step of 2.5 nm were simulated using SICSTEM [16], to test whether the Eshelby twist could be observable directly by optical sectioning. The results indicate that a clockwise rotation-like displacement of the atomic columns, in the direction of the twist, around the core of the dislocation occurs, as displayed in Figure 2 where the position of the intensity maxima of an atomic column at an approximate distance of 1.7 nm from the dislocation core is shown.

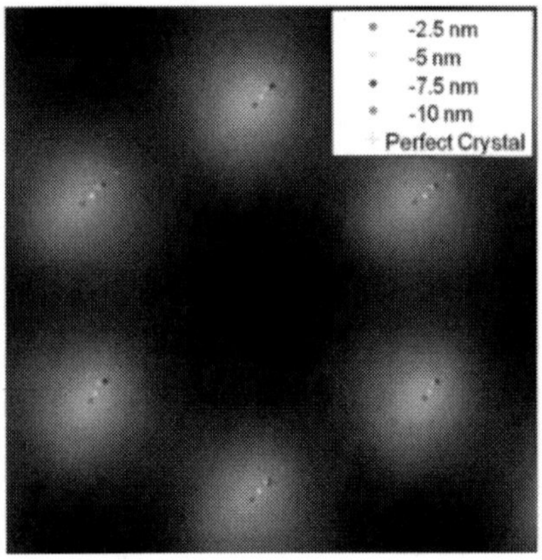

Figure 2: Simulated HAADF image of GaN (0001) indicating the positions of the intensity maxima of different atomic column for different defocus values, showing a clockwise rotation of the lattice in the direction of the Eshelby twist.

Series of HAADF STEM optical sectioning experiments were performed in the aberration corrected JEOL ARM 200F operating at 200 kV. The samples investigated consisted of a 4.2μm thick GaN layer, grown on sapphire substrates by metal-organic chemical vapour deposition (MOCVD). The threading dislocation density was low, in the order of 4-5×10^8 cm^{-2}. Out of the three types of dislocations, pure screw ones have the lowest density, but they can be identified as they are known to form nanopipe structures around them, as the one shown in Figure 3a. An open core screw dislocation can then be seen in an area close to the centre of the nanopipe, and images were recorded from this region at defocus step of 1 nm and subsequently drift corrected. The results indicate that the effect of Eshelby twist is observable as a clockwise rotation of the GaN lattice around the dislocation core. Following the arguments given by Hirth and Lothe [14], for a right handed screw dislocation looking from the surface into the sample, the rotation due to the Eshelby twist to annul the shear stress due to the screw at the entrance surface must be anti-clockwise, looking in the same direction. Since the observed image rotation is clockwise as the focal plane goes away from the surface and deeper into the crystal, the dislocation is a right-handed screw. The experimental rotation angle as function of defocus for a given column (indicated by a dotted line in Figure 2b) is 0.074 $^{\circ}nm^{-1}$, which is not far from the 0.116 $^{\circ}nm^{-1}$ measured in the simulated stack of images.

3. Conclusions

We have shown that is possible to detect a depth dependent strain associated to the different types of dislocations in GaN using optical sectioning in HAADF STEM, in this case arising from the Eshelby-Stroh twist. The presence and sign of the depth dependent strain provides a method of detecting dislocation Burgers vectors parallel to the electron beam.

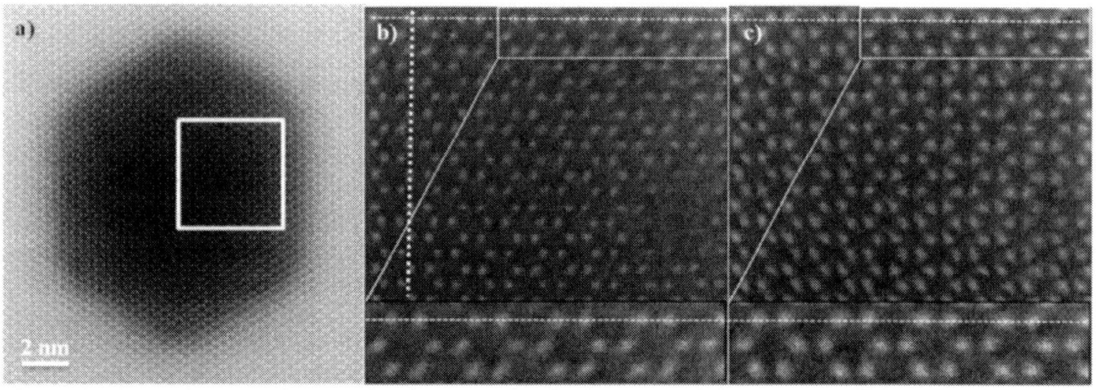

Figure 3. a) HR-STEM micrograph of a nanopipe in GaN containing a screw dislocation (b) and (c) HR-STEM micrographs of the area marked with a white square taken at defocus values 4 nm apart. The insets show show magnified areas. In (b), the dashed line runs through the centre of the atomic columns. In (c), the dashed line runs through the centre of the columns on the right hand side of the image (see inset), while on the left side of the image, the atomic columns are above the dashed line. This indicates a clockwise rotation of the lattice in the image as the focus has changed.

4. References

[1] Wright A F and Grossner U 1998 *Appl. Phys. Lett.* **73**, 2751
[2] Hino T, Tomiya S, Miyajima T, Yanashima K, Hashimoto W and Ikeda M 2000 *Appl. Phys. Lett.* **76**, 3421
[3] Belabbas I, Béré A, Chen J, Petit S, Akli M, Belkhir M, Ruterana P and Nouet G 2007 *Phys. Rev. B* **75**, 115201
[4] Belabbas I, Nouet G and Komninou Ph 2007 *J. Crystal Growth* **300,** 212
[5] Eshelby J D and Stroh A N 1951 *Philosophical Magazine Series* 7 **42**, 1401
[6] Groger R, Dudeck J K, Nellist P D, Vitek V, Hirsch P B and Cockayne D J 2011 *Philosophical Magazine* **91**, 2364
[7] Cosgriff E C, Nellist P D, Hirsch P B, Zhou Z, and Cockayne D J 2010, *Philosophical Magazine* **90**, 4361
[8] Arslan I, Bleloch A, Stach E A, Ogut S and Browning N D 2006 *Philosophical Magazine* **86**, 4727
[9] Rez P and Amali A 1997 *Microsc. Microanal.* **3**, 28
[10] Nellist P D, Beham G, Kirkland A I, and Hederington C J D 2006 *Appl. Phys. Lett.* **89**, 124105
[11] Wang P, Hashimoto A, Takeguchi M, Mitsuishi K, Shimojo M, Zhu Y, Okuda M, Kirkland A I and Nellist P D 2012 *Appl. Phys. Lett.* **100**, 213117
[12] Wang P, Behan G, Takeguchi M, Hashimoto A, Mitsuishi K, Shimojo M, Kirkland A I, and Nellist P D 2010 *Phys. Rev. Lett.* **104**, 200801
[13] Van Benthem K, Lupini A R, Kim M, Suck Baik H, Doh S, Lee J H, Oxley M P, Findlay S D, Allen L J, Luck J T, and Pennycook S J 2005 *Appl. Phys. Lett.* **87**, 034104
[14] Hirth J P and Lothe J 1968, *Theory of Dislocations* (McGraw-Hill, New York,), pp. 59-62.
[15] Arslan I and Browning N D 2003, *Phys. Rev. Lett.* 91, 165501
[16] Pizarro J, Galindo P L, Guerrero E, Yáñez A, Guerrero M P, Rosenauer A, Sales D L and Molina S I 2008 *Appl. Phys. Lett.* **93**, 153107

Helium ion microscopy and energy selective scanning electron microscopy – two advanced microscopy techniques with complementary applications

C Rodenburg[1], M A E Jepson[2], Stuart A. Boden[3], Darren M. Bagnall[3],

[1]Department of Engineering Materials, University of Sheffield, Mappin Street, Sheffield, S1 3JD, UK
[2]Department of Materials, Loughborough University, Loughborough, LE11 3TU, UK
[3]Electronics and Computer Science, University of Southampton, Highfield, Southampton, SO17 1BJ, UK

c.rodenburg@shef.ac.uk

Abstract. Both scanning electron microscopes (SEM) and helium ion microscopes (HeIM) are based on the same principle of a charged particle beam scanning across the surface and generating secondary electrons (SEs) to form images. However, there is a pronounced difference in the energy spectra of the emitted secondary electrons emitted as result of electron or helium ion impact. We have previously presented evidence that this also translates to differences in the information depth through the analysis of dopant contrast in doped silicon structures in both SEM and HeIM. Here, it is now shown how secondary electron emission spectra (SES) and their relation to depth of origin of SE can be experimentally exploited through the use of energy filtering (EF) in low voltage SEM (LV-SEM) to access bulk information from surfaces covered by damage or contamination layers. From the current understanding of the SES in HeIM it is not expected that EF will be as effective in HeIM but an alternative that can be used for some materials to access bulk information is presented.

1. Introduction

The field of scanning electron microscopy (SEM) has seen major changes in the past decade in particular in terms of resolution improvements for low accelerating voltages, e.g. monochromated LVSEM resulting in sub-nanometer resolution [1]. Also a novel microscope was introduced, the Helium Ion microscope, HeIM, (in which the electron gun is replaced by a helium ion source), with an ultimate expected focused spot size of 0.25nm [2]. With such progress in lateral resolution, new application areas for both SEM and HeIM are opened up and it is increasingly recognized that interpretation of contrast in SEM images needs more attention.

While there have been some successful attempts to quantify SEM contrast for specific applications microscopes, it is also recognised that the contrast obtained from different instruments varies widely [3]. Often this has been blamed on the variation of secondary electron yield with the surface condition, i.e. the presence of oxide or carbon contamination layers. Of course this has to be considered as a possible explanation, especially since it has been found in both HeIM [4] and LVSEM [5] that organic mono-layers lead to strong contrast, albeit the exact mechanism is a matter of debate with several attempts to link contrast based on SE yield to differences in functional groups, surface potential, surface energy and molecular ordering. While very useful for the imaging of self-assembled organic mono-layers (SAMs), it raises the issue of uncertainty in SEM contrast interpretation as result of surface layers. Further doubt on any attempts of quantitative SEM image contrast arises from the observation that contrast is very strongly influenced by the particular detector used in one instrument or different instruments. Given this situation one could come to the conclusion that there is no hope in understanding SEM contrast because of the limited information collected in the SEM.

One solution is to move away from the sole focus of SE yields (intensities) and start to consider the additional information contained within the SE emission spectrum (SES). Therefore, here SES are considered in more detail and the ways that they can be exploited for a better informed interpretation of image contrast are explored. In addition, differences in SES found in HeIM and LVSEM and the consequences for accessing bulk information or obtaining the best possible resolution are considered.

2. The secondary electron spectrum caused by low energy ELECTRON impact – accessing bulk information

Unlike electron energy loss spectra (EELS), secondary electron spectra (SES) are not widely known to have very distinctive features, perhaps due to the fact that SEMs are often operated at accelerating voltages of 20kV. Nevertheless, coincidence spectroscopy between EELS and SES has revealed a clear link between them [6]. It is therefore not surprising that SES collected using LVSEM (<5kV), where collision cascades are short and localised, differ sufficiently enough for fingerprint identification of inorganic materials based on the LVSEM SES spectra [7]. A more topical example demonstrating the importance of SES is that of graphene covered Ni in comparison to bare Ni. While the yield of the graphene covered Ni and the bare Ni were found to be similar, very marked differences in the LV SES for these two cases were reported [8]. Further experiments of varying accelerating voltage (250eV-900eV) on the graphene covered Ni show very distinctive changes in the SES, whereby the low energy peak starts to dominate with increasing accelerating voltage, reminiscent of the electron cascade, that leads to SE generation in the bulk resulting in SE with low energy when reaching the surface [8]. Detailed Monte Carlo calculations of the SES of this system were not reported but Monte Carlo modelling of SiO_2 also demonstrates that the high energy end of the SES is dominated by SEs generated at the surface while a substantial fraction of low energy SEs are likely to have been excited closer to the bulk (deeper than 10nm in the case of SiO_2) rather than at the surface [9]. Figure 1 (field of view, FV 6 μm) demonstrates that this increased escape depth for low energy SEs can be used to recover bulk information on a damaged surface, using dopant contrast as an example.

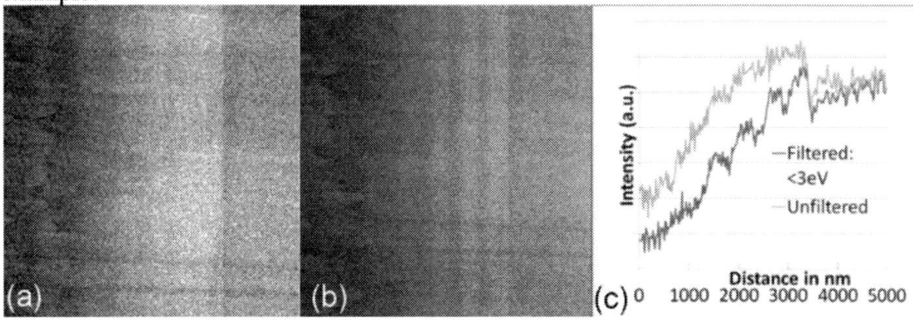

Figure 1 LV SEM (a) p-doped layers in Si not separated after Ga ion exposure, (b) as (a) but SE < 3eV only (c) Intensity profiles, from (a) and (b)

In LVSEM p-doped silicon layers show clear contrast on cleaved specimens but individual layers are not resolved when the sample surface is damaged during focused ion beam (FIB) preparation using 10 kV Gallium ions at grazing incidence. This can be understood as a result of the lateral displacement of dopant atoms, originally confined to the doped layers, by up to ~25nm, according to TRIM simulations. EF- imaging of such a Si surface does not improve the situation because the maximum escape depth, even for low energy SEs, is still located within the damage layer. However, if the same specimen is kept in air or exposed to an oxygen containing plasma, oxidation occurs from the surface, which also contains the most heavily ion beam damaged layer (with majority of damage located within of ~5nm from the surface, as determined by TRIM). The escape depth of the low energy SEs is increased in grown SiO_2 compared to Si and now exceeds 10nm, deeper than the majority of the Ga-ion damage. This makes no difference to dopant contrast of a standard (unfiltered) SE image which is strongly dominated by SEs originating from the damaged layer. Hence the individual doped layers still cannot be distinguished (Figure 1a). However, when only low energy SEs (<3eV) are detected, the individual layers can be resolved again, as the image is now predominantly formed from SEs originating from below the heavily ion beam damaged material. Figure (1c) contains intensity profiles that demonstrate the above more clearly.

This filtering approach also works for other materials systems as demonstrated in Figure 2 (a,b). The material is a triblock copolymer. In the standard SEM image in Figure (2a) the presence of details is obscured and contrast is dominated by topography as result of microtome damage. When only SE with energies <8.3 eV are used to form the image, fine detail appears (Figure 2b), best seen in the higher magnification inset. This demonstrates again the power of LVSEM in combination with EF in accessing bulk information. We note that the best suited cut-off energy depends on the material investigated based on the experimental or modelled SES.

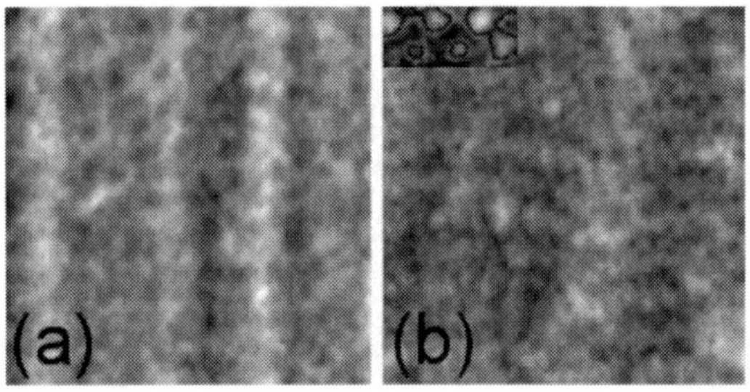

Figure 2 LV SEM images taken at 1 keV primary beam, Tube bias 250V, Working distance 3mm: (a) stained tri-block polymer (FV 500nm), the vertical lines are sample preparation artefact (microtome damage), (b) as (a) but energy cut off for SE set at 8.3eV, the damage related features are much less pronounced, instead horizontal chains become visible. The inset shows chain at higher magnification (width 94 nm).

3. The secondary electron spectrum caused by HELIUM ION impact – accessing surface information

The HeIM is renowned for its high chemical contrast and high lateral resolution. Figure 3a is a HeIM image from the same material, shown in Figures 2. While Figure 2b suffers from a limited lateral resolution, Figure 3a has limited chemical information, in-spite of the expected better lateral resolution. It is assumed that this lack of chemical information is a result of the extreme surface sensitivity of HeIM. Evidence for this can be found by comparing the contrast at p-n junctions with varying concentrations of dopant atoms in the HeIM and LVSEM. This contrast is related to the potential difference V_{pns} between the p and n doped regions at the surface. V_{pns} is expected to reach a value determined by the doping concentration differences through the Fermi level alignment, V_{pnFA} if no surface states were present. In the presence of surface states (expected for silicon cleaved in air), V_{pns} only gradually reaches the value of V_{PnFA}, when moving from the surface into the specimen. The depth of this surface band bending region becomes shallower, with increased doping levels and can explain a sharp increase in doping contrast in the LVSEM for doping levels $>10^{18}$ cm^{-3}[10,11]. The sharp increase in contrast is a result of the surface band bending depth becoming smaller than the SE escape depth. This phenomenon can also be observed in the HeIM, but at a higher doping level of 10^{19} cm^{-3}[11], and hence a shallower band bending region. The latter suggest a shallower escape depth for SEs generated by He ion impact when compared to those generated by electron impact, leading to extreme surface sensitivity.

Further evidence for the enhanced surface sensitivity is found by the collection of SES generated by 30kV He ion impact [12]. These SES show a shift of the peak energy to smaller energies as well as a reduction in the width of the SES peak. Based on the description of the relationship between the depth of SE origin and SE energy in section 2, it can be concluded that the energy of the most energetic SEs generated at the surface by HeIM impact differs from that resulting from electron impact. If we assume this also applies to the internally He-ion generated SEs then the escape depth is affected and for carbon it is only ~0.5 of that for electron impact according to the data in [13].

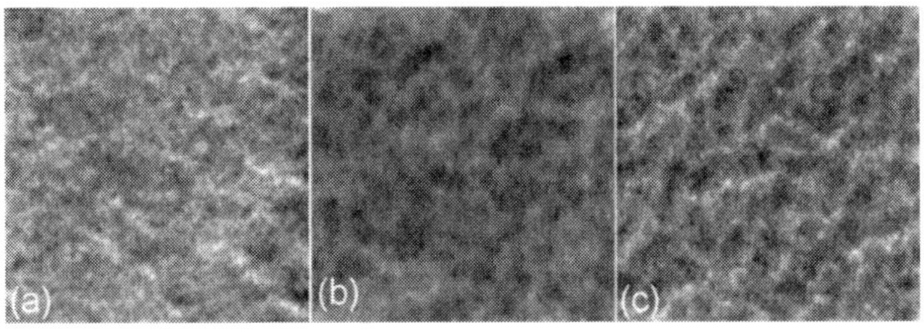

Figure 3 HeIM images): (a) stained tri coblock polymer field of view 300nm, (b) P3HT:PCBM polymer blend as received field of view 150nm, (c) same specimen as (b) after 3 min of plasma cleaning (process gas is air).

Based on the above discussion, HeIM will have less scope for making use of energy filtering for the purpose of accessing bulk information covered by damage layers due to the smaller maximum SE escape depth involved. Of course, we should recognise that it is precisely the latter that contributes to the excellent lateral resolution and surface sensitivity of HeIM SE imaging.

An alternative solution to access bulk information on some material in the HeIM may be the use of a plasma to etch away surface layers as demonstrated in [14] and in Figure 3(b,c). Figure 3b shows the as-received surface of a P3HT based bulk heterojunction solar cell material imaged in the HeIM at 30keV but the contrast becomes clearer when P3HT rich surface layer is removed through plasma etching.

4. Conclusion

Many development efforts in microscopy have been devoted to improving lateral resolution, but a practical limitation often encountered for advanced SEM is how to gain information representative of the bulk material in the presence of limited contrast due to damage or contamination layers. It has been demonstrated here that the filtering of the relatively broad SES in LVSEM can address this challenge effectively, while HeIM is best suited for applications where extreme surface sensitivity is paramount. Energy filtering in HeIM provides less scope for access to bulk information as a result of a reduced SE escaped depth but under some circumstances, a plasma can be used to remove surface layers.

Acknowledgments

The authors thank M. Tureau & T. H. Epps (College of Engineering University of Delaware) for providing the tri block polymer samples and A. J. Pearson and D. G. Lidzey Department of Physics and Astronomy, University of Sheffield for providing the P3HT:PCBM polymer blend sample.

References

[1] Michael J R 2011 Scanning, 33: 147–154
[2] Ward B W *et al* 2006 J. Vac. Sci. Technol, B **24** 2871
[3] Griffin B J 2011 Scanning, 33: 162–173[4]
[4] George A *et al* 2012 Langmuir. 28, 3045
[5] Srinivasan C *et al* ACS Nano 2007 1 (3) 191-201
[6] Pijper F J and Kruit P 1991 Phys. Rev. B 44(17) 9192
[7] Joy D C *et al* 2004 Journal of Microscopy 215: 77–85
[8] Riccardi P *et al* 2012 APL 101, 183102
[9] Fitting H-J *et al* A 2001 J. of Electron Spectroscopy and Related Phenomena 35–47
[10] Chee K W A *et al* 2010 J. Phys.: Conf. Ser. 209 012054
[11] Jepson M A E *et al* 2009 EPL 86 26005
[12] Petrov Y and Vyvenko O 2011 Proc. SPIE 8036, doi:10.1117/12.886347
[13] Ramachandra R *et al* 2009 Ultramicroscopy 109 748–757
[14] Pearson A J *et al* 2011 Nano Lett.11 (10) 4275–4281

Development of a high-efficiency DF-STEM detector

T. Kaneko[1], A. Saitow[1], T. Fujino[1], E. Okunishi[1] and H. Sawada[1]

[1] JEOL Ltd. 1-2 Musashino 3-Chome, Akishima, Tokyo 196-8558, Japan

E-mail: takekane@jeol.co.jp

Abstract. We experimentally evaluated the signal intensity produced by an electron on three types of scintillators. A powder scintillator showed high efficiency at a low accelerating voltage, whereas a single crystal scintillator showed high efficiency at higher accelerating voltages. On the basis of these characteristics, for a dark-field (DF) scanning transmission electron microscope (STEM), we have developed a new hybrid type scintillator, which consisted of a powder deposited on a single crystal substrate. The luminescent quantum efficiency of the hybrid scintillator was measured to be twice as large as that of the single crystal type detector at 60 kV and was about 8 times higher than that of the powder type detector at 300 kV. The developed detector has advantages of powder and single crystal type detectors, and covers the observation at the accelerating voltages from low to high voltages. Especially, it is useful for low voltage observations of carbon-based materials consisted of few atomic layers that produces weak scattering of electron.

1. Introduction

High-resolution dark-field (DF) imaging with an aberration corrected scanning transmission electron microscope (STEM) has been widely used for structural analysis in materials science. A conventional DF-STEM detector consists of a scintillator, a glass light pipe, and a photomultiplier tube (PMT). The scintillator of the DF-STEM detector is selectable either a powder scintillator or a single crystal scintillator. Y_2SiO_5 doped with Ce (P47) is used for the powder scintillator. And, the single crystal scintillator is made of yttrium aluminum garnet (YAG) or yttrium aluminum perovskite (YAP). Faruqi and Tyrell investigated the relationship between the thickness of the powder scintillator and the accelerating voltage [1]. In their study, a powder scintillator was reported to show a maximum of efficiency at a particular accelerating voltage. The accelerating voltage showing the maximum efficiency depends on the thickness of the powder scintillator. A YAG single crystal scintillator has been used for the charge-coupled device camera of transmission electron microscopy [2-4]. The YAP scintillator is mostly used for the DF-STEM detector, because its wavelength of the luminescence is sensitive for a conventionally commercialized PMT [5, 6].

Most recently, observations at low accelerating voltages have been widely popular for carbon-based materials such as carbon nanotubes or graphenes [7-9], to reduce knock-on damage due to irradiation of an electron beam. These materials generally produce a small amount of scattered

[1] To whom any correspondence should be addressed.

electrons for the DF-STEM detector because they are composed of a few atomic layers. A DF-STEM detector of high-efficiency is longed to obtain an STEM image with a better signal-to-noise ratio. In this study, we explain the characters of a new highly-efficient DF-STEM detector. And we show the DF-STEM image taken at a lower accelerating voltage using the detector.

2. Experimental and methods

Three scintillators (P47 powder, YAP single crystal, and hybrid scintillators) were tested, as described below. The intensity of the signal per electron was measured, on each scintillator at accelerating voltages ranging from 30 kV to 300 kV. An electron beam was focused and scanned on the scintillator screen by adjusting the excitations of several lenses to see the map of the signal intensity. The probe current was measured using a Faraday cage, and the number of electrons per a unit time was calculated using the elementary charge of an electron. The Gatan DigiScan™ system was employed to obtain an average of the digitally counted intensity on each scintillator. The intensity per electron on each scintillator was calculated as a quotient of averaged the digital counts on the image of the scintillator divided by the number of electrons.

Figure 1(a) shows the schematic of the detector, and an intensity map of the scintillator, which shows a flatness of detection, is displayed in Fig. 1(b). In the conventionally in-use DF-STEM detectors, the image of the scintillator lacked uniformity around the center hole of the detector [10-16]. Whereas, for the developed DF-STEM detector, we reduced this non-uniformity by redesigning the structure of the DF-STEM detector, thereby achieving a relatively uniform signal intensity over the scintillator, as shown in Fig. 1(b).

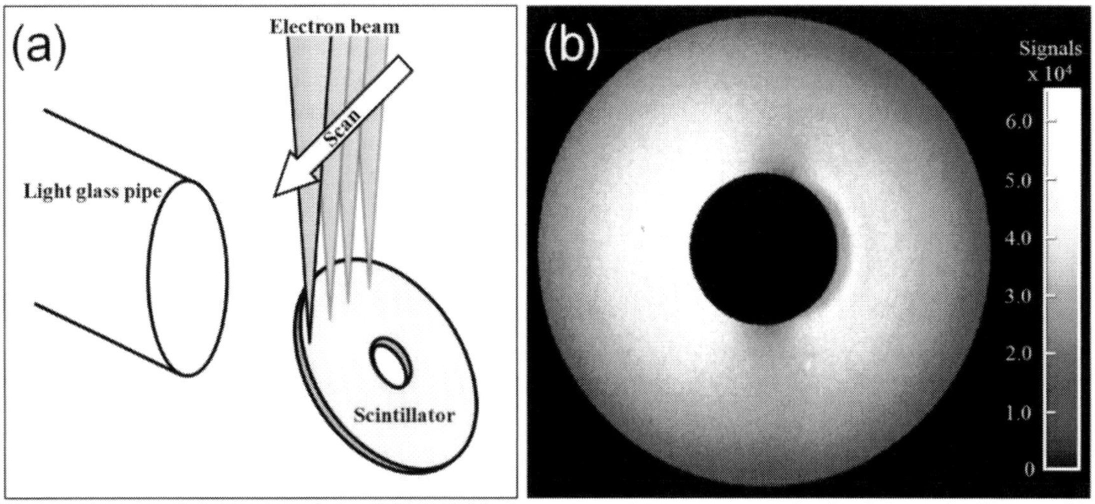

Figure 1. (a) Schematic of the DF-STEM detector and scanning method of the scintillator. (b) Scanning image of the scintillator.

3. Results and discussion

3.1. Development of the hybrid detector

Scanning images of the scintillators were obtained at several accelerating voltages, and their efficiencies were measured. Figure 2 shows the plotted signal intensity per electron as a function of the accelerating voltage for the three scintillators. The YAP single crystal and the P47 powder scintillators are used in the conventional detectors. The thicknesses of scintillators for the YAP single crystal and the P47 powder were 0.5 mm and 30 μm, respectively. The efficiency (signal intensity per electron) of the YAP single crystal scintillator increases as the accelerating voltage increases. The efficiency of the YAP single crystal scintillator is higher than that of P47 powder scintillator at accelerating voltages greater than 100 kV. On the other hand, the efficiency of the P47 powder scintillator is relatively uniform with respect to that of the YAP single crystal scintillator. P47 shows a maximum efficiency at 60 kV. For accelerating voltages less than 60 kV, the signal intensity of the P47 powder scintillator is higher than that of the YAP single crystal scintillator.

These characteristics are attributed to the forms and thicknesses of the materials of scintillators. In the YAP single crystal scintillator, the electron produces photons until the electron energy becomes lower than the threshold energy to produce a photon. We can say that the number of photons produced by electron is roughly proportional to the accelerating voltage. The thickness of YAP single crystal scintillator is sufficiently thick for electrons of high energy, resulting in more photons produced in a thick YAP single crystal scintillator than thin P47 powder scintillator at accelerating voltages higher than 60 kV. On the other hand, in scintillator made of a thin layer of P47 powder, emitted photon tends to encounter to the surface of the powder particle with high incident angle, because of small curvature of powder particles. The high incident angle leads, the higher escape probability of photons than thick and scintillation crystal that has wide and flat surfaces, resulting in higher emission of photons for powder scintillator. And most of energy of incident electrons less than 60 kV is consumed in the layer. Thus, the scintillator of powder is superior than that made of single crystal scintillator at accelerating voltage below 60 kV.

By combining the advantage of these scintillators, we built a hybrid scintillator, which is made of deposited a 30-μm-thick P47 powder layer on a 0.5-mm-thick YAP single crystal substrate. The efficiency of the developed hybrid scintillator is also plotted in Fig. 2, as a function of the accelerating voltage. When an electron hits the P47 powder scintillator at the surface, the hybrid scintillator produces luminescence, and the emitted photons are transmitted through the transparnent YAP single crystal scintillator. Moreover, the transmitted electrons or X-rays passing through P47 produce another luminescence in the YAP single crystal. Thus, the total efficiency is the sum of these two scintillators, resulting in the higher efficiency of the developed detector with respect to the two conventional scintillators at each accelerating voltage. The signal intensity of the hybrid scintillator was about 8 times as large as the intensity of the P47 powder scintillator at 300 kV. Furthermore, the luminescent quantum efficiency of the hybrid scintillator was twice as large as that of the YAP single crystal scintillator at 60 kV.

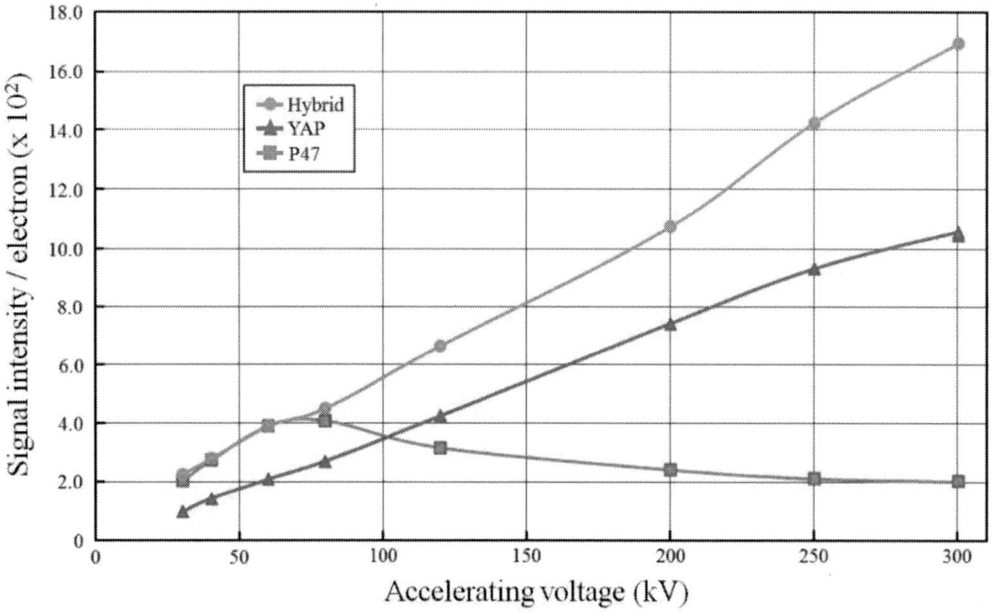

Figure 2. Signal intensity per electron of three scintillators plotted as a function of the accelerating voltage.

3.2. Observation of the STEM image of graphene

Using the developed hybrid detector, we performed STEM observation of graphene at 80 kV, using an aberration-corrected microscope (JEM-ARM200F) equipped with a cold field emission gun. Figure 3 shows the DF-STEM images with a probe current of 22 pA and a convergence semi-angle of 33 mrad. The detecting angle for the DF-STEM detector was set to range from 54 mrad to 216 mrad in this observation. The dwell time for a pixel and the number of pixels for the STEM image were 19.3 μs and 512 × 512 pixels, respectively. Single carbon atoms in mono atomic sheet were clearly observed as a bright dot. The intensity profile taken from the area indicated by a dotted rectangle is shown in Fig. 3(b). C-C atoms separated by 0.14 nm were imaged with sufficient dips between them.

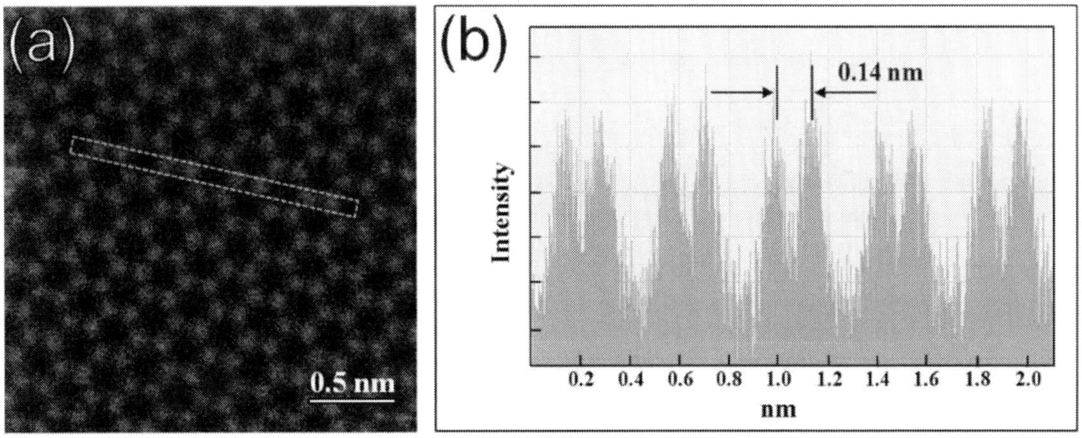

Figure 3. (a) DF-STEM image of graphene. (Raw data) (b) Intensity profile of dotted rectangular area.

4. Summary

A DF-STEM detector was developed by combining the powder and transparent crystalline scintillators. Experimental measurements showed that the developed hybrid detector had higher efficiency than the conventional detectors at accelerating voltages ranging from 30 kV to 300 kV. A single carbon atom in a mono atomic sheet of graphene was successfully imaged with new hybrid detector at 80 kV, using an atomic resolution electron microscope (JEM-ARM200F).

Acknowledgements

This work was supported by JST under the Research Accelerating Program (2012–2016).
The authors thank Dr. U. Mirsaidov of National University of Singapore for providing us the graphene sample.

References

[1] Faruqi A R and Tyrell G C 1999 *Ultramicroscopy* **76** 69
[2] Daberkow I, Herrmann K H, Liu L and Rau W D 1991 *Ultramicroscopy* **38** 215
[3] Nishi R, Yoshida K, Takaoka A and Katsuta T 1996 *Ultramicroscopy* **62** 271
[4] Zuo J M 2000 *Microscopy Research and Technique* **49** 245
[5] Haider M 1989 *Ultramicroscopy* **28** 240
[6] Kirkland E J and Thomas M G 1996 *Ultramicroscopy* **62** 79
[7] Suenaga K and Koshino M 2010 *Nature* **468** 1088
[8] Krivanek O L, Dellby N, Murfitt M F, Chisholm M F, Pennycook T J, Suenaga K and Nicolosi V 2010 *Ultramicroscopy* **110** 935
[9] Zan R, Ramasse Q M, Bangert U and Novoselov K S 2012 *Nano Lett.* **12** 3936
[10] LeBeau J M and Stemmer S 2008 *Ultramicroscopy* **108** 1653
[11] Findlay S D and LeBeau J M 2013 *Ultramicroscopy* **124** 52
[12] Katz-Boon H, Rossouw C J, Dwyer C and Etheridge J 2013 *Ultramicroscopy* **124** 61

Simulating Neutron Radiation Damage of Graphite by In-situ Electron Irradiation

Brindusa E Mironov[1], H M Freeman[1], R M D Brydson[1], A V K Westwood[1], A J Scott[1]
[1] Institute for Materials Research, University of Leeds, Leeds, UK, LS2 9JT

E-mail: pmbem@leeds.ac.uk

Abstract. Radiation damage in nuclear grade graphite has been investigated using transmission electron microscopy (TEM) and electron energy loss spectroscopy (EELS). Changes in the structure on the atomic scale and chemical bonding, and the relationship between each were of particular interest. TEM was used to study damage in nuclear grade graphite on the atomic scale following 1.92×10^8 electrons nm^{-2} of electron beam exposure. During these experiments EELS spectra were also collected periodically to record changes in chemical bonding and structural disorder, by analysing the changes of the carbon K-edge. Image analysis software from the 'PyroMaN' research group provides further information, based on (002) fringe analysis. The software was applied to the micrographs of electron irradiated virgin 'Pile Grade A' (PGA) graphite to quantify the extent of damage from electron beam exposure.

1. Introduction

Around 90% of the UK's current nuclear reactors are graphite-moderated generation II Advanced Gas Cooled Reactors. As well as moderating the energies of fission neutrons, the graphite core provides structural support, contains the fuel and control rods and allows for coolant flow. Graphite blocks in these reactors are subject to high levels of radiation and as a result they experience chemical and physical property changes which can affect other reactor components. The lifetime of such reactors is therefore primarily governed by the performance of the graphite moderator so an accurate estimation of its condition is essential for economic success and plant safety.

Since the 1960s much effort has been made to understand the behaviour of irradiated graphite [1-5]. Despite the wide knowledge base achieved to date there are significant gaps in understanding at the atomic level. The bulk properties of damage features have been thoroughly experimented upon and theoretical models have been derived. Although this has allowed behavioural changes in the irradiated bulk to be accounted for to some extent, in current and future graphite -based reactor designs, the mechanisms of such processes on the nanoscale still remain uncertain. This work investigates the effect of electron irradiation on nuclear grade graphites to understand the fundamental processes involved in radiation damage.

2. Experimental Details

Virgin PGA graphite sourced from the University of Manchester was chosen for inspection. Samples were crushed using a marble pestle and mortar and mixed with acetone before being dispersed onto a holey carbon coated copper grid (3mm).

Journal of Physics: Conference Series **522** (2014) 012051 doi:10.1088/1742-6596/522/1/012051

TEM investigations were performed on an FEI CM200 field emission TEM operated at 197 kV with a tip extraction bias of 3.21 kV routinely providing an electron flux of approximately 3.2×10^5 electrons $nm^{-2}\ s^{-1}$ and an electron energy loss spectroscopy (EELS) energy resolution of 0.7–0.8 eV. For EELS, the microscope was operated in diffraction mode with a collection semi-angle of 6 mrad and a convergence semi-angle of approximately 1 mrad, using a 115mm camera length (corresponding to the magic angle [6]). Digital images and energy loss spectra were captured using a Gatan imaging filter (GIF) 200 with a 1 megapixel slow scan CCD array. Data from the array (i.e. images and spectra) were processed using Gatan's Digital Micrograph (DM) software.

3. Results and Discussion

3.1 TEM

PGA graphite was exposed to the electron beam for 600 seconds, corresponding to a flux of 1.92×10^8 electrons nm^{-2}. Electron micrographs and energy loss spectra were recorded periodically throughout. It can be seen in figure 1 that electron radiation causes damage to the atomic structure of graphite. The tortuosity (or curvature, τ) of planes increases and fringe length decreases.

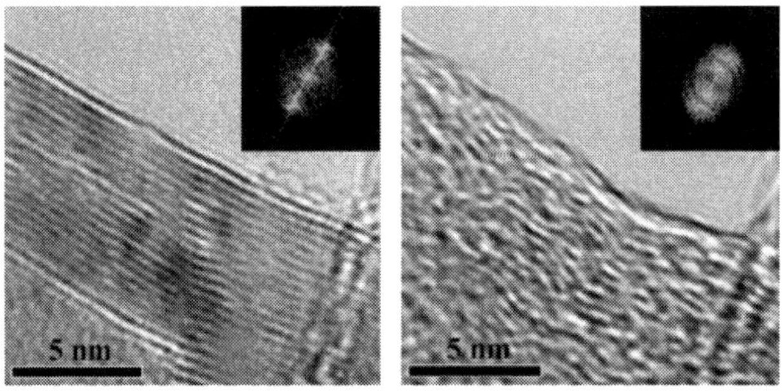

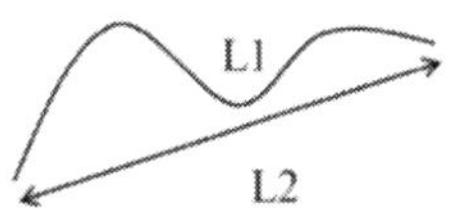

$$\tau = \frac{L1}{L2}$$

Figure 1 Electron micrographs and their corresponding Fourier transforms of PGA graphite at 0 seconds (a) and 600 seconds (b) of beam exposure, receiving 3.2×10^5 electrons $nm^{-2}\ s^{-1}$ ($\pm$ 5%).

The use of image analysis software (provided by the 'PyroMaN' research group [7]) revealed that 600 seconds of beam exposure broke up fringes; reducing fringe length by approximately 68% (figure 2, figure 6). Based on 002 fringe analysis, the software masks a manually selected area and applies a grey scale pixel selector to detect fringes. Once detected, information about their length and tortuosity can be extracted.

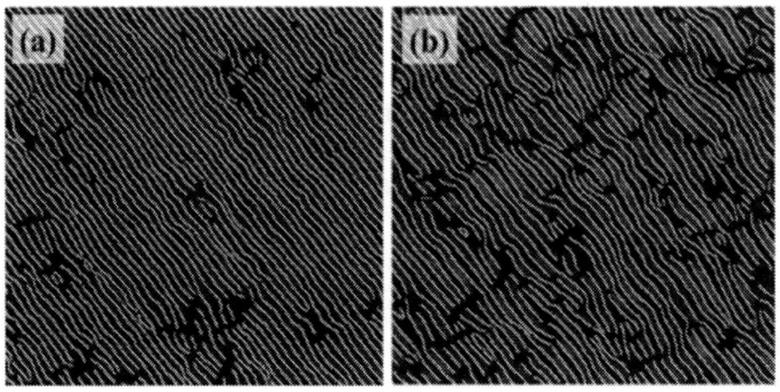

Figure 2 Application of PyroMaN image analysis software provides information on tortuosity (a measure of curvature) and fringe length before (a) and after (b) 600 s of beam exposure.

3.2 EELS

The carbon K-edge was collected for the area presented in figure 1(a) corresponding to the EELS spectrum shown in figure 3. The image was recorded after approximately 1 second of beam exposure, and represents highly ordered graphite. The on-set energy of the carbon K-edge for all the collected EELS spectra was around 285 eV, and a spectrum of the carbon K-edge was collected each 90-100 seconds (figure 4 (a)-(e)) along with the zero loss peak.

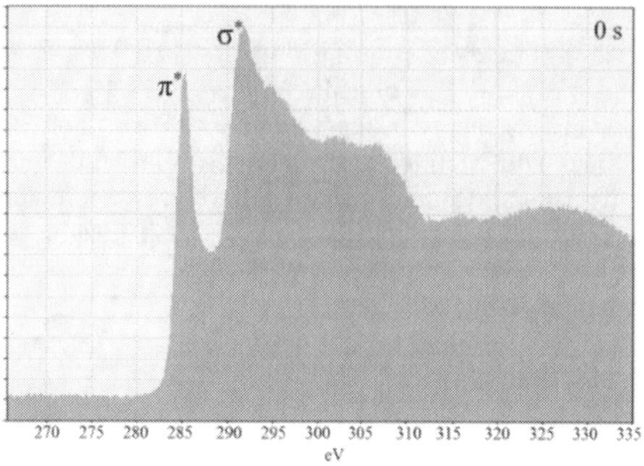

Figure 3 EELS spectrum for highly ordered graphite.

To characterise the ratio of order to disorder (sp^2 bound carbon to total carbon) the "two window method" was used [8]. It relates graphite's two core-loss peak intensities, for $\pi^*(sp^2)$ and $\sigma^*(sp^3)$ peaks, as follows:

$$\%sp^2 = [(I_{\pi^*} / I_{(\pi^*+\sigma^*)}) / (I_{\pi^*} / I_{(\pi^*+\sigma^*)})_{gr}] \text{ x } 100\%$$

where the intensity values from the EELS of the highly ordered virgin graphite were used to calculate $(I_{\pi^*} / I_{(\pi^*+\sigma^*)})_{gr}$ since this provides the best available approximation to perfect graphite (100% sp^2). The degree of disorder changed approximately linearly with respect to electron beam exposure time (figure 5).

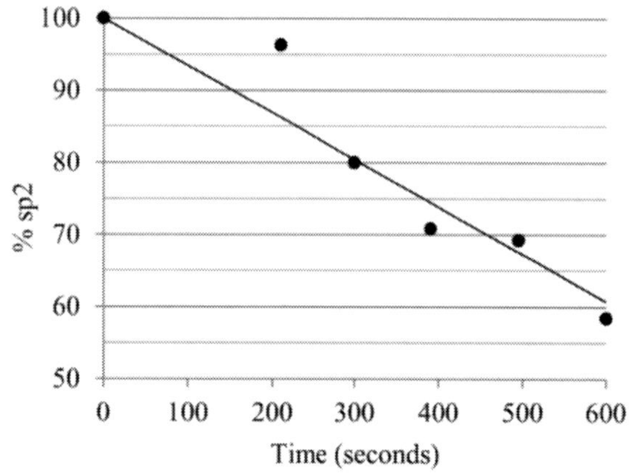

Figure 5 Correlation between the sample's degree of order (%sp^2) and electron irradiation time.

The sp^2 percentage in figure 5 dropped from 100% to below 60% after 600 seconds of beam exposure, showing that electron irradiation introduces atomic disorder. In figure 6 it can be seen that the rate in maximum fringe length reduction follows a similar path.

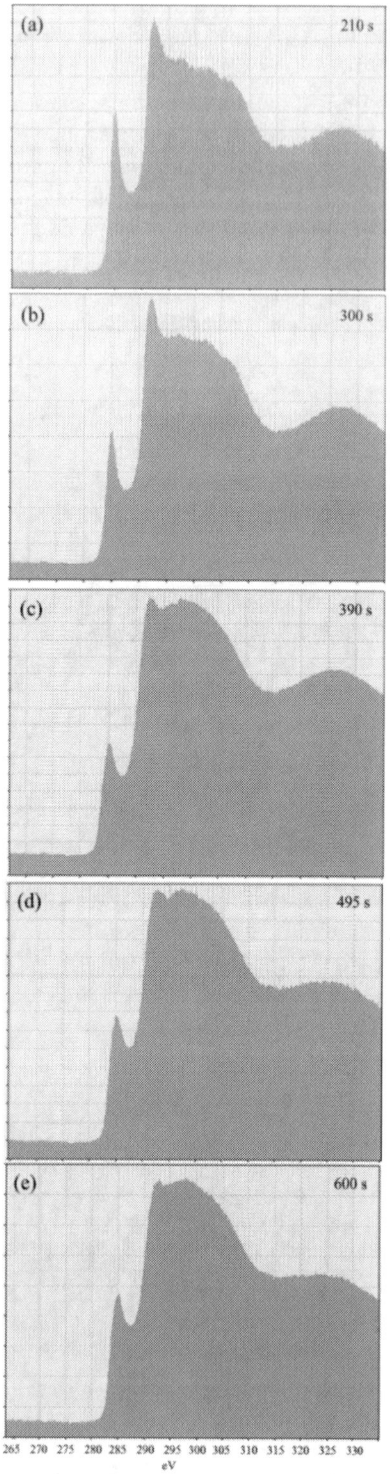

Figure 4 EELS spectra following various durations of electron beam exposure at a flux of 3.2 x 10^5 electrons nm^{-2} s^{-1}.

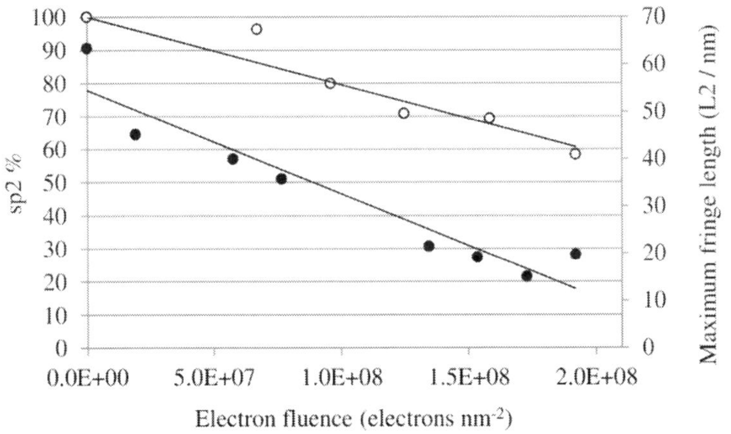

Figure 6 Correlation between the change in maximum fringe length (L2) and atomic order (% sp^2) with respect to electron fluence.

4. Summary

It has been shown that in-situ electron irradiation at room temperature damages PGA graphite at the atomic level. This can be used to better understand the behaviour of graphite moderators inside a nuclear reactor. The electron flux in the TEM can be easily controlled allowing for the observation of radiation damage mechanisms. Future work is in progress using in-situ heating to observe the effects of electron beam irradiation at elevated temperatures. Samples of neutron irradiated graphites are also being investigated.

Acknowledgements

The authors would like to thank the National Nuclear Laboratory and EPSRC (grants EP/J502042/1 and EP/I003312/1) for research funding. Thanks must also be given to J P Da Costa, B Farbos and J M Leyssale of the 'PyroMaN' research group for useful discussions and image analysis software availability (University of Bordeaux, Laboratoire des Composites Thermo Structuraux UMR 5801, 33600 Pessac, France).

References

[1] Hennig G R 1962 Vacancies and Dislocation Loops in Graphite *J. Appl. Phys. Lett.* **1** 55
[2] Nightingale R E 1962 *Nuclear Graphite.* vol. 1 (New York: Academic Press)
[3] Reynolds W N and Thrower P A 1963 The Flux Level Effect in Radiation Damage in Graphite *J. Nucl. Mater.* **10** 209
[4] Bollmann W and Hennig G R 1964 Electron Microscope Observations of Irradiated Graphite Single Crystals *Carbon* **1** 525
[5] Gray B S, Brocklehurst J E, Kelly B T, Sorenson H and Dietrich O W 1969 Irradiation Damage in Solids and Reactor Materials (Vienna: IEAEA)
[6] Daniels H, Brown A, Scott A, Nichells T, Rand B and Brydson R M D 2003 Experimental and Theoretical Evidence for the Magic Angle in Transmission Electron Energy Loss Spectroscopy *Ultramicroscopy* **96** 523
[7] Raynal P I, Monthioux M, Da Costa J P, Dugne O 2010 Multi-Scale Quantitative Analysis of Carbon Structure and Texture: iii. Lattice fringe imaging analysis. *Proc. Int. Carbon conf. (USA)*
[8] Zhang Z, Brydson R M D, Aslam Z, Reddy S, Brown A, Westwood A and Rand B 2011 Investigating the structure of non-graphitising carbons using electron energy loss spectroscopy in the transmission electron microscope *Carbon* **49** 5049

Electron Microscopy and Analysis Group Conference 2013 (EMAG2013) IOP Publishing
Journal of Physics: Conference Series **522** (2014) 012052 doi:10.1088/1742-6596/522/1/012052

Exit wave reconstruction of radiation-sensitive materials from low-dose data

C. Huang, K. B. Borisenko and A. I. Kirkland

Department of Materials, University of Oxford, Parks Road, Oxford, OX1 3PH, UK.

Email: chen.huang@materials.ox.ac.uk

Abstract. We report exit wave reconstruction using a focal series of low-dose images for ZSM-5, a zeolite which has a wide range of industrial applications. The exit wave phase was successfully reconstructed showing the structure of ZSM-5 at a near-atomic resolution even though the dose in the individual images in the focal series was as low as 26 electrons/Å^2. This implementation shows the method's potential for application to other radiation-sensitive materials.

1. Introduction

With recent developments in both the instrumentation and the off-line data processing techniques it is possible to obtain structural and chemical information about materials at ever increasing level of detail. Spherical aberration, one of the main factors that limited the resolution of electron microscopes in the past, has now been used to improve the imaging contrast [1,2]. In addition, exit wave restoration has evolved into a standard *a posteriori* method that can be conveniently carried out and provide information on the complex electron wave at the exit plane of the specimen [3], thanks to the automatic data acquisition and the increasing availability of reconstruction software packages [4,5]. The retrieved exit wave should in theory contain no artifacts caused by various kinds of aberrations. Even in the era of aberration corrected TEM, exit wave reconstruction still has the potential of revealing the information that is lost with the phase during the image formation.

Radiation damage is known to be largely responsible for the limited resolution achieved in materials that are radiation sensitive. Attempts of minimising the influence of radiation damage can be roughly divided into two categories.

The first approach is to increase the resistivity of the samples against radiation damage through special sample preparation techniques. For example, in cryo-electron microscopy (cryo-EM) the damage rate of the sample is reduced by keeping the specimen at the liquid nitrogen temperature. This allows a higher total imaging dose to be used.

The second approach is associated with the properties of the illuminating electron beam, namely accelerating voltage and electron dose. There are two main mechanisms of the radiation damage, knock-on damage and radiolysis, both of which are dependent on the accelerating voltage, however, in the opposite ways. Using lower accelerating voltage generally decreases the knock-on damage, but increases the radiolysis, and vice versa. Therefore, these two kinds of damage cannot be completely eliminated solely by adjusting the accelerating voltage. Instead, optimal beam conditions need to be chosen according to the chemical bonding type in the studied materials to minimise radiation damage [6].

Regardless of what voltage the microscope is operated at, for light-element inorganic and organic materials, low-dose image acquisition is nearly a prerequisite for any high resolution images to be taken [7]. The inherent problem of low-dose imaging is that it substantially reduces the signal-to-noise ratio (SNR) as the electron dose decreases, resulting in blurred images, loss of the detailed structural information, and restricted interpretation of the experimental data. Although exit wave restoration from a focal/tilt series of images is an established off-line method for improving resolution and data interpretation for high-dose images, there are very few examples of exit wave reconstruction from low-dose images and the tested materials are not particularly easily damaged by electron beam [8]. In the present work we test the low-dose exit wave restoration on the radiation sensitive ZSM-5 aluminosilicate zeolite.

2. Experimental

2.1. Sample preparation
ZSM-5 is an MFI-type zeolite, belonging to the space group Pnma, with the cell parameters of a = 20.090 Å, b = 19.738 Å, c = 13.142 Å. The sample of ZSM-5, with a Si/Al ratio of 70 was first crushed into very fine powder in liquid nitrogen to make the particles more brittle and easier to generate sharp edges with thin areas. The powder was then dispersed in ethanol and deposited onto the TEM sample grid.

Prior to the actual TEM imaging, the specimen was left in the high vacuum column of the microscope for about 50 hours. It is believed that the residual ethanol and any adsorbed water had been eliminated from the material by the time of the TEM imaging.

2.2. Dose calibration
The imaging was carried out at 200 kV in the aberration-corrected JEOL 2200MCO TEM, equipped with a Faraday cage for the measurement of electron beam current, i. The gain of the CCD camera, g, can be calculated by the following equation,

$$g = it/ce \qquad (1)$$

where t is the exposure time, c is the total counts in an image that contains the whole beam illumination, and e is the elementary charge. The calibration shows that every primary electron would generate 16.7 counts on average when the microscope is operated at 200 kV. Knowing the pixel area on the detector, a^2, the dose, D, for each image is calculated as:

$$D = gc/a^2 \qquad (2)$$

2.3. Low-dose focal series image acquisition
The low-dose focal series of 21 images used in this paper was taken with the defocus ranging from -8 nm to 32 nm, with a 2 nm focal step between the adjacent images. The dose for a single image was controlled at 26 electrons/Å^2 over an exposure time of 3 s. There was a 1 s delay between each two images when the sample was still exposed to radiation. The total dose in this particular focal series was therefore approximately 720 electrons/Å^2.

The radiation damage was evident after recording several more focal series in the same area with the total dose of ~7000 elections/Å. In the FFT of the very first image the (14 0 0) spot is visible, corresponding to a distinguishable lattice spacing of 1.44 Å. In the last image of the final series, the visible spot with the highest spatial frequency was the (6 0 0) spot, corresponding to a lattice spacing of 3.35 Å.

3. Results and discussion

3.1. Exit wave reconstruction from the low-dose focal series

The low-dose high resolution image and the atomic model of ZSM-5 along with two exit wave phases reconstructed by FTSR and the real-space optimisation codes are shown in Figure 1. The multi-ring structure of ZSM-5 can be discerned in exit waves obtained by both methods.

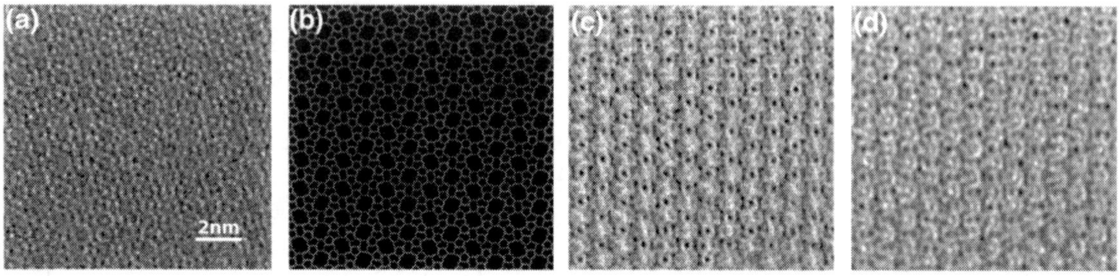

Figure 1. (a) An HRTEM image from the low-dose focal series with -2 nm defocus; (b) atomic model of ZSM-5 zeolite; (c) exit wave phases reconstructed by the FTSR plugin and (d) by the MATLAB least-squares optimisation code. The phase range is between 0 (black) ~ 0.88π (white).

3.1.1. FTSR. FTSR [4] is a commercially available DM plugin for exit wave reconstruction. It employs a phase correlation function/phase contrast index function (PCF/PCI) method for the image registration and focus determination. The registered images can then be used to restore the exit wave through a linear Wiener filter. The expected structure of ZSM-5 has larger ten-membered rings of about 9 Å in diameter, each surrounded by ten smaller rings of about 4-5 Å in diameter. The retrieved exit wave phase (Figure 1c) clearly resolved this multi-ring structure in the zeolite sample with a much lower noise level, as compared to the original image (Figure 1a).

3.1.2. MATLAB code. The real-space least-squares optimisation MATLAB code [9] uses non-linear iterative reconstruction method which integrates GPU computing to improve the performance. In the refinement process it minimises the sum of squared differences between the observed images and the images computed from the restored exit wave. The restored wave phase (Figure 1d), after filtering out the high frequency noise with a band-pass filter, gives a promising estimation of the sample structure. This method is intended for quantitative comparison with simulations. It is constrained by the quality of the initial guess from where the refinement starts and the usually finite computing power.

3.2. Image registration

A successful restoration of the exit wave needs the images in the focal series to be precisely aligned. Most of the alignment tools of EM images are based on the cross-correlation function (CCF). The FTSR software uses an enhanced phase correlation function (PCF) which includes the modulus normalisation and phase compensation for image registration. In practice, the registration result changes considerably when choosing different image slices of the series as the reference. The high noise level in the low-dose TEM images can greatly compromise the effectiveness of the image registration and is regarded as the main restriction to the quality of the exit wave restoration.

The real-space code employed a least-squares optimisation approach to the registration of images in the focal series. In addition, the corrections to the initial registration were performed after each optimisation cycle of the exit wave during the restoration.

3.3. Total dose in focal series acquisition

It must be noted that the total dose on the sample is more than just the number of images times the dose rate multiplied by the acquisition time. A considerable dose of electron irradiation the sample has been exposed to during searching and focusing processes needs to be taken into account. Although looking for a suitable area can be done at a lower magnification and the initial adjustment of focus and aberrations can be completed in adjacent areas, even then an experienced and skillful microscope operator needs time for tilting the sample to the right orientation and moving the region of interest to the optimal defocus. In reality, these two steps are not entirely independent from each other and have to be done right at the area of interest. Considering the focal series imaging, which typically takes about 1 min to collect at a constant dose, the whole preparation process before the image recording may well exceed the series imaging process in time and may be comparable in dose. Measuring this additional dose accurately will require knowledge of different fixed dose rates for searching, adjusting, and imaging, with each step precisely timed. This will be included in the future experimental design.

3.4. Defocus in the low-dose focal series

It is very challenging to identify the optimum defocus at very low doses. Therefore, even only for the purpose of getting the best contrast out of single TEM images, it is a better strategy to simply take a focal series after the coarse adjustment to the zero focus. As far as an image series for exit wave restoration is concerned, it is not a problem to have the images not equally distributed on both sides of the zero focus. The exit wave restoration software needs to be able to determine the defocus value automatically and also refine the defocus value and the focal step for more accurate restoration. These steps were implemented in the MATLAB least-squares optimisation code.

4. Conclusions

We have shown that exit wave reconstruction using low-dose focal series images is possible on an example of a radiation-sensitive ZSM-5 zeolite even when the imaging dose is as low as only 26 electrons/$Å^2$ per image.

Further improvement in image registration of the low-dose focal series and restoration algorithms are in progress to increase the accuracy of restoration to a level suitable for more detailed and direct comparison with the simulations.

References

[1] Haider M, Uhlemann S, Schwan E, Rose H, Kabius B and Urban K 1998 *Nature* **392** 768–9
[2] Chang L-Y, Chen F-R, Kirkland A I and Kai J-J 2003 *J. Electron Microscopy* **52** 359–64
[3] Kirkland A I and Meyer R R 2004 *Microsc Microanal* **10** 401–13
[4] Meyer R, Kirkland A I and Saxton W 2002 *Ultramicroscopy* **92** 89–109
[5] Ishizuka K 2006 *Microsc Microanal* **12** 1460
[6] Yoshida K and Sasaki Y 2013 *Microscopy* **62** 369–75
[7] Carlson D B and Evans J E 2012 Low-Dose Imaging Techniques for Transmission Electron Microscopy ed K Maaz (InTech) p 392
[8] Barton B, Jiang B, Song C, Specht P, Calderon H and Kisielowski C 2012 *Microsc Microanal* **18** 982–94
[9] Borisenko K B, Moldovan G, Kirkland A I, Wang A, Van Dyck D and Chen F-R 2012 *Journal of Physics: Conference Series* **371** 012057

The influence of emitter conditioning on the performance of a tungsten <111> cold field emission gun operating at 300kV

I M Ross, W Li and T Walther

Kroto Centre for High Resolution Imaging & Analysis, Department of Electronic and Electrical Engineering, University of Sheffield, Mappin Street, Sheffield S1 3JD, UK

i.ross@shef.ac.uk

Abstract: In this contribution, we examine the influence of emitter conditioning for a <111> tungsten cold field emission gun on the emission and beam characteristics of a double aberration corrected electron microscope. By varying the post flash build-up parameters we can control the effective emitter tip radius. A sharp emitter yields an energy resolution of 0.31eV but relatively low beam current whereas an increased tip radius results in a reduction in energy resolution to 0.4eV but much higher potential beam current. Consequently, careful control of the build-up parameters can be used as a means of tailoring the emission to suit specific instrumental requirements.

1. Introduction

Recent developments in commercial field emitter design have seen a renaissance of cold-field emitters (c-FEG) in modern aberration corrected TEM and STEM instruments to optimise brightness and energy resolution [1]. The performance of these emitters is strongly dependent on their geometry and the level of vacuum around the tip. Nevertheless, even with reasonable gun vacuum levels of 10^{-9} Pa or better there is still a need to periodically recondition or "flash" the electron source to remove residual gas absorbed on the tip surface [2]. Conventional high-flash were the tip is heated to ~2400 K results in the removal of the absorbed surface gas but can effectively blunt the tip and significantly reduce the brightness of the electron source for a given extraction potential. The geometry of the emitter tip can however be restored by a build-up process using a thermal field treatment [3] in which tip is heated under the influence of a strong electric field. In the case of a tungsten <111> single crystal, the build-up process results in the diffusion of surface atoms at the apex of the tip to form {110} and {211} side facets the latter of which ultimately intersect to produce a sharp pyramid truncated by the (111) plane.

In this report, we investigate the effect of varying the post flash build-up parameters applied to a <111> tungsten c-FEG on the emission characteristics, extraction potential, probe current stability and primary energy spread.

2. Experimental methods

The W <111> c-FEG utilised in this study is fitted to a JEOL Z3100F-R005 double aberration corrected TEM/STEM normally operating at 300kV. The periodic "flashing" procedure involves an initial high-flash (HF) in which the tip is raised to a high temperature by passing a current of 2.2A for 3 seconds. The subsequent build-up (BU) treatments investigated are detailed in Table 1. In each case the emission current and corresponding probe current was measured as a function of extraction voltage (A1 kV) with a constant second anode (A2) potential of 4.6kV. The emission current was determined by indirect measurement of the potential across the appropriate test point of the emission feedback control board. This was necessary as the emission current displayed in the microscope operating software (TEM-COM) has a minimum increment of 1μA and indicates zero for emission currents below 1μA. For consistency with our previous studies the corresponding probe current was measured in CBD mode α1 with a nominal spot size of 1.2nm and a 150μm condenser aperture using a Keithley 2635 test meter via the small diameter viewing screen, previously calibrated using a Faraday cup at the specimen plane. Emission patterns were recorded using a TV rate camera positioned at the gun inspection port focused onto the surface of the perforated second anode.

Table 1: Emitter conditioning parameters investigated in this study.

	Current (A)	A1 Bias (kV)	Time (Sec)
High Flash	2.2	-	3
BU 1	2.2	-2.70	10
BU 2	2.1	-2.60	10
BU 3	2.05	-2.55	10
BU 4	2.0	-2.50	10

To assess the influence of tip conditioning on the energy spread in the primary beam an energy-loss spectrum was acquired immediately after build-up in each case for a series of equivalent probe current by recording the full width half maximum (FWHM) of the zero energy loss spectrum using a Gatan Tridium GIF with an entrance aperture of 1mm, dispersion of 0.01ev/pixel and an exposure of 0.1 second. The relative drift in beam current and gun vacuum was determined as a function of time immediately after build-up (BU-1) treatment.

3. Results and Discussion

A series of emission patterns for the different build-up conditions studied is given in Figure 1 with the respective details of the extraction potential, emission current and corresponding probe current shown in parenthesis in each case. In figure 1(a) we observe the classic three-fold symmetry of the <111> W field emission pattern.

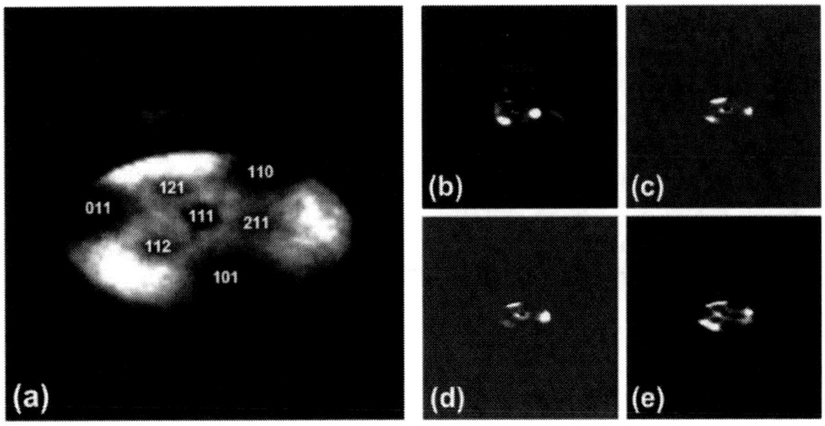

Figure 1: C-FEG emission patterns for the different conditions immediately after build-up (a) high-flash only (2.00kV, 4.08μA, 6.25nA) with the low order emitting facets indexed, (b) BU-1 (1.05kV, 5.05μA, 397pA), (c) BU-2 (1.40kV, 5.03μA, 1.35nA), (d) BU-3 (1.5kV, 4.94μA, 4.3nA) and (e) BU-4 (1.79kV, 4.26μA, 1.3nA). Note - pattern (a) is x6 magnified with respect to the subsequent patterns to show detail.

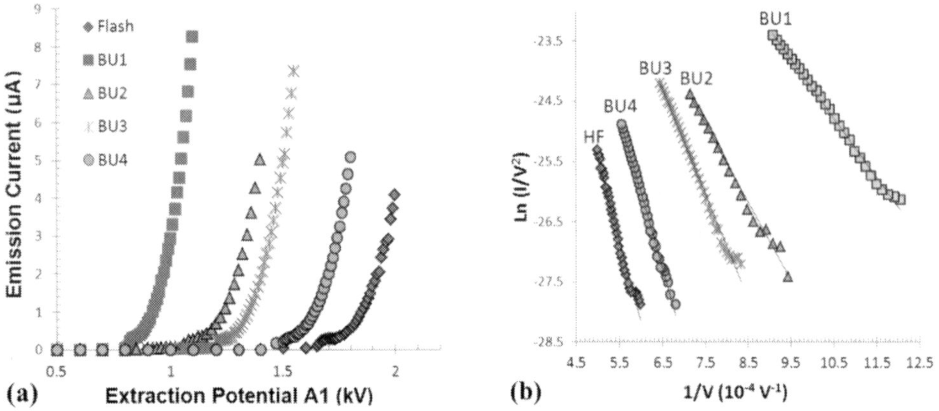

Figure 2: (a) Emission current as a function of extraction voltage, (b) the corresponding Fowler-Nordheim plots.

Plots of emission current versus extraction potential are given in Figure 2(a) with the corresponding Fowler-Nordheim (F-N) plots shown in Figure 2(b). An estimate of the tip radius can be directly extracted from the F-N field emission theory which shows the slope of the $\ln(I/V^2)$ versus $1/V$ plot to be $-6.8 \times 10^7\ \Phi^{3/2}\ \alpha kr$ [4,5] where, Φ is the work function (eV) (4.47eV for <111> tungsten), α is the Nordheim image correction term (which in this instance is assumed to be unity [4]), k is the field reduction factor and r is the tip radius (in cm). The exact value of k can vary significantly and depends on the actual tip geometry with reported vales between the limits of 3 for blunt tips to 35 for sharp tips [6]. Table 2 shows the estimated tip radii from the F-N plots in Figure 2(b). The value of kr is presented to provide a relative comparison, while the actual estimated tip radius is given for the limits k=3 and k=35. The estimates clearly illustrate the sharpening of the tip with increasing heating current and applied reverse potential.

Table 2: Estimation of the tip radii from the linear fits of the Fowler-Nordheim plots shown in Fig 2b.

	BU1	**BU2**	**BU3**	**BU4**	**Flash**
kr (nm)	160	175	278	431	505
kr (k=3) (nm)	53	58	92	144	168
kr (k=35) (nm)	4.6	5.0	8.0	12.0	14.4

The beam current at the specimen as a function of extraction potential is given in Figure 3(a). From Table 2 it can be observed that as the tip radii decreases, the required extraction potential decreases however, we find that for sharper tips the total beam current is limited by the emission current, Figure 2(a) since the latter is limited within the instrument software to 20µA and hence the extractor cannot simply be increased indefinitely. The influence on the energy spread is presented in Figure 3(b). With the exception of the high-flash only condition, the value of the FWHM of the zero energy loss peak is relatively consistent for a given value of the beam current, although the best energy resolution (0.31eV) is still achieved using the sharpest tip (BU1) while a ΔE of 0.45eV can be attained for less sharp tips (BU3, BU4) while maintaining significantly higher beam currents.

Figure 4(a) shows the gun vacuum as a function of beam-on time. While there is some slight initial degradation, likely due to out-gassing of the second anode aperture, the major component relates to the differential pressure between the gun and column. Nevertheless, the increase in gun pressure tends to plateau after 2-3 hours, remaining better than 6.5×10^{-9} Pa. Analysis of the F-N plots for the tip described in Figure 4(b) after prolonged periods without flashing reveals that the relative value of kr increases to ~180nm after 266 hours. The corresponding beam current was 290pA, approximately half the original beam current generated after initial build-up. The application of a low-flash (1.460A for 45 seconds) restored the probe current to ~750pA (for an equivalent emission current of 4μA) while the value of *kr* differed only slightly at ~190nm, indicating the successful removal of adsorbed gas but marginal change in emitter shape under low-flash conditions.

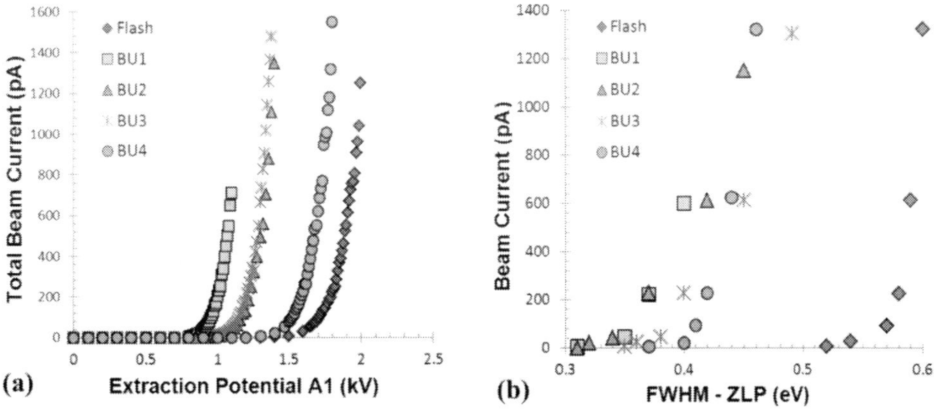

Figure 3: (a) Plot of beam current as a function of A1 with (b) the corresponding energy spread in the primary beam as a function of build-up condition for a given beam current.

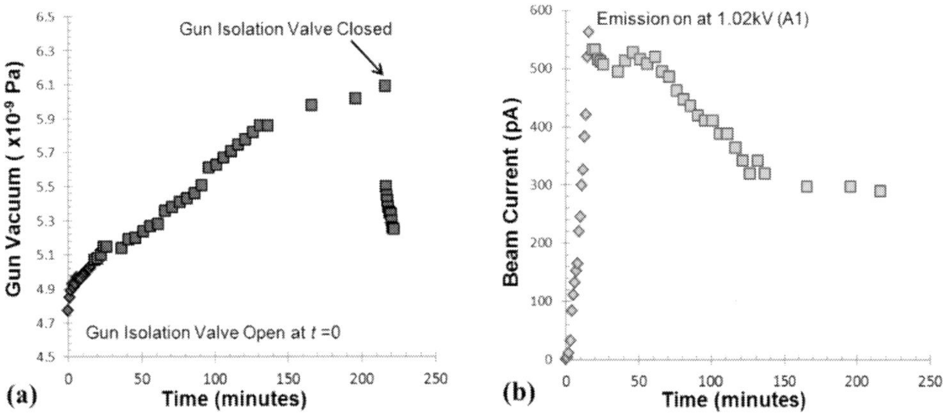

Figure 4: Plot of drift in (a) Gun vacuum and (b) beam current after high-flash and BU-1 build-up.

4. Summary

We demonstrate that the post flash build-up parameters can strongly influence the resulting emission and beam characteristics. Hence, careful choice of these parameters is required to fully optimise instrument performance for specific applications.

References

[1] Kohno Y, Okunishi E, Tomita T, Ishikawa I, Kaneyama T, Ohkura Y, Kondo Y, Isabell T, 2010 Microscopy and Analysis, **142** S9-S13
[2] Crewe A V, Eggenberger D N, Wall J, Welter L M, 1968 *Rev. Sci. Instrum.* **39** 576
[3] Kurata H, Isoda S, Tomita T, 2006 *Proceedings **IMC16** Sapporo* **2** 583
[4] Gomer R, *Field Emission and Field Ionization*, 1961 *Harvard Monographs in Applied Science* (9) Harvard University Press.
[5] Binh V T, Marien J, 1988 *Surf. Sci. Letts.* **202** L539
[6] Lucier A S, 2004 *Preparation and characterisation of tungsten tips*, MSc Thesis, McGill University.

The energy spread of a LaB_6 cathode operated in the virtual source mode

T Wells[1] and M El-Gomati[2*]
[1]York Probe Source Ltd, York, UK, www.yps-ltd.com
[2]Department of Electronics, University of York, York, YO10 5DD, UK

Abstract. The LaB_6 cathode has been the brightest thermionic source used in microprobe applications requiring longer lifetime [1-2]. It is x100 lower in brightness than thermal field emitters (TFE) ca Zr/W(100) [3]. There are attractive similarities between these cathodes in terms of work function and operating temperature that are worth considering. Major differences include their respective source sizes (>10μm vs 30nm) and energy spread of 1-2 eV vs 0.6-0.7eV for the LaB_6 and TFE, respectively [4,3].

We report here on the experimental measurement of the energy spread of a LaB_6 cathode operated in the virtual source mode. The cathode used has an end-form measuring 15μm. Total energy spread values obtained using a dedicated electron energy analyser shows values of 0.4eV-0.7eV, significantly lower than typical values in the thermionic mode of 1-2eV.

1. Introduction

The Lanthanum Hexaboride (LaB_6) electron cathode have been recognised as the highest brightness thermionic source since its introduction in the early 1970s [4]. This cathode is preferred to the conventional tungsten heated filament due to its relatively smaller source size but much higher brightness with respect to tungsten by about x10. The latter property has improved the microscopes' spatial resolution at lower electron beam energies. However, a fundamental impediment with all thermionic electron sources in comparison to the more recent point source cathodes lies in the configuration of the electron source (i.e. gun) part of the microscope. This, for most thermionic types, has been used with a Wehnelt electrode to aid focussing and control of the electron emission, see Figure 1. The Wehnelt electrode consists of a cylindrical cap with a small central hole where the cathode is located behind it. In operation of the source, a focussed cloud of electrons is formed in front of the Wehnelt which typically measures about 10-20μm in diameter. This first cross-over of the probe causes a broadening in the velocities of the emitted electrons, due to Coulomb interaction, that in turn causes a broadening in the final beam diameter via the chromatic aberration component of the electron beam. Values of the energy spread of the emitted electrons in such configuration of greater than 2eV can result at moderate to higher beam currents. This is a much worst value than the natural spread of thermally emitted electrons at such temperatures (estimated to be in the order of 0.4eV).

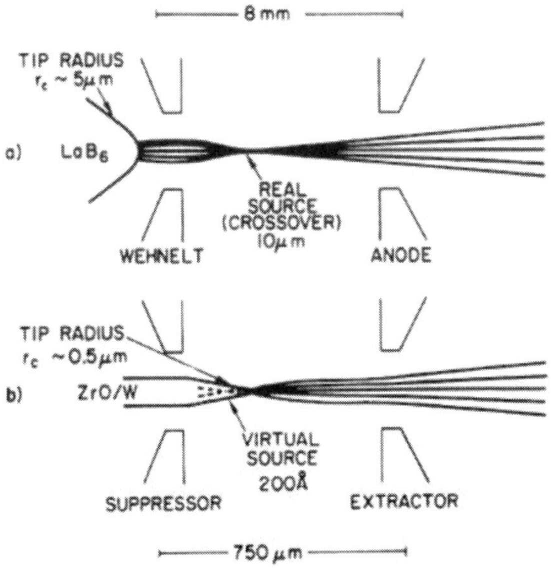

Figure 1 Two electron gun cathode configurations **a)** Thermionic LaB_6 with real source and **b)** TFE ZrO/W in the virtual source mode [3].

These two properties have been improved upon by using a point source such as the Zr/W(100) Schottky electron emitters (also referred to as thermal field emitters, TFE) [5](Fujita 2010). In this case, , see Figure 1, the projection of the emitted electrons appears to come from a virtual source of a much smaller size than the diameter of the TFE cathode and is normally assumed to be in the range of 20-30nm. The LaB_6 and the TFE cathodes, however, share two fundamental properties that are worth exploiting in probe forming systems. These are similar work function (2.6-2.8eV) and the same operating temperature of 1800K. The work reported here is a measurement of the energy spread of a LaB_6 cathode when operated in the virtual source mode and shows that by operation in this mode one achieves an energy spread similar to that of the typical TFE cathodes.

2. Energy spread measurements

Operation of the LaB_6 cathode was performed in the virtual source mode using a YPS M20 minimodule which is comprised of an extractor and suppressor electrodes with a cathode in the virtual source configuration shown in Figure 1. The M20 was designed to fit a standard Kimball physics ES-423E LaB_6 emitter, with a 19-20μm flat and a 90degree cone angle, see Figure 2, with a 1mm suppressor aperture that allows the LaB_6 crystal to protrude by up to 0.2mm above the suppressor electrode. The M20 module was then loaded into a single lens electrostatic mini-electron column, comprising of a grounded anode electrode and a lens electrode, for beam formation.

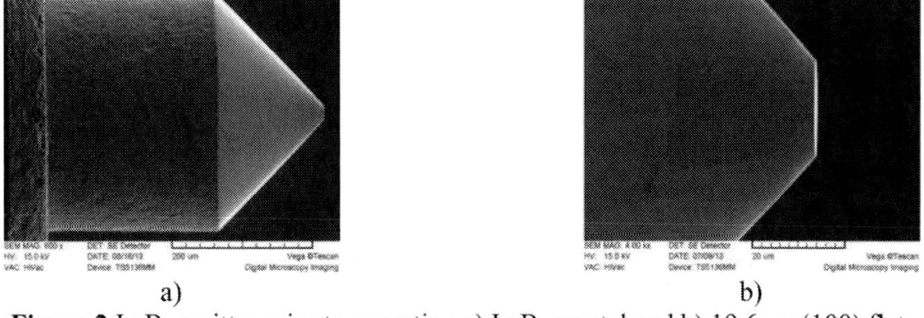

a) b)

Figure 2 LaB_6 emitter prior to operation a) LaB_6 crystal and b) 19.6μm (100) flat.

With a 50μm aperture in the extractor electrode of the M20 a beam current of up to 1μA could be obtained from the mini-electron column. The maximum beam current is limited by the total emission current from the LaB_6 emitter, > 3mA, that could be delivered by the power supply.

The energy spread measurements were performed using a **VG CLAM100** hemispherical electron energy analyser with an electrostatic lens on the entrance optics. A beam of electrons was fired at a Ag coated sample with the subsequent electron spectra collected in the CLAM100. The elastic scattered electrons could then be used to represent the electron source energy spread.

3. Results

Filament Temperature (K)	1600	1625	1650	1675	1700	1725	1750	1800
Δe FWHM Ve = 600V (eV)	0.41	0.43	0.44	0.45	0.47	0.48	-	-
Δe FWHM Ve = 300V (eV)	-	-	0.42	-	0.45	-	0.48	0.54

Table 1 **Energy spread with filament temperature at Ve = 600V and Ve = 300V.**

Operation of the LaB_6 emitter at a fixed extractor voltage showed an increase in the energy spread as the temperature of the emitter was increased, see Table 1. This is as expected with TE emission and TFE, but the increase in the energy spread is in the high energy tail as can be seen in Figure 3 which is in keeping with a TE emission mechanism.

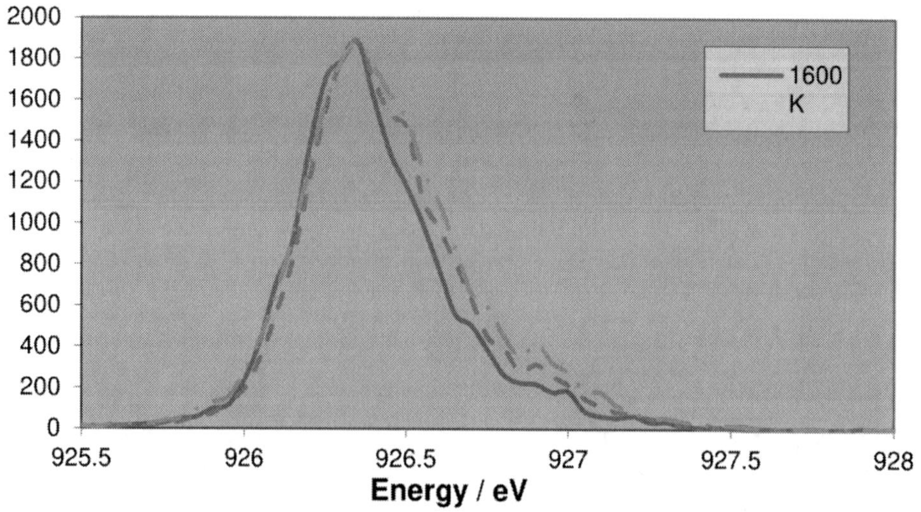

Figure 3 Energy spread with increasing filament temperature at Ve = 600V.

From Table 1 it can be seen that the temperature dependence of the energy spread is repeated when the extraction voltage is dropped to 300V. The change in the extraction voltage in Table 1 shows a limited effect on the energy spread. Figure 4 shows the energy spread spectra as the extraction voltage is increased from 200-800V. It is important to note that the LaB_6 emitter used in this experiment is different than the one discussed above, although of the same type. The energy spread obtained shows a

FWHM value which is constant at 0.66±0.01eV over these extraction voltages and with a beam current of 250-300μA.

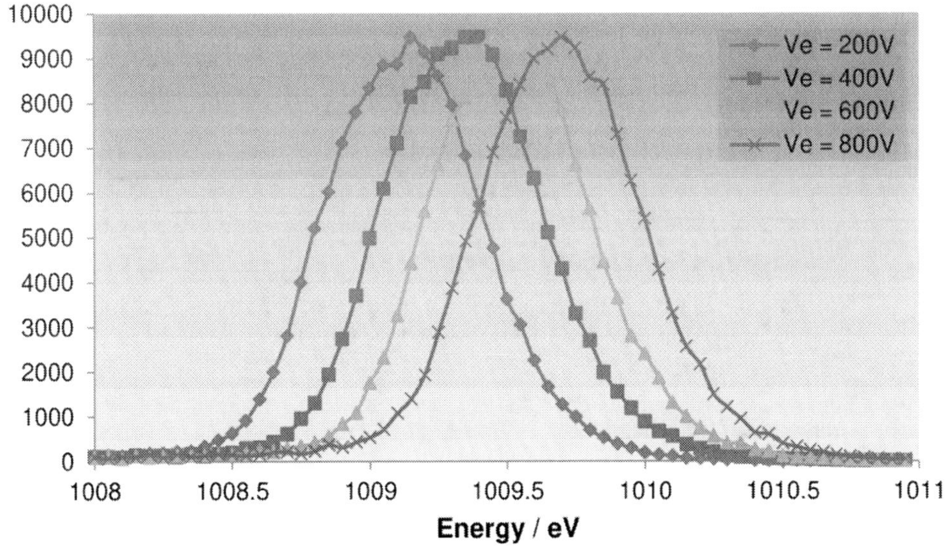

Figure 4 Energy spread with increasing extraction voltage: *note the peaks have been offset 0.2eV from each other for ease of viewing.*

4. Discussion and Conclusion

From the Richardson theory of electron behaviour, a 0.04eV increase in the energy spread is predicted for a 125K increase in the temperature as opposed to the 0.07eV increase measured in Figure 3. This, however, may have been caused by a small electron interaction and is also within the experimental errors for such measurements. What is more important here is that the measured energy spread reflects the expected increase in the high energy tail of the peak due to thermionic electron emission. The results shown in Figure 4 demonstrates that there are no increase in the energy spread with increasing the electric field around the LaB_6 emitter. This also points to a pure thermionic electron emission mechanism.

In conclusion, the results obtained here show that operating the LaB_6 in the virtual source mode results in a much smaller electron energy spread than normally obtained in conventional operation of this type of electron cathode. This should aid one to obtain a superior spatial resolution at lower beam voltages. Operating the source in this mode also results in a much smaller source size thus, estimated to be at least x10 smaller than its thermionic mode, thus positively contributing to superior source electron optical characteristics.

References

[1] J. Electron Microscopy. Suppl., 1989, (40th Anniversary of Japanese Society of EM) Vol.38
[2] Bell A E and Swanson L W, 1979, Phys. Rev. B Vol. 19, p.3353
[3] Swanson L W and Schwind G A, 1997, Handbook Book of Charged Practical Optics, CRC press, New York.
[4] Broers A N, Electron Gun Using Long-Life Lanthanum Hexaboride Cathode, 1967, J. Appl. Phys., Vol. 38, p. 1991
[5] Fujita S, Enhanced angular current intensity from Schottky emitters, 2010, Journal of Microscopy, vol. 239, issue 3, p. 215

The use of transmission electron microscopy in the quantification of nanoparticle dose

N Hondow, R Brydson and A Brown

Institute for Materials Research, School of Process, Environmental and Materials Engineering, University of Leeds, Leeds LS2 9JT, UK

n.hondow@leeds.ac.uk

Abstract. There are an increasing number of potential applications for nanoparticles in clinical medicine, including targeted drug delivery and contrast agents for biomedical imaging. Current *in vitro* studies are concerned with the biological impact of nanoparticles, with electron microscopy commonly employed to image their intracellular location. It is critical to quantify the absolute nanoparticle dose internalized by cells in a given exposure, and to understand the factors which affect this. In this work we are aiming to develop a full quantitative description of quantum dot uptake by an *in vitro* cell line. Transmission electron microscopy of thin cell sections provides the location and number of cellular vesicles per 2-D cell slice plus the number of quantum dots per vesicle. These results can then be correlated to other techniques to quantify the internalized nanoparticle dose distribution for whole cells.

1. Introduction

Quantum dot nanoparticles are fluorescent semiconductor nanocrystals commonly 2 to 15 nm in length that are finding numerous biological applications [1]. In particular they are being used in the development of nanomedicines as diagnostic nanoprobes or to label drug molecules or nanocarriers [2]. One of the challenges of nanomedicine is to understand how nanoparticles will be distributed in the body, making it critical to measure the absolute nanoparticle dose internalized by cells and tissues for a given exposure [3].

Predicting internalized dose of nanoparticles is complicated because the dispersion and surface activity can alter in different liquid environments (e.g. water, ionic or physiological media). We have already addressed the representative analysis of dispersions of nanoparticles by TEM following sample preparation by plunge freezing and have specifically quantified the dispersion of quantum dots [4] and other nanoparticles such as silica and zinc oxide [5] in cell culture media. The advantage of TEM analysis is that we can measure and quantify the dispersion and the condition of the primary nanoparticles, identifying the number, size and even the compositional distribution of primary and agglomerated particles.

TEM is commonly employed to image the internal location of nanoparticles taken up by cells and tissues. *In vitro*, these results are often qualitatively compared to cell line responses measured by standard viability and genotoxicity assays, and are essential to confirm uptake. Optical techniques can be used to track nanoparticles inside live cells, if for example the particles fluoresce such as quantum dots, however resolution is limited to identifying intracellular features such as membrane-bound vesicles loaded with nanoparticles and not individual particles [6]. Only with the improved resolution afforded by TEM can the fundamental unit of dose, individual nanoparticles, be monitored and measured. In this work we present TEM results of human osteosarcoma (U2-OS) cells loaded with quantum dots, in which analytical techniques (energy dispersive X-ray (EDX) spectroscopy and energy filtered TEM (EFTEM)) have been used to confirm the identity of the quantum dots. We have also performed extensive TEM imaging to allow for quantification of nanoparticle uptake [7].

2. Experimental

U2-OS cells were loaded with quantum dots (Qtracker 705, ~10 nm in length, Invitrogen Ltd) as described previously [7]. After 1 hour exposure, the cells were harvested and processed for electron microscopy by fixation, dehydration, resin embedding and thin sectioning [7]; however no conventional heavy metal stains (uranyl acetate or lead citrate) were used. TEM was conducted on two electron microscopes; an FEI Tecani F20 FEG-TEM operating at 200 kV and fitted with a Gatan Orius SC600A CCD camera and an Oxford Instruments X-Max SDD EDX detector, and a JEOL 2100 TEM operated at 120 kV and fitted with a Gatan Tridium energy filter.

3. Results and Discussion

TEM imaging of thin sections of human osteosarcoma cells exposed to Qtracker 705 quantum dots (figure 1a) indicates nanoparticles enclosed in membrane bound vesicles (membranes are lightly stained by the osmium tetroxide fixative), confirming uptake by endocytosis. In some instances, nanoparticles can even be imaged entering a cell by endocytosis [4]. The crystalline nature of the quantum dots is confirmed with lattice imaging (figure 1e).

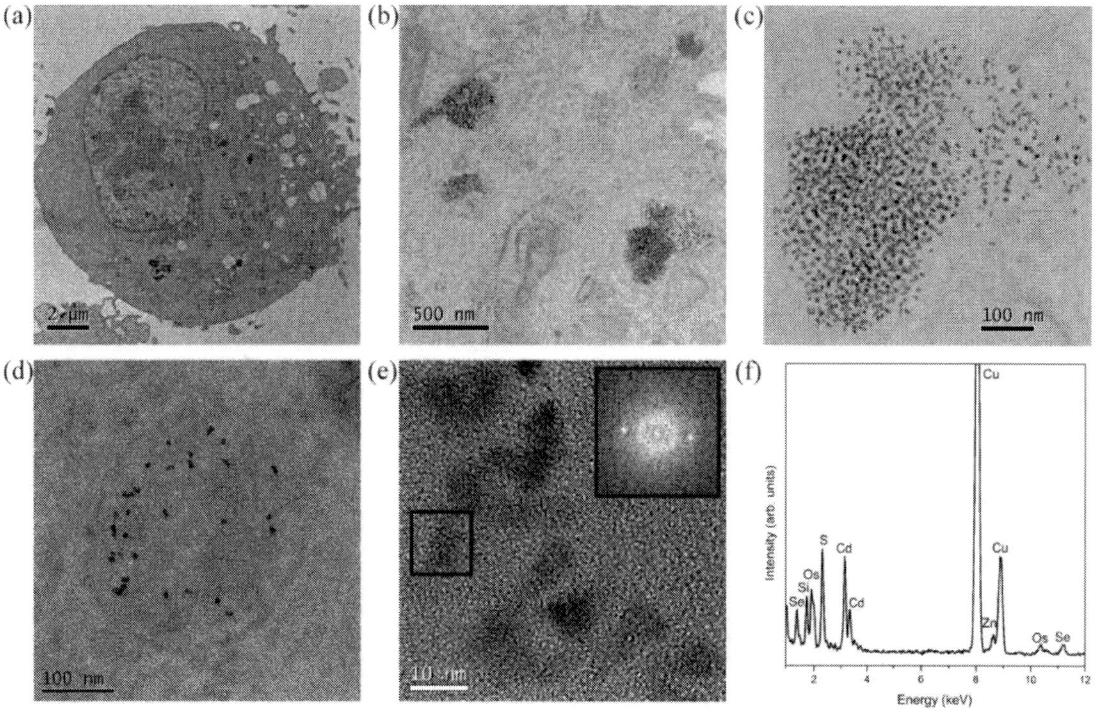

Figure 1. Electron microscopy of thin sections of U2-OS cells exposed to quantum dots. a) Low magnification image of a cell; b) higher magnification image from the centre of the cell; c) and d) higher magnification images of vesicles in (b) containing quantum dots; e) lattice image of quantum dots inside a vesicle with FFT inset; f) EDX spectrum from a group of quantum dots inside a vesicle.

Previous studies have highlighted the importance of distinguishing the identity of nanoparticles internalized by cells *in vitro* from the stains decorating cellular nano-structures [8]; contrast enhancing heavy metal stains (uranyl acetate or lead citrate are not in this study) and even the osmium tetroxide fixative (used here), are nanoparticulate themselves. In the first instance, we have used energy dispersive X-ray (EDX) spectroscopy from a collection of quantum dots inside a vesicle (figure 1f) to measure composition; selenium, sulfur, cadmium and zinc are all detected (with silicon, copper and osmium resulting from the TEM grid and the sample preparation). In addition, high magnification imaging (figure 1e) reveals the (100) atomic planes of hexagonal CdSe, absolutely confirming the identification of CdSe-ZnS dots.

The elemental distribution over a larger area has been mapped using EFTEM, mapping for cadmium and sulfur (figure 2). This technique also confirms that the particles identified are quantum dots and we have shown it is possible to identify individual dots within the cytoplasm using a combination of HAADF-STEM and elemental analysis [9]. An advantage of not using any post-embedding stain to highlight cellular features is more straightforward nanoparticle identification.

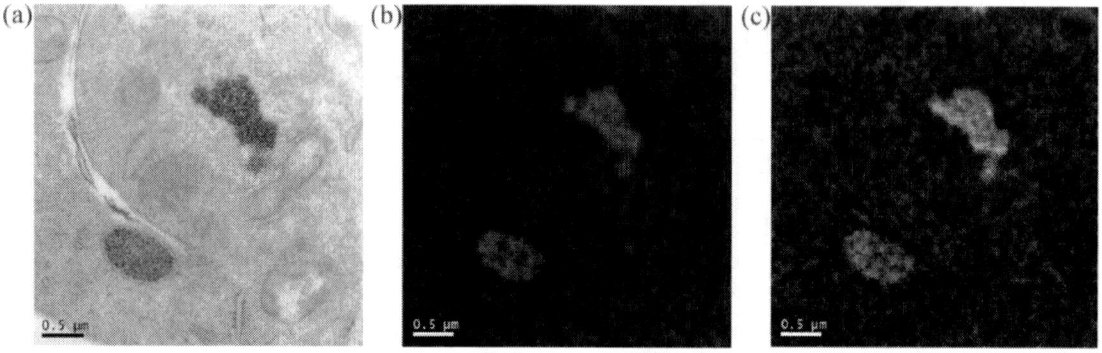

Figure 2. EFTEM analysis of U2-OS cells exposed to quantum dots. a) zero-loss energy filtered TEM image of an area within a cell containing several quantum dot filled vesicles; b) red false-coloured Cd $M_{4,5}$ map of (a); c) green false-coloured S $L_{2,3}$ map of (a).

Having established confidence in the accurate identification of quantum dots, counting of dots in vesicles has been undertaken. Images have been taken at a magnification appropriate for individual dots to be identified within in a vesicle (figure 3a). Numerous images have been subjected to analysis (figure 3b) to quantify the number of quantum dots per vesicle [7]. Performing this analysis on a thin section of many different cells (~100), with all quantum dot loaded vesicles quantified, reveals the distribution of quantum dot uptake by endocytic vesicles (figure 3c).

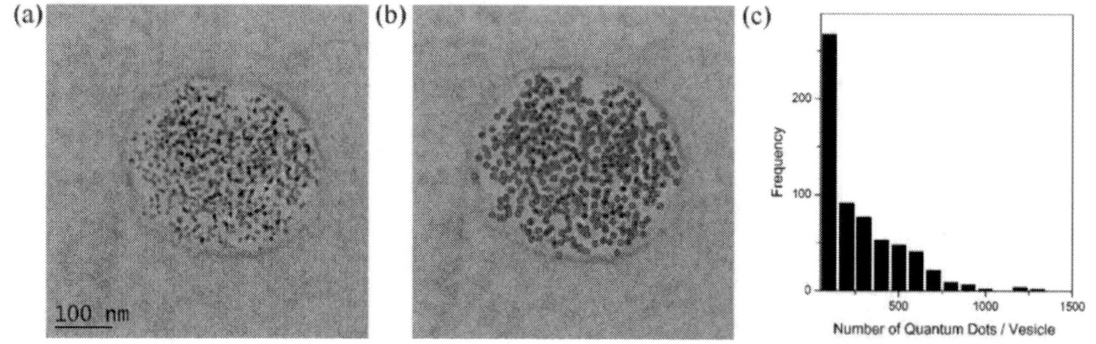

Figure 3. Quantitative TEM imaging to measure the distribution of nanoparticle uptake. a) A membrane bound vesicle containing numerous quantum dots; b) automated counting of quantum dots; c) frequency plot of number of quantum dots per vesicle.

To obtain the internalized quantum dot dose distribution for a cell population, we have correlated this low-throughput, high-resolution TEM analysis to high throughput, low resolution optical imaging by flow cell cytometry with an imagestream [7]. It is however also possible to use this quantitative TEM imaging method to monitor the quantum dot dose distribution per endocytic vesicle over the cell cycle; this sample is fixed immediately after a 1 hour exposure of the quantum dots to U2-OS cells but the distribution (figure 3c) will be compared to another time point in the cell cycle (for example, fixing 24 hours after exposure), potentially examining endosome maturation or merging and even inheritance of quantum dots after cell division. These distributions (figure 3c) and the correlation to whole cell populations [7] suggest that cells exposed to nanoparticles targeted for endocytosis receive a considerably variable dose of nanoparticles by number. This has major implications for the use of nanoparticles as drug delivery vehicles.

4. Conclusions

The use of transmission electron microscopy is essential in the measurement of nanoparticle number or dose internalized by cells and tissues as it can resolve the fundamental unit of dose, the nanoparticle. Analytical TEM techniques, such as EDX spectroscopy and EFTEM, can provide unequivocal evidence that particles imaged are indeed the target particles, and with extensive imaging and image analysis numbers of nanoparticles per membrane bound vesicle or per cell section can be determined. This can then be extrapolated to three dimensions and correlated to other techniques to develop full dose models, or could be repeated at different time points in the cell cycle to examine nanoparticle inheritance.

Acknowledgments

This work has been supported by the EPSRC (EP/H008578/1) and the AXA Research Fund (AXA fellowship to NH). We would like to thank: Prof. Huw Summers, Prof. Paul Rees and Dr M. Rowan Brown (Swansea University) for samples and image analysis; Prof. Martin Saunders (University of Western Australia) for help with EFTEM studies; and K. Brady (Kings College London) for sample preparation.

References

[1] Jamieson T, Bakhshi R, Petrova D, Pocock R, Imani M and Seifalian A M 2007 *Biomaterials* **28** 4717

[2] Wang Y and Chen L 2011 *Nanomed.-Nanotechnol.* **7** 385

[3] Sanhai W R, Sakamoto J H, Canady R and Ferrari M 2008 *Nat. Nanotechnol.* **3** 242

[4] Hondow N S, Brydson R, Wang P, Holton, M R, Rees P, Summers H D and Brown A 2012 *J. Nanopart. Res.* **14** 977

[5] Hondow N S, Wang P, Brydson R, Holton M D, Rees P, Summers H D and Brown A 2012 *J. Phys. Conf. Proc.* **371** 012020

Wallace R, Brown A P B, Brydson R, Milne S J, Hondow N and Wang P 2012 *J. Phys. Conf. Proc.* **371** 012080

[6] Summers H D, Rees P, Holton M D, Brown M R, Chappell S C, Smith P J and Errington R J 2011 *Nat. Nanotechnol.* **6** 170

[7] Summers H D, Brown M R, Holton M D, Tonkin J A, Hondow N, Brown A P, Brydson R and Rees P 2013 *ACS Nano* **7** 6129

[8] Bradenburger C, Clift M J D, Vanhecke D, Mülfeld C, Stone V, Gehr P and Rothen-Rutihauser B 2010 *Particle Fibre Toxicol.* **7** 15

[9] Brown A and Hondow N 2013 *ElectronMicroscopy of Nanoparticles in Cells* in Summers H D (Ed.) *Frontiers of Nanoscience: Nanomedicine* (Elsevier)

Studying the interface between cyanobacteria and biotite mineral surfaces using FIB and TEM

M B Ward and A P Brown

Institute for Materials Research, School of Process, Environmental and Materials Engineering, University of Leeds, LS2 9JT, UK

m.b.ward@leeds.ac.uk

Abstract. Recent analysis of the bioweathering of minerals has highlighted the challenges for investigating the interface between fungi or bacteria and the surface of the mineral that they live on. Transmission electron microscopy (TEM) with its ability to gather imaging information and collect elemental data at high spatial resolution is the ideal technique to analyse such interfaces. Further to this, a dual beam scanning electron and focused ion beam (FIB) microscope is an ideal instrument to prepare specimens for TEM because of its ability to simultaneously cut through hard and soft materials from specific sites of interest. There are however precautions that must be taken when analysing such mineral systems. The electron beam sensitive nature of most sheet silicate minerals means that consideration has to be made as to whether the structure and/or chemistry of the material is being altered during (S)TEM analysis. Here, results from a study of cyanobacteria grown on the surface of biotite are discussed. Particular reference is given to the methods used to determine an electron beam intensity threshold, below which STEM-EDX analysis could be performed without detrimental alteration to the mineral.

1. Introduction

The bioweathering of minerals by organisms such as fungi and bacteria requires a suitable analysis method to characterize the interface chemically and structurally at a high spatial resolution. Transmission electron microscopy (TEM) with Focused ion beam (FIB) sample preparation has been shown to be a useful combination of techniques for the productive study of such interfaces [1, 2].

Dual beam SE-FIB microscopy has the advantage of being a versatile, site-specific technique, and is routinely used for the preparation of thin specimens for TEM [3]. Further to this, TEM analysis of geological samples such as zoned minerals, allows for accurate determination of chemistry/structure at high spatial resolutions, within the limits of the materials' resistance to damage from the electron beam [4]. For example, the weak interlayer bonds of sheet silicate minerals make them prone to radiation damage from the electron beam. Irradiation results in the mobilisation of certain atomic species, mass loss and structural rearrangement, that have a rate dependency on the crystal orientation to the electron beam [5].

The identification and minimization of electron beam damage is therefore an important part of any electron microscope studies that are to be carried out on such mineral materials. Ideally the amount and rate of any beam damage can be quantified to enable the pristine structure of the mineral to be inferred. This can be achieved by the identification of a characteristic or threshold electron dose, below which there is minimal specimen alteration. Here, we have established an electron intensity threshold for the sheet silicate mineral biotite, irradiated with 200 kV electrons by STEM. .

2. Methods

The filamentous cyanobacterial strain of *Hassallia byssoidea* was isolated from biotite granite in the Czech Republic and incubated on a freshly cleaved biotite surfaces in modified BBM medium (BBM-mod3) for 95 days. 60W fluorescent lamps were used to provide 16/8 hour intervals of light/dark, and a temperature of 22± 1°C was maintained throughout [1].

TEM samples were prepared by FIB using a FEI Nova200 Nanolab dual beam SEM/FIB fitted with a Kleindiek micromanipulator for ex situ lift-out. The instrument was operated at 30 kV, and at beam currents between 0.1 and 5 nA. Lamella cross sections were taken across the cyanobacteria so that the contact interface with the mineral beneath could be studied (figure 1). Samples were attached to copper support grids (Omniprobe) ready for TEM examination. TEM was carried out using a FEI Tecnai F20 FEGTEM (200 kV) and a Philips CM200 FEGTEM (200kV) fitted with an Oxford Instruments UTW EDX detector and a Gatan GIF200 imaging filter.

Figure 1. A secondary electron SEM image showing a TEM lamellar cross-section specimen from a cyanobacterial filament attached to a biotite surface, prepared by FIB and prior to lift-out onto a TEM grid.

3. Results and discussion

Scanning (S)TEM combined with Energy dispersive X-ray spectroscopy across the cyanobacteria/mineral interface was used to characterize elemental diffusion between the bacteria and minera]. HAADF images were used to identify the interface and measure the size of the electron probe (as will be described later), and STEM-EDX linescans were used to measure any compositional variations at distance from the interface, however the linescans induced significant damage of the biotite [1]. This damage results from the relatively high electron dose within a very focussed electron probe of high current density and with relatively long dwell times that is required for high spatial resolution EDX analysis by STEM.

The TEM samples were oriented such that the viewing direction was normal to the cyanobacteria/mineral interface (Figure 2). Several STEM probes were scanned across the interface using different beam conditions (extraction voltage, spot size and C2 aperture) resulting in linescans of different probe dimensions and beam currents, and thus different fluence rates. The interface scans could be seen to mark the biotite (figure 2 lines A, B and C). Beam currents were measured using brightness readings taken from the phosphor viewing screen following the procedure described in Pan et al [6], and the probe dimensions were measured using a HAADF image taken of the interface using each respective probe setting. For these images, the sample was oriented such that the viewing direction was normal to the interface, resulting in the 002 planes of the biotite being on axis. This gave an atomically abrupt mineral surface, and so any blurring of the interface in the image would be a result of the probe width. By taking an integrated intensity profile across the image of the interface, the probe width could be measured (figure 3). The total fluence per dwell point measured for probes A,B and C in figure 2 were 1.2×10^4, 3.0×10^4 and 1.3×10^4 electrons nm^{-2} respectively [1].

A clear transition can be seen between probes of increasing fluence A, B and C; from contamination build-up (A) to hole-drilling (B, C) (figure 2). We suggest that below this electron intensity threshold the rate of hydrocarbon contamination induced by electron beam heating exceeds the rate of ionization induced sputtering or mass loss, providing a coating on the specimen that inhibits further mass loss [7]. Thus STEM-EDX data can be collected using probe setting A without significant alteration of the structure or chemistry of the biotite and has shown potassium depletion of biotite by fungal hypha [2] but no significant elemental alteration of biotite by cyanobacteria [1].

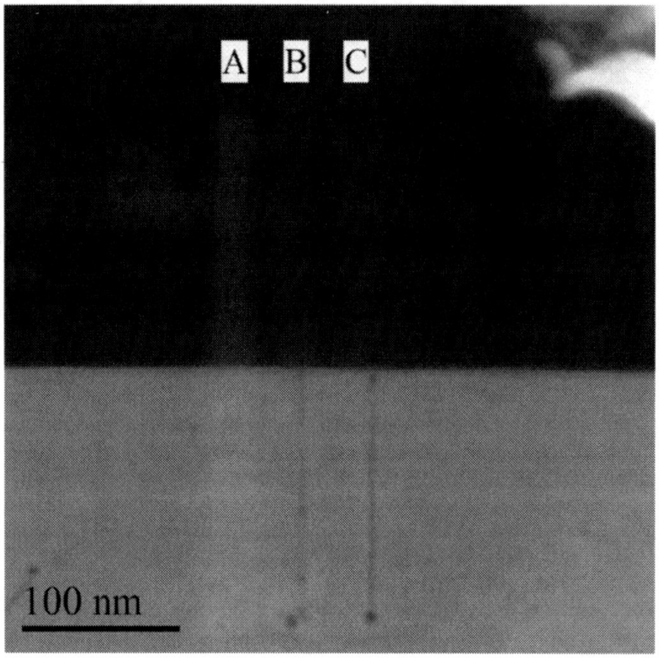

Figure 2. A HAADF image of a cyanobacteria/mineral interface showing three electron probe linescans produced using different beam conditions. A transition from hydrocarbon contamination build-up to hole-drilling in the mineral can be seen between line A and the higher electron fluence lines B and C.

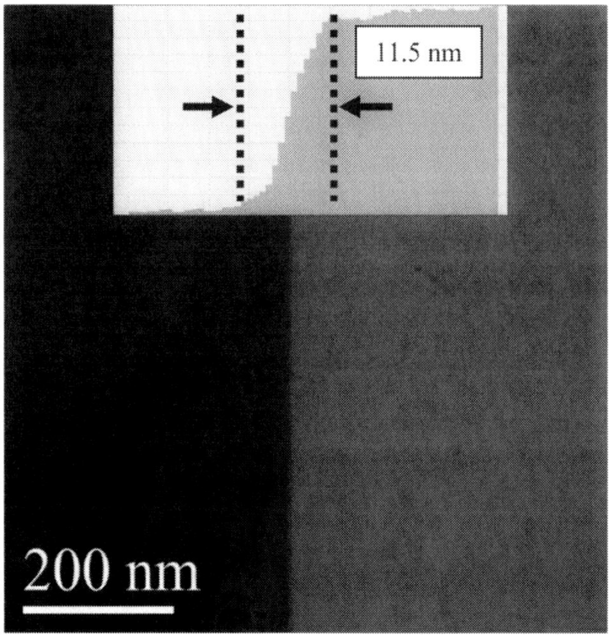

Figure 3. An example HAADF image of the cyanobacteria/mineral interface with an intensity profile across the interface inset. The intensity profile was integrated across 200 pixels, and the width of the profile transition was taken to be the electron probe width for the beam conditions the image was taken at.

4. Conclusions

This paper describes a method to characterize electron beam damage induced in biotite by 200kV STEM analysis. The purpose was to determine STEM operating conditions for EDX elemental analysis of biotite without the artefactual loss of material from the specimen.

An electron probe intensity for a transition between hydrocarbon contamination build-up and mass-loss was demonstrated. Representative analytical STEM studies were conducted below this electron intensity threshold for biotite and reported elsewhere. In general analytical STEM analysis of similar layered minerals should only be carried out once a thorough investigation into beam damage has been performed.

References

[1] Ward M B, Kapitulčinová D, Brown A P, Heard P J, Cherns D, Cockell C S, Hallam K R and Vala Ragnarsdóttir 2013. *Micron.* **47** 10.
[2] Bonneville S, Smits M M, Brown A, Harrington J, Leake J R, Brydson R and Benning L G 2009. *Geology.* **37** 615.
[3] Wirth R 2009. *Chem. Geol.* **261** 217.
[4] Lee M R 2010. *Mineral. Mag.* **74** (1). 1-27.
[5] Ma C, FitzGerald J D, Eggleton R A and Llewellyn D J 1998. *Clay. Clay Miner.* **46** (3) 301.
[6] Pan Y, Brown A, Brydson R, Warley A, Li A and Powell J 2006. *Micron.* **37** 403.
[7] Egerton R F, McLeod R, Wang F and Malac M 2010. *Ultramicroscopy.* **110** 991.

Sample preparation and EFTEM of Meat Samples for Nanoparticle Analysis in Food

L. Lari[1,2,3] **and A. Dudkiewicz**[4, 5]

[2]Department of Physics, University of York, Heslington, YO10 5DD, York, UK.
[3]The York JEOL Nanocentre, York Science Park, Heslington, YO10 5BR, York, UK
[4]Food and Environment Research Agency, YO41 1LZ, York,
[5]Department of Environment, University of York, Heslington, YO10 5DD, York, UK

E-mail: leonardo.lari@york.ac.uk

Abstract. Nanoparticles are used in industry for personal care products and the preparation of food. In the latter application, their functions include the prevention of microbes' growth, increase of the foods nutritional value and sensory quality. EU regulations require a risk assessment of the nanoparticles used in foods and food contact materials before the products can reach the market. However, availability of validated analytical methodologies for detection and characterisation of the nanoparticles in food hampers appropriate risk assessment. As part of a research on the evaluation of the methods for screening and quantification of Ag nanoparticles in meat we have tested a new TEM sample preparation alternative to resin embedding and cryo-sectioning. Energy filtered TEM analysis was applied to evaluate thickness and the uniformity of thin meat layers acquired at increasing input of the sample demonstrating that the protocols used ensured good stability under the electron beam, reliable sample concentration and reproducibility.

1. Introduction

The new emerging trend in the food industry exploits nanotechnology for versatile developments. One example is silver nanoparticles (NPs) used for food application. For their antimicrobial properties, silver nanoparticles (AgNPs) are used in food supplements and various food contact surfaces (e.g. packaging, cutting boards, cutlery)[1]. AgNPs are reported to be cytotoxic not only to bacterial but also human cells [2]. One of the reasons for this increased toxicity of AgNPs is associated with their size [2]. However, NPs size undergoes dynamic changes after spiking in complex matrices e.g. agglomeration/ aggregation, dissolution, deagglomeration [3-5]. Therefore The European Food Safety Authority emphasizes that for appropriate risk assessment of NPs in foods and feed, a particle size measurement in the hosting food products is necessary [6]. Of special interest among the methods allowing NP size measurement is electron microscopy (EM) coupled spectrometry methods. However, standard sample preparation protocol for solid food samples e.g. meat involves such methods as resin embedding or freezing and sectioning of frozen material [7]. Both methods are laborious, require sophisticated equipment and technical skills. Additionally the volume of the sample that can be

[1] To whom any correspondence should be addressed.

analysed in EM at once is very limited, making the method only useful for samples containing very high numbers of particles. In this study we test an alternative sample preparation protocol based on sedimentation of homogenized and highly diluted samples onto TEM grids. This approach allows a reduction in sample preparation time down to 1-2 hours and does not require skills necessary for thin sectioning of frozen or embedded material. The samples are dried prior to analysis and therefore it is possible to increase the analysed sample volume when compared to traditional approaches. In this work we report stability of the sample in the electron beam and summarize data regarding the sample layer thickness and its uniformity at increasing sample concentration. These parameters have a crucial meaning for the size measurement and quantification of NPs in food samples by EM. The study demonstrates application of powerful tools, energy filtered transmission electron microscopy (EFTEM) and electron energy loss spectrometry (EELS) for the measurement of the meat sample thickness.

2. Material and methods

Meat emulsion was obtained using the equipment and the methodology as described in [8]. However, at the final stage, the cryo-milled meat was not frozen and no AgNPs were added. The material was preserved using Proclin 150™ at a concentration of 1.1 g/ kg and aseptically packed in 50 ml amber vials. The density of the meat emulsion was 1.006 g/ ml. Aqueous dispersion of AgNPs stabilized with polyvinylpyrrolidone at concentration of 0.1% m/m and nominal size of 42±10 nm was originally obtained from Nanogap (Milladoiro, Spain). AgNPs were mixed with the meat emulsion in a 1:1 ratio using a potter type homogenizer until visual homogeneity of the resulting viscous liquid was obtained.

TEM samples were prepared by dilution, homogenization and subsequent sedimentation of the AgNPs/meat viscous liquid on formvar-carbon coated, 400 mesh Cu TEM grids (Agar Scientific, Stansted, UK). A borate buffer at pH 8.0 (0.05M H_3BO_3, 0.05M KCl, 0.004M NaOH) was used for the dilution step for three dilution factors, 200 fold m/m (Sample *A*), 500 fold m/m (Sample *B*) and 2000 fold m/m (Sample *C*) respectively. The diluted emulsions were dispersed in the buffer using potter type homogenizer and subsequently homogenized with ultrasonic probe (Misonix, USA) in vials for 1 min at 100 W. During homogenization, vials were kept cool by immersion in water with ice. Lastly, 0.5 g of sample at each dilution level were transferred to polyallomer centrifuge tubes and topped up with the buffer to the level of 2-3 mm from the tube rim. The tubes were equipped with solid Agar 100 resin fillers (Agar Scientific, Stansted, UK) at the bottom to ensure flat support for the TEM grids. Samples were sedimented on the TEM grids in Beckman XL-100 ultracentrifuge (Beckman, California, USA), equipped with SW40Ti rotor and operating at RCF=100.000 g at 20^{0}C for 1h. Characterization by EFTEM was undertaken using a JEM-2200 FS (JEOL Ltd., Tokyo, Japan) field emission TEM operating at 200 kV, equipped with an in-column Omega-type energy filter, and CEOS image and probe 3rd order aberration correctors. Sample thickness was measured using the Digital Micrograph (DM) (Gatan, Pleasanton, USA) EFTEM thickness map routine.

3. Results and discussion

Figure 1 shows typical sample thickness analysis for the three samples. Bright field images in Fig. 1a), 1b) and 1c) show nanoparticles embedded in the dried emulsions. t/λ maps in Fig. 1e), 1f) and 1g) (where t is the sample thickness and λ is the electron mean free path for inelastic scattering) were obtained through acquisition and application of the log-ratio method [9] of an unfiltered bright field (BF) image followed by an elastic image with 10eV energy slit centred onto the zero loss peak (512x512 pxl, 3s acquisition). AgNPs were found positioned at different heights within the sample regardless of the dilution applied, demonstrating that the sample preparation method provided layers of significant thickness. The sample drift during the map acquisition was evaluated and removed using AgNPs as a reference for image correlation. Figure 1d)-f) show t/λ maps respectively for sample *A*, *B* and *C*. Cross correlation results from each map acquisition show an average drift of 2.4pxl (Sample *A*), 1.6 pxl (Sample *B*) and 4 pxl (sample *C*) corresponding to 3.2nm (0.5% of the field of view), 2.2nm (0.3%) and 5.4nm (0.8%). This shows that during the images acquisition, and regardless of the level of sample dilution, the samples were stable under the electron beam, *i.e.* no significant sample drift or shrinkage could be observed.

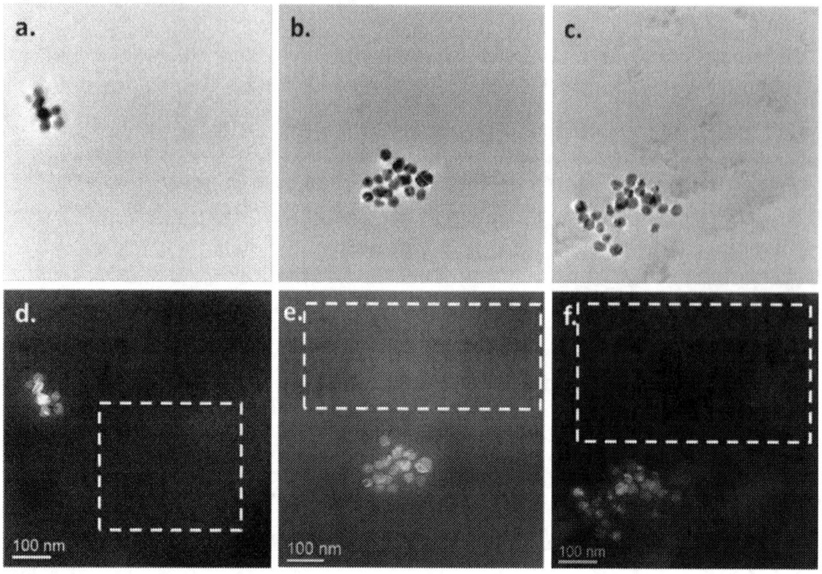

Figure 1. EFTEM analysis of *Ag NPs in 200 (a) and (d) (Sample A), 500 (b) and (e) (Sample B), and 2000 fold (c) and (f) (Sample C) with BF TEM images a.)-c.) and thickness maps d)-f);dashed boxes indicates typical areas of τ/λ mean values measurement .*

The sample thickness was calculated from t/λ measurements on particle free areas and the results are reported in Table 1. λ was estimated using the following equations:

$$\lambda = \frac{106\,F\,\frac{E_0}{E_m}}{\ln(2\,\beta\,\frac{E_0}{E_m})} \quad (1)\,[9]; \qquad E_m = 7.6\,Z_{eff}^{0.36} \quad (2)\,[10]; \qquad Z_{eff} = \frac{(\sum Z_i\,f_i)^{1.3}}{(\sum Z_i\,f_i)^{0.3}} \quad (3)\,[9],$$

where F is-a relativistic factor (0.618 for 200kV electrons), E_0 the incident beam energy, E_m the mean energy loss (calculated using eq.(2)). The effective atomic number Z_{eff} is calculated according to fractions (f) of the (i) elements in the sample of atomic number Z as described in equation (3).

β indicates the collection aperture, in our case no aperture was used and a value of 15mrad was chosen to feed into eq.(1) according to [10]. The sample (+support) composition was estimated by EELS spectroscopy (see Figure2a). A spectrum was acquired within a region comprising the main elements present in the meat (C, O, N) sample. Quantification was done using calculated cross sections in DM quantification routine producing the following elemental composition: C 88at%, N 7at% and O 5at%. λ value was estimated to be 149nm.

Table 1. EFTEM measurements of t/λ and mean values of calculated thickness

Dilution factor	t/λ ± s.d.	t (nm) ± sd
(A) 1 in 199	1.75 ± 0.31	260 ± 46
(B) 1 in 499	0.79 ± 0.26	118 ± 39
(C) 1 in 1999	0.50 ± 0.10	74 ± 15

Standard deviations (s.d.) are given from measurements in 5 different areas per sample.

The mean sample thickness (see Table 1) varies with the sample dilution and increases with decreasing dilution factor. The measurements over 5 areas across the TEM grid point out good homogeneity of the sample thickness over the grid, potentially allowing quantification of homogenously distributed NPs in meat emulsion. However, the calculated sample thickness included the thickness of the formvar-carbon film initially present on the TEM grid. To estimate the film's thickness we used a linear fit of the measured sample+support thickness versus the quantity of sample applied in milligrams (see Figure 2b). The support thickness can be evaluated at the intercept on the x-axis (no sample applied) of the fitting function, which corresponds to a value of ~ 40 nm. This value is in good agreement with the value provided by the manufacturer: 30-40 nm of formvar and ~10 nm of carbon. Therefore the thickness for sample A, B and C is respectively about 220, 80, and 30nm. We have calculated that same samples in the hydrated state (based on meat emulsion density) would create 34±9 times thicker layer on the TEM grid. This figure also defines the level of the sample pre-concentration compared to the cryo- and resin embedding preparations where the volume changes in comparison to hydrated sample are minimal.

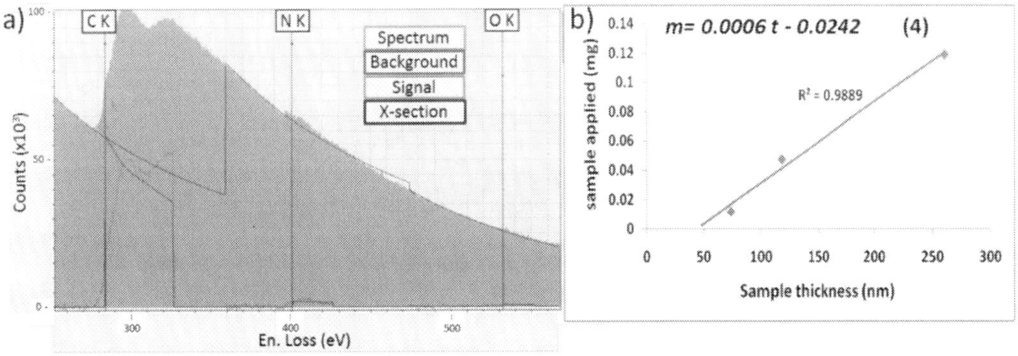

Figure 2. a) *Quantified EELS spectrum showing C, O and N K-edges. Red, green and blue lines represent background, net signal, and calculated cross section for quantification. b) Linear fit of sample applied versus sample thickness with linear fit equation (4) inset .*

4. Conclusions

In this paper we have evaluated a new sample preparation approach for potential quantification and measurement of NPs in food. EFTEM and EELS were successfully applied to measure the thickness of meat sample containing AgNPs. Using the described sample preparation protocol, we were able to obtain meat emulsion layers where the weight input of meat emulsion was linearly related to the TEM sample thickness and allowed to predict the support thickness for null input. This protocol provides a quicker and easier sample preparation method with respect to traditionally used solutions, such as resin embedding and cryo-sectioning, and allows increasing the amount of sample volume analysed.

Acknowledgements

The authors would like to acknowledge Prof. Kristian Mølhave from Danish Technical University (Lyngby, Denmark) for advice regarding experimental design, as well as Dr T. Linsinger and Dr R. Grombe at IRMM, JRC (Brussels, Belgium) for providing materials. Financial support from EU Programme NanoLyse (FP7/2007-2013) under grant agreement n° 245162 is gratefully acknowledged

References

[1] "PEN The Project on Emerging Nanotechnologies." http://www.nanotechproject.org/, 2013.
[2] Liu W. *et al.* 2010 Nanotoxicology **4** 319
[3] Luo P. *et al.*, J. 2013 Microsc **250** 32
[4] Zook J. M. *et al.* 2011 Anal. Bioanal. Chem. **401** 1993
[5] Peters R. *et al.* 2012 Acs Nano **6** 2441
[6] EFSA Scientific Committee, 2011 EFSA Journal **9** 1
[7] Dudkiewicz A. *et al.*, 2011 Trends in Analytical Chemistry **30** 28
[8] Grombe R. *et al.* "Production of Reference Materials for Analytical Method Development for Nanoparticles in Food. Part 2" submitted Anal. Bioanal. Chem.
[9] Egerton R. F., *Electron energy-loss spectroscopy in the electron microscope* (Plenum Press, New York, 1996).
[10] Malis T *et al.* 1988 J. Electron Microscopy Techniques **8** 193

Measuring *in vitro* cellular uptake of nanoparticles by transmission electron microscopy

A P Brown, R M D Brydson and N S Hondow

Institute for Materials Research, School of Process, Environmental and Materials Engineering, University of Leeds, Leeds LS2 9JT, UK

a.p.brown@leeds.ac.uk

Abstract. Biomedical application of engineered nanoparticles (NPs) is a growing area of research and development. Uncertainty remains as to the mode of action of many NP types and TEM is a tool capable of addressing this if used in conjunction with standard cellular response assays. We will demonstrate imaging of thin sections of fixed, plastic embedded cells by analytical TEM to identify: superparamagnetic iron oxide NP translocation into cell compartments such as endosomes; amorphous silica NP penetration through a cell membrane without membrane encapsulation and zinc oxide NP degradation in cell compartments. We will then discuss how the *in vitro* cellular responses to a dose of NPs exposed to cell lines can be correlated to the internalized dose per cell section noting however that quantification of the latter requires random sampling procedures or correlation to higher throughout techniques to measure a population of whole cells. Similarly, analytical TEM measures of NP degradation within intracellular compartments will require a more appropriate sample preparation such as cryo-fixation.

1. Introduction

Engineered nano-particles (NPs) have been the focus of much recent research with a particular emphasis on application in the field of medicine as imaging probes for cells and tissues, for drug delivery and as sensors of target molecules [1-3]. Efficacious and safe implementation of NPs requires a complete understanding of cellular response to NP exposure. Knowing precisely how an exposure dose relates to the response is therefore critical to the development of both NP toxicology and medicinal applications. Electron microscopy has sufficient spatial resolution to achieve this correlation and has recently been used to monitor and quantify NP dispersion and dose internalised by *in vitro* cell lines [4,5].

In this work we will draw on the imaging and spectroscopic capabilities of (scanning) transmission electron microscopes (S/TEMs) to identify cellular uptake pathways and the final location of internalised particles. The key challenges for this work are representative sample preparation (that does not detrimentally alter the cells or NPs, or obscure NP location with contrast enhancing heavy metal stains) and suitable control of the electron fluence applied during imaging and, more importantly, spectroscopic analysis such that the particle structure (and ideally the cellular material) remains unaltered during investigation [6]. We will demonstrate the use of analytical TEM to identify: superparamagnetic iron oxide NP translocation into cell compartments such as endosomes; amorphous silica NP penetration through a cell membrane without membrane encapsulation and zinc oxide NP degradation in cell compartments.

2. Experimental

Cells from the human lymphoblastoid cell line (MCL-5) were exposed to dextran coated iron oxide nanoparticles (Liquids Research, Wales) as described in [7]. Cells from the human lung alveolar carcinoma cell line (A549) were exposed to amorphous silica nanoparticles (Ludox SM-30) and aliphatic polyether, polymer-coated zinc oxide nanoparticles (Nanotek, Alfa Aesar) as described in [8] and [9] respectively.

Preparation of NP exposed cells for TEM followed the conventional wet chemical approach to view cellular ultrastructure: aldehyde and then osmium tetroxide fixation, dehydration in ethanol and infiltration with and polymerization of epoxy or acrylic resins. After embedding, thin sections (70 – 90 nm) were cut by an ultramicrotome, floated on a water bath and mounted on TEM grids [7-9]. No bulk and post embedding, heavy-metal staining steps were applied. TEM was conducted on two microscopes; an FEI CM200 field emission gun (FEG-)TEM running at 197 kV equipped with an Oxford Instruments energy dispersive X-ray (EDX) spectrometer (X-Max, silicon drift detector) and a Gatan Imaging Filter (GIF-200) and an FEI Tecnai F20 FEG TEM operating at 200 kV fitted with a Gatan Orius SC600A camera and an Oxford Instruments energy dispersive X-ray (EDX) spectrometer (X-Max, silicon drift detector). Tilt series were collected through a tilt range of +46 to -46 degrees with an image recorded every one degree on the Gatan Orius SC600A camera and processed using the software ImageJ, with the TomoJ plugin. Images in the tilt series were defocused (underfocus) by 180 μm from minimum contrast to enhance the contrast of the NPs in the unstained cell sections.

3. Results and Discussion

Dextran coated superparamagnetic iron oxide NPs are used as MRI contrast enhancement agents however partial oxidation of these particles has been shown to dramatically alter cellular internalization, with the more oxidised form exhibiting the most uptake and inducing significant DNA damage [7]. TEM confirms cellular internalization of these iron oxide NPs and suggests that uptake is dominated by encapsulation within vesicles, indicative of endocytosis (figure 1). The cellular response (including DNA damage) has been correlated to the uptake and the probable release of Fe^{2+} following dissolution of the NPs in lysosomes or late endosomes.

(a) (b) (c)

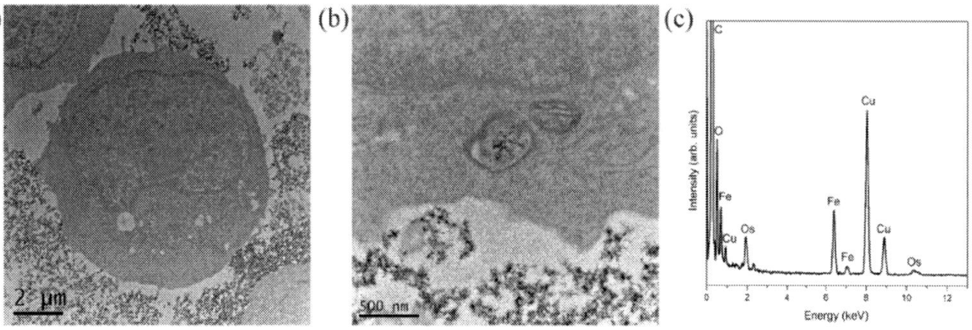

Figure 1 a) BF TEM image of MCL-5 cells exposed to iron oxide NPs, b) BF TEM image of an agglomerate of NPs in a vesicle within a cell (confirmed by tilting, [10]), and c) EDX spectrum from particles in the cell confirming the presence of Fe (the Cu signal is generated by the support grid) [6].

Amorphous silica NPs have potential for drug delivery however significant levels of induced DNA damage and reduction in cell viability (at higher exposure doses) are observed upon *in vitro* exposure to these NPs [8]. A TEM tilt-series indicates that cellular internalization of NPs has occurred without membrane encapsulation and the observation of irreversible binding to model lipid membranes suggest this is driven by an adhesive interaction with cell membranes. NP internalization without membrane encapsulation is confirmed following cellular exposure at 4 °C, when endocytic uptake pathways are suppressed (figure 2). The identification of a non-active mode of entry directly into the cytoplasm supports the idea that these particles might be suitable for drug delivery and gene therapy, at relatively low exposure doses.

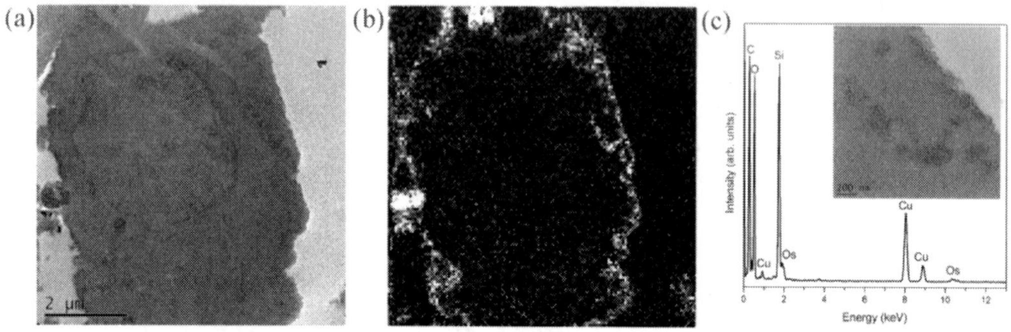

Figure 2 a) BF TEM image of A549 cells exposed to silica NPs at 4 °C, b) STEM Si-X-ray map showing membrane adsorption *and* localized penetration of silica NPs into the cell, without membrane encapsulation (BF TEM image inset in (c)) and c) EDX spectrum from particles in the cell confirming the presence of Si [6].

Zinc oxide (ZnO) NPs are used extensively in sun creams and whilst it is now generally accepted that these NPs do not penetrate beyond the stratum corneum of skin, the risks to human health through inhalation and ingestion at the manufacturing stage are still a concern [9]. Potential toxicity of ZnO is due to the compound's solubility in aqueous media, releasing Zn^{2+}. *In vitro*, exposure of cells to polymer-coated ZnO NPs results in reductions in cell viability that are more acute than exposure to uncoated ZnO NPs and almost comparable to the response upon exposure to soluble zinc salt solutions at similar Zn concentration. Monitoring of intracellular Zn^{2+} concentrations and non-reversible interactions with model membranes suggests the polymer-coated NPs are internalized before dissolution of the NPs is complete. TEM confirms NP internalization in membrane encapsulated compartments with clear evidence of porosity within the NPs, indicating partial dissolution (figure 3). Thus to limit toxicity, ZnO NPs with coatings that only weakly interact with cellular membranes and that inhibit pH driven dissolution of the ZnO in aqueous media are recommended.

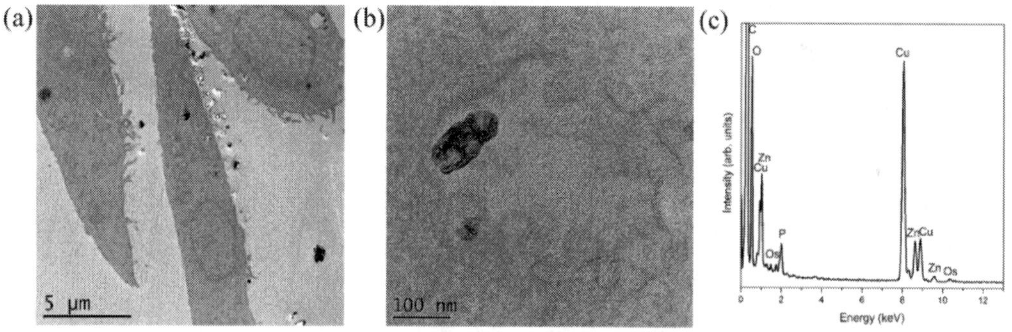

Figure 3 a) BF TEM image of A549 cells exposed to zinc oxide NPs, b) BF TEM image of ZnO NP in a vesicle within the cell, particle porosity suggests partial dissolution of the ZnO, and c) EDX spectrum from the particle in the vesicle confirming the presence of Zn [6].

In general, employing TEM sample preparation by resin embedding with only chemical-fixation (including osmium tetroxide) does enable visualization of the main cellular structure without contrast from the NPs themselves being obscured by heavy metal stains. This approach enables the easy localization of NPs within a cell and the use of a corresponding analytical capability is essential to confirm the elemental identity of the NPs [11]. Looking forward, quantification of NP uptake within thin sections of cells is also possible, however random sampling procedures or correlation to higher throughout techniques are required to measure a population of whole cells [5, 6]. Also quantification, by analytical TEM, of potential degradation of NPs within intracellular compartments (such as in figure 3), requires a more appropriate sample preparation to avoid re-diffusion of ionic species during fixing and processing. Analytical measures of the ionic content of whole cells prepared for EM by cryo-fixation and freeze drying have been shown to be a sensitive measure of cell viability [12]. Therefore, thin sections of cells prepared in this way should enable quantification of NP composition and ionic content within intracellular compartments.

4. Conclusions

Biomedical application of engineered nanoparticles is a growing research field in which some uncertainty remains as to the mode of action of many nanoparticle types. To eliminate this, *in vitro* cellular responses to a dose of NPs exposed to cell lines need to be correlated to the internalized dose per cell. In addition to the position and number of NPs within a cell, identification of any degradation of the NPs within intracellular compartments and its impact on cell viability is required. TEM is capable of identifying and quantifying cellular uptake of NPs however appropriate sample preparation and additional elemental analysis are required.

References

[1] Lee D E, Koo H, Sun I C, Ryu J H, Kim K and Kwon I C 2012 *Chem. Soc. Rev.* **41** 2656.
[2] Doane T L and Burda C 2012 *Chem. Soc. Rev.* **41** 2885
[3] Saha K, Agasti S S, Kim C, Li X and Rotello V M 2012 *Chem. Rev.* **112** 2739
[4] Hondow N S, Brydson R, Wang P, Holton M D, Brown M R, Rees P, Summers H D and Brown A 2012 *J. Nanopart. Res.* **14** 977
[5] Summers H D, Brown M R, Holton M D, Tonkin, J A, Hondow N, Brown A P, Brydson R and Rees P 2013 *ACS Nano* **7** 6129.
[6] Brown A and Hondow N 2013 *Electron Microscopy of Nanoparticles in Cells* in Summers H D (Ed.) *Frontiers of Nanoscience: Nanomedicine* (Elsevier)
[7] Singh N, Jenkins G J S, Nelson B C, Marquis B J, Maffeis T G G, Brown A P, Williams P M, Wright C J and Doak S H 2012 *Biomaterials* **33** 163
[8] Mu Q, Hondow N S, Krzemiński Ł, Brown A P, Jeuken L J C and Routledge M N 2012 *Particle Fibre Toxicol.* **9** 29
[9] Mu Q, David C A, Galceran J, Rey-Castro C, Krzemiński Ł, Wallace R, Bamiduro F, Milne S J, Hondow N S, Brydson R, Vizcay-Barrena G, Routledge M N, Jeuken L J C and Brown A P 2013 *Particle Fibre Toxicol.* Submitted
[10] Hondow N, Harrington J, Brydson R, Doak S H, Singh N, Manshian B and Brown A 2011 *Nanotoxicology* **5** 215
[11] Brandenberger C, Clift M J D, Vanhecke D, Mühlfeld C, Stone V, Gehr P and Rothen-Rutishauser B 2010 *Particle Fibre Toxicol.* **7** 15
[12] Fernandez-Segura E and Warley A 2008 *Electron Probe X-ray Microanalysis for the Study of Cell Physiology* in Terence D A (Ed.) *Methods in Cell Biology 88* (Academic Press) 19

Microstructural characterization of low and high carbon CoCrMo alloy nanoparticles produced by mechanical milling

TA Simoes[1], AE Goode[2], AE Porter[2], MP Ryan[2], SJ Milne[1], AP Brown[1], RMD Brydson[1]

[1]Institute for Materials Research, SPEME, University of Leeds, LS2 9JT, UK
[2]Imperial College London, London, SW7 2AZ, UK

pmtase@leeds.ac.uk

Abstract. CoCrMo alloys are utilised as the main material in hip prostheses. The link between this type of hip prosthesis and chronic pain remains unclear. Studies suggest that wear debris generated in-vivo may be related to post-operative complications such as inflammation. These alloys can contain different amounts of carbon, which improves the mechanical properties of the alloy. However, the formation of carbides could become sites that initiate corrosion, releasing ions and/or particles into the human body. This study analysed the mechanical milling of alloys containing both high and low carbon levels in relevant biological media, as an alternative route to generate wear debris. The results show that low carbon alloys produce significantly more nanoparticles than high carbon alloys. During the milling process, strain induces an fcc to hcp phase transformation. Evidence for cobalt and molybdenum dissolution in the presence of serum was confirmed by ICP-MS and TEM EDX techniques.

1. Introduction

Hip replacement surgery is one of the most successful medical operations and has been a good treatment for arthritis and other joint diseases in terms of chronic pain[1]. Metal on metal (MoM) devices can provide longer lifetimes for these implants, and cobalt-chromium (CoCr) based alloys are increasingly utilized due to the characteristics of good longevity[2] and low wear[3], especially in the second generation of these implants.

Both high and low carbon CoCrMo alloys are utilised in MoM hip implants. The carbon plays an important role in the strength, ductility and wear resistance of the implants. Nevertheless, the carbon can induce the formation of carbides and these may become sites for localized corrosion (dissolution) [4]. As a consequence, ions and/or particles released in this process may interact with the surrounding tissue and lead to an inflammatory response.

This paper investigates ball milling of low and high carbon CoCrMo alloy gas atomized powders in order to simulate the effects of mechanical wear of the alloy during implant use. The characterization of the size, shape, composition and phase structure of these nanoparticles gives relevant information about the physicochemical behaviour in relevant biological media and any preferential dissolution of the elements. CoCrMo powders have been milled in bovine serum albumin achieving a particle size reduction to the nanometre scale. The nanoparticles generated are comparable in terms of size and composition to some of the nanoparticle wear debris produced by real hip implants[5].

The samples were analyzed by Transmission Electron Microscopy (TEM) (in terms of size and shape), Energy-Dispersive X-ray spectroscopy (EDX) (composition) and High-Resolution

Transmission Electron Microscopy (HRTEM) (phase identification). These results are compared to bulk techniques such as X-ray Diffraction (XRD), Dynamic Light Scattering (DLS) and Inductively Coupled Plasma Mass Spectrometry (ICP-MS), demonstrating the effect of milling fluid on the size and stability of CoCrMo nanoparticles when in the presence of serum proteins.

2. Experimental techniques

CoCrMo gas atomized powders (ASTM75), both high and low carbon, were purchased from Sandvik Osprey. These starting powders had a particle size of around 22 μm and 16 μm, respectively. Bovine Serum Albumin (BSA), heat shock, pH 7.0 was acquired from Fisher Scientific, and used in solution during milling as a process control agent. The solution was composed of 0.4g of BSA diluted in 10 mL of deionized water. The powders were ball milled for 150 min in a high energy Spex 8000M mill using zirconia ceramic vial and balls. Before milling, the vial was fully filled with an inert gas (Argon) to avoid atmospheric contamination. A ball-to-powder weight ratio of 3:2 was used. After collection, samples were centrifuged in a Heraeus Megafuge for 10 min, at 25°C and a centrifugal force of 8000G. This procedure removes part of the BSA and coarse particles, maintaining most of the fine particles in dispersion.

Characterization by XRD was carried out using a Philips/PANalytical X'pert Diffractometer with Cu Kα radiation (λ = 1.5418 Å). The suspension was dropped onto a Si wafer and dried. TEM samples were analyzed using a Phillips CM200 FEGTEM operated at 200kV equipped with an Oxford Instruments UTW SDD EDX detector. Samples were drop cast onto lacey carbon coated Cu TEM grids. HRTEM images were acquired on an image-corrected FEI Titan 80-300 (S)TEM operated at 300 kV. For DLS, the samples were placed in an ultrasonic bath for 20 min. The refractive index and absorption coefficient used for hydrodynamic size analysis were 2.064 and 3.595, respectively.

3. Results and Discussion

3.1 Particle size and shape

The nature of the milling process could be comparable to the way that real wear debris particles are mechanically generated. The process includes the action of scratching and/or pounding of two CoCrMo surfaces, creating damage on the surface of the alloy. Shear bands are characteristic of this process and are often observed in particles – see Fig.2 (f).

The raw CoCrMo atomized powder shows perfectly round particle morphologies [Fig. 2 (g) and (h)]. After milling, DLS analysis [Fig.2 (a)] showed a significant decrease in particle size: a reduction down to 10-600nm for the high carbon and 10-800nm for the low carbon alloy powders. For both milled powders, the majority of the particles are found in the range of 10-200nm, and these results agree with TEM analysis in Fig. 1(b). By TEM, the low carbon alloy generated almost 50% more nanoparticles in the range of 10-200nm than the high carbon alloy. This could be related to carbon improving the ductility of the alloy, resulting in the absorption of more impact energy during milling.

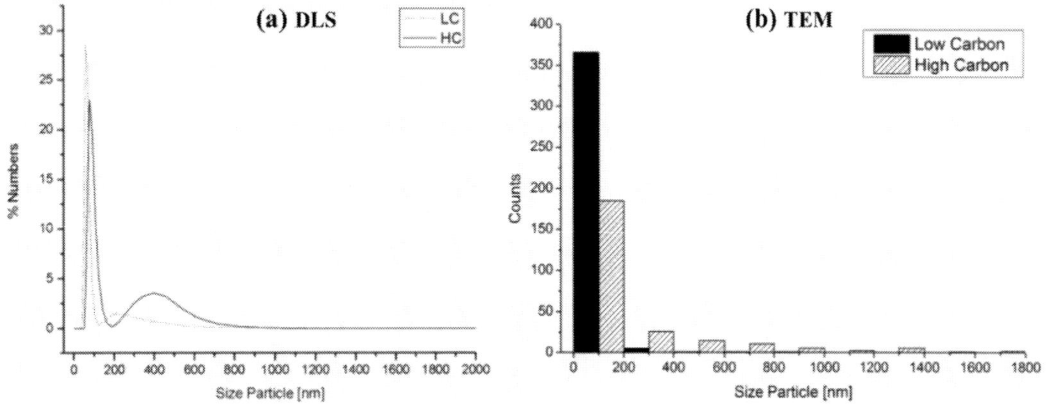

Figure 1 – (a) DLS and (b) TEM particle size distributions for the high and low carbon milled material.

3.2 Phase analysis

XRD of the raw material showed that both high [Fig.2 (a)] and low [Fig.2 (b)] carbon alloy powders, are dominated by the presence of the CoCrMo face-centred cubic phase (fcc, JCPDS 04-16-6869). Traces of the CoCr hexagonal close-packed phase (hcp, JCPDS 04-002-1030) are also evident. With the mechanical action of milling, post-milled material showed a significant increase in the hexagonal phase. This phase transformation is a strain induced fcc-hcp transformation (SIT), where temperature usually plays an important role, however in this case, stress is the most important contribution. The presence of carbon substantially alters the retention of the fcc phase after milling which is expected because carbon is known to stabilize this phase. S(TEM) analysis reveals coarse particles dominated by the fcc phase [Fig.2 (i)], whereas fine particles (less than 200nm) are predominantly hcp [Fig.2(e)]. This effect probably occurs due to the higher energy transferred to produce fine particles. In addition, XRD showed the appearance of zirconia traces (tetragonal, JCPDS 01-083-0113) as a result of contamination by erosion of balls/vial of the mill, possibly by hydrothermal degradation.

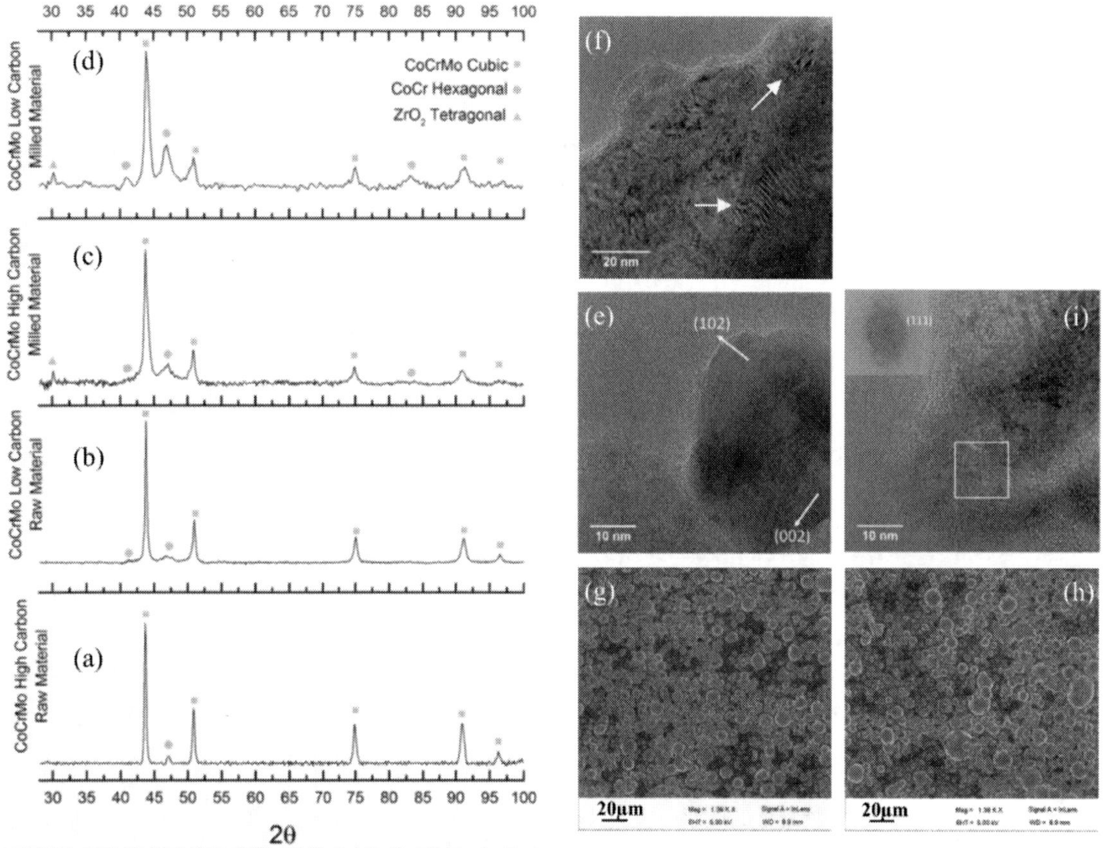

Figure 2 – Phase identification by XRD: (a) high carbon raw material, (b) low carbon raw material, (c) high carbon milled material and (d) low carbon milled material. TEM shows shear bands characteristic of milling process (f). TEM hcp and fcc phase identification in (e) and (i). SEM of the raw material: (g) low carbon and (h) high carbon powder.

3.3 Composition

The raw bulk materials were analysed in terms of composition (atom %) by SEM Energy Dispersive X-ray Spectroscopy (EDX). TEM-EDX [Fig.3 (b)] of individual particles showed that the low carbon milled material contained 75% of the particles with a cobalt composition below that of the raw material, accompanied by a corresponding increase in chromium. Following milling, high carbon alloy particles presented a wide distribution in composition. Some of the very small particles (about 10nm) were composed of almost 100% cobalt. However, 50% of the high carbon particles were found with a cobalt composition below that of raw material. In both high and low carbon alloy powders, ca. 50% of the particles showed no presence of molybdenum (or a level below the detection limits of EDX), with the remaining 50% showing 10% to 20% Mo, above that found in the raw material. ICP-MS results [Fig.3 (a)] strongly suggest that cobalt and molybdenum undergo dissolution during the milling process in serum in agreement with the TEM results. This behaviour is apparently independent of the amount of carbon in the alloy. The difference in the absolute degree of dissolution between the high and low carbon powders shown in ICP is likely related to the Ostwald-Freundlich relation, where the particle solubility is inversely proportional to particle size, TEM having revealed that the low carbon alloy produced more nanoparticles. Moreover, ICP indicates that chromium is resistant to the dissolution process.

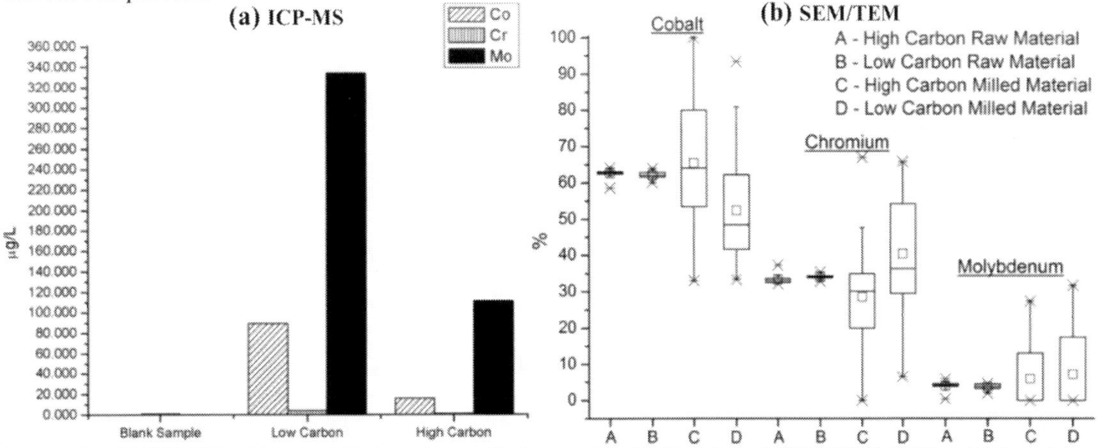

Figure 3 – (a) ICP-MS of low and high carbon milled material. (b) Boxplot† of TEM/SEM EDX of high and low carbon, raw and milled material.

4. Conclusions

Mechanical milling of CoCrMo-C alloy atomized powders has been shown to decrease particle size to between 10-200nm which is comparable with that seen in real wear debris released by hip implants. During milling, low carbon alloy powders generated more nanoparticles than the high carbon alloy, which is directly related to the mechanical properties and the stabilization of the fcc phase by carbon. ICP-MS together with SEM/TEM EDX gave an indication of cobalt and molybdenum dissolution in the presence of serum for both high and low carbon alloy milled powders. TEM shows a good agreement with bulk techniques, and allows differentiation of results in terms of particle size.

Acknowledgements
The authors gratefully acknowledge funding from the Capes Foundation (Ministry of Education of Brazil): BEX-5515/10-6 and EPSRC Science and Innovation Programme: EP/D063329.

References

[1] I.D. Learmonth, C. Young, C. Rorabeck, *The Lancet* **2007**, *370*, 1508.
[2] B.G. Weber, *Clinical Orthopaedics and Related Research* **1996**, *329*.
[3] H.-P. Sieber, C.B. Rieker, P. Köttig, *Journal of Bone & Joint Surgery, British Volume* **1999**, *81-B*, 46.
[4] E. Bettini, T. Eriksson, M. Boström, C. Leygraf, J. Pan, *Electrochimica Acta* **2011**, *56*, 9413.
[5] P.F. Doorn, P. a Campbell, J. Worrall, P.D. Benya, H. a McKellop, H.C. Amstutz, *Journal of Biomedical Materials Research* **1998**, *42*, 103.

† The bottom of each box is the 25th percentile; top is the 75th percentile; line in the middle is the medium value; square inside the boxes represents the average; "Whiskers" above and below each box represent 1.5 of the average values; crosses are the outliers.

EELS from organic crystalline materials

R Brydson[1], M D Eddleston[2], W Jones[2], C R Seabourne[1] and N. Hondow[1]

[1] Institute for Materials Research, SPEME, University of Leeds, Leeds, LS2 9JT, UK
[2] Department of Chemistry, University of Cambridge, Lensfield Road, Cambridge, CB2 1EW, UK

E-mail: mtlrmdb@leeds.ac.uk

Abstract. We report the use of the electron energy loss spectroscopy (EELS) for providing light element chemical composition information from organic, crystalline pharmaceutical materials including theophylline and paracetamol and discuss how this type of data can complement transmission electron microscopy (TEM) imaging and electron diffraction when investigating polymorphism. We also discuss the potential for the extraction of bonding information using electron loss near-edge structure (ELNES).

1. Introduction

Polymorphism is the ability of a compound to crystallise with different packing arrangements and is critically important in several industrial sectors as many of the solid-state properties of a compound are dependent on the crystal form. For example, in the pharmaceutical industry it is a requirement that the polymorphic behaviour of a drug compound is thoroughly investigated and understood as different crystal phases dissolve at different rates affecting the adsorption of the compound *in vivo*, making it essential to control which crystal form is dosed to patients.

Recently, TEM has been applied to the study of polymorphism in organic materials such as pharmaceuticals and has been shown to have advantages over more routinely used analytical techniques [1, 2]. Using a combination of imaging and diffraction each of the crystals in a sample can be analysed one-by-one during TEM analysis and different polymorphs of a compound can be distinguished, or new polymorphs of a compound identified, on the basis of crystal morphology and/or by indexing electron diffraction patterns. A further advantage is the ability to solve crystal structures of beam sensitive organic compounds using a single crystallite with sub-micron sized dimensions, by combining electron diffraction and crystal structure prediction, where standard single crystal or powder X-ray approaches would not be applicable. For example, a new polymorph of the pharmaceutical compound theophylline has been identified from analysis of a single crystallite having an estimated mass of 3 pg, despite this phase being present as a minor component in a mixture of crystal forms [3]. Here we investigate the potential use of EELS to provide additional chemical and structural information from single crystalline particles of organic material.

2. Experimental

Four different materials were studied: theophylline Form II; a 1:1 caffeine:theophylline co-crystal; para-terphenyl and Paracetamol Form I. Theophylline is a xanthine derivative with the chemical formula $C_7H_8N_4O_2$ which is used as a treatment for asthma; it bears a structural and pharmacological similarity to caffeine ($C_8H_{10}N_4O_2$). para-terphenyl ($C_{18}H_{14}$) consists of a central benzene ring substituted with two phenyl groups. Paracetamol (or acetaminophen) ($C_8H_9NO_2$) is a mild analgesic. These organic crystals are classified as molecular crystals where the molecules retain their character and pack and interact via weak van der Waals and hydrogen bonds. Preparation of thin organic crystals for TEM is a significant issue; para-terphenyl crystals were prepared by evaporation from an organic solvent at the interface with water, paracetamol crystals were prepared by crystallisation from the melt, theophylline crystals were prepared by cooling a saturated solution in nitromethane, and the 1:1 caffeine:theophylline co-crystal was prepared by ball milling.

Samples were dispersed and drop cast onto holey carbon coated TEM grids and analysed in a FEI CM200 field emission TEM/STEM operated at 200 kV equipped with a Gatan GIF200 Electron Energy Loss Imaging Filter and an Oxford instruments energy dispersive X-ray (EDX) detector for elemental analysis. An extraction voltage of 3.25 kV, spot size 6 and a magnification of 5000 times were employed. The value of probe current was estimated using an FEI calibration curve based on the measured brightness (i.e. exposure time) incident on the viewing screen – this was independently checked by measuring the current incident on the drift tube of the GIF. For measurement, the beam was typically spread over a circular area of > 20 micron radius giving an electron fluence rate of ca. 4 x 10^{-5} A/cm^2 and a total electron fluence during spectral acquisition of typically 1500-2000 electrons/nm^2. All electron energy loss (EELS) spectra were collected in diffraction mode from crystalline areas of ca. 0.8 micron diameter. Low loss EELS spectra were collected with a collection semi-angle of 6 mrads (typically 0.1 second acquisition time) and core loss spectra with a collection semi-angle of 30 mrads (typically 20 second acquisition time). Beam damage (presumed to be by a radiolytic mechanism) was monitored by observation of the diffraction pattern; typical lifetimes of crystals were of the order of minutes. Interestingly, progressive amorphisation by the electron beam proceeded via progressive loss of short range order (fading of the outer diffraction spots first) as observed by Glaeser [4] possibly due to breaking of hydrogen bonds and small molecular rearrangements during the initial stages of damage.

3. Results

3.1. EELS Low loss spectra

Small variations were apparent between EELS plasmon peak energies for the different organic crystals (see Table 1). The most noticeable difference was between theophylline Form II and the 1:1 caffeine:theophylline co-crystal where an increase in energy of over 1 eV was consistently measured. This would correlate with an increase in physical density between the two crystals and hence, as the two molecules are chemically similar, an increase in valence electron density (N) which is known to influence the plasmon energy as $\sqrt{N}$. A further effect in the low loss was the tentative observation of intensity in the region between 6-8 eV which would be associated with π to $\pi*$ transitions or a π plasmon in the conjugated systems such as para-terphenyl and paracetamol (figure 1).

Table 1. Plasmon peak energies (average and standard deviation of three measurements) compared to physical density derived from single crystal X-ray structure determination.

Sample	Plasmon Energy/ eV	Density/ gcm^{-3}
Theophylline Form II	22.6 ± 0.2	1.493
1:1 Caffeine:theophylline co-crystal	23.8 ± 0.1	1.551
para-terphenyl	22.9 ± 0.1	1.233
Paracetamol Form I	23.3 ± 0.1	1.296

3.2. EELS Core loss spectra

In principle EELS is ideal to quantify light elements in organic crystals such as C, N and O. However at the low electron fluences required to prevent irradiation damage, core loss spectra suffered from low signal to noise ratios (SNR) due to dominance of detector readout noise and dark current over the inherent statistical noise in the data (see figure 2). Despite this current limitation, it was possible to derive some elemental ratios which are listed in Table 2. Here the values are averages of three measurements and we estimate the accuracy of the ratios to be at best 20%. The core loss EELS spectrum of theophylline is shown in figure 2 and shows the presence of carbon, nitrogen and oxygen with a derived [C]/[N] atom ratio close to the expected value but provides an overestimate of the level of oxygen. Data for paracetamol showed the presence of carbon and oxygen in the sample in reasonable agreement with the expected value, however no nitrogen K-edge was observed above the baseline noise owing to the relatively low levels present.

Table 2. Values of the measured and theoretical carbon to nitrogen and carbon to oxygen atom ratios. Values determined using a 30 eV integration window and Hartree-Slater cross-sections.

Sample	EELS [C]/[N] atom ratio	EELS [C]/[O] atom ratio	Theoretical [C]/[N] atom ratio	Theoretical [C]/[O] atom ratio
Theophylline Form II	1.7	2.6	1.8	3.5
Paracetamol Form I	N not detected	3.4	8.0	4.0

With such low SNR, the core loss edge shape (ELNES) was difficult to extract reliably. However there were indications that the C K-ELNES was different between Theophylline and para-terphenyl consistent with the presence of conjugated π bonding in the latter. As an indication of the potential of ELNES measurements to differentiate between organic crystal polymorphs, figure 3 shows calculations using the CASTEP density functional theory (DFT) code of the predicted N K-ELNES for two polymorphs of para-aminobenzoic acid $C_7H_7NO_2$ [5 - 6]. These calculations were carried out using appropriately converged cell sizes and calculation parameters, using 0.5 electron core-holes for the α and β polymorphs, with an energy-dependent final state lifetime broadening applied [7-10]. The calculations show clear differences in the main peak some 10 eV above the edge onset which is split in the α phase but not in the β polymorph.

4. Future strategies for EELS measurements

A major problem remains that, at the low electron fluences required to prevent irradiation damage in these samples, core loss spectra suffered from low signal to noise ratios (SNR) due to dominance of detector readout noise and dark current over the inherent statistical noise in the data. We outline strategies to improve EELS analysis of organic crystals which include: (1) cooling the sample to allow, in principle, higher electron fluences and fluence rates to be used so as to improve spectral SNR; (2) optimisation of the SNR induced by the CCD detector via reduction of the number of detector pixels used in the non-energy dispersive direction on the CCD, as well as the use of new fast readout, low noise detectors; (3) the use of thinner crystals for analysis and (4) the use of a scanned beam as in STEM.

Assuming that EELS can be optimised further to give reliable quantitative information about chemical composition it would be useful when studying polymorphism by TEM. For example, when analysing a sample of a given compound by TEM, if a crystal giving diffraction patterns that do not index to a known crystal form of this compound (or with an unexpected crystal habit) was observed, EELS analysis would help in determining if this crystal was a new polymorph of the compound or a chemical impurity. Also, when performing electron diffraction analysis on samples containing several different compounds, such as a drug formulation, EELS analysis would be important for determining which of the compounds was being analysed when diffraction patterns were recorded, making the process of indexing the diffraction patterns to a particular crystal form significantly easier.

References

[1] Li Z Get al. 2002 *Microsc. Microanal.*8, 134
[2] Eddleston M D et al. 2010 *J. Pharm. Sci.* 99, 4072
[3] Eddleston M D et al. 2013 *Chem. Eur. J.* 19, 7874
[4] Glaeser R M 1971 *J. Ultrastructure Res.* 36, 466
[5] Clark S J et al. 2005 *Z. Kristallogr.* 220, 567
[6] Gao, S P et al. 2008 *Phys. Rev. B* 77, 115122.
[7] Athimoolam S and Natarajan S 2007 *Acta Cryst.* C63, o283
[8] Gracin S and Fischer A 2005 *Acta. Cryst.* E61, o1242
[9] Stevens J S et al. 2013 *PCCP* in preparation
[10] Moreau P et al. 2006, *Phys. Rev. B* 73, 195111.

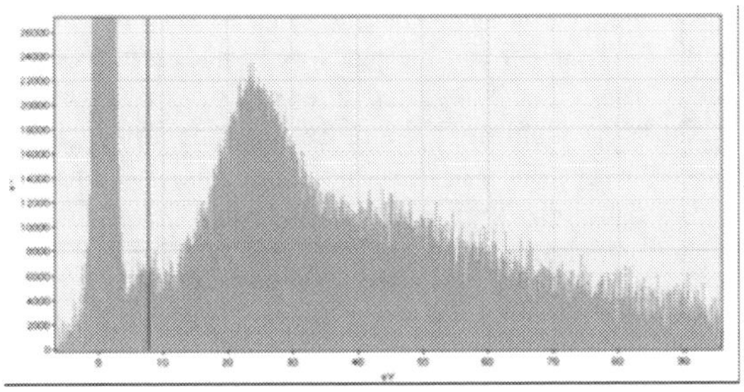

Figure 1. Low loss EELS from a crystal of paracetamol showing evidence for a π plasmon/transition below 10 eV (total acquisition time a few seconds at 4 x10^{-5} A/cm^2).

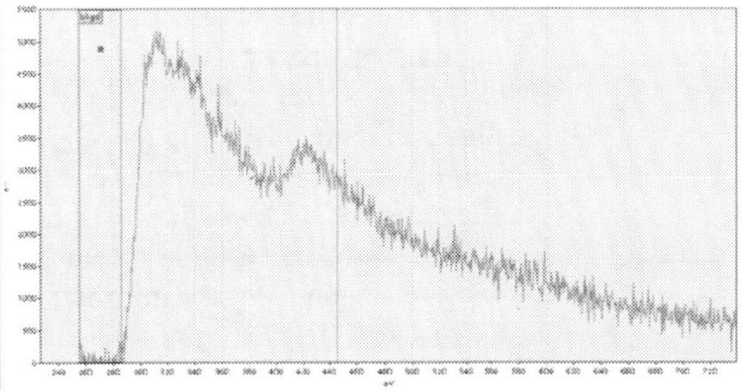

Figure 2. EELS core loss spectra from a theophylline crystal showing the nitrogen K-edge (ca. 400 eV) and the oxygen K-edge (ca. 530 eV) superimposed on the tail of the background subtracted carbon K-edge at ca. 285 eV (total acquisition time 20 seconds at 4 x10^{-5} A/cm^2).

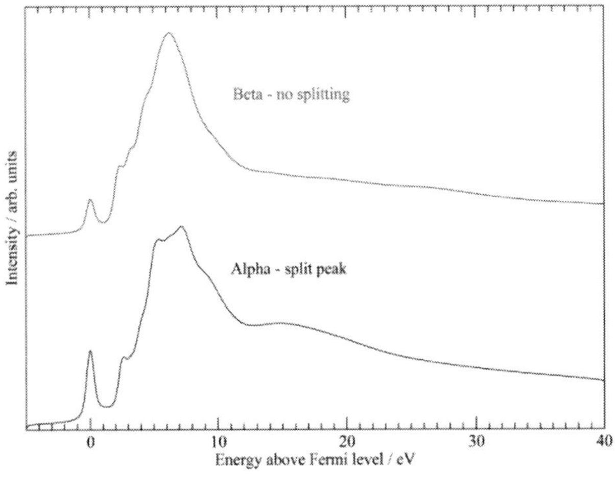

Figure 3. Calculations of the N K-ELNES for two polymorphs of para-aminobenzoic acid using the CASTEP DFT code.

Structural quantification of nanoparticles by HAADF STEM

K E MacArthur[1], L B Jones[1], S Lozano-Perez[1], D Ozkaya[2], P D Nellist[1]
[1]University of Oxford, Department of Materials, Parks Road, Oxford, OX1 3PH, UK
[2]Johnson-Matthey Technical Centre, Blounts Court, Sonning Common, Reading, RG4 9NH, UK

E-mail: katherine.macarthur@materials.ox.ac.uk

Abstract. A new method of quantifying the HAADF STEM signal on absolute scale, building on the z-contrast nature of the technique, has been developed to address the problem of characterising nanoparticles to an atomic scale. Experimental images are scaled to a fraction of the incident beam intensity from a detector map. The integrated intensity of each individual atomic column is multiplied by the pixel area yielding a more imaging-parameter robust quantity, termed a scattering cross-section. Using this cross-section approach, and simulated reference data we show how it is possible to count the number of atoms in individual columns, in order to compare against theoretical 3-dimensional structures. All this is with the aim of trying to obtain as much information as possible from a single image of these beam sensitive samples.

1. Introduction

Platinum nanoparticles are used for catalysing a wide range of reactions however their high cost has driven research into reducing particle size and trying to optimise active sites by controlling particle shape. Further improvement of these systems can only be made by careful investigation of the atomic structure of these particles, including surface facets and steps. These catalysts nanoparticles provide a severe characterisation challenge due to their minute size and beam sensitivity. The necessary aim, therefore, is to extract as much information from one image as possible

Much quantification work has been previously carried out on more large scale bulk materials or model systems. The new scattering cross-section technique [1] for high angle annular dark-field (HAADF) scanning transmission electron microscopy (STEM) quantification shows real promise for analysing nanoscale objects where statistical counting [2] and unit cell averaging are less applicable [3] due to the low total number of atomic columns within an image and therefore poor column statistics.

2. Method

The samples investigated were platinum supported on an amorphous 3-dimensional carbon support; received in powder form which was dusted onto a copper holey carbon grid. All experimental images were taken on an FEI Titan[3] G2 60-300 (S)TEM operating at 300kV with convergence angle of 20.2mrad and a camera length of 145mm giving detector collection angles of 34.9-190mrad.

Experimental images were taken and converted to an absolute scale through a detector calibration, Figure 1. [4] The accuracy of this approach was further refined by taking into account the variation in detector sensitivity as a function of scattering angle leading to an up-scaling of the experimental results. A peak finding routine was then carried out on the images. Once the location of peaks are know it was then possible to carry out a locally varying background subtraction behind the particle. This is carried out using the remainder of the image which contains no peaks carrying out iterative in-painting over the remainder of the image. The image was then subsequently segmented into Voronoi polygons; each pixel was assigned to its nearest peak up to a maximum value, to prevent excessive expansion at the surface. Conversion to scattering cross-sections was carried out by integrating within each Voronoi cell and then multiplying by the pixel area. The results were then constructed into a cross-section map, Figure 1, allowing the cross-section of each atomic column to be related back to it location within the particle structure. A program for automated processing of the images as outlined here is freely available online. [5]

A library of scattering cross-sections was obtained by simulating a platinum crystal oriented down the <110> crystallographic direction with increasing number of atoms in the z-direction. These simulations were carried out using a full frozon phonon calculation (QUEP) with 8 phonon configurations. The simulations were carried out to match those experimental conditions used for imaging; voltage of 300kv, Cs of 1.84µm, C5 of -1.25mm, convergence angle of 20.2mrad and detector collection angles of 34.9-190mrad. A 'super-cell' was constructed in CrystalMaker™ with a cross-section of 30 atoms and vacuum around the edge to act as a buffer. The cross-section of each of the central 12 atomic columns were averaged to give the tabulated result. For the calculated probe it was estimated that the intensity had fallen to 1% of the maximum value by a radius value of 0.85Å or by 1.58Å at 8.5nm depth into a crystal due to simple geometrical broadening of the probe, this will be slightly larger with a sample due to scattering. For the Pt crystal structure this means that the cross-section of each column can be evaluated independently.

Using the simulated library, Figure 2, each atomic column in the image was assigned to its nearest integer number of atoms. We can then check consistency of proposed nanoparticle models with the atom counting.

3. Results and Discussion

From the library simulations, Figure 2, it is possible to see the strength of the scattering cross-section technique as the channelling effects are much less strong than those presented by the peak intensity with increasing number of atoms. The peak intensity measurement will also be affected by the pixel sampling and whether or not this is coincident with the maximum intensity of the 2-dimensional Gaussian profile. It is still, however, vital to carry out direct comparison with simulation due to the non-linearity of the cross-section value with increasing thickness.

Figure 3, demonstrates a 3-dimensional reconstruction which is in agreement with the atom counting results. From this information it would then be possible to extract important catalytic information such as what surface planes predominate and the distribution of surface atoms with low coordination, which are thought to be the more catalytically active.

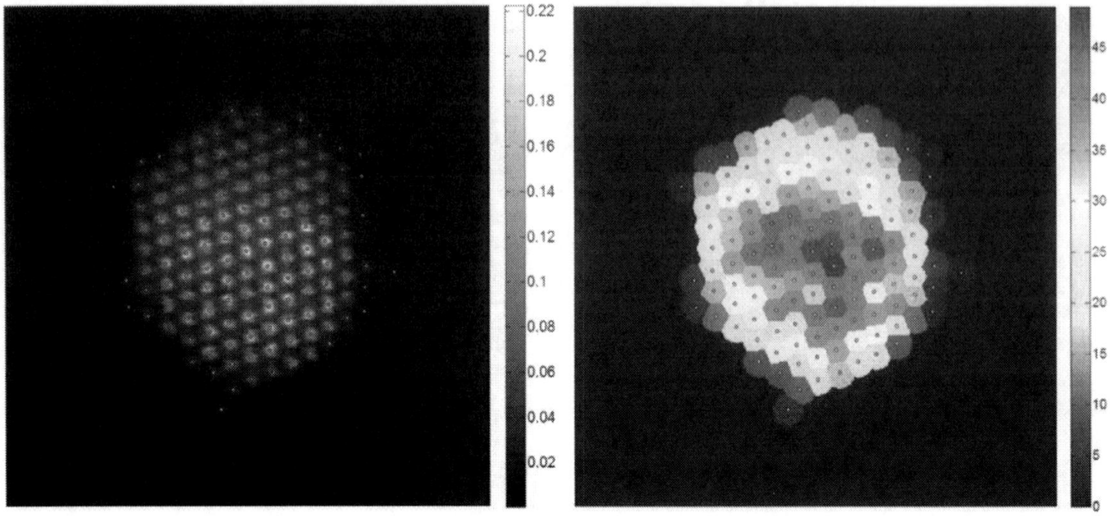

Figure 1. Left: The greyscale HAADF image of a Pt nanoparticle viewed down the <110> direction, scaled as a fraction of the incident beam intensity. Right: A scattering cross-section map of the same particle where each coloured region represents the Voronoi polygons over which the cross-section for that atomic column was integrated.

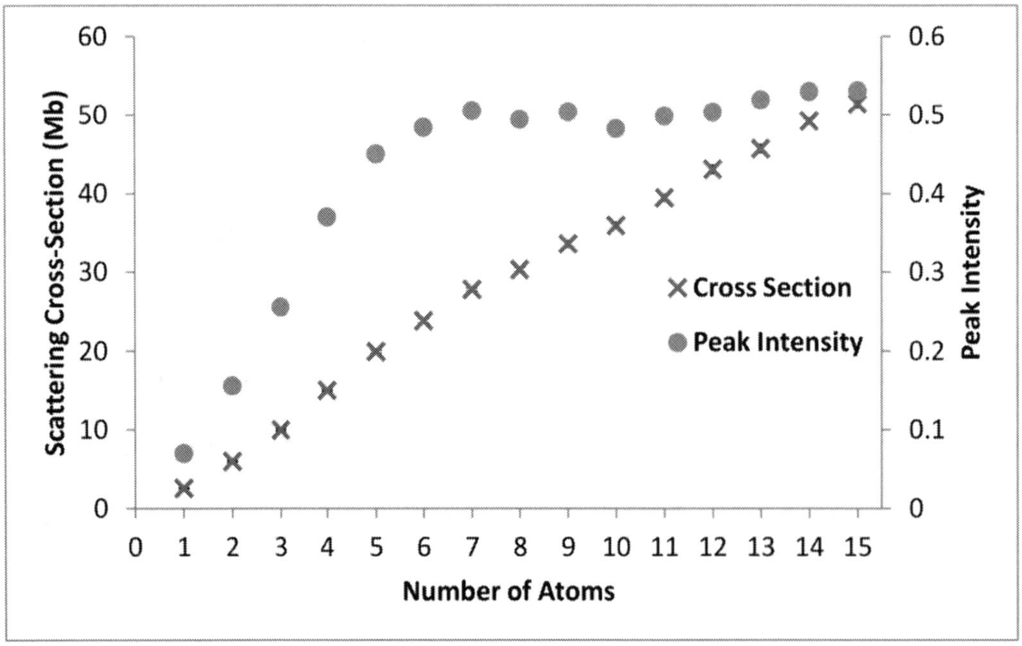

Figure 2. The scattering cross-section and peak intensity plotted against number of atoms in the column, along the <110> direction. This is the library used to assign columns for the reconstruction.

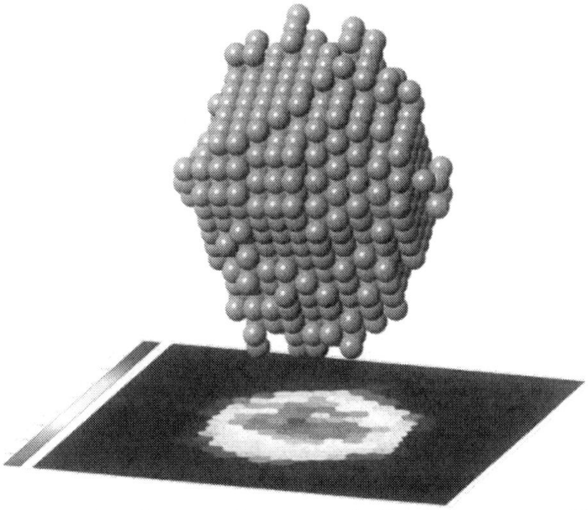

Figure 3. Shows a 3-dimensional image of an example final reconstructed Pt nanoparticle based on the atom counting results.

4. Conclusion

The scattering cross-section technique produces a much more reliable method for direct comparison with simulations, producing a library which varies monotonically with thickness. Using the new scattering cross-section technique it is now possible to estimate the atomic structure of a single element nanoparticle from one HAADF image. This has particular advantages for samples of high beam sensitivity, such as nanoparticles, where a tomographic series is not feasible.

5. References

[1] H. E, K.E. MacArthur, T.J. Pennycook, E. Okunishi, A.J. D'Alfonso, N.R. Lugg, L.J. Allen, P.D. Nellist, Ultramicroscopy (2013) In Press.

[2] S. Van Aert, J. Verbeeck, R. Erni, S. Bals, M. Luysberg, D. Van Dyck, G. Van Tendeloo, Ultramicroscopy 109 (2009) 1236–1244.

[3] A. Rosenauer, T. Mehrtens, K. Müller, K. Gries, M. Schowalter, P.V. Satyam, S. Bley, C. Tessarek, D. Hommel, K. Sebald, M. Seyfried, J. Gutowski, A. Avramescu, K. Engl, S. Lutgen, Ultramicroscopy 111 (2011) 1316–27.

[4] M. Retsky, Optik 41 (1974) 127–142.

[5] http://www.lewysjones.com/integrator/integrator_features.shtml

6. Acknowledgments

The authors gratefully acknowledge Dr Armand Béché, EMAT Antwerp, for acquiring the images and Dr Adrian D'Alfonso discussion related to the simulation codes. They would also like to thank EPSRC for financial support, and Johnson-Matthey for financial support and providing the catalyst samples.

The research leading to these results has received funding from the European Union Seventh Framework Programme under Grant Agreement 312483 - ESTEEM2 (Integrated Infrastructure Initiative–I3).

Hydrothermal synthesis, off-axis electron holography and magnetic properties of Fe_3O_4 nanoparticles

Trevor P. Almeida,[1*] Adrian R. Muxworthy[1], Wyn Williams[2], Takeshi Kasama[3] and Rafal E. Dunin-Borkowski[4]
[1]Department of Earth Science and Engineering, South Kensington Campus, Imperial College London, London, SW7 2AZ, UK.
[2]School of GeoSciences, University of Edinburgh, Edinburgh, EH9 3JW, UK.
[3]Centre for Electron Nanoscopy, Technical University of Denmark, DK-2800 Kongens Lyngby, Denmark
[4]Ernst Ruska-Centre for Microscopy and Spectroscopy with Electrons and Peter Grünberg Institute, Forschungszentrum Jülich, D-52425 Jülich, Germany.

Abstract. The hydrothermal synthesis of Fe_3O_4 nanoparticles (NPs) (< 50 nm) from mixed $FeCl_3$ / $FeCl_2$ precursor solution at pH ~ 12 has been confirmed using complementary characterisation techniques of transmission electron microscopy and X-ray diffractometry. Off-axis electron holography allowed for visualisation of their single domain (SD) nature, as well as inter-particle interactions, with the latter attributed to explain the pseudo-SD/multi-domain behaviour demonstrated by bulk magnetic measurements.

1. Introduction

The ability of a rock to reliably record the geomagnetic field depends on the geometry and size of the constituent magnetic particles, whereby large or closely packed grains that have large magnetostatic interactions between them will affect their recording fidelity [1]. Generally, particles in the single domain (SD) grain size range (< 100 nm) are regarded as ideal palaeomagnetic recorders because of their strong remanence and high magnetic stability, with potential relaxation times greater than that of the age of the Universe [2]. However, rocks rarely contain ideal assemblages of mono-dispersed SD particles and techniques used to determine the magnetic domain state of remanence carriers often reveal the presence of very small superparamagnetic (SP) particles, i.e., nanometric, plus pseudo-single domain (PSD) and multi-domain (MD) particles. As the magnetic minerals in rocks commonly show broad variations in particle size, shape and spacing, synthetic analogues are considered more ideal for study.

Magnetite (Fe_3O_4) is arguably the most important naturally occurring magnetic naturally mineral on Earth [3]. Understanding its grain size behaviour and recording fidelity is crucial to palaeomagnetism. Previously nano-sized Fe_3O_4 have been prepared by various methods such as sol–gel processing, hydrothermal synthesis, forced hydrolysis, ball milling and microemulsion. In this paper, we consider the approach of hydrothermal synthesis (HS), which in particular, offers effective control over the size and shape of Fe_3O_4 nanoparticles (NPs) at relatively low reaction temperatures and short reaction times, providing for well-crystallised reaction products with high homogeneity and definite composition [4]. Off-axis electron holography permits nanometre-scale imaging of magnetic induction within and around materials as a function of applied field and temperature [3, 5]. In this context, the HS of Fe_3O_4 NPs is presented, and comparison of their magnetic properties investigated using off-axis electron holography and more conventional rock magnetism technique was made.

[*] Author to whom correspondence should be addressed: t.almeida@imperial.ac.uk.

Journal of Physics: Conference Series **522** (2014) 012062 doi:10.1088/1742-6596/522/1/012062

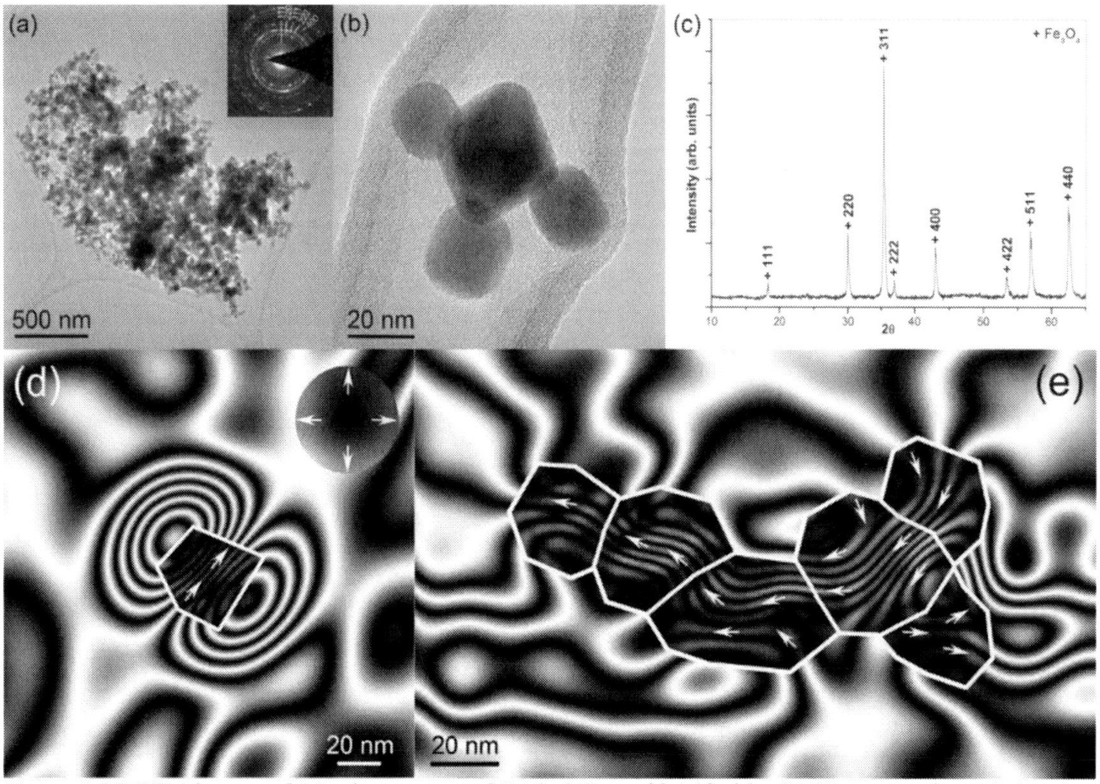

Figure 1 BF TEM images of (a) a large group of Fe_3O_4 NPs, as identified by SAED pattern (inset, indexed to Fe_3O_4 (JCPDS 88-0315)); and (b) a few individual Fe_3O_4 NPs. (c) Associated XRD pattern of the synthesised NPs, confirming the Fe_3O_4 phase. (d) Magnetic induction map of an individual Fe_3O_4 NP in the SD size range (~ 45 nm) acquired in magnetic field-free conditions and determined from the magnetic contribution to the phase shift reconstructed from off-axis electron holograms, revealing the SD nature of the particle. Magnetisation direction is depicted in the colour wheel (inset). (e) Magnetic induction map of a group of Fe_3O_4 NPs, showing magnetic contours flowing through the SD particles. The contour spacings are (d) 0.029 and (e) 0.049 radians.

2. Experimental

For the HS of Fe_3O_4, 0.2 ml $FeCl_3$ aqueous solution (45% pure $FeCl_3$; Riedel-de Haen, Germany) and 150 mg $FeCl_2$ powder (98.0% $CoCl_2$; Sigma-Aldrich, UK) were further diluted in 40 ml distilled water and mechanically stirred in a 125 ml Teflon-lined steel autoclave. The precursor solution was adjusted to pH ~ 12 with the drop-wise addition of aqueous NaOH solution, then sealed and inserted into a temperature controlled furnace at 200°C for 3 hours. The autoclave, once removed from the furnace, was allowed to cool down to room temperature naturally.

The synthesised reaction products were centrifuged for 5 minutes at 5000 rpm, cleaned with acetone and deposited onto single crystal silicon substrates for the purpose of structural characterisation using a PANalytical X'Pert PRO Diffractometer (Cu*Kα* radiation; θ/2θ diffraction geometry). For the purpose of transmission electron microscopy (TEM) investigation, the acetone cleaned HS suspensions were dispersed using an ultrasonic bath before deposition onto lacey carbon / copper mesh support grids (Agar Scientific Ltd, UK). Conventional bright field (BF) imaging was performed using a FEI Tecnai TEM operated at 200 kV (Centre for Electron Nanoscopy (CEN), Technical University of Denmark) and selected area electron diffraction (SAED) allowed for phase identification. For off-axis electron holography, a FEI Titan TEM fitted with a C_S corrector on the condenser lens, and a Lorentz lens and biprism, was used at 300kV (CEN).

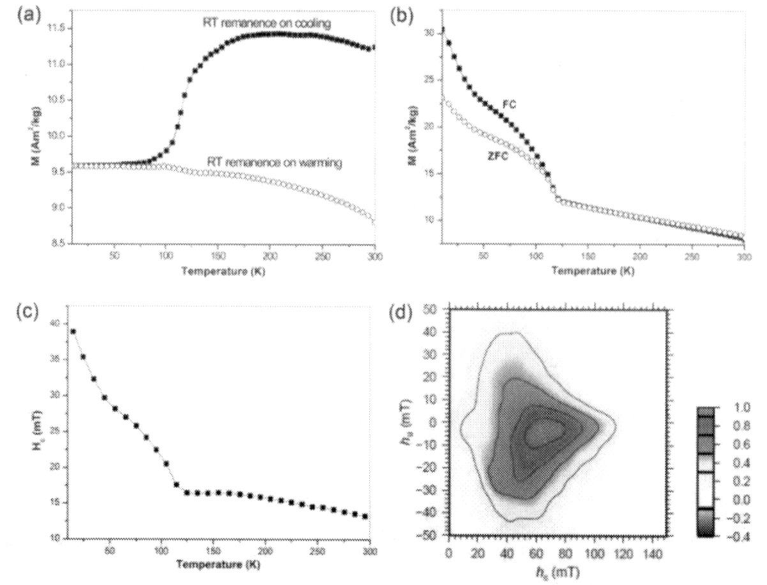

Figure 2 (a) RT-SIRM cycling curves; and (b) ZFC-FC magnetisation measurements of the Fe_3O_4 NPs; (both measured from 10 – 300 K). (c) H_c versus temperature (15-295 K) measurement and (d) normalized ZFC FORC diagram acquired at 15 K (smoothing factor = 2).

For the purpose of magnetic measurements, a dry powdered sample was inserted, compressed and sealed in a plastic pellet and examined at the Institute for Rock Magnetism (I.R.M) at the University of Minnesota, Minneapolis, USA. Low-temperature hysteresis measurements from 15 – 295 K (in 10 K intervals), and first-order reversal curve (FORC), at 15 K, were made using a Princeton Measurements Vibrating Sample Magnetometer (VSM). A Quantum Designs Magnetic Properties Measuring Systems (MPMS) cryogenic magnetometer was used to measure cooling and warming curves of room temperature (RT) induced saturation isothermal remanence magnetisation (SIRM) to identify low-temperature crystallographic transitions. In addition, warming curves of SIRM at 10 K after field cooling (FC) at 2.5 T and zero-field cooling (ZFC) from 300 K to 10 K, were made.

3. Results

The hydrothermal processing of mixed $FeCl_3$ / $FeCl_2$ precursor solution led to the synthesis of a nano-structured Fe_3O_4 reaction product, the evidence for which is now presented in detail.

Fig. 1 provides information on the morphology, size and structure of the synthesised Fe_3O_4 NPs. The BF TEM image of Fig. 1a presents the HS reaction product and is shown to comprise crystalline Fe_3O_4 NPs, as identified by SAED (inset), which exhibit some faceting but are generally isotropic and < 50 nm in size (Fig. 2b). The peaks in the XRD pattern of Fig. 1c are in good agreement with the formation of Fe_3O_4 (Joint Committee on Powder Diffraction standards (JCPDS) reference 88-0315).

The magnetic induction map of an individual Fe_3O_4 NP (~ 45 nm) presented in Fig. 1d, determined from the magnetic contribution to the phase shift reconstructed from electron holograms, depicts the magnetic field contours to be concentrated to – and flow diagonally right and upward through – the centre of the Fe_3O_4 NP, as well as a stray magnetic field expected for a SD particle. Fig. 1e shows magnetic induction map of a group of Fe_3O_4 NPs, displaying the magnetic contours to again flow through the SD particles, and the expected stray magnetic field.

Fig. 2 provides information on the magnetic properties of the Fe_3O_4 NPs. The low-temperature SIRM cycling curves (Fig. 2a) displays behaviour typically reported for PSD magnetite with a Verwey transition of ~ 120 K; however, theoretical models also predict such behaviour for SD particles at this temperature [6]. The drop in the remanence at the Verwey transition suggests that part of the magnetic remanence is a carried by interacting SD and/or small PSD (vortex) domain states.

Low-temperature warming of SIRM to 10K (Fig. 2b), clearly reveals the Verwey transition, and is indicative of near stoichiometric magnetite. The hysteresis parameter of coercivity (H_C) is plotted as a function of temperature (15 K to 295 K) in Fig. 2c and displays H_C behaviour typical of SD or small PSD magnetite [7], i.e., on warming H_C drops at 15K, until the Verwey transition, followed by a gradual drop from 16 to 13 mT on warming to room temperature; such behaviour has been previously predicted [8]. The FORC diagram of Fig. 2d was measured at 15 K after ZFC and exhibited behaviour characteristic of SD grains displaying moderate inter-particle magnetostatic interactions [9].

4. Discussion

This TEM and XRD investigation of the HS of mixed $FeCl_3$ / $FeCl_2$ precursor solution, at pH ~ 12, has provided evidence for the development of Fe_3O_4 NPs that are generally isotropic and < 50 nm in size. A HS reaction temperature of 200°C was observed sufficient to promote the formation of the thermodynamically stable Fe_3O_4 iron oxide phase as the sole reaction product, whilst it is considered that the Fe_3O_4 NP growth mechanism may have involved the growth and dissolution of intermediate iron (oxy)hydroxide phases. Kinematically accessible amorphous $Fe(OH)_3$, poorly crystalline 'hydroxychloride green rust 1': $Fe_{3.6}Fe_{0.9}(O, OH, Cl)_9$ (JCPDS 13-88) sheets and α-FeOOH nanorods are known to precipitate during HS at relatively low temperatures in iron-chloride salt solutions at pH ~ 12 [4, 10], and their dissolution with increasing temperature is considered to supply elemental growth constituents for formation of the more thermodynamically stable Fe_3O_4.

Complementary hysteresis and thermomagnetometry combined with off-axis electron-holography measurements, provided information on the magnetic properties of the Fe_3O_4 NPs. The magnetic induction map of an individual Fe_3O_4 NP (Fig. 1d) demonstrates its SD behaviour, as evidenced by the uniaxial magnetic contours concentrated to the NP centre and stray magnetic field. This is reinforced by the FORC diagram that exhibits behaviour characteristic of interacting single-domain grains. However, the H_c, RT-SIRM and ZFC-FC cycling data reveals behaviour more indicative of PSD/MD domain-states. Considering the HS reaction product is in a powdered form, the PSD/MD-like behaviour is suggested to be caused by inter-particle interaction of the Fe_3O_4 NPs during measurement [11], which is supported by the visualised interactions between the SD particles shown in Fig. 1e and the FORC data (Fig. 2d).

In summary, complementary TEM and XRD have confirmed the HS of SD Fe_3O_4 NPs (< 50 nm in diameter). Off-axis electron holography showed their SD nature and explained the PSD/MD behaviour demonstrated by the bulk magnetic measurements through visualisation of inter-particle interactions.

References

[1] Krasa D, Wilkinson C D W, Gadegaard N, Kong X, Zhou H, Roberts A P, Muxworthy A R and Williams W 2009 *J. Geophys. Res.* **163,** 243.
[2] Dunlop D J and Ozdemir O, 1997 *Rock Magnetism: Fundamentals and Frontiers,* (Cambridge Univ. Press, Cambridge, U.K.)
[3] Harrison R, Dunin-Borkowski R E and Putnis A 2002 *PNAS* **99**, 16556-16561.
[4] Zheng Y, Cheng Y, Bao F and Wang Y 2006 *Mater. Res. Bull.* **41**, 525-529.
[5] Dunin-Borkowski R E, McCartney M R, Frankel R B, Bazylinski D A, Posfai M and Buseck P R 1998 *Science* **282**, 1868-1870.
[6] Muxworthy A R and Williams W 2006 *J. Geophys. Res.***111**, B07103.
[7] Schmidbauer E and Keller R 1996 *J. Magn. Magn. Mater.*,**152**, 99–108.
[8] Carter-Stiglitz B, Jackson M and Moskowitz B 2002 *Geophys. Res. Lett.* **29**, NO. 7, 10.1029/2001GL014197.
[9] Muxworthy A, Heslop D and Williams W 2004 *Geophys. J. Int.* **158,** 888–897.
[10] Almeida T P, Fay M W, Zhu Y Q and Brown P D 2012 *J. Phys. Conf. Ser.* **371,** 012074.
[11] Muxworthy A R, Williams W and Virdee D 2003 *J. Geophys. Res.* **108,** 2517.

Acknowledgements

This project was primarily funded by NERC grant NE/H00534X/1, with additional funding from Institute for Rock Magnetism, University of Minnesota, USA, which is funded by the W. M. Keck Foundation, The National Science Foundation, and the University of Minnesota.

Quantitative electron tomography investigation of a TiO_2 based solar cell photoanode `

G Divitini[1], A Abrusci[2], F Di Fonzo[3], H Snaith[2], C Ducati[1]
[1] Department of Materials Science & Metallurgy, University of Cambridge, Cambridge, CB2 3QZ, UK
[2] Clarendon Laboratory, University of Oxford, Oxford, OX1 3PU, UK
[3] IIT - Istituto Italiano di Tecnologia, Via Pascoli 70/3, I-20133 Milan, Italy

E-mail: gd322@cam.ac.uk

Abstract. The development of efficient thin film solar cells requires a deep knowledge of the nanoscale morphology of the active layers. While conventional investigation is usually limited to 2D information, here we use electron tomography to unravel a complex particle network in a non-ambiguous, 3D reconstruction. We present our study of a dye sensitised solar cell, based on a nanostructured TiO_2 photoanode produced by pulsed laser deposition (PLD) and displaying a hierarchical, quasi-1D arrangement. We prepare the sample for electron tomography using focused ion beam (FIB) milling to obtain a micro-pillar, instead of a conventional TEM lamella. This approach has the advantage of allowing higher quality tomographic reconstructions of complex morphologies due to the increased tilt range available and the constant thickness of the section. We analyse the resulting reconstruction to quantitatively investigate the geometry of the TiO_2 network. We compare the findings with a photoanode based on a conventional TiO_2 paste, determining the anisotropy of the PLD-grown film. To complement our nanoscale TEM characterization, we also employ FIB tomography, to obtain a complete structural characterisation of the photoanode at different length scales.

1. Introduction

The use of solar power is being widely investigated to address the ever-increasing need for energy of modern society in a sustainable way. Research groups all over the world are looking into developing photovoltaic devices which could compete with the established silicon-based technology while offering some benefits, particularly in the energy footprint of the manufacturing, a lower cost or more versatility of use. Dye sensitised and polymer heterojunction cells, in particular, are experiencing a new surge of interest, trigged by the development of novel cobalt-based redox couples which led to an increase in power conversion efficiency above 12% [1, 2]. In order to push the conversion efficiency even further, the structure of the electrode needs to be finely tailored at the nanoscale: the interplay between surface area, crystal structure, defects and electron transport properties is of paramount importance for the behaviour of the macroscopic device [3]. The optical absorption of the device can also be enhanced by exploiting light scattering induced by mesostructures in an appropriate size range (normally hundreds of nm). A deep understanding of the material morphology can be obtained with investigation through electron microscopy, with electron tomography being the technique of choice for obtaining three-dimensionally resolved information on assemblies of nanoparticles.

We investigate the potential of electron tomography on a prototype of a dye sensitised solar cell with a polymer hole transporter (P3HT). The electron acceptor, TiO_2, is produced by Pulsed Laser Deposition (PLD) [4], where a TiO_2 target is ablated by nano-second Nd:Yag laser pulses (4 J/cm^2 energy density) in a low-pressure oxygen atmosphere (7 Pa). The material experiences a supersonic expansion, followed by deposition on a substrate. The resulting films retain a high porosity and self-assemble in quasi-1D hierarchical structures of branched columns. Similar structures have been demonstrated to be promising when employed as photoanode for a dye sensitised solar cell [5]. For comparison, we also analysed a prototype with a TiO_2 paste prototype from commercial nanoparticles (Dyesol 18NR-T). We will refer to the PLD-deposited sample as "H" (Hierarchical) and to the TiO_2 paste sample as "P" (Paste).

The morphology of the photoanode was studied using two tomographic approaches - electron tomography in a transmission electron microscope and slice & view tomography in a Focused Ion Beam (FIB), providing information on different size ranges.

2. Methods

Electron tomography was carried out in a FEI Tecnai F20 TEM using HAADF STEM imaging on a sample prepared via Focused Ion Beam (FIB) milling. The specimen was cut into a micropillar geometry in order to provide a high quality reconstruction [6-8], since in this case the projected thickness is constant at all tilt angles. We used a FEI Helios NanoLab for the specimen preparation, shown in Figure 1. For each sample, a tilt series was acquired from -76° to +76°, using a Fischione 2010 on-axis holder (sample H) and a Fischione 2020 tomography holder (sample P).

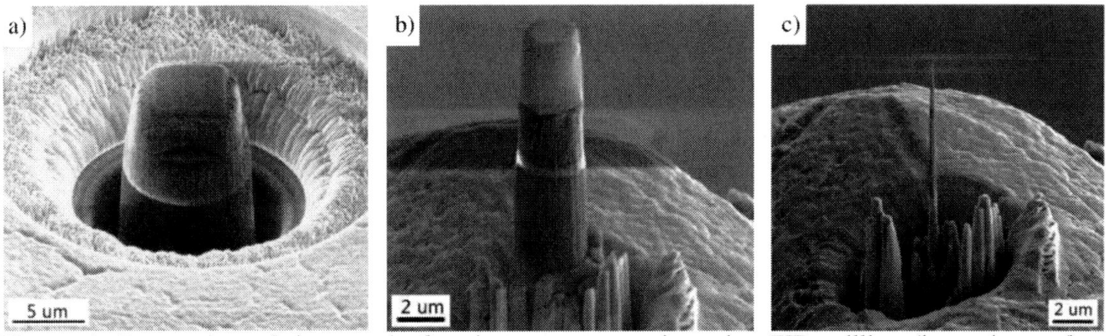

Figure 1. Preparation of the sample for electron tomography using FIB milling (here shown for sample H). A cylindrical part of the specimen is isolated (a) and moved to a mount compatible with a sample holder (a tomography post or an Omniprobe grid) (b). The needle is thinned by progressive milling of annular regions until electron transparency, corresponding to 2-300 nm for this material (c).

The tilt series were aligned and the original volume was reconstructed using FEI Inspect3D. ImageJ was used to analyse the result, extracting information on particle size, porosity and network structure [8].

In order to obtain morphological information on a larger length scale, the electron tomography characterisation was integrated with "slice & view" FIB tomography carried out in a FEI Helios NanoLab. The volume under analysis was 3.32 μm (sample P) and 4.85 μm (sample H) wide and 1.52 mm (P) and 2.56 mm (H) deep in the direction parallel to the "slicing". The larger volume in sample H was chosen to better sample the morphology, which is less homogeneous than in sample P. Avizo 7 was used to align the SEM images and segment the resulting volume, which was then processed using ImageJ.

3. Discussion

3.1 Electron tomography

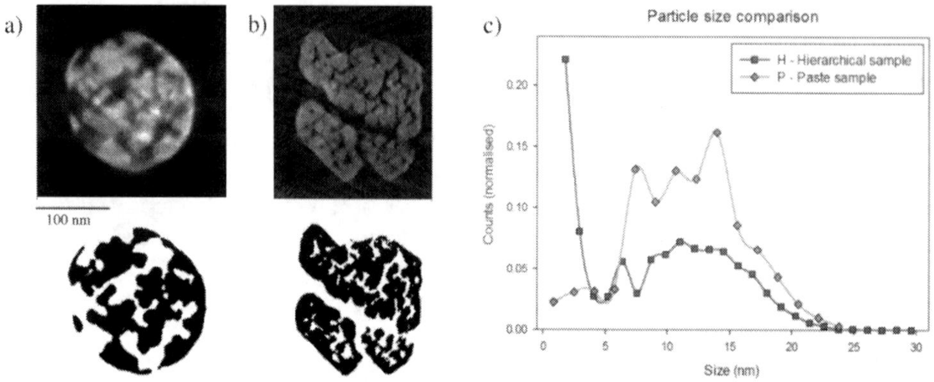

Figure 2. Orthoslices before and after segmentation for samples P (a) and H (b) from the reconstructed volumes, taken along the micropillar axis. The width of the pillar is here 180 nm (P) and 210 nm (H). Channels can be seen in sample H between different columns. (c) Comparison of particle size, obtained through iterated erosion in 3D. The high counts for small particles in sample H are attributed to noise in the reconstructed volume.

Although the two systems present similar particle size (see Figure 2) and crystallinity, their arrangement is different. While sample P presents a homogeneous network of spherical particles, features in sample H resemble conical columns which tend to align at a constant angle to the growth axis of the structure. Some gaps between different structures or between branches of the same structure, as shown in figure 2, run through the film. Porosity and surface area measured from the reconstruction are shown in Table 1, along with the light conversion efficiency measured for the complete device: sample P has higher porosity and surface area. This suggests that the combination of lower porosity and the hierarchical ordering favour electron transport, and enhanced scattering increases light absorption, compensating for the lower surface area and resulting in a better light conversion efficiency.

Table 1. Comparison of performance and morphology for the two solar cell prototypes.

Specimen	Porosity	Surface area	Light conversion efficiency (η)
Paste	0.483	88.5 m^2/g	2.64 %
Hierarchical	0.354	41.2 m^2/g	2.83 %

3.2 FIB Slice & View

FIB tomography was carried out on both samples. Differences at the length scale of 100-1000 nm can be seen. As expected, sample P has a homogeneous structure over several microns, while sample H shows a semi-regular arrangement of columnar hierarchical structures with side branches. To quantify the properties of the pore space in the two samples, the area coverage (defined as the ratio between TiO_2 and total area) and the number of pores per slice parallel to the electrode surface have been calculated (Figure 3). Sample H presents a higher area coverage. Sample P seems to become more compact away from the FTO substrate, although the effect is small (a change from 0.90 to 0.93) and might be due to artefacts or to the film becoming more compact upon polymer infiltration. The sharp decrease in coverage far from the substrate corresponds to the top surface of the nanostructured film. The number of pores is constant through the thickness of sample P, confirming its homogeneity. Sample H has a more complex behaviour: close to the electrode (thickness < 200 nm) the number of pores decreases as the bases of the hierarchical columns are progressively more densely packed; then the film presents an intermediate region in which the structures are closely packed and the number of pores is roughly constant. Finally, in the 500-800 nm close to the surface, side branches open out, with an obvious increase in pore number.

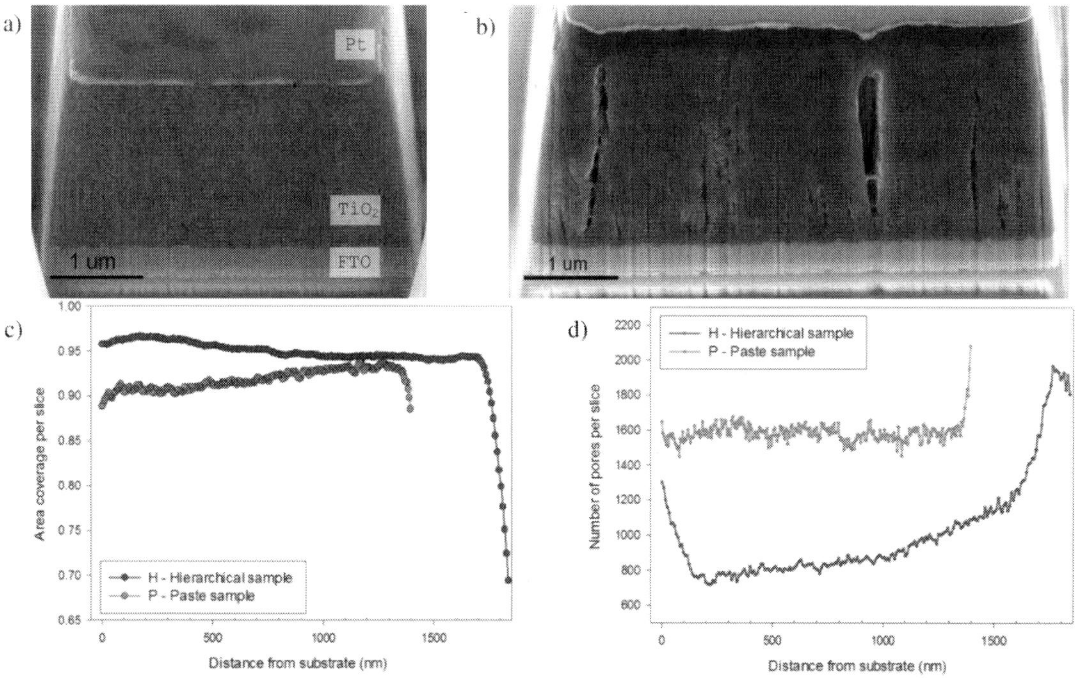

Figure 3. FIB tomography. (Top) Comparison of slices from sample P (a) and H (b). The substrate is FTO-coated glass, the active layer is protected by platinum for FIB processing. (Bottom) Area coverage (c) and number of pores (d) in function of the distance from the FTO layer.

4. Conclusions

The combination of electron and FIB tomography is a very powerful tool for the characterisation of increasingly popular composite structures, where both nano- and meso-scale need to be analysed quantitatively. The approach shown here for solar cell photoanodes can be extended to other fields of nanoscience, such electrochemical cells and photocatalytic devices.

Acknowledgements

The authors acknowledge funding from the ERC under grant number 259619 PHOTO EM.

References

[1] Hardin B E, Snaith H J and McGehee M D 2012 *Nature Photonics* **6** 133
[2] Yella A, Lee H-W, Tsao H N, Yi C, Chandiran A K, Nazeeruddin M, Diau E W-G, Yeh C-Y, Zakeeruddin S M and Grätzel M 2011 *Science* **334** 629
[3] Barbé C J, Arendse F, Comte P, Jirousek M, Lenzmann F, Shklover V and Grätzel M 1997 *Journal of the American Ceramic Society* **80** 3157
[4] Di Fonzo F, Casari C S, Russo V, Brunella M F, Li Bassi A and Bottani C E 2009 *Nanotechnology* **20** 015604
[5] Sauvage F, Di Fonzo F, Li Bassi A, Casari C S, Russo V, Divitini G, Ducati C, Bottani C E, Comte P and Grätzel M 2010 *Nano Letters* **10** 2562
[6] Biermans E, Molina L, Batenburg K J, Bals S and Van Tendeloo G 2010 *Nano Letters* **10** 5014
[7] Kawase N, Kato M, Nishioka H and Jinnai H 2007 *Ultramicroscopy* **107** 8
[8] Divitini G, Stenzel O, Ghadirzadeh A, Guarnera S, Russo V, Casari C S, Li Bassi A, Petrozza A, Di Fonzo F, Schmidt V and Ducati C 2013 - *submitted*

Three-dimensional electron diffraction mapping for cyclic twinned nanostructures

Xin Fu[1] **and Jun Yuan**[2]

[1] General Research Institute for Nonferrous Metals, Beijing, 100088, P.R. China
[2] Department of Physics, University of York, York YO10 5DD, United Kingdom

Email: fuxints@gmail.com

Abstract. In recent years, the information retrieval of three-dimensional (3D) intensity distribution in reciprocal space has attracted much attention in nanometrology research. However, the development of 3D analysis of electron diffraction intensity including crystal structural identification and morphology determination of nanocrystals have so far focused on single crystalline nano-objects. Using boron carbide five-fold twinned nanowires as a model system for nanostructured nano-objects, we demonstrated that such method can also identify the internal structure of cyclic twinned nanostructures. Here, the methodology of the 3D mapping of electron diffraction from such nanowires will be presented and the approach is compared with the conventional method of axial rotation electron diffraction analysis to demonstrate its comprehensive, quantitative and non-destructive nature. We believe that such approach can be easily extended to the investigation of internal structures of other polycrystalline nanostructures.

1. Introduction

Cyclic twinned nanostructures, such as five-fold twinned nanoparticles and nanowires, have drawn much research interests due to the unusual internal structure and the unexpected mechanical properties [1-4]. A direct method of characterization for these nanostructures is through cross-sectional observation by transmission electron microscopy (TEM) [2]. However, it typically requires preparation of ultrathin cross-sectional samples either by microtome or focused ion beam (FIB), both of them inevitably introduce sample damage and creating difficulty in identifying the intrinsic structural defects. The real-space electron tomography is non-destructive but suffers from a 'missing wedge' problem [5], and it cannot be routinely used to determine the crystallography of the internal structure. Besides the real-space imaging methods, electron diffraction analysis has also been employed to investigate the structural feature of the complex nanostructures. The conventional approach is to acquire a set of high symmetry zone-axis related electron diffraction patterns from some specific orientations [3, 4]. However, there is a wealth of structural information contained in the relative position and intensity of diffraction spots. 3D analysis of electron diffraction intensity distribution aims to fully exploit such information and has been applied in crystal structure identification and morphology determination of nanocrystals, so far all focusing on single crystalline nano-objects [6, 7]. Here, we demonstrated the approach and advantages of the 3D electron diffraction mapping method for revealing the internal structure of cyclic twinned nanostructures and contrast this with the conventional approach using the axial rotation electron diffraction analysis.

2. Axial rotation electron diffraction analysis

For boron-rich five-fold twinned nanowire model system, an ideal structure has been proposed [3, 4]. It consists of five identical crystalline segments sharing a common $[001]_r$ (*r* refers to the rhombohedral representation of the unit cell) axis along the nanowire growth direction and joined together by $\{100\}_r$ twinned planes. Fig. 1a shows the cross-sectional view where five constituent segments are marked with different colors and identified from T1 to T5 for convenience. To identify this twinning structure experimentally, at least three different characteristic diffraction patterns, 18 degrees apart, must be recorded in the conventional method of axial rotation electron diffraction analysis, as shown in Fig. 1. This is consistent with the five-fold pseudo-symmetry of such nanostructures (Fig. 1a and 1e).

For example, if the nanowire is initially oriented in such a way that a mirror-symmetric diffraction pattern can be obtained as shown in Fig. 1c, then the nanowire can be rotated along its own long axis (RT in Fig.1a) either clockwise or anticlockwise by the angle of 18°, and two additional characteristic diffraction patterns with one of the crystalline segments at a $[1\bar{1}0]_r$ zone-axis orientation should be seen. Comparison with simulated diffraction pattern using the idealized structural model suggests that in this case T2 is oriented along the $[1\bar{1}0]_r$ zone-axis in Fig.1b, and T5 is approximately oriented along the $[1\bar{1}0]_r$ zone-axis in Fig.1d. If the nanowire is oriented such that one segment is already close to the $[1\bar{1}0]_r$ zone-axis, as shown in Fig. 1g which is in the same orientation as that shown in Fig. 1b and Fig. 1d, then the nanowire can be rotated clockwise and anticlockwise by 18° along its long axis to observe two mirror-symmetric diffraction patterns respectively. Because of the rhombohedral structure, further tilting the nanowire by approximate 7°, along the direction perpendicular to its long axis, will be required to reveal the mirror-symmetric diffraction patterns with one segment oriented to $[\bar{1}\bar{1}1]_r$ zone-axis as shown in Fig. 1f and Fig. 1h. As a result, such diffraction analysis is complex when applied to an unknown structure. In addition, the large angle of rotation involved is also practically inconvenient for most analytical TEMs where the tilting ranges of the specimen holders are restricted to ±20°or ±30° due to the limited space inside the pole piece of the objective lens. If the initial orientation of the nanowires is too far off the orientation as shown in Fig. 1c or Fig. 1g, not all three characteristic diffraction patterns, Fig. 1b-1d or Fig. 1f-1h, may be accessible experimentally. In addition, the diffraction information at a large range of orientations is not effectively used because of their complex and low-symmetry features.

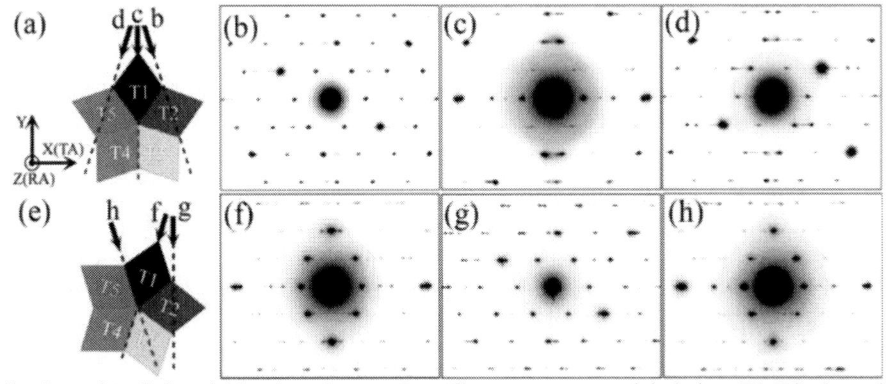

Fig.1 Identification of B_4C five-fold twinned nanowires (five segments are labeled as T1-T5) through electron diffraction analysis under the axial rotation conditions. (a) and (e) illustrate the 18° orientation relationship of the nanowire with the incident electron beam (black arrows) viewed down the nanowire's axis (i.e. the $[001]_r$ direction). The corresponding diffraction patterns shown in (b)-(d) respectively are obtained by successive 18° rotations along the nanowire's axis (RT as shown in (a)). The same approach is used to generate (f)-(h), except that (f) and (h) has been tilted additionally approximate 7° along an axis perpendicular to the nanowire's axis after axial rotation, because of the low symmetry rhombohedral unit cell, to reveal $[\bar{1}\bar{1}1]_r$ zone-axis diffraction from T1 and T3 segments respectively.

3. Three-dimensional (3D) electron diffraction mapping

The essence of 3D electron diffraction mapping is to recreate the 3D intensity distribution of specific reflections in reciprocal space. Experimentally, this is achieved by recording a series of systematic tilting two-dimensional electron diffraction patterns and then using those as the basis for the following 3D reconstruction of the specific reflection.

For the B_4C five-fold twinned nanowire, we start by orienting the nanowire such that the diffraction pattern shown in Fig. 1c is generated and a series of systematic tilting diffraction pattern is collected by either tilting the nanowires or the incident electron beam along the tilting axis (TA, i.e. X axis shown in Fig. 2a), with step of 1° and an overall tilting range about ±10°. The recorded diffraction patterns share a common feature of a set of diffraction spots along an invariant line (perpendicular to the mirror symmetry plane of the nanowire) which is parallel to the tilting axis (TA) as shown in Fig 2c. The invariant line consists of a systematic row of spots due to diffraction by a set of $(1\bar{1}0)_r$ planes of T1 and a set of $(100)_r$ planes ($(0\bar{1}0)_r$ planes) of T3 (T4) (Fig. 2b). As shown in Fig. 2c, each of the recorded diffraction image represents the intersection of the plane-like segment of the Edward sphere, associated with the 200kV electron beam, with the reciprocal lattice structures of the crystalline components of the nanowires and as such that the raw data of the series of systematic tilting electron diffraction patterns samples the 3D reciprocal space, although with irregularly shaped volume element.

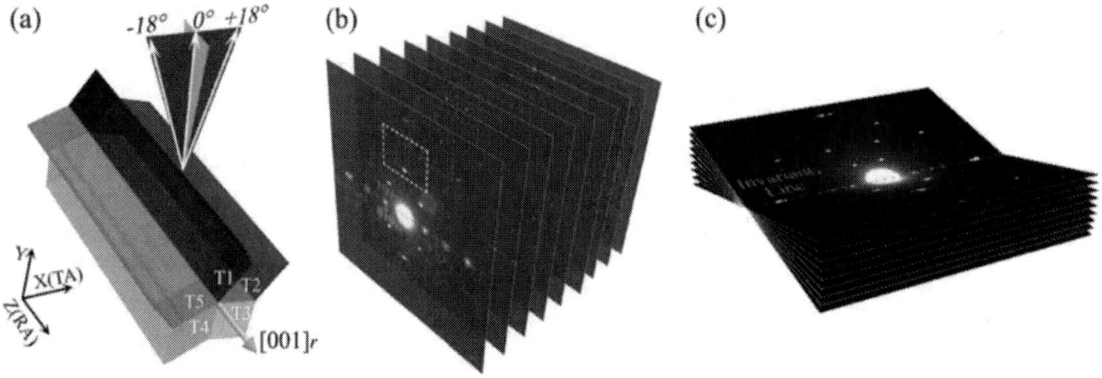

Fig. 2 (a) Schematic view of the orientation relationship between a B_4C five-fold cyclic twinned nanowires and the incident electron beam. The grey plane represents the tilting range of the incident electron beam in the 3D electron diffraction mapping. The arrows in the brown plane indicate the direction of the incident electron beams for the axial rotation electron diffraction. (b) The stack of the collected systematic-tilting diffraction patterns. The dashed rectangle indicates the data extracted area which contains the intensity information of $(112)_r$ and $(113)_r$ reflections. (c) The orientation relationship of the collected diffraction patterns for 3D reconstruction of specific reflections in reciprocal space.

To represent the diffraction intensity properly in the reciprocal volume, we need to convert the experimental raw dataset to a 3D representation in reciprocal space with uniform voxels. The schematics of the post-processing involved are shown in Fig. 3a-3b. We extracted a series of 2D intensity data from a rectangular area with the same pixel size and offset the same distance from the invariant line of every diffraction pattern (Fig.2b). In reciprocal space, the extracted 2D slices form a volume with a sectorial cross section viewing along the invariant line as shown in Fig.3a. We obtained a 3D dataset with uniform voxels as illustrated in Fig.3b through interpolation.

Following this approach, a 3D mapping of diffraction intensity distribution, including the $(112)_r$ and $(113)_r$ reflections of the B_4C five-fold twinned nanowire, has been reconstructed [8]. Fig. 3c presents the reciprocal space volume containing the five $(113)_r$ reflections of the B_4C five-fold twinned nanowire. This intensity distribution directly exhibits the cyclic twinning structure, confirming the overall crystallographic feature of the ideal model. However, detailed analysis [8] shows that the five reflections are not identical in their intensity distribution nor in their angular positions, both reflecting the inhomogeneous distribution of strain and microtwin defects within the five segments. As the nanowire has not been mechanically strained in this method, we have revealed the inhomogeneous nature of the intact B_4C five-fold twinned nanowire. The quantitative nature of the resulting diffraction intensity distribution also means that similar dataset of sufficient fidelity together with the use of high coherent field emission source also opens up the further possibility of diffractive imaging [7] to reveal the real-space distribution of the internal nanostructures.

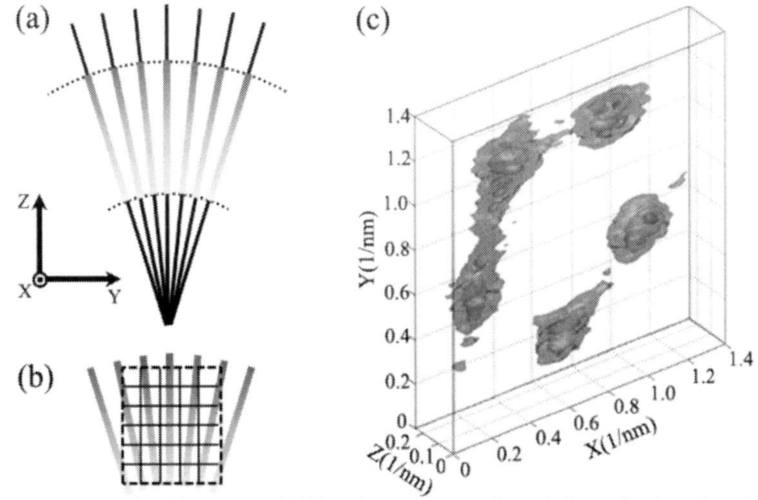

Fig. 3 (a)Schematic view of the tilt series of diffraction patterns aligned along the invariant line (parallel to the X axis, i.e. the tilting axis). The black line with a colored section represents a slice of diffraction pattern. The colored section symbolically indicated the area of interest in the raw data. (b) illustrated the mathematical process applied to raw dataset within the area of interest. (c) The 3D intensity distribution of $(113)_r$ reflections of a B_4C five-fold twinned nanowire.

4. Conclusions

We demonstrated the approach of 3D electron diffraction mapping for cyclic twinned nanostructures. In contrast to the conventional axial rotation electron diffraction analysis, this approach can effectively use the electron diffraction information within a narrow tilting range, so can be used in most analytical microscopes. Additionally, the retrieved quantitative 3D diffraction intensity distribution can provide comprehensive information about the internal structure of such nancomplex, such as the twinning nature, strain relaxation mechanism and the defect structures.

Acknowledgement

This research is supported by National Natural Science Foundation of China (No. 51201015).

References

[1] Wu B, Heidelberg A, Boland J J, Sader J E, Sun X M and Li Y D 2006 *Nano Lett.* **6** 468-472
[2] Chen H Y, Gao Y, Zhang H R, Liu L B, Yu H C, Tian H F, Xie S S and Li J Q 2004 *J. Phys. Che. B* **108** 12038-12043
[3] Jiang J, Cao M H, Sun Y K and Yuan J 2006 *Appl. Phys. Lett.* **88** 163107
[4] Fu X, Jiang J, Liu C and Yuan J 2009 *Nanotechnology* **20** 365707
[5] Midgley P A and Dunin-Borkowski R E 2009 *Nat. Mater.* **8** 271-280
[6] Mugnaioli E, Gorelik T and Kolb U 2009 *Ultramicroscopy* **109** 758-765
[7] Dronyak R, Liang K S, Tstai J-S, Stetsko Y P, Lee T-K and Chen F-R 2010 *Appl. Phys. Lett.* **96** 221907
[8] Fu X and Yuan J 2013 *Nanoscale* first published online DOI:10.1039/c3nr01839c

Characterisation of Nanoparticle Structure by High Resolution Electron Microscopy.

Robert D. Boyd[1], Rickard Gunnarsson, Iris Pilch and Ulf Helmersson

Plasma and Coatings Physics Division, Department of Physics, Chemistry and Biology (IFM), Linköping University, Linköping, SE-581 83, Sweden

E-mail: robbo@ifm.liu.se

Abstract. Whilst the use of microscopic techniques to determine the size distributions of nanoparticle samples is now well established, their characterisation challenges extend well beyond this. Here it is shown how high resolution electron microscopy can help meet these challenges. One of the key parameters is the determination of particle shape and structure in three dimensions. Here two approaches to determining nanoparticle structure are described and demonstrated. In the first scanning transmission electron microscopy combined with high angle annular dark field imaging (HAADF-STEM) is used to image homogenous nanoparticles, where the contrast is directly related to the thickness of the material in the electron beam. It is shown that this can be related to the three dimensional shape of the nano-object. High resolution TEM imaging, combined with fast Fourier transform (FFT) analysis, can determine the crystalline structure and orientation of nanoparticles as well as the presence of any defects. This combined approach allows the physical structure of a significant number of nano-objects to be characterised, relatively quickly.

1. Introduction

Nanoparticles are generally defined as those particles which have at least one dimension in the 1 - 100 nm range. Such materials have been produced and used for many hundreds of years. Medieval red stainless glass windows are dependent on colloidal gold nanoparticles and the distinctive dark hair colour dye used by the ancient Egyptians had nanoparticles of lead sulfide as its core ingredient. Interest in engineered nanoparticles as the basis for new and improved products has increased strongly since their emergence in the 1970s [1]. They effectively act as a bridge between the bulk properties of materials and the behaviour of individual atoms and molecules and as such often display unique properties. For example, the optical properties of gold nanorods are highly dependent on their aspect ratio (with colours ranging from red to dark purple) and could be incorporated into thin films for new optical devices. [2]. Recently concerns have been raised regarding the short and long term health and environmental impact of such materials [3].

Parameters which define the properties of nanoparticle production batches include the size distribution, shape, crystal structure (different outermost crystal planes will behave differently) and chemical composition and homogeneity. To ensure the continued innovation and safety of nanoparticles and their related products nanoparticle production batches must be fully and accurately characterised.

Here it will be demonstrated how all of these parameters can be determined in an effective way using conventional and scanning transmission electron microscopy, with a sample of titanium dioxide nanoparticles produced in the gas phase.

2. Experimental

A novel method of producing nanoparticles in the gas phase has recently been developed as described in reference [4]. Briefly titanium dioxide nanoparticles were prepared by sputtering a titanium hollow cathode in a reactive atmosphere containing oxygen. These particles were deposited directly onto a silicon wafer coated with gold (200 nm) with an adhesion layer of titanium (2.5 nm). Once deposited onto the substrate the particles were transferred onto gold TEM grids coated with a thin film of amorphous carbon by gently wiping the substrate and grid together.

All transmission electron microscopy (TEM) and energy dispersive X-ray spectroscopy (EDS) measurements were taken using a FEI Tecnai G^2 TF20 UT microscope equipped with a field emission gun operating at a voltage of 200 kV with an emission current of 38 μA. Scanning transmission electron microscopy images were taken with a high angle annular dark field (STEM-HAADF) detector with a camera length of 220 mm. All the measurements were taken in a single experimental session within a 2 hour period.

3. Results and Discussion

The nature of nanoparticles means that determining the size distribution is critical for understanding their overall properties. Due to the high resolution capabilities of transmission electron microscopy, it is now routinely used to determine the size distribution of nanoparticle samples However the contrast in conventional bright field is strongly affected by the crystalline orientation of the sample. This can lead to incorrect results when analysing polycrystalline particles. This is not the case for HAADF-STEM, where contrast is primarily affected by the average atomic number and thickness of the sample. A HAADF-STEM image of the produced particles is shown in Figure 1(a). The particles are widely dispersed across the TEM foil with the individual particles clearly visible. Rather than the traditional view of nanoparticles as being spherical, here the particles appear highly faceted with either four or six visible faces. The latter also appear to be anisotropic. Using automatic image analysis using the ImageJ software package the size distribution can be determined [5]. A common method of doing this is to record the minimum and maximum Feret dimensions (Figures 1(b and c)), which is the maximum and minimum calliper distances respectively for each particle [6]. The minimum Feret distribution shows a high monodisperse distribution with an average particle diameter of 25 nm. The larger particle sizes measured (above 60 nm) are primarily due to agglomerated particles being detected as a single particle. The maximum Feret distribution is more widely dispersed, reflecting the anisotropicity of some of the particles. Another advantage of the HAADF-STEM approach for homogeneous materials is the fact the image intensity is directly related to the thickness, for very small or thin speciemens. This provides a route for determining the thickness distribution for the particles without the need for expensive and time consuming tomography measurements [7]. The HAADF image intensity for each particle is plotted against the minimum Feret dimension and shown in Figure 1(d). It can be seen that there is a strong linear relationship between the two, indicating that the particle grow uniformly and that they have a uniform chemical composition.

Journal of Physics: Conference Series **522** (2014) 012065 doi:10.1088/1742-6596/522/1/012065

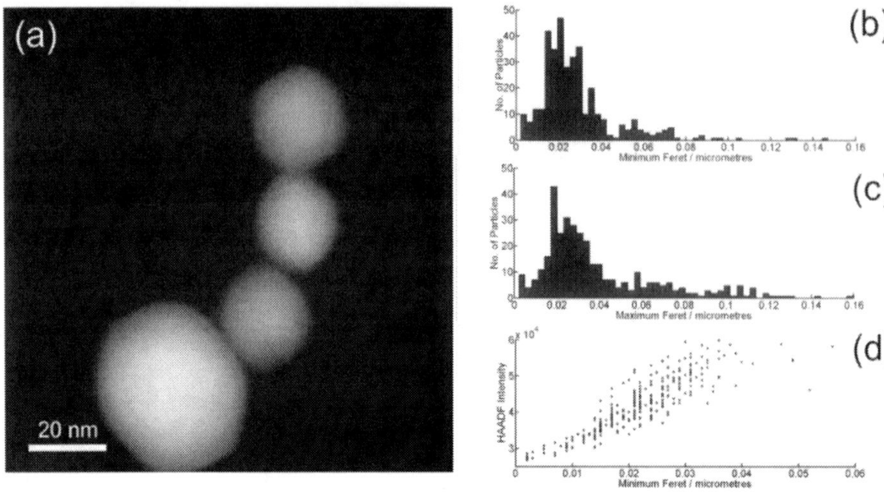

Figure 1: (a) HAADF-STEM image of obtained nanoparticles. (b) and (c) minimum and maximum Feret diameter distributions respectively and (d) the variation HAADF image intensity with Feret dimension.

High resolution electron microscopy images of the particles were used to determine the crystalline structure. Typical high resolution images of particles with different orientations are shown in Figures 2(a) and (c). The crystalline lattice is clearly resolvable. Processing of the high resolution images with 2D fast Fourier analysis produces a pattern which can be indexed to various crystalline structures Figures 2(b) and (d). A close fit was obtained to the rutile titanium dioxide structure, indicating the primary composition of the particles. This was confirmed with X-ray diffraction (data not shown). If a significant number of the particles contain a number of crystalline defects then this will also be revealed. This is shown in Figure 2(e) where a particle with the same orientation as that shown in Figure 2(a) is shown, containing a large number of either twin or stacking faults.

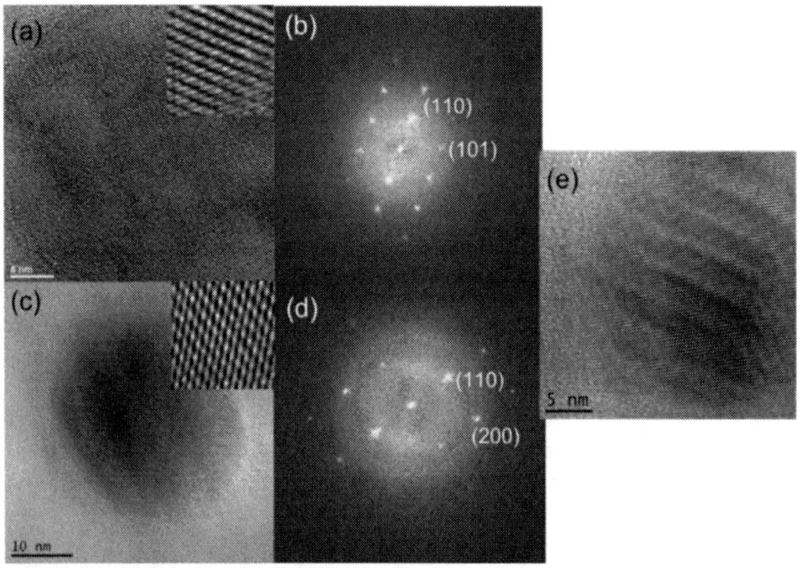

Figure 2: Bright field TEM analysis of individual particles. (a) and (c) high resolution TEM images and (b) and (d) their respective FFT patterns which could be fitted to a rutile TiO_2 structure as indicated and (e) a high resolution image showing crystalline defects within the particle.

As expected from the previous measurements, chemical analysis of the particles by EDS revealed the presence of titanium and oxygen (Figure 3(a)). Chemical line profiles across several particles (Figure 3(b)) confirms that both (and only) of these elements are associated with the particles. This allows the composition of individual particles to be determined and, as expected, remains consistent.

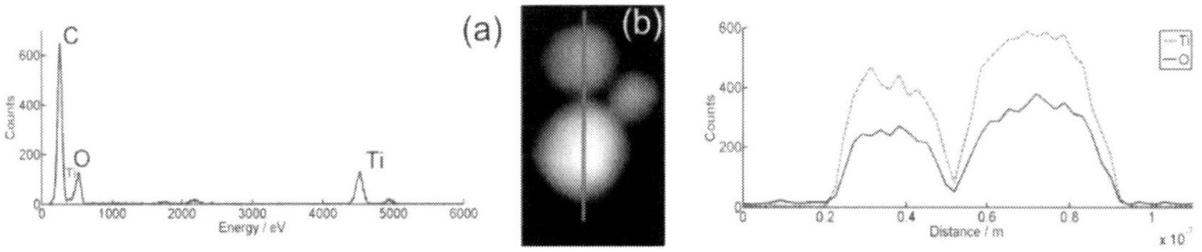

Figure 3: EDS analysis of the nanoparticles. (a) Spectra showing the presence of oxygen and Titanium along with carbon from the substrate and contamination. (b) HAADF image of several nanoparticles and the associated EDS line scan.

4. Conclusions

A nanoparticle sample produced by a novel process has been analysed using (scanning) transmission electron microscopy. In a single experimental session enough data was obtained to determine their dimensional distributions (length, width and thickness). In addition, two particular shapes of particles were described. The crystalline structure was determined and confirmed by chemical analysis which also confirmed the homogeneity of the particles.

5. Acknowledgements

This work was financially supported by the Knut and Alice Wallenberg Foundation under contract KAW 2012.0083.

References

[1] Kreuter J *Int J Pharm.* 2007 **331(1)** p1
[2] Pérez-Juste J, Pastoriza-Santos I, Liz-Marzán LM, Mulvaney P. *Coordination Chemistry Reviews* 2005 **249(17–18)** p1870–1901.
[3] Herve-Bazin B. *Ann Chim-Sci Mater.* 2006 **31(3)** p339
[4] Pilch I, Söderström D, Brenning N, Helmersson U. *Appl Phys Lett.* 2013 **102(3)** p033108.
[5] Abramoff M, Magelhaes P, Ram S. *Biophotonics Int.* 2004 **11(7)** p36.
[6] Allen T. 1999 *Particle size measurement. 5th ed.* (Dordrecht : Kluwer)
[7] Boyd RD, Young TJ, Stolojan V. *J Nanoparticle Res* 2012 **14(4)** p797

TEM and EELS studies of microwave-irradiation synthesis of bimetallic platinum nanocatalysts

N R Mathe[1, 2], N J Coville[2], J E Olivier[3], M R Scriba[1]

[1] National Centre for Nano-Structured Materials, CSIR, P.O. Box 395, Pretoria, 0001, South Africa.
[2] Molecular Sciences Institute, School of Chemistry, University of the Witwatersrand, Private Bag 3, Wits, 2050, Johannesburg, South Africa.
[3] Centre for HRTEM, Nelson Mandela Metropolitan University, P.O. Box 77000, Port Elizabeth, 6031, South Africa
Email: nmathe@csir.co.za

Abstract: Microwave-irradiation (MW) synthesis of nanostructured materials provides for the synthesis of metal nanoparticles, using fast and uniform heating rates. This procedure affords better control of the shape and size of the nanoparticles when compared to conventional methods. In this work, microwave-irradiation was used to produce platinum-cobalt (Pt-Co) and platinum-nickel (Pt-Ni) nanoparticles for use as electrocatalysts in the methanol oxidation reaction. High resolution TEM imaging and EELS studies revealed that these bimetallic nanoparticles form islands or hetero-structures.

1. Introduction

Nanoparticles are of interest due to their enhanced electronic, magnetic and catalytic properties compared to bulk structures, mainly due to their relatively larger surface to volume ratio and consequent increased surface activity [1]. The preparation of nanoscale metal clusters and colloids with desirable properties presents a significant scientific challenge. In order to understand the architecture in bimetallic nanoparticles, it is important to synthesize the different structures, i.e. heterostructures, alloys and core-shell structures. Conventional synthesis procedures involve the use of volatile organic solvents with harmful effects on the environment and human health [2]. Therefore, the search for alternative synthesis procedures using uniform heating methodologies has become a high priority.

New synthesis approaches include microwave (MW)-irradiation [3]. Microwave heating results from dipole-dipole interactions between the polar molecules in the solution, and the electromagnetic field. The main advantages of the microwave-assisted reactions over conventional methods are; (a) the kinetics of the reaction be can increased by the 1 – 2 orders of magnitude, (b) novel phases can be formed, (c) the initial heating rate is rapid and can lead to energy saving, and (d) the selective formation of specific phases [4]. This MW synthesis approach is a versatile, cost-effective, energy efficient {Grace, 2007 #1} method with potential for mass production. In terms of the use of MW irradiation for nanoparticles, Bensebaa et al.[5], synthesized polymer-stabilized Pt-Ru nanoparticles using this approach. In this work, Pt-Ni and Pt-Co nanoparticles were synthesized using a MW process, which have both shown good activity towards the methanol oxidation reaction (MOR) in fuel cells.

Journal of Physics: Conference Series **522** (2014) 012066 doi:10.1088/1742-6596/522/1/012066

2. Synthesis Procedure

Materials: The samples were prepared using metal precursor salts ($H_2PtCl_6.6H_2O$, M = $NiCl_2.6H_2O$ and $CoCl_2.6H_2O$) and ethylene glycol as both the solvent and reducing agent. Polyvinylpyrrolidone (PVP) was used as the capping agent and Pt:M was added at a molar ratio of 1:1.75. All the chemicals were obtained from Sigma Aldrich$^{(R)}$, and used as received.
Synthesis: Microwave synthesis was performed using an Anton Paar M300 reactor (temperature = 200 °C, power = 750 W, time = 30 min). The Pt-Co nanoparticles were synthesized using a seeding method, i.e. Pt seeds were synthesized first, filtered and dried. Then the Co shells were grown on the surface of the platinum. Pt-Ni nanoparticles were synthesized using a co-reduction procedure in the microwave oven, i.e. the two metals salts are reduced simultaneously in solution.
Characterization: The morphology of the Pt-Co and Pt-Ni nanoparticles were determined using transmission electron microscopy (TEM) on JEOL JEM-2100 microscope. Electron energy loss spectroscopy (EELS) and scanning transmission electron microscop (STEM) were performed on a JEOL ARM 200F, both at 200 kV.

3. Results

3.1. TEM analysis

In the lower magnification TEM images (figure 1 a-b), the Pt-Co nanoparticles have a homogeneous, spherical structure with observed lattice fringes. However, as indicated by the arrows in figure 1(b), lattice information was observed on the material underneath the nanoparticles. This "layer" structure was suggested as Co sheets, also observed by Shin et al. [6] for Co nanoparticles synthesized in supercritical methanol. The crystalline nature of the nanoparticles is seen in figure 1c, with the cubic lattice structure confirmed by the Fast Fourier Transform (FFT) spectrum; figure 1(d), of a single nanoparticle (dashed box in figure 1(c).

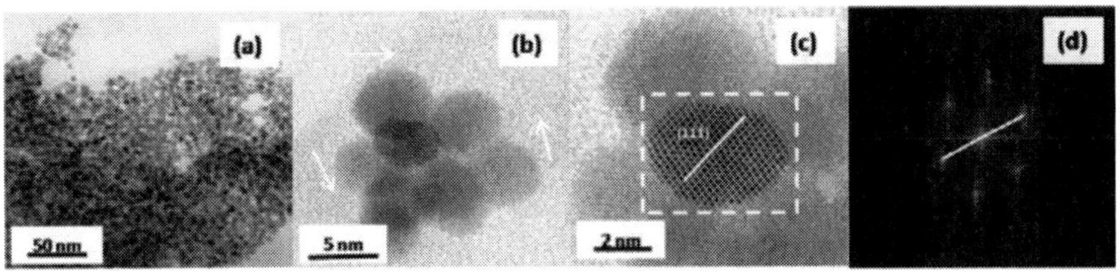

Figure 1. TEM images of microwave synthesized Pt-Co nanoparticles using Pt-seeds, (a) low magnification, B-) STEM images and (d) the FFT spectrum of a single nanoparticle.

The absence of any contrast in the particles indicates that the structure is not an intermetallic, as the difference in electron densities between the two metals was not observed. Further, the d-spacing of 2.30 Å is typical of the (111) direction of a Pt fcc crystal.

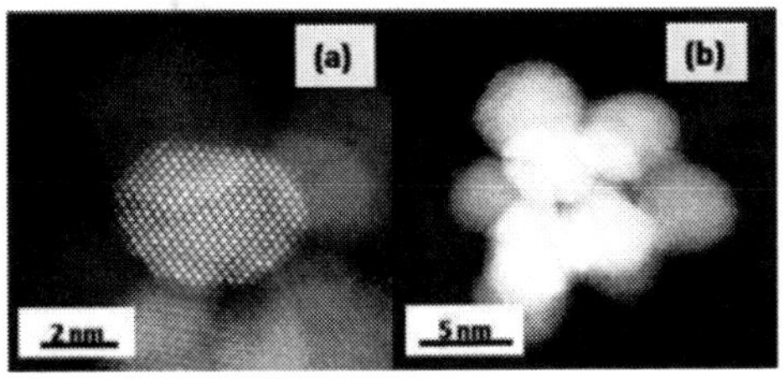

Figure 2. HAADF images of microwave synthesized Pt-Co nanoparticles, (a) single and (b) cluster of nanoparticles.

As shown in figure 3a, the Pt-Ni nanoparticles are crystalline (figure 3b) with two distinctive shapes, i.e. spherical and hexagonal. The FFT spectrum of figure 3a (dashed box), is illustrated in figure 3b, showing a crystalline pattern with a calculated d-spacing of 2.38 Å for fcc (111). Figure 3d indicates a possible hexagonal structure (from figure 3c) with a d-spacing of 2.24 Å for the hcp (100) direction. This could represent Ni because it can form a combination of fcc (NiO) and hcp (Ni) structures [7].

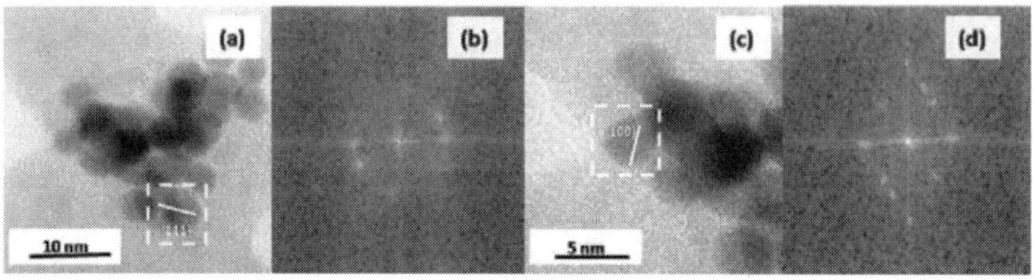

Figure 3. STEM images of microwave synthesized Pt-Ni nanoparticles (a,c) taken at different magnifications and (b,d) FFT spectra of single nanoparticles.

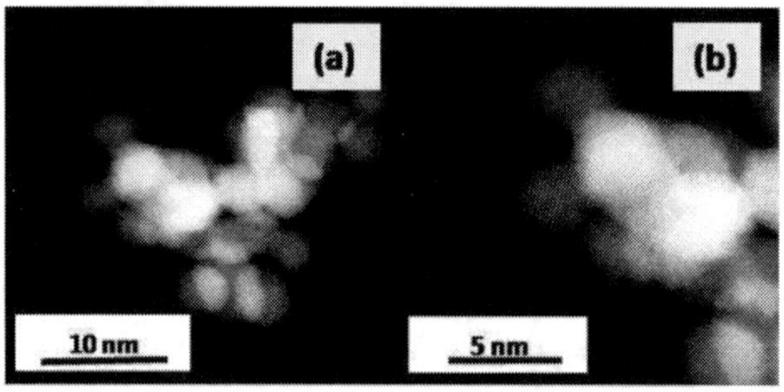

Figure 4. HAADF images of the Pt-Ni nanoparticles.

The dark-field images of the Pt-Ni nanoparticles (figure 4) also showed no clear phase in the nanoparticles. Therefore to confirm this observation, EELS was performed on both samples.

3.2. EELS analysis

EELS maps of Pt-Co nanoparticles synthesized by adding the $CoCl_2$ precursor solution to Pt seed nanoparticles followed by microwave heating, are shown in figure 5.

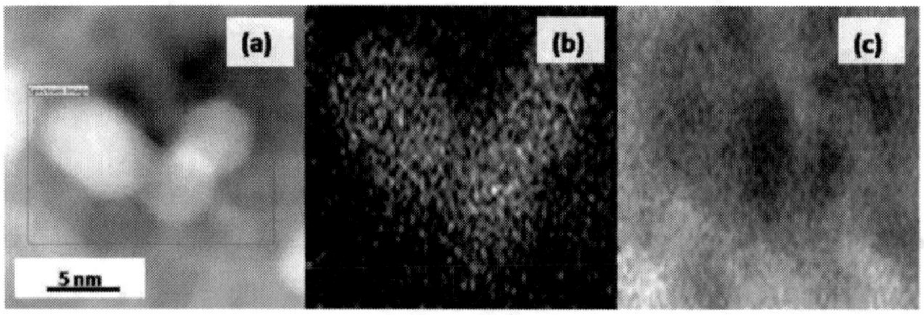

Figure 5. Energy-filtered EELS maps;(a) survey image,(b) Pt map and (c) Co map Pt-Co nanoparticles.

For the Pt and Co sample, contrast was observed in the images of the maps in figure. 5b, c. Furthermore the Pt seeds remain unaffected by the microwave process and the Co form sheets/layers of material underneath the Pt instead of growing as a shell around the Pt seeds. This is illustrated particularly in figure 5c where the Co is present as a homogeneous layer, whereas in figure 5b the Pt map illustrates the presence of Pt only in the nanoparticle structure.

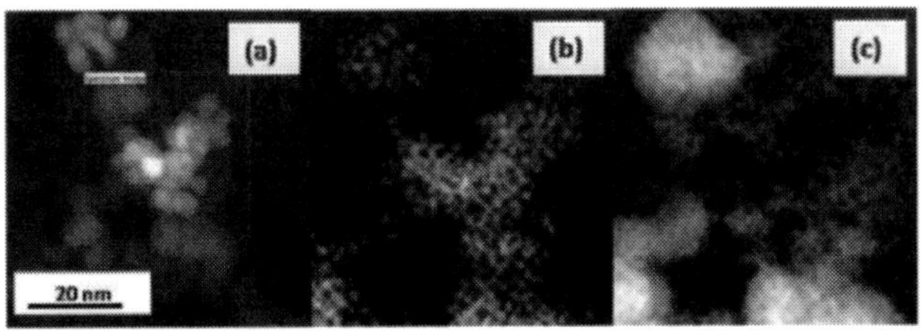

Figure 6. Energy-filtered EELS maps; (a) survey image, (b) Pt map and (c) Ni map, of microwave synthesized Pt-Ni nanoparticles.

The EELS maps of Pt-Ni nanoparticles (figure 6) illustrate the presence of two types of structures in the nanoparticles mixture, i.e. Pt-rich (figure. 6b) and Ni-rich (figure. 6c) areas. This suggests that an island/heterostructure is formed instead of an alloy structure. The Pt has been shown to grow independently from the alloying metal under certain conditions[8], therefore there is a possible island structure for this bimetallic systems.

4. Conclusion

The structures of the Pt-Co and Pt-Ni nanoparticles produced by microwave irradiation were investigated using electron microscopy. It was determined that for Pt-Co and Pt-Ni, the structures are not intermetallic or core-shell as expected, but that heterostructures were formed. In the case of Pt-Co, Pt nanoparticles are supported on Co layers, while in the Pt-Ni there is a presence of Pt-rich and Ni-rich areas. It has thus been shown that the seeding approach does not offer a route to the creation of core-shell structures, but instead results in hetero- or island structures. This stems from the differences that exist in the thermal decomposition rates of the Pt, Co and Ni.

Acknowledgements
The authors will like to acknowledge the CSIR NIC/NCNSM for financial assistance, NMMU Centre for HRTEM JEOL ARM 200F, JEOL UK and ALS for sponsoring the trip to EMAG 2013.

References
[1] Grace A.N. and Pandian K. 2007 *Materials Chemistry and Physics* **104** 191 - 8
[2] Zhu Y.-J., Wang W.-W., Qi R.-J. and Hu X.-L. 2004 *Angewandte Chemie* **116** 1434 - 8
[3] Tu W. and Liu H. 2000 *Chemistry of Materials* **12** 564 - 7
[4] Komarneni S., Li D., Newalkar B., Katsuki H. and Bhalla A.S. 2002 *Langmuir* **18** 5959 - 62
[5] Bensebaa F., Patrito N., Page Y.L., L'Ecuyer P. and Wang D. 2004 *Journal of Materials Chemistry* **14** 3378 - 84
[6] Shin N.C., Lee Y.-H., Shin Y.H., Kim J. and Lee Y.-W. 2010 *Materials Chemistry and Physics* **124** 140 - 4
[7] Garcia-Cerda L.A., Bernal-Ramos K.M., Montemayor S.M., Quevedo-Lopez M.A., Betancourt-Galindo R. and Buenos-Baques D. 2011 *Journal of Nanomaterials* **2011**
[8] Langlois C., Alloyeau D., Bouar Y.L., Loiseau A., Oikawa T., Mottet C. and Ricolleau C. 2008 *Faraday Discussions* **138** 375 - 91

Novel bilayer graphene structures produced by arc-discharge

Peter J.F. Harris

Centre for Advanced Microscopy, J.J. Thomson Building, University of Reading, Whiteknights, Reading RG6 6AF, UK.

Email: p.j.f.harris@reading.ac.uk

Abstract. A new form of carbon is described, which consists of hollow, three-dimensional shells bounded by bilayer graphene. The new carbon is produced very simply, by passing a current through graphite rods in a commercial arc-evaporation unit. Characterisation of the carbon using high resolution transmission electron microscopy is described, and the possible formation mechanism discussed.

1. Introduction

Many interesting structures, including C_{60} and carbon nanotubes, can be produced using a graphite arc-evaporation apparatus [1 - 4]. Recently, yet another kind of carbon has been discovered in graphite samples which have been subjected to arc-discharge [5 - 7]. This new carbon apparently consists of hollow graphitic shells bounded by curved and faceted planes, typically made up of two graphene layers. The hollow structures, which can be several 100s of nm in size, are frequently decorated with nano-scale carbon particles, or short nanotubes. In some cases, nanotubes are found to be seamlessly joined to the thin shells, indicating that the formation of the shells and the nanotubes is intimately connected. In this paper, this new carbon material is characterised using high resolution transmission electron microscopy, and evidence is presented that the structures are three-dimensional rather than flat. The possible mechanism whereby graphite could be transformed into the three-dimensional shells is considered, and possible applications for the new material discussed.

2. Experimental Methods

The graphene material described in this paper was prepared in a commercial arc-evaporator, the Quorum Q150T ES, which is normally used for carbon-coating specimens for electron microscopy. In this unit the electrodes are 3 mm graphite rods, one of which is thinned to a diameter of approximately 1.4 mm and held in contact with the other electrode with a spring mechanism. The chamber is pumped by a turbomolecular pump to a pressure of approximately 3 x 10^{-4} mbar. Before carrying out the "evaporation", the rods are out-gassed by passing a current of about 30A for 1 minute. For evaporation, a current of 75A is passed for 3 s. Following evaporation, the thinner carbon rod was found to have slightly shortened, and a small deposit was formed in the area where the two rods made contact. This was collected and prepared for TEM by grinding in an agate mortar under isopropanol, mixing in an ultrasonic bath and depositing onto lacey carbon TEM grids. The microscope used was a JEOL 2010, with a point resolution of 0.19 nm, operated at an accelerating voltage of 200kV. Samples from the fresh graphite rods were also imaged, for comparison with the carbon collected after arcing.

Journal of Physics: Conference Series **522** (2014) 012067 doi:10.1088/1742-6596/522/1/012067

3. Results

A typical image of material from the fresh graphite rod is shown in Fig. 1 (a). As expected, this consists mainly of flat crystallites, ranging from a few 100 nm to about 5 μm in size, containing up to 100 layers. The crystallites were often folded and buckled, and were covered with small amounts of finely-divided material. However, nanotubes or other fullerene-related structures were not seen in the fresh graphite. The carbon collected from the graphite rods following arcing contained some "normal" graphite, but this was accompanied by many regions which had a very different appearance. One of these areas is shown in Fig. 1 (b). Here, the outline of the structure is much more irregular than in the fresh graphite, with many curved and unusually-shaped features. The material is decorated with numerous short nanotubes and nanoparticles.

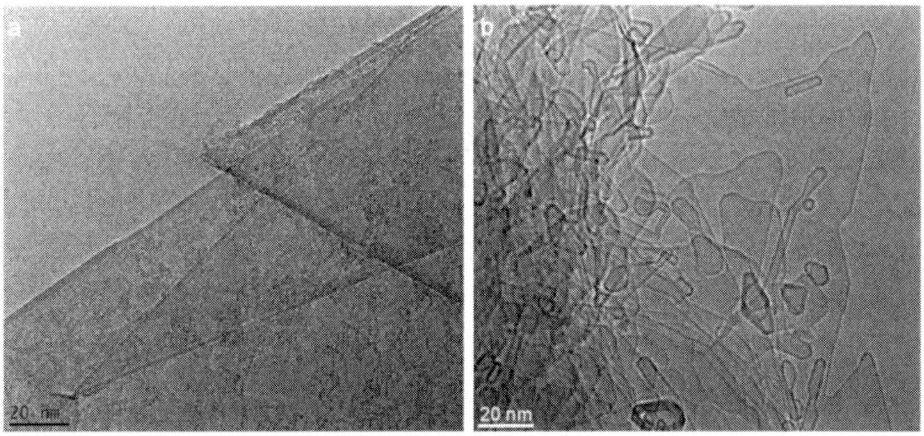

Figure 1: (a) Low magnification micrograph of carbon from fresh graphite rod. (b) Micrograph at same magnification showing transformation in structure following arc-discharge.

A higher magnification image of the material following arcing is shown in Fig. 2 (a). It can be seen that this largely consists of bilayer graphene. A striking feature of the graphitic material was that nanotubes were often observed to be seamlessly joined to the larger regions. An example can be seen in Fig. 2 (b). The tubes were generally 4 – 6 nm in diameter. This strongly suggests that the large bilayer structures are three-dimensional rather than flat, since it is difficult to envisage a way in which nanotubes, with their circular cross-section, could be connected to flat, bilayer, graphene without being seriously distorted, at least in the vicinity of the junction.

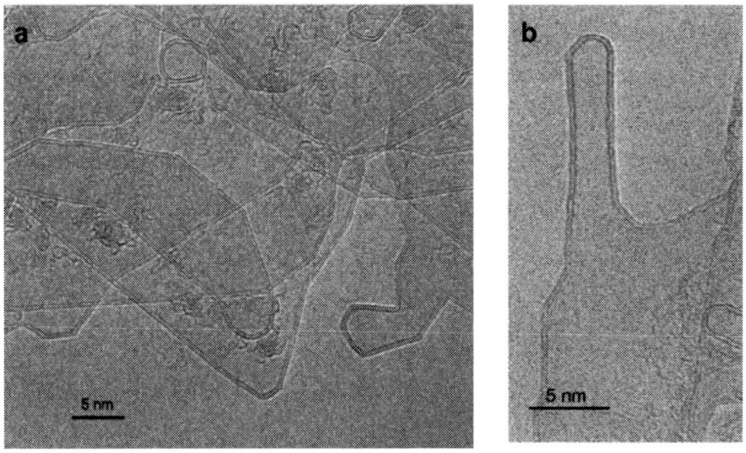

Figure 2: Higher magnification micrograph of carbon following arc-discharge. (b) Micrograph showing bilayer nanotube joined to larger region.

Another reason for believing that this novel carbon material is three-dimensional rather than flat is that nanotubes or nanoparticles can sometimes be found encapsulated inside the larger structures, showing that they are hollow. An example is shown in Fig. 3. Here, a bilayer nanoparticle can be seen apparently inside a larger bilayer structure. The way in which the contours of the nanoparticle follow those of the larger structure strongly suggests that it is encapsulated inside, rather than supported on the surface.

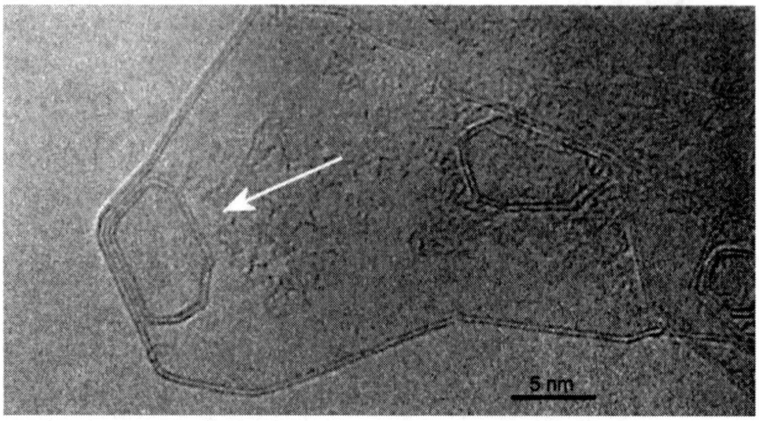

Figure 3: Bilayer graphene structure with nanoparticle apparently encapsulated inside (arrowed).

4. Discussion

The structural transformation of graphite as a result of the passage of an electric current (i.e. by Joule heating), has now been observed by a number of groups [8 - 12]. For example, Huang *et al.* [9], reported experiments in which samples of graphene prepared from highly orientated pyrolytic graphite (HOPG) were Joule-heated inside a transmission electron microscope. This produced a radical restructuring of the graphene rather similar to that described here, except that the final structures were made up of single-layer rather than bilayer graphene. Like other authors in this field, Huang *et al.* discussed the process in terms of the sublimation and edge reconstruction of flat graphene. However, as argued in this paper, there are good reasons for believing that the material produced by this transformation is in fact three-dimensional. In particular, the presence of nanotubes seamlessly joined to larger regions, and the observation of nanoparticles encapsulated inside larger graphene structures, are difficult to explain in terms of a "flat graphene" model. Obtaining clear proof that the structures are three-dimensional will require the use of electron tomography, and such experiments are planned.

If we accept that the transformation involves the growth of three-dimensional structures from initially two-dimensional graphite, what could be the mechanism? It seems likely that the key to this lies in the edge structure of graphite. It is well established that graphite planes often have "closed" edges, so that the layers resemble folded sheets [e.g. 13 - 15], as in the micrograph shown in Fig. 4 (a).

a b

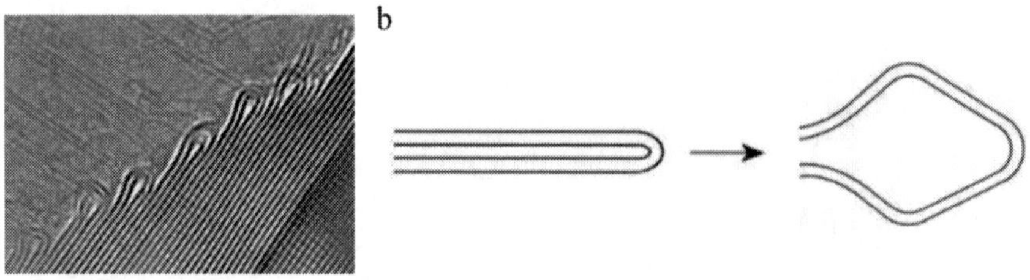

Figure 4: (a) Folded bilayer edges in graphite, reproduced with permission from ref. [15] (Suppl. data). (b) Schematic illustration of transformation of folded graphene sheets into hollow structure.

In a sense, therefore, these adjacent layers already represent closed structures, and the transformations induced by Joule-heating may simply involve an opening up of the layers to produce hollow particles, as illustrated schematically in Fig. 4(b). Such a process might be initiated at pentagonal rings which occur where zig-zag and armchair edges meet [15]. In the case of the HOPG experiments [9], we have noted that these result in the formation of structures bounded by single layers instead of the bilayers observed in the present paper. This suggests that the edge structure in HOPG differs from that in "normal" synthetic graphite, with folded monolayer rather than bilayer edges.

Acknowledgements
I thank Kazu Suenaga for discussions in the early stages of this work.

References

[1] Krätschmer W, Lamb L D, Fostiropoulos K and Huffman D R 1990 *Nature* **347** 354
[2] Iijima S 1991 *Nature* **354** 56
[3] Iijima S and Ichihashi T 1993 *Nature* **363** 603
[4] Bethune D S *et al.* 1993 *Nature* **363** 605
[5] Harris P J F 2009 *J. Phys.: Condens. Matter* **21** 355009
[6] Harris P J F 2011 *Phil. Mag.* **91** 2355
[7] Harris P J F 2012 *Carbon* **50** 3195
[8] Jia X *et al.* 2009 *Science* **323** 1701
[9] Huang JY *et al.* 2009 *PNAS* **106** 10103
[10] Qi L *et al.* 2010 *Carbon* **48** 2354
[11] Barreiro A *et al.* 2012 *Nano Letters* **12** 1873
[12] Cruz-Silva E *et al.* 2013 *ACS Nano* **7** 2834
[13] Buseck P R, Bo-Jun H and Keller L P 1987 *Energy & Fuels* **1** 105
[14] Ugarte D 1992 *Chem. Phys. Lett.* **198** 596
[15] Liu Z, Suenaga K, Harris P J F and Iijima S 2009 *Phys. Rev. Lett.* **102** 015501

Journal of Physics: Conference Series **522** (2014) 012068 doi:10.1088/1742-6596/522/1/012068

Aberration Corrected (S)TEM and Nano Beam Diffraction (NBD) of Pt and Pd on Al_2O_3 Catalysts

M R Ward[1,2], **E D Boyes**[1,2,3], **P L Gai**[1,2,4]

[1] Department of Physics, University of York, York, YO10 5DD, UK
[2] York JEOL Nanocentre, University of York, York, YO10 5DD, UK
[3] Department of Electronics, University of York, York, YO10 5DD, UK
[4] Department of Chemistry, University of York, York, YO10 5DD, UK

E-mail: mrw502@york.ac.uk

Abstract. Nano-beam electron diffraction (NBD) has been used in addition to HAADF-STEM and HRTEM analyses of Pd-Al_2O_3 and Pt-Al_2O_3 catalysts. NBD and its STEM nanodiffraction counterpart are not widely used for nanoparticle catalysts but the more common SAD method does not offer the same level of structural information for individual ultrafine (< 10 nm) nanoparticles, even with aberration correction. Here, NBD has been applied to commercially relevant catalysts for added information and to investigate defect structures in sintered particles.

1. Introduction

Pt nanoparticles loaded onto γ-Al_2O_3 are popular catalytic systems used in automobile emission treatment catalysts [1,2] and various other industrial processes [3,4]. Although there have been many studies of the Pt-Al_2O_3 system, including *in-situ* and *ex-situ* HRTEM, there seem to be relatively few that concentrate on the nanostructures of the individual nanoparticles. This may be related to the complex structure of γ-Al_2O_3 which can often coexist with δ, θ [5,6] and γ' phases [7]. HAADF-STEM is the ideal choice for the study of small clusters and atomically dispersed Pt because it's Z (atomic number) image contrast is ideally suited to the system [8,9]. However, the defective spinel structure of Al_2O_3 and the weakly scattering nature of Al and O often result in images of commercial Al_2O_3 being poorly resolved. With HRTEM, it is possible to gather structural information from nanoparticles on Al_2O_3 but clusters and atomically dispersed metal remains elusive in the image.

Here, we apply nano-beam diffraction (NBD) in addition to HRTEM and HAADF-STEM to Pd and Pt-Al_2O_3 systems. To investigate potential metal-support interactions under an O_2 atmosphere, we also examine catalysts which have been treated at elevated temperatures in air. We found that on these catalysts there is no fixed epitaxy between the support and nanoparticles. NBD also provided insights into structural intricacies in nanoparticles containing defects which would be unlikely to be detectable in a normal SAD pattern.

2. Technique

The Pt-Al_2O_3 and Pd-Al_2O_3 catalysts were provided by Johnson Matthey. A powder and pellet variant of the Pt-Al_2O_3 catalyst was provided in addition to a Pd-Al_2O_3 powder catalyst.

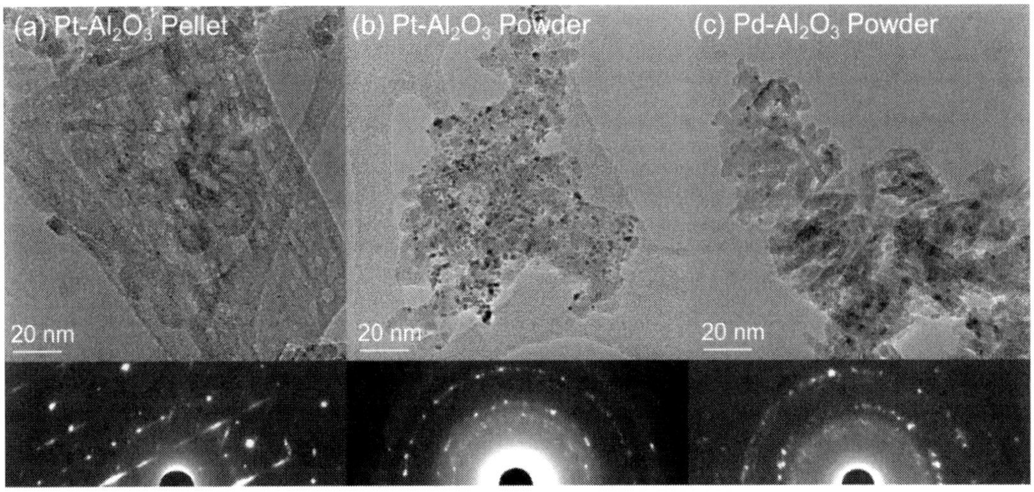

Figure 1. TEM images with corresponding SAD diffraction patterns of the labelled untreated fresh catalysts. The nanoparticles on the Pt-Al_2O_3 pellet specimen can be barely seen and the Al_2O_3 crystals are much larger than in the powder catalysts. The diffraction pattern in (a) is δ-Al_2O_3 $[1\,1\,0]$ whereas the other two SAD patterns are dominated by the γ-Al_2O_3 $\{4\,0\,0\}$ and $\{4\,4\,0\}$ rings

Portions of the fresh catalyst powders and pellets were heat treated in a furnace (held in a Al_2O_3 crucible) in air at 250 °C and 500 °C for 3 h and 15 h in addition to 750 °C for 3h prior to (S)TEM specimen preparation. (S)TEM specimens were prepared by suspending some of the powder in ethanol. The pellets required crushing. The powder suspensions were deposited using 2.5 μl drops onto a holey-C film suspended by a Cu grid.

The microscope used for these studies was a double aberration corrected JEOL 2200FS. Aberrations were corrected using a reference sample (Au nanoparticles on 2.5 nm amorphous C) in HRTEM mode. Nano-beam diffraction (NBD) mode was used for acquiring diffraction patterns from individual nanoparticles. A beam diameter of 2 nm was used with the smallest condenser aperture. The resulting diffraction patterns were indexed like a SAD pattern.

3. Results and Discussion

Figure 1 shows a montage of TEM images of the untreated fresh catalysts along with a section of their SAD patterns. Using HAADF-STEM, the average nanoparticle diameters were determined to be as follows. Pt-Al_2O_3 pellet: 1.57 ± 0.02 nm, Pt-Al_2O_3 powder: 2.69 ± 0.04 nm and Pd-Al_2O_3 powder: 3.43 ± 0.04 nm. These randomly orientated ultrafine particles gave virtually no contribution to the SAD patterns as shown in Figure 1. Figure 2 shows a NBD pattern, HRTEM image and its FFT from a nanoparticle located on the edge of a Al_2O_3 crystal. It is known that HRTEM contrast from thin oxide crystals can be ambiguous [10] and the crystal quality varies considerably in the presented Al_2O_3 powder catalysts. The NBD pattern shows the nanoparticle approximately 2° from the exact $[1\,1\,2]$ zone axis. Intricacies in the diffraction pattern can be found on the upper reflections. Here, an array of spots can be seen near the main reflections. Such intensity phenomena have been shown to originate from the effect of the crystal's finite size [11]. The additional reflections, in the 2D projection extend in the $[\bar{3}\,1\,1]$ direction, so the nanoparticle is likely a thin wedge based on the expectations of size effects from a cube (see, for example [11]). Such intricacies would be missed in a SAD pattern and FFT. Work continues to deduce epixatial relations between the support and nanoparticles.

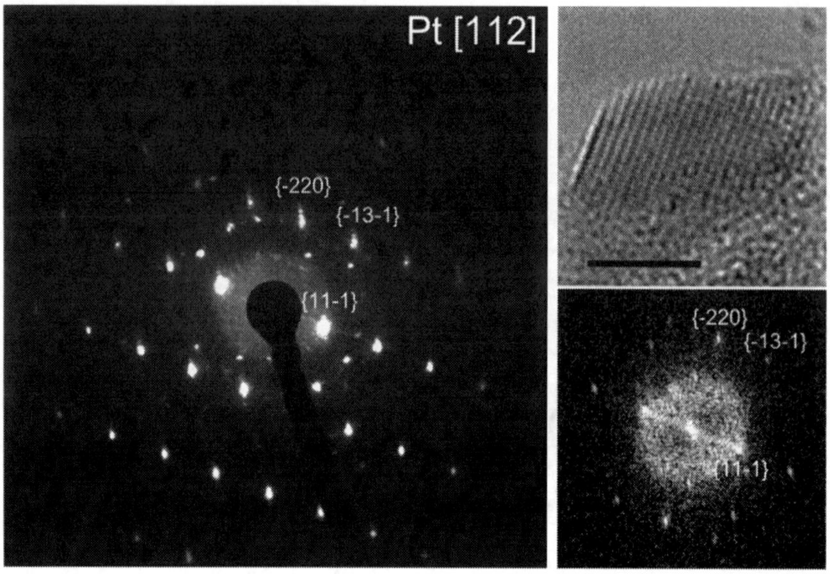

Figure 2. NBD pattern with HRTEM image and FFT of a small Pt nanoparticle on the Pt-Al_2O_3 powder catalyst. Scale bar is 2 nm

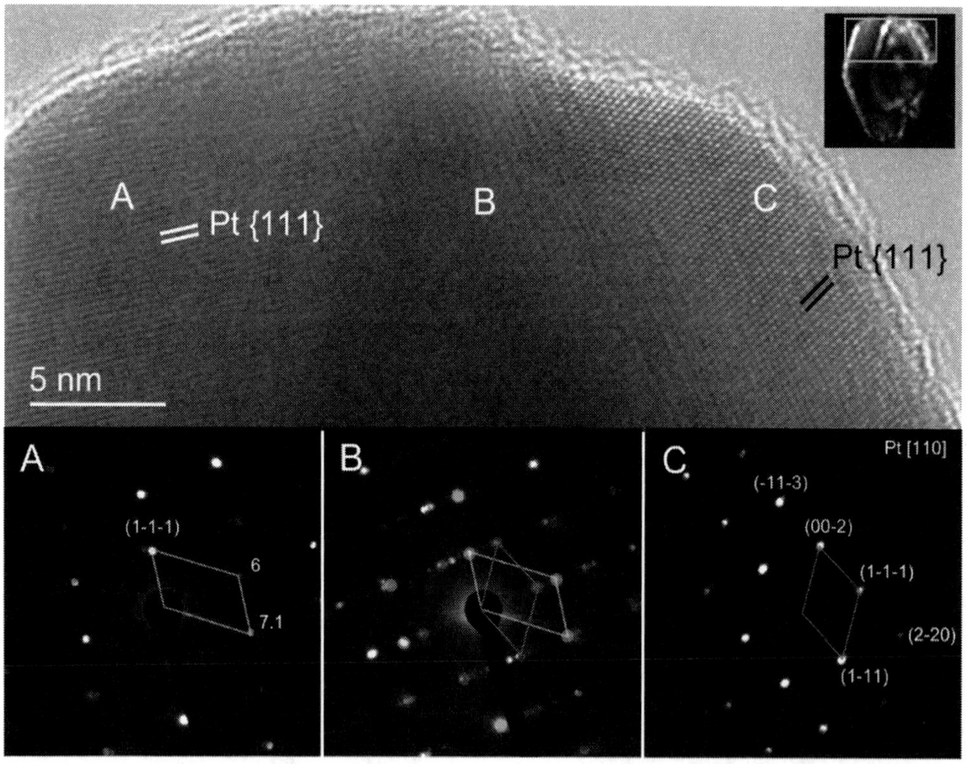

Figure 3. Example of a large structure on the Pt-Al_2O_3 catalyst treated for 3 h at 750 °C. Nano- area diffraction reveals the presence of additional diffraction phenomena caused by the presence of a series of stacking faults. Measurements in nm^{-1} in diffraction pattern A

Figure 3 shows a series of NBD patterns from a large Pt structure on the Pt-Al_2O_3 powder catalyst which has been treated at 750 °C in air for 3 h. Although defects have been studied in nanoparticles by dark field TEM they have generally been analysed in very large particles. With NBD it is possible to obtain diffraction patterns from defects on the nano-scale. The diffraction pattern from region A shows what appears to be a highly distorted Pt lattice. The pattern resembles the [2 1 3] zone axis but the spatial frequencies of most of the reflections are wrong for this zone axis. A dark field image is shown as an inset in the HRTEM image which implies that the structure observed is an overlap between two differently aligned nanoparticles approximately each 20 nm in diameter at their widest point measured from left to right in the dark field image. We believe the pattern in A is a highly distorted Pt [1 1 2] zone axis where the distortion is facilitated by the presence of a tilted grain boundary in region B which may have induced defects between the two nanoparticles.

The diffraction pattern from region B shows near continuous intensity along the $[1\,\bar{1}\,\bar{1}]$ direction (with respect to C) and this type of contrast is consistent with the presence of stacking faults but it may also be caused by the presence of a tilted gain boundary which may induce its own shape effects into the diffraction pattern. The main reflections in B originate from an overlap of the diffraction patterns in A and C. In diffraction pattern C, Pt [1 1 0] can be seen. A key feature shown in these diffraction patterns is the alignment of different directions in each crystal, namely $[1\,\bar{1}\,\bar{1}]$ in A with $[\bar{1}\,1\,\bar{3}]$ in C.

4. Conclusions

Our NBD investigation on Pt and Pd-Al_2O_3 has provided additional structural insights of small particles as well as larger particles that contain defects. Such features are likely to be missed using SAD patterns. For the smallest particles, such as those on the Pt-Al_2O_3 pellet catalyst shown here, nanodiffraction, pioneered by Cowley and co-workers [12] would be the most practical diffraction technique. Recent developments at the York-JEOL Nanocentre regarding in-situ environmental microscopy [13] could allow diffraction techniques to be combined with traditional imaging techniques that could allow dynamic structural changes to be investigated in more detail than before.

Acknowledgments

M. R. Ward thanks the EPSRC for PhD funding.

References

[1] Gandhi H S, Graham G W and McCabe R W 2003 *Journal of Catalysis* **216** 433–442
[2] Russell A and Epling W S 2011 *Catalysis Reviews* **53** 337–423
[3] Che M and Bennett C O 1989 The influence of particle size on the catalytic properties of supported metals (*Advances in Catalysis* vol 36) ed DD Eley H P and Weisz P B (Academic Press) pp 55 – 172
[4] Shi B and Davis B 1995 *Journal of Catalysis* **157** 359 – 367
[5] Lippens B C and de Boer J H 1964 *Acta Crystallographica* **17** 1312–1321
[6] Levin I and Brandon D 1998 *Journal of the American Ceramic Society* **81** 1995–2012
[7] Paglia G, Buckley C E, Rohl A L, Hart R D, Winter K, Studer A J, Hunter B A and Hanna J V 2003 *Chemistry of Materials* **16** 220–236
[8] Nellist P D and Pennycook S J 1996 *Science* **274** 413–415
[9] Sohlberg K, Rashkeev S, Borisevich A Y, Pennycook S J and Pantelides S T 2004 *ChemPhysChem* **5** 1893–1897
[10] Jefferson D A 2000 *Philosophical Transactions of the Royal Society of London. Series A: Mathematical, Physical and Engineering Sciences* **358** 2683–2692
[11] Shevchenko V Y, Madison A E and Smolin Y I 2002 *Glass Physics and Chemistry* **28** 326–333
[12] Cowley J 2004 *Micron* **35** 345 – 360
[13] Boyes E D, Ward M R, Lari L and Gai P L 2013 *Annalen der Physik* **525** 423–429

Probing the core-shell-shell structure of CdSe/CdTe/CdS type II quantum dots for solar cell applications

E A Lewis[1], R C Page[2], D J Binks[3], T J Pennycook[4, 5], P O'Brien[1, 2], S J Haigh[1]

[1]School of Materials, University of Manchester, Manchester, M13 9PL, UK
[2]School of Chemistry, University of Manchester, Manchester, M13 9PL, UK
[3]Photon Science Institute & School of Physics and Astronomy, University of Manchester, Manchester, M13 9PL, UK
[4] Department of Materials, Oxford University, Parks Road, Oxford, OX1 3PH, UK
[5]SuperSTEM , STFC Daresbury Laboratories, Warrington, WA4 4AD, UK

Email: sarah.haigh@manchester.ac.uk

Abstract. A greater understanding of multiple exciton generation in heterostructured colloidal quantum dots can be achieved through detailed modelling, and used to optimise their design for solar cell applications. However, such modelling requires an accurate knowledge of the physical structure of the quantum dots. Here we report the use of high angle annular dark field (HAADF) scanning transmission electron microscope (STEM) imaging to study the size and shape of CdSe/CdTe/CdS type II quantum dots at each of the three stages of their synthesis.

1. Introduction
Multiple exciton generation[1] (MEG) is the process, in colloidal quantum dots, where the absorption of a single photon results in the formation of two or more excitons. This phenomena is of great interest for solar cell applications, enabling more efficient use of the solar spectrum and allowing cells to be designed which exceed the Shockley Queisser limit.[2] Controlling exciton dynamics and interactions by using Type II heterostructures has been suggested as a means of improving the quantum yield of multiple exciton generation and thereby increasing the efficiency of solar cells.[3] Changes in the heterostructure of quantum dots affect excitonic interactions, including MEG, in a complex way. Detailed modelling of these processes will enable a greater understanding of MEG in quantum dots to be obtained, and allow the design of optimised heterostructures for enhanced MEG. However, for models to be validated by comparison with experimental results the quantum dot heterostructures must be characterized in detail. This paper reports the use of HAADF STEM imaging to accurately determine the core and shell dimensions of CdSe/CdTe/CdS type II quantum dots.

2. Experimental details
The CdSe/CdTe/CdS quantum dot samples were synthesised using a step-wise route based on the synthesis of CdSe quantum dots reported in reference 4.[4] The solutions of quantum dots were then cleaned by centrifugation to remove excess capping ligands, before being redispersed in fresh toluene. The resulting quantum dot solutions were drop cast onto holey carbon support grids which were then rinsed with methanol to remove residual solvent.

HAADF STEM imaging was performed using a probe side aberration corrected FEI Titan G2 80-200kV with an X-FEG electron source operated at 200kV (convergence angle of 26mrad and a HAADF inner angle of 52mrad). Image analysis was performed using FEI TIA and Gatan DigitalMicrograph software. STEM imaging and electron energy loss spectroscopy (EELS) was also performed using a Nion UltraSTEM 100 operated at 60kV, equipped with a cold field emission electron source, an aberration corrector capable of neutralising up to fifth order aberrations and an Enfina EEL spectrometer.[5] EELS compositional analysis revealed the presence of the expected elements (Cd, Se, Te, and S) within the complete nanocrystals. However, the quantum dots were found to be electron beam sensitive at both 60kV and 200kV, shrinking under prolonged imaging. Consequently EELS line scans through individual quantum dots exhibited a poor signal to noise ratio which prevented detection of the core-shell structure via EELS analysis. Instead an alternative approach was employed to quantifying the core-shell-shell dimensions by imaging the dots at each of the three different stages of their sequential growth (CdSe cores, CdSe/CdTe core-shell structures and CdSe/CdTe/CdS core-shell-shell structures).

3. Results and discussion

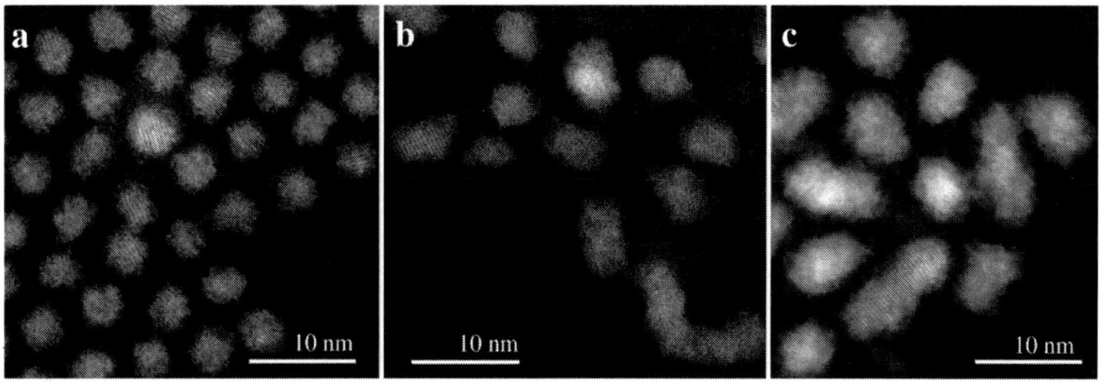

Figure 1. 3 HAADF STEM images from the Titan G2 operated at 200kV showing the sequential growth stages used in the synthesis of the colloidal nanoparticles. (a) CdSe cores (b), CdSe/CdTe core-shell, and (c) CdSe/CdTe/CdS core-shell-shell quantum dots.

Representative HAADF STEM images of the quantum dots at the different stages used in the colloidal synthesis of the CdSe/CdTe/CdS core-shell-shell structures are shown in Figure 1. Atomic resolution lattice fringe information was observable for ~50% of the final CdSe/CdTe/CdS quantum dots imaged in this study. Of these, the majority appeared to be single crystal, with less than 30% containing visible defects in the crystal structure. It was also apparent that the structure at the surfaces of the dots was less crystalline and relatively unstable during exposure to the electron beam, consistent with previous studies.[6] Lattice separations observable in the images matched those of the zinc blende crystal structure at all stages of growth, in agreement with x-ray diffraction data.

Quantum dot size analysis was performed using the TIA Imaging software to measure 100 particles at each growth stage. Overlapping dots were excluded from the analysis. For each growth step the dot length was estimated as the maximum diameter visible for each dot. This was compared to the width (particle diameter measured in the perpendicular direction) as shown in Figure 2a. The aspect ratio of each dot was calculated from the ratio of the length and width.

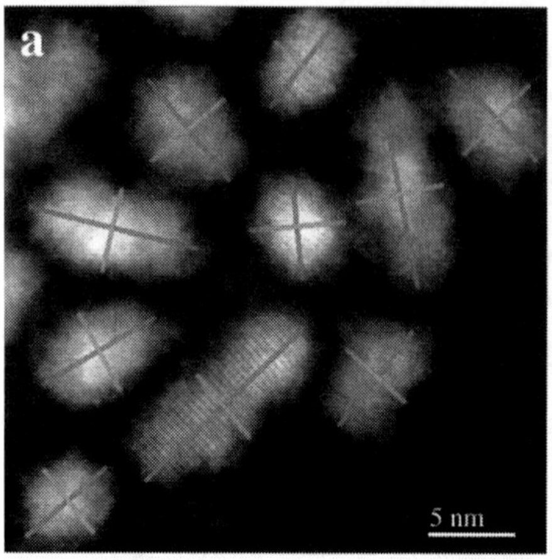

b	CdSe	CdSe/CdTe	CdSe/CdTe/CdS
Mean length	3.72 nm	5.54 nm	7.70 nm
Relative standard deviation	9.5%	21.7%	25.7%
Mean width	3.23 nm	4.14 nm	5.63 nm
Relative standard deviation	9.8%	15.2%	15.0%
Mean aspect ratio	1.16	1.37	1.40
Relative standard deviation	12.3%	26.5%	29.9%

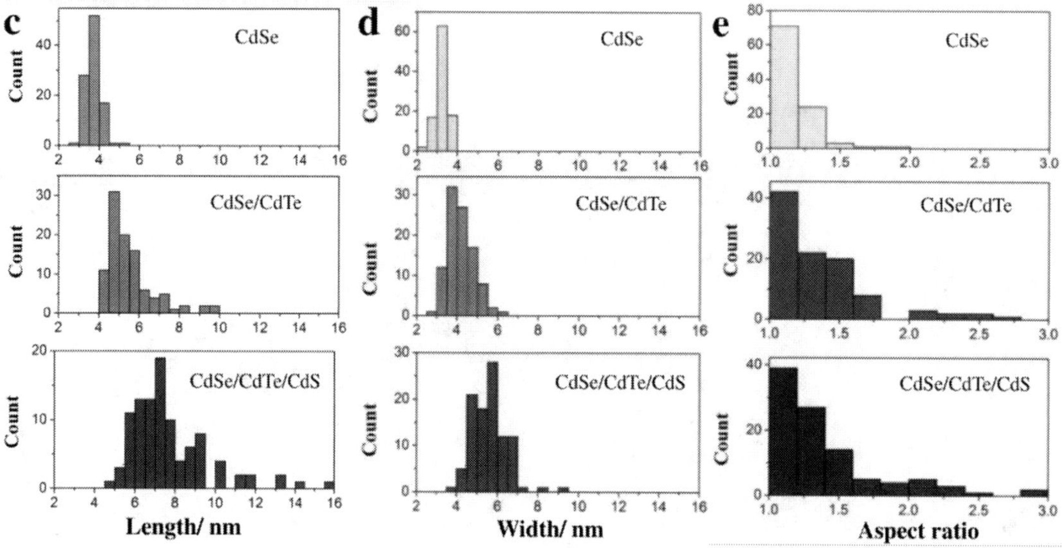

Figure 2. (a) Illustrating the measurement of the individual CdSe/CdTe/CdS quantum dot lengths (red lines) and widths (green lines). (b) Table listing the mean and standard deviations in the lengths, widths and aspect ratios (length/width) measured for 100 dots at each growth stage (CdSe cores, CdSe/CdTe core-shell, and CdSe/CdTe/CdS core-shell-shell). The data used to calculate these values is presented as histograms in (c), (d), and (e).

The mean and standard deviation of the measured dimensions for each stage of growth are shown in Figure 2b. The CdSe seed cores were found to be approximately spherical in shape, with a mean aspect ratio of 1.2. The cores had a mean width of 3.2 nm and narrow size distribution with the standard deviation of both length and width values less than 10%, less than a single Cd-Cd atomic spacing (zinc blende CdSe has a lattice constant of 0.61 nm and a Cd-Cd separation of 0.43 nm). For comparison a standard deviation of less than 5% is typically considered monodisperse.[7] The uniform nature of the CdSe dots favoured their assembly into periodic 2D rafts as shown in Figure 1a. The uniformity of the population decreases significantly after shell growth such that regular packing arrangements were no longer observed for either the CdSe/CdTe core-shell or the CdSe/CdTe/CdS core-shell-shell structures. As expected, the size increased after the shell growth stage with the CdSe/CdTe and CdSe/CdTe/CdS particles having mean widths of 4.1 nm and 5.6 nm respectively. These measurements are consistent with size estimates obtained using ultraviolet-visible spectroscopy (with band edges of 555 nm, 617 nm, and 637 nm recorded for CdSe, CdSe/CdTe, and

CdSe/CdTe/CdS quantum dots respectively).[8] The increase in the mean size of the dots is accompanied by a corresponding increase in polydispersity as illustrated by the histograms in Figure 2c-e. The mean aspect ratio of both the CdSe/CdTe and the CdSe/CdTe/CdS is ~1.4. High aspect ratio particles are especially prevalent in the complete core-shell-shell sample, where 11% of particles were found to have having aspect ratios >2. The increasing polydispersity of the dots at the later growth stages could indicate that monomer addition favours certain crystal facets such that shell growth is not uniform in all directions. An alternative explanation is that dot-dot coalescence plays an important role in the growth process and that elongated dots are a result of two of more cores fusing during the shelling reaction. There is a growing acceptance that the latter mechanism plays an important role in many nanocrystal growth processes, provided in part by in-situ electron microscope imaging.[9]

4. Conclusion

We have measured the size distribution of CdSe/CdTe/CdS core-shell-shell quantum dots at each stage of their growth. Our results show the quantum dots have a mean CdSe core diameter of 3.2±0.3 nm, an inner CdTe shell with a mean thickness of 0.5±0.4 nm and an outer CdS shell with a mean thickness of 0.7±0.5 nm. The initial cores are close to spherical with a mean aspect ratio of ~1.2. At later stages of synthesis the aspect ratio increases to ~1.4, with the complete CdSe/CdTe/CdS dots having the largest shape variation. These results provide essential experimental input for models which will enable a greater understanding of MEG in heterostructured nanocrystals to be gained, allowing optimisation of their design for solar cell applications.

References

[1] Nozik A J 2008 *Chem. Phys. Lett.* **457** 3-11
[2] Shockley W, Queisser H J 1961 *J. Appl. Phys.* **32** 510-519
Hanna M C, Nozik A J 2006 *J. Appl. Phys.* **100** 074510
[3] Binks D J 2011 *Phys. Chem. Chem. Phys.* ***13*** 12693-12704
[4] Mohamed J A U, Tonti M B, Al-Salman D, Chemseddine A, Chergui A 2005 *J. Phys. Chem. B* **109** 10533-10537
[5] Krivanek O L, Corbin G J, Dellby N, Elston B F, Keyse R J, Murfitt M F, Own C S, Szilagyi Z S, Woodruff J W 2008 *Ultramicroscopy* **108** 179-195
[6] Pennycook T J, McBride J R, Rosenthal S J, Pennycook S J, Pantelides S T 2012 *Nano Lett.* **12** 4415-4415
[7] Murray C B, Kagan C R, Bawendi M G 2000 *Annu. Rev. Mater. Sci.* **30** 545-610
[8] Yu W W, Qu L H, Guo W Z, Peng X G 2003 *Chem. Mat.* **15** 2854-2860
Tyrrell E J, Smith J M, 2011 *Phys. Rev. B* **84** 165328
[9] Liao H G, Cui L K, Whitelam S, Zheng H M 2012 *Science* **336** 1011-1014

EBSD study of substrate-mediated growth of hexagonal boron nitride

J Dias[1], P R Kidambi[2], S Hofmann[2], C Ducati[1]
[1] Department of Materials Science and Metallurgy, University of Cambridge, Cambridge, CB3 0FS, UK
[2] Department of Engineering, University of Cambridge, Cambridge, CB3 0FA, UK

E-mail: cd251@cam.ac.uk

Abstract. Hexagonal Boron Nitride (h-BN) is a promising insulating material to complement and enable graphene electronics. Given the good lattice match to graphite, graphene/h-BN heterostructures may be grown with negligible amounts of strain and defect states, resulting in high carrier mobilities approaching values for suspended graphene. Chemical vapour deposition (CVD) has emerged as one of the preferred routes for the synthesis of 2D materials for electronic applications. Here we report on the growth of h-BN by low pressure CVD, using borazine as a precursor. Electron backscattered diffraction (EBSD) in conjunction with topographic imaging in the scanning electron microscope are used to investigate the change in crystal structure and orientation of three metallic catalyst substrates: Co, Ni and Cu, by high temperature processing and the growth of nanoscale h-BN domains. The behaviour of the metal foils is interpreted in light of the prevalent growth models. EBSD and imaging conditions are optimized to allow efficient acquisitions for these composite and nanostructured specimens.

1. Introduction

Hexagonal boron nitride (h-BN) is a promising insulating material to complement graphene in two dimensional electronics. The ultra-flat morphology and insulating properties of h-BN allow carrier mobilities to approach values reported for suspended graphene [1]. Growth of h-BN was demonstrated using chemical vapour deposition (CVD) on metallic substrates, with recipes conceptually analogous to those employed for the synthesis of graphene. In particular, here we adopted borazine $(HBNH)_3$ as B-N precursor in a low pressure CVD (LPCVD) approach, because the gas already carries a stochiometric balance of the constituent elements, with a low pressure of H_2. In the literature, a range of macroscopic characterization techniques is employed to analyse the initial stages of BN growth, including Raman spectroscopy and electron backscattered diffraction (EBSD) in the scanning electron microscope (SEM). To gain control over the deposition in view of applications, it is important to understand how the specific CVD conditions (p, T, precursor dilution) affect the re-structuring of the metallic substrate, which in turns affects the growth in term of nucleation and the quality of the formed layers.

Here we present a survey of the effect LPCVD processing has on grain size and crystal orientation of three common metallic foils used as substrates for the growth of h-BN. Topographic secondary electron imaging (topo-SEI) complemented by electron backscattered diffraction (EBSD) are used for characterising the morphology of the h-BN phase and the orientation of the metallic substrate, grain by grain.

2. Experimental methods

LPCVD growth of boron nitride was performed using a borazine $(HBNH)_3$ precursor at 800-1000°C for 30 minutes to produce h-BN nuclei on Co (μm Alfa Aesar purity > 99.9%), Ni (μm Alfa Aesar purity > 99.999%), and Cu (25 μm Alfa Aesar purity > 99.999%) foils. The crystallographic phase of BN was confirmed by various characterization techniques including Raman spectroscopy post transfer from the metallic substrate .

Topo-SEI and EBSD were performed using a FEI Helios FEG SEM/FIB microscope. The Channel 5 software suite distributed by Oxford Instruments with the EBSD Nordlys II detector was used for data acquisition and processing. The metallic foils were mounted on SEM stubs and loaded in the SEM chamber and imaged at a working distance of about 4 mm. This was found to be a good compromise because it allowed sufficient spatial resolution for topo-SEI of the boron nitride islands, as well as efficient detection of the EBSD signal for a specimen tilt angle of 60°. The acceleration voltage was set to 2 kV for topo-SEI to improve the visibility of the overlayers, to 20 kV for low mag topo-SEI, and to 20 kV for EBSD analysis. The optimum beam current for these acquisitions was found to be 0.69-1.4 nA. MonteCarlo simulations were used to calculate the maximum escape depth of the backscattered electrons for the three metals ($Z = 27$, 28 and 29 for Co, Ni, and Cu, respectively) using the software CASINO v. 3.2.0.1. This was found to be of the order of 200 nm and therefore much smaller than the average size of the metallic grains within the foils.

For each location we acquired a low magnification topo-SEI micrograph and corresponding EBSD map, followed by a set of high magnification, low kV, micrographs of the metallic grains in the map to identify h-BN growth habit as a function of crystal orientation. The main factors limiting mapping efficiency were found to be surface coverage and macroscopic surface roughness, the latter causing contrast inversion that prevents automatic indexing during mapping.

3. Results and discussion

3.1 Cobalt

The grains in the Co foil have an average grain diameter of about 20 μm, and a wide range of crystallographic orientations. A unique feature observed on Co is the presence of heavily stepped grains along-side smooth ones. By correlating topo-SEI and EBSD phase maps, e.g. in figure 1(a-b), it appears that stepped surfaces correspond to the hexagonal phase, while smooth surfaces, decorated with BN dendrites, correspond to the cubic phase. We note that the indexing rate on stepped surfaces is lower than on smooth surfaces, and seems to be influenced by the relative orientation of the steps, possibly due to differences in the take-off angle of the backscattered electrons. Beyond this observation, statistically significant correlations between Co grain orientation and h-BN growth were hard to find. This is evident from Figure 1(a) and 1(c) [2].

3.2 Nickel

Ni foils have an average grain diameter of 50 μm, and display a strong preponderance of grains for which the [100] is perpendicular to the foil surface, with a few that are [111]-oriented. Figure 2(a) shows an example of BN nuclei on a [111] Ni grain, from which it is clear that the growth is partly 3D and is attributed to conditions specific to the LPCVD recipe.in this case. Figure 2(b) shows a low mag image of the Ni foil from which the EBSD map in (c) was acquired.

Epitaxial growth of h-BN on Ni(111) was first demonstrated by Nagashima and co-workers [3], and arises from a good lattice match between the two materials. The 0.5% mismatch gives rise to corrugations of the h-BN according to Auwarter et al. [4], and indeed figure 2(a) shows a strong interplay of the two phases, much more than observed in Co. Dendritic h-BN growth can also be observed on a (101)-like facets. Two growth mechanism on reconstructed (101) surfaces are suggested [5], 1x6 being energetically favourable because corrugation of the film can compensate mismatch. 7x5 structure produces negative mismatch that introduces strain in film.

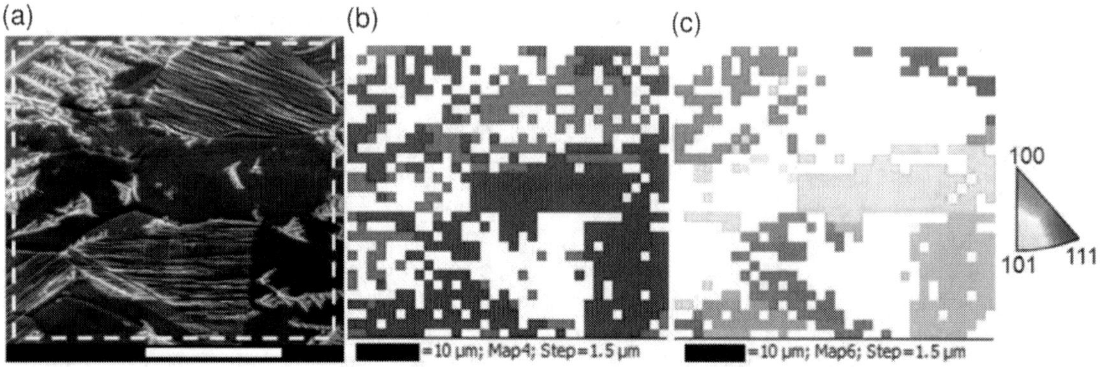

Figure 1. (a) Secondary electron SEM image of Co foil after processing, the scale bar corresponds to 20 µm (20 kV acceleration voltage). The dashed rectangle corresponds to the area of the EBSD maps shown in (b) and (c). The EBSD phase map in (b) displays the hexagonal (green) and cubic (blue) grains in the foil. The map in (c) shows the relative orientation of the cubic grains, colour coded according to the legend shown on the right.

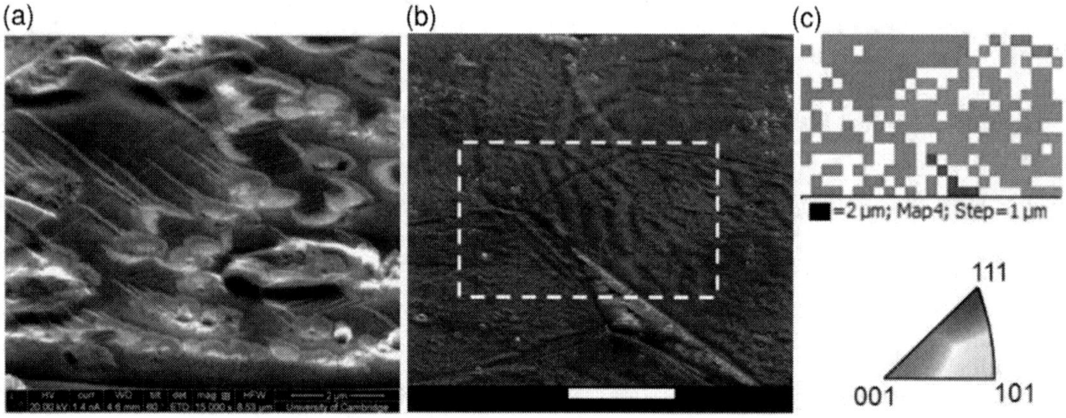

Figure 2. (a) Secondary electron SEM image of the Ni foil after processing. Boron nitride islands are visible in the topographic image acquired at 20 kV acceleration voltage. The scale bar corresponds to 2 µm. (b) Low mag SE image of the Ni foil. The scale bar corresponds to 10 µm. The dashed rectangle corresponds to the area of the EBSD maps shown in (c), which shows the relative orientation of the cubic grains, colour coded according to the legend shown on the right.

3.3 Copper

Copper foil responds to the processing by forming the largest grains, up to 200 µm, but with a large size distribution, visible in figure 3(b). A large fraction of the grains is oriented such that the [100] axis is parallel to the surface normal. The EBSD map of figure 3(c) shows that only a small fraction of the foil surface is (111), but the morphology of these regions is peculiar, i.e. the domains form long, thin strips. In the literature h-BN growth was only reported on (100) and (111)-like facets, but in this study we observed nuclei on (110) too, as shown in the image in figure 3(a).

Growth of h-BN (001) on Cu (111) surfaces is likely to be epitaxial, as reported by Preobrajenski [6] and Kim [7]. This is due to the fact that the two have the same symmetry and lattice spacings, where residual misfit is possibly accommodated by the h-BN. Growth on Cu (100) was only reported by Guo [8], no clear mechanism was reported for it. It is interesting to note that h-BN nuclei stretch across grain boundaries.

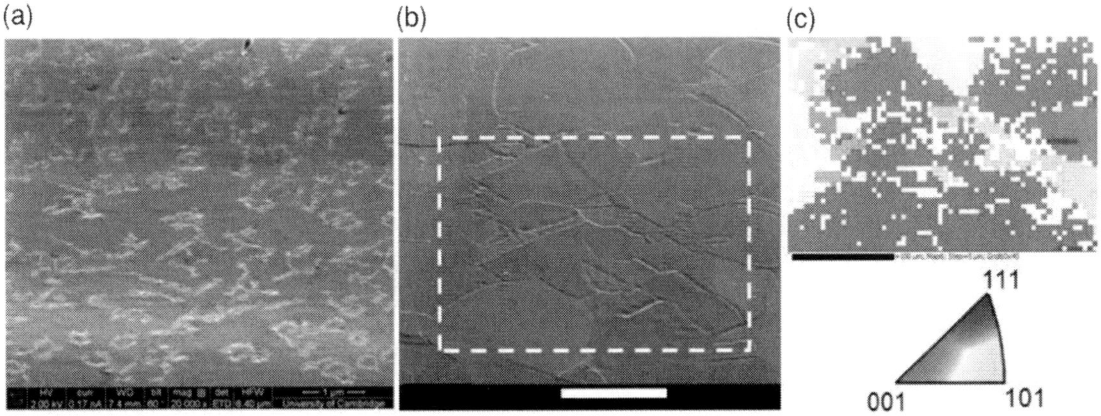

Figure 3. (a) Secondary electron SEM image of the Cu foil after processing. Boron nitride islands are visible in the topographic image acquired at 2 kV acceleration voltage. The scale bar corresponds to 1 μm. (b) Low mag SE image of the Ni foil. The scale bar corresponds to 100 μm. The dashed rectangle corresponds to the area of the EBSD maps shown in (c), which shows the relative orientation of the cubic grains, colour coded according to the legend shown on the right.

4. Conclusions

The synthesis of 2D h-BN by CVD methods requires a detailed understanding of the role of transition metal foils both as catalysts and as substrates. In this study, our aim was to observe to what extent a specific high temperature LPCVD process affected the grain size and crystallographic orientation of cobalt, nickel, and copper foils. EBSD mapping revealed that for Ni and Cu foils the [100] is predominantly parallel to the surface normal. Low voltage topographic secondary electron imaging of the foils showed boron nitride nuclei on all surfaces, some with a marked dendritic character, e.g. on cubic Co grains, and some with a more irregular shapes, e.g. with Cu (101) surfaces.

References

[1] Butler SZ et al., 2013 *ACS Nano*, **7**, 2898-2926
[2] Orofeo CM et al., 2013 *Nano Research*, **5**, 335-347
[3] Nagashima A et al. 1996 *Surface Science* **357**, 307
[4] Auwarter W, Suter HU, Sachdev H, and Greber T., 2004 *Chem. Mater.*, **16**, 343-345
[5] Greber T. et al. 2006 *e-Journal of Surface Science and Nanotechnology* **4**, 410
[6] Preobrajenski AB et al. 2005 *Surface Science* **582**, 21–30
[7] Kim KK et al. 2012 *Nano Letters* **12**, 161–166
[8] Guo et al. 2012 *Nanotechnology* **23**, 415605

Acknowledgements

The authors acknowledge funding from the ERC under grant number 259619 PHOTO EM

Nanoscale characterisation of hybrid photovoltaic cells based on C_{61} capped CdSe QDs

S Cacovich[1], G Divitini[2], E Capria[3], A Fraleoni Morgera[3], V Lughi[1] and C Ducati[2]

[1]Department of Engineering and Information Science, University of Trieste, via Valerio 6/a, 34127 Trieste, Italy
[2]Department of Materials Science and Metallurgy, University of Cambridge, Cambridge CB3 0FS, UK
[3]Elettra-Sincrotrone Trieste SCpA-SS 14.5, km 163.5, 34149 Basovizza (TS), Italy

E-mail: cd251@cam.ac.uk

Abstract. Hybrid solar cells based on 1,2 methanofullerene (C_{61}) capped CdSe and poly (3-hexylthiophene) (P3HT) were been investigated through a range of techniques. High resolution transmission electron microscopy (HRTEM) was used to characterize size, morphology and crystal structure of as-grown and C_{61}-capped CdSe quantum dots. Cross sectional lamellar specimens were prepared from full photovoltaic devices using a focused ion beam milling approach. The sections were analysed by high angle annular dark field imaging in scanning TEM mode to determine the morphology of the device, in particular the intermixing of P3HT and capped quantum dots.

1. Introduction

In recent years organic photovoltaic cells based on semiconducting polymers have achieved good power conversion efficiencies, but an improvement of the devices has been demonstrated using polymer blended with fullerene derivates [1,2] or with quantum dots [3]. We have studied a ternary system that includes P3HT, methanofullerene and CdSe quantum dots (QDs). The ratios of C_{61}-capped CdSe to P3HT and C_{61} to CdSe were varied and the macroscopic optoelectronic properties were investigated in solution, using UV-VIS absorption and normalized PL spectra. It was found that incorporation of C_{61}-capped QDs into P3HT increases the total absorbance through the cumulative contributions of P3HT and CdSe. Photoluminescence measurements show that the C_{61} capping induces the quenching of the emission peak typical of CdSe QDs, which suggests a charge-transfer event which effectively competes with radiative recombination of electron and hole in the QD. This is an important prerequisite for photovoltaic behavior, as photo-generated electron-hole pairs can be transferred to the outer circuit through the interconnected carbonaceous network. Prototype solar cells were prepared on ITO coated glass substrates, and tested in a solar simulator. To complement the macroscopic characterization of this novel hybrid solar cell, we performed a detailed investigation of the active components of the device using a range of transmission electron microscopy techniques, aimed at verifying the nanoscale architecture.

2. Experimental methods

CdSe quantum dots were synthesized according to the procedure of *Peng et al.*[4]. QDs growth was carried out by colloidal chemical approach, using dihydrate cadmium acetate as precursor and the native capping agent was substituted with 1,2 methanofullerene C_{61} [5]. Methanofullerene (1,2 mg) was dissolved in 8 ml of 1,2 dichlorobenzene and the solution was added to the oleic acid capped CdSe. The mixture was held at 50 °C for 6 hours, stirred magnetically and sonicated. The exchange capping procedure was monitored by absorption and photoluminescence spectra. High resolution transmission electron microscopy (HRTEM) was used to characterize size, morphology and crystal structure of CdSe QDs and capped QDs, drop-cast from solution on holey carbon Cu grids. HRTEM images were taken using a JEOL 4000 EX (400 kV acceleration voltage).

Photovoltaic devices were prepared using drop casting method from a 1, 2 dichlorobenzene solution, on a clean ITO glass substrate in a nitrogen atmosphere, and Al contacts were deposited by thermal evaporation. The hybrid solar cells had the following configuration: ITO/TiO_2/P3HT-CdSe C_{61}/Al. A thin cross sectional sample of the solar cell was prepared by Ga ion focused ion beam milling in a FEI Helios FEG SEM/FIB microscope. A 200 kV FEI Tecnai F20 scanning transmission electron microscope (S)TEM was employed to characterize the x-sectional device.

3. Results and discussion

Figure 1(a) shows a typical high resolution TEM image of CdSe QDs. Fast Fourier Transform analysis of the QDs revealed that the CdSe crystallized in the zincblende structure. Some defects in the crystalline structure of CdSe, e.g. twinning, probably due to imperfect growth, are visible. The diameter distribution of CdSe QDs was measured from about 100 nanoparticles, and the resulting histogram is shown in Figure 1(b). From macroscopic absorption spectra obtained from QDs in solution, it is possible to determine the diameter average size of the QDs using the equation proposed by *Peng et al.* [4]. The absorption peak for the QDs was observed at 561 nm (2.21 eV). The two approaches give comparable results: $D_{HRTEM} = 3.12 \pm 0.48$ nm and $D_{Peng} = 3.28$ nm.

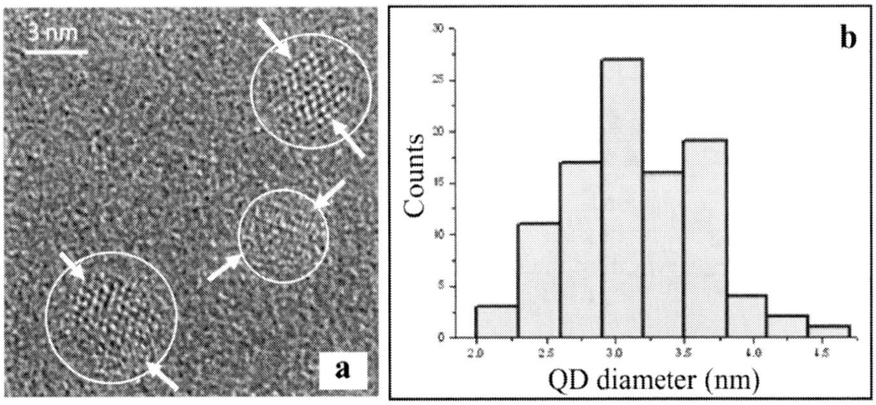

Figure 1. a) Inverted contrast HRTEM image of CdSe QDs, b) Particle size distribution calculated from about 100 QDs.

Figure 2(a) shows a high resolution TEM image of the capped CdSe QDs. The presence of fullerenes is evident on the surface of CdSe QDs. The diameter distribution measured from HRTEM images is displayed in Figure 2(b). The difference between average diameter CdSe QDs and capped CdSe QDs (Fig. 1(b) and 2(b)), is 0.70 nm, which is comparable with the diameter of a C_{60} molecule, and also compatible with (twice) the length of the 1,2 methanofullerene radical. As seen from the photoluminescence spectra in Figure 2(c), CdSe QDs show an emission peak at 571 nm, while capped CdSe QDs do not show any emission, confirming that radiative recombination is suppressed by the capping. This suggests that charge transfer has taken place.

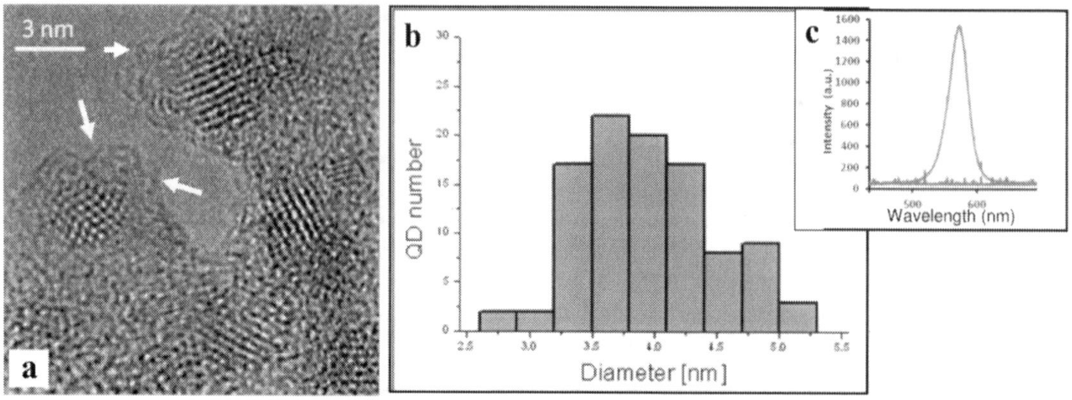

Figure 2. a) HRTEM image of the capped CdSe QDs, with white arrows highliting some of the fullerene molecules, b) Particle size distribution, and c) PL spectra of CdSe QDs and C_{61} capped QDs.

To verify the dispersion of QDs in the polymeric blend, a cross section of the device was prepared by FIB milling. The lamella was mounted on an Omniprobe grid, and thinned to a final thickness of about 150 nm, using low Ga ion currents to reduce Ga implantation. The section, shown at low magnification in Figure 3, was investigated using HAADF STEM imaging. The thickness of the various layers was measured at multiple locations along the section, and the average values are given in Table 1.

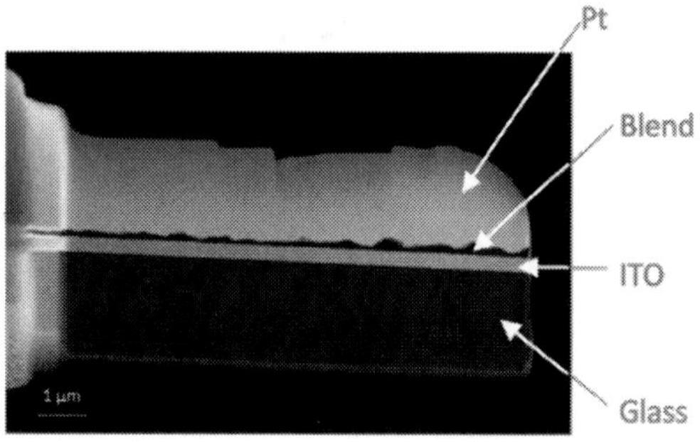

Figure 3. FIB lamella obtained from a complete hybrid solar cell. The Pt layer was deposited to protect the device from the Ga ion beam.

Table 1. Average thickness of the different layers measured from the cross section of the solar cell.

Layer	Thickness (nm)
ITO	300 ± 3
TiO_2	20 ± 2
Blend	140 ± 44

In this type of hybrid photovoltaic cell, CdSe and P3HT are used to absorb light and generate excitons, while a C_{61} network provides a conduction path for the electrons (holes are transported by the P3HT). The distribution and concentration of the nanoparticles within the P3HT matrix play a fundamental role in the functioning of the cell. The best compromise between carrier generation and transport was estimated for the ratios: CdSe:C_{61} 1:3 and CdSe@C_{61}:P3HT 1:1.

Figure 4(a) shows a high magnification HAADF STEM image of the active area of the solar cell obtained for these concentrations. The atomic number (Z) contrast allows us to identify the CdSe QDs easily. The QDs are quite uniformly dispersed in the polymer, with an average distance between particles of 5.57 nm measured in the 2D projections. By evaluating the maximum intensity of the HAADF signal for the QDs in Figure 4(a), we can estimate the projected CdSe thickness within the section, and therefore calculate the number of overlapping CdSe particles in this projection, as shown in Figure 4(b-c). QDs are assumed to have the same average atomic number. It is interesting to notice that, assuming the lowest intensity corresponds to a single quantum dot of diameter D_{HRTEM}, the highest intensities correspond at most to the overlap of two QDs (2 x D_{HRTEM}) over the thickness of the lamella, which confirms an excellent dispersion of QDs within the polymer matrix.

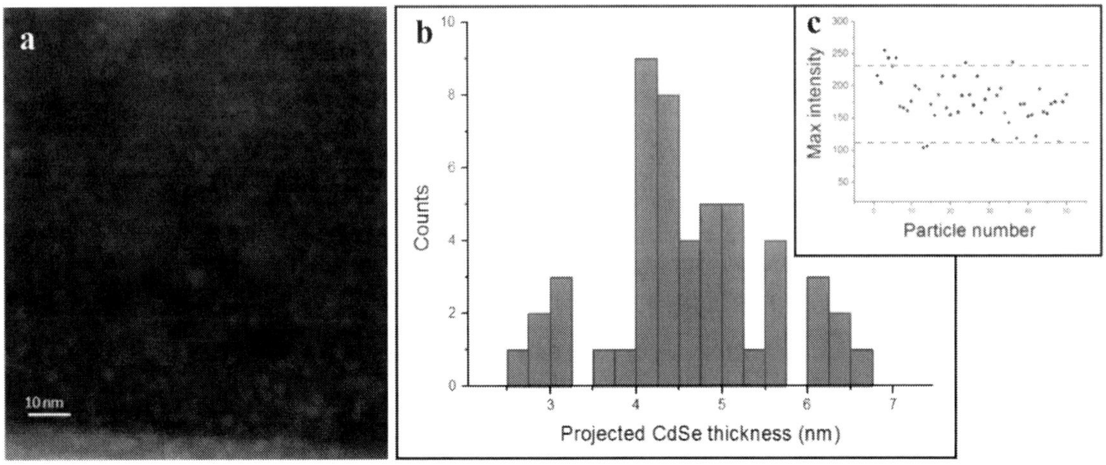

Figure 4. a) HAADF (S)TEM image of the ternary system CdSe/C_{61}/P3HT b) histogram of projected thickness obtained from the intensity distribution of CdSe in the HAADF image, shown in (c).

4. Conclusions

The incorporation of CdSe QDs in a conjugated polymer improves light absorption, as demonstrated by the presence of two absorption peaks in UV - VIS spectra for the composite. Charge transfer processes occur at the interface QDs/ C_{61} and P3HT/C_{61}, as suggested by the quenching effect shown by PL spectra. (S)TEM images demonstrate that individual zincblende CdSe QDs are effectively coated in fullerenes and uniformly dispersed in P3HT, which represents a good compromise between creating large contact area, promoting photon absorption, and aiding charge transport through the device. Recombination processes are not fully understood yet, due to the complexity of the system. Further studies are required to optimized the efficiency of the devices and to improve the knowledge of the physical mechanisms that determine the photovoltaic behaviour of this kind of nanocomposite.

References

[1] Gunes S, Neugebauer H and Sariciftci S N 2007 *Chem. Rev.*, **107** 1324-1338
[2] Chen D, Nakahara A, Wei D, Nordlund D, Russell T 2011 *Nano Lett.*, **11** 561- 567
[3] Huynh W U, Dittmer J J, Libby W C, Whiting G L, Alivisatos A P 2003 *Adv. Funct. Mater.*, **13** 73-79
[4] Yu W, Qu L, Guo w and Peng X 2003 *Chem. Mater.*, **15** 2854-2860
[5] Lee J, Kim J, Yang C 2011 *Nanoscale Res. Lett.*, **6** 23

Acknowledgements

The authors acknowledge funding from the ERC under grant number 259619 PHOTO EM and the Erasmus Placement Programme.

TEM and HRTEM of Soot-in-oil particles and agglomerates from internal combustion engines

M W Fay[1,2], A La Rocca[2] and PJ Shayler[2]
[1] Nottingham Nanotechnology and Nanoscience Centre, University of Nottingham, Nottingham NG7 2RD, UK
[2] Department of Mechanical, Materials and Manufacturing Engineering, Faculty of Engineering, University of Nottingham, University Park, Nottingham, NG7 2RD, UK.

E-mail: michael.fay@nottingham.ac.uk

Abstract. Over time, the performance of lubricating oil in a diesel engine is affected by the build-up of carbon soot produced by the combustion process. TEM and HRTEM are commonly used to investigate the characteristics of individual and agglomerated particles from diesel exhaust, to understand the structure and distribution of the carbon sheets in the primary particles and the nanostructure morphology. However, high resolution imaging of soot-in-oil is more challenging, as mineral oil is a contaminant for the electron microscope and leads to instability under the electron beam. In this work we compare solvent extraction and centrifugation techniques for removing the mineral oil contaminant, and the effect on particle size distribution.

1. Introduction

Most of the particulate matter produced by the combustion process in a diesel engine is expelled with the exhaust gases into the exhaust system, but a small proportion is transferred to the lubricating oil. Soot-in-oil differs from exhaust soot in that it has not been subject to oxidation processes to the same extent. Hence, the outer shell structure is more likely to remain intact. The nanostructure of the soot surface is of interest because of the potential influence on its reactivity and oxidation rates. The length and orientation of the graphene layers on the soot surface determine soot graphitisation and consequently the reactivity of the soot particles. Shorter layers will result in more edge sites, and therefore more reactivity. Soot in oil does not agglomerate as much as exhaust soot.

High resolution TEM is commonly used to examine exhaust soot [1, 2], but the presence of mineral oil in soot-in-oil samples complicates TEM analysis. Cryogenic vitrification and imaging have been used to analyse soot-in-oil in the past, but issues related to viscosity of the typical used engine oil lead to localised thick layers [3,4].

Solvent extraction has been used to prepare suitable samples for TEM, although significant engine oil remains that is an issue due to contamination in the TEM. Subsequent centrifugation has been reported to produce samples suitable for HRTEM; however there are indications in the literature that the increased stresses involved in this process have modified the characteristics of the soot agglomerates.

[1] To whom any correspondence should be addressed.

2. Experimental

Samples of soot extracted from engine oil were prepared using a solvent extraction process, diluting the oil at a dilution ratio of 1:60 in heptane. This produces a heptane solution containing a much lower oil content, and also at a suitable low viscosity to allow deposition onto a carbon coated TEM grid. Following deposition, the solvent evaporates rapidly to leave soot particles of varying sizes and aggregations. This process puts little strain on the soot aggregates, so is expected to reveal a structure typical of the soot as it was in the engine oil. The centrifugation process is performed on the diluted oil/heptane solution by five stages at 14,000 rpm for 90 minutes at 25°C, following the work of Clague et al [5].

For TEM analysis, after preparation, 3 microlitres of the soot suspension in heptane were dispersed onto amorphous carbon or graphene oxide TEM support films obtained commercially from Agar Scientific. Graphene oxide films, which have only recently become commercially available, are particularly advantageous for work of this nature, as the reduced contribution to the image from the supporting film makes identification of the structures in the smaller soot primary particles more straightforward. The solvent extraction samples were given a 30 second O2/Ar plasma clean in a Fischione Model 1020 Plasma Cleaner prior to imaging in the TEM.

Imaging of solvent extraction samples was performed at liquid nitrogen temperatures using a Gatan 914 Cryo-tomography holder. Imaging of centrifugation samples was performed using a standard JEOL single tilt holder. TEM analysis was performed in a JEOL 2100F TEM equipped with a Gatan Orius CCD camera.

3. Results and Discussion

Samples of soot in oil after solvent extraction only were found to be suitable for conventional TEM (Figure 1), showing the overall structure of the soot particles and the level of agglomeration. The soot agglomerates were found to be comprised of 12-40nm sized primary particles. These formed a mixture of chain-like structures and clusters of spherules. The presence of 2nm particles was also observed; these are also found in samples of the clean oil (not shown), and are ascribed to be an oil additive. However, high resolution TEM of samples prepared by solvent extraction was not practical due to the rapid build-up of carbon contamination in most areas of the TEM grid investigated, rapidly obscuring the structure of interest, even at liquid nitrogen temperatures.

In comparison, the five-step centrifugation process was found to produce samples suitable for HRTEM analysis, at room temperature, without the need for a plasma cleaning stage subsequent to dispersal onto the TEM grid (Figure 2). The nanostructure of the primary particles from soot samples shows tortuous graphitic segments constituting the outer shell, with defects and incomplete layers present, indicating that these particles will be reactive. However, the centrifugation process was found to have altered the structure of soot aggregates, although it does not appear to have altered that of the primary particles [6].

Dispersion of centrifuged samples onto graphene oxide grids was found to give superior imaging of the soot structures in comparison to dispersion onto conventional amorphous carbon film grid. The reduced contrast from the support film is particularly beneficial when observing small, light element structures such as the soot primary particles in this study.

Soot-in-oil agglomerates were found to be composed of primary spherical particles of 12-40nm. The core-shell structure is clearly visible. Primary particles from used oil samples show the structure as it was at the time of adsorption into the engine oil, which prevents further oxidation. The classic core shell structure is present in all particles from soot in oil samples, with an outer shell of 6-12nm thick.

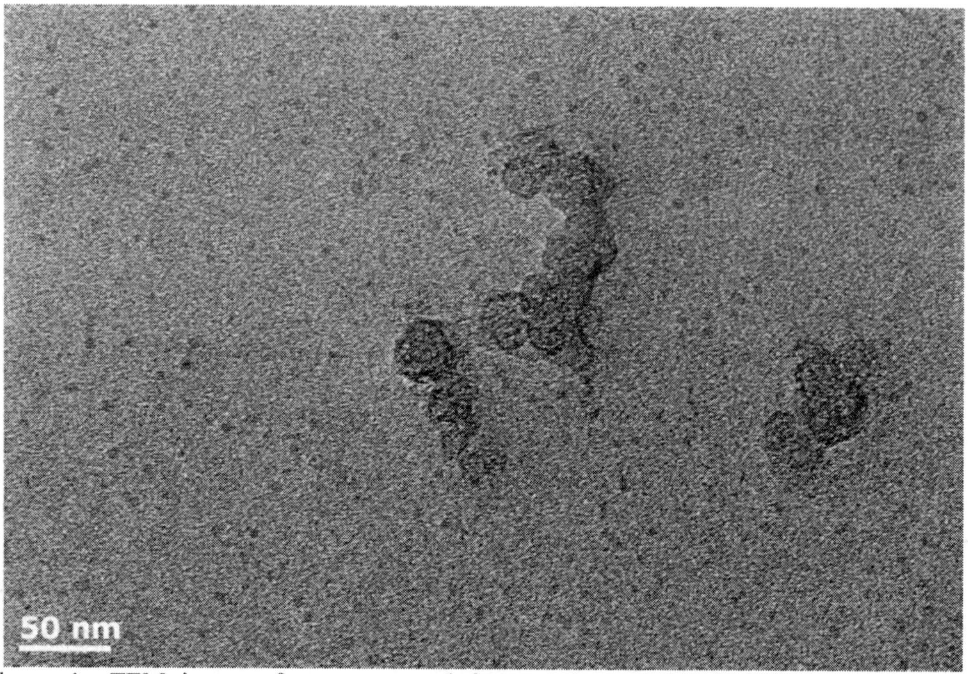

Figure 1 –TEM image of soot extracted from engine oil following solvent extraction, deposited on a conventional amorphous carbon film.

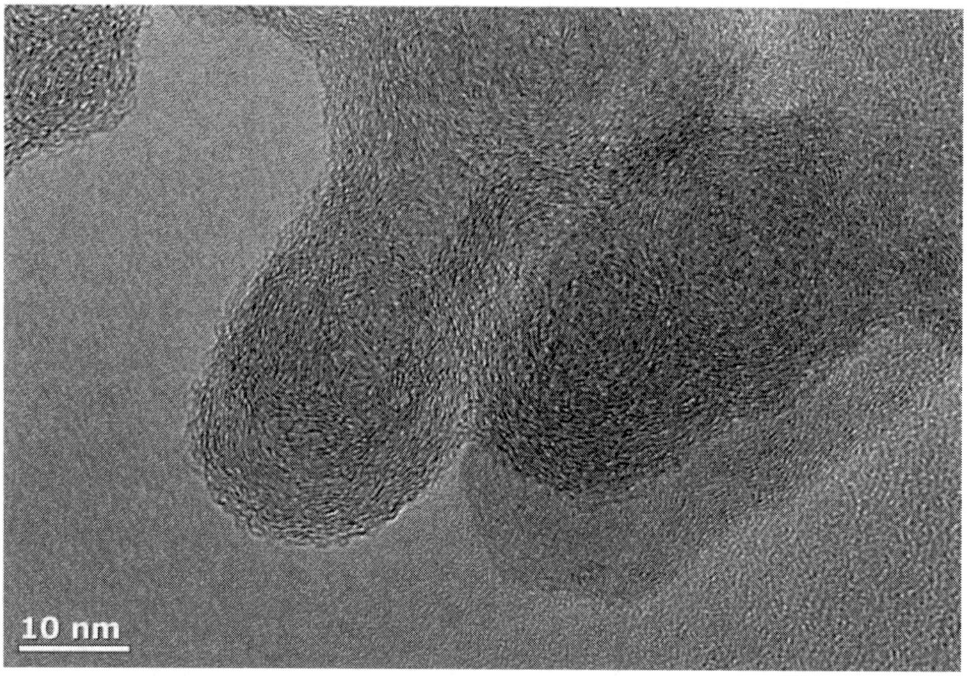

Figure 2 – HRTEM image of soot particles extracted from engine oil following solvent extraction and ultra-centrifugation. This sample was dispersed onto a graphene oxide support grid to give a reduced contrast from the supporting film.

4. Conclusion

Samples produced by solvent extraction were found to be sufficiently clean of mineral oil to analyse the aggregate structure of soot-in-oil. A centrifugation process subsequent to solvent extraction was found to produce samples with minimal remnant oil and hydrocarbon contamination, suitable for high resolution TEM analysis, but is found to produce a differing particle size distribution to samples produced using the solvent extraction process. Analysis of samples using both methods can be combined to obtain a full understanding of the soot-in-oil structure. Graphene Oxide support grids are beneficial for HRTEM analysis of such nanoscale, light element structures.

References

[1] Park A, Kittelson DB, Zachariah MR, McMurry JC 2004 Journal of Nanoparticle Research **6** 267–72.
[2] Ishiguro T, Takatori Y, Akihama K, 1997 Combustion and Flame, **108**, 231-4
[3] Kawamura M, Ishiguro T, Fujita K, and Moromoto H. 1998 Wear, **123,** 269–80.
[4] Liu C, Nemoto S, and Ogano S. 2003 Tribology Transactions **46**, 12–8.
[5] Clague ADH, Donnet JB Wang TK and Peng JCM, 1999 Carbon **37** 1553-65
[6] La Rocca A, Di Liberto G, Shayler PJ, Fay MW, 2013 Tribology International, **61** 80-87

In situ heating and tomography of gold nanoparticles on carbon structures

M W Fay[1,3], A La Torre[2,4], M Giménez-López[2], C Herreros Lucas[2], A N Khlobystov[2], P D Brown[2]

[1] Nottingham Nanotechnology and Nanoscience Centre, University of Nottingham, Nottingham NG7 2RD, UK
[2] School of Chemistry, University of Nottingham, Nottingham NG7 2RD, UK
[3] Department of Mechanical, Materials and Manufacturing Engineering, Faculty of Engineering, University of Nottingham, University Park, Nottingham, NG7 2RD, UK
[4] Institut de Physique et Chimie des Materiaux de Strasbourg, France

michael.fay@nottingham.ac.uk

Abstract. A series of thermally mediated dynamic (in situ) experiments in the transmission electron microscope (TEM) have been performed, in order to obtain improved understanding and control of the ripening and migration processes of gold nanoparticles on multi-wall carbon nanotubes (MWNTs). In particular, post-heating tomography was used to appraise the resultant dispersion of the nanoparticles within these three-dimensional (3D) composite structures.

1. Introduction

Supported gold nanoparticles may be used for a wide range of applications, including low temperature catalysis and light absorption [1]. As the functional properties of gold nanoparticles are strongly affected by their size and shape, it is necessary to obtain a full understanding of the thermal stability of supported gold nanoparticles and how it is affected by the material support.

In this work, we performed in-situ heating of gold nanoparticles on a support material, followed by post-anneal tilt analysis to understand the ripening and migration processes of the nanoparticles, and how the relate to the structure of the support material.

The commercial availability of graphene oxide films is advantageous for work of this nature. This substrate has benefits for both the heating and tomographic analysis of these composite materials. Firstly, the substrate provides better heat conductivity than traditional amorphous carbon films, allowing the temperature at the sample to be more reliably calculated from knowledge of the heating holder crucible temperature. Secondly, the low contrast background of the graphene oxide film assists greatly in tomographic reconstruction, producing minimal background to the sample even at high tilt angles.

2. Experimental

Arc-discharge produced multi-wall nanotubes (AD-MWNTs, MER corporation) were treated with nitric acid (16 M for 24 h) at the reflux point and the resulting mixture was then diluted with deionised water (200 mL). This was filtered using a 0.45 μm pore size PTFE membrane filter, washed thoroughly with water until neutral pH and ethanol (50 mL), and dried under vacuum to yield a black solid (192 mg). 1 mg of Gold nanoparticles were suspended in hexane and mixed, via sonification, with 1 mg of AD-MWNTs [2]. The resultant composite material was again filtered through PTFE and washed with ethanol and acetone. The composite was dispersed in ethanol and spotted onto graphene oxide TEM support films (Graphene Supermarket).

Imaging for in situ heating and tomographic analysis has been performed largely in scanning TEM mode. This rastering mode of imaging reduces beam heating effects [3], ensuring that the ripening of the nanoparticles during in situ heating is predominantly an effect of thermal heating from the support stage, with no observable further ripening occurring during post-heating tomographic analysis. High angle annular dark field (HAADF) imaging is particularly suitable for analysing metal nanoparticles on 3D carbon supports as the atomic number sensitivity makes few-nm sized metallic nanoparticles visible against the background for a significantly greater volume of graphitic material. As the orientation of the crystalline nanoparticles does not strongly affect the intensity of the HAADF image, this mode produces an image series that is particularly suitable for tomographic reconstruction.

TEM investigations were performed using a Jeol 2100F equipped with JEOL and Gatan digital STEM detectors and a Gatan Orius CCD camera for conventional imaging. In situ heating was performed using a Gatan 652 double tilt heating holder. Post-annealing tilt series were acquired using a Gatan 916 room temperature tomography holder. Tilt series acquisitions was performed using SerialEM software, with tomographic reconstruction performed using IMOD 4.3

The relationship between the temperature of the nanoparticles on a graphene oxide film to the recorded temperature of the heating holder crucible was calibrated prior to this work using the evaporation temperature of Ag nanoparticles dispersed onto graphene oxide film, as a reference point [4,5].

3. Results and Discussion

Samples of Au nanoparticles on acid treated MWNTs were first investigated at room temperature to locate a suitable region of the grid and nearby identifying marker features. The size of Au nanoparticles on the structure was found to range from approximately 1.5nm to 6nm in diameter prior to annealing. The holder temperature was then rapidly raised to 500 °C, with the sample location and focus tracked manually using the HAADF STEM mode on high scan rate. Once the temperature had reached 500 °C, images were taken in HAADF STEM mode every five minutes, for a total duration of 150 minutes (Figure 1). The particles had largely reached a stable size after 135 minutes.

After the in-situ heating, the sample was cooled to room temperature in vacuum before transferring to the tomography holder for high-tilt image series acquisition on two axis. A stereogram showing typical images of the post-anneal structure is shown in figure 2. Finally, HRTEM analysis was performed on the exterior gold nanoparticles (figure 3).

Subsequent to annealing, gold nanoparticles within the carbon nanostructure were found to be predominantly between 1 and 2.5nm in diameter. Particles on the exterior of the carbon support ranged from 2nm to 7nm, with a greater number at the higher end of the range than prior to annealing. Coalescence of some gold nanoparticles on the exterior of the support into non-spherical shapes was observed.

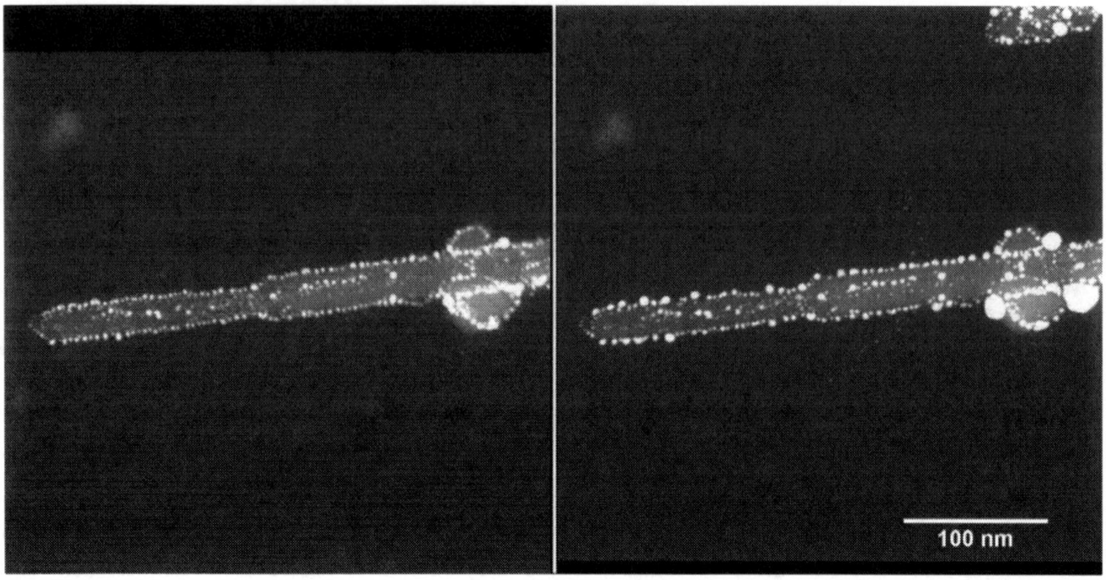

Figure 1 HAADF images of Au nanoparticles on acid treated multi-walled carbon nanotubes, annealed in situ at 500 °C for 135 minutes, illustrating the ripening of the nanoparticles. Left: t=0 min, right t=135min.

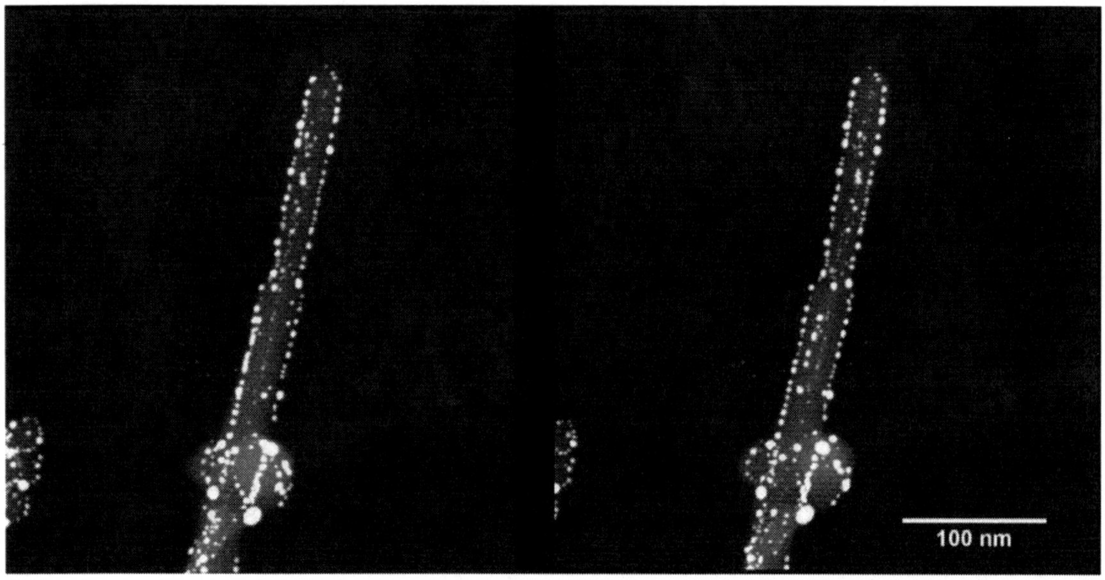

Figure 2 HAADF stereogram images of gold nanoparticles on multi-walled carbon nanotubes after in situ annealing.

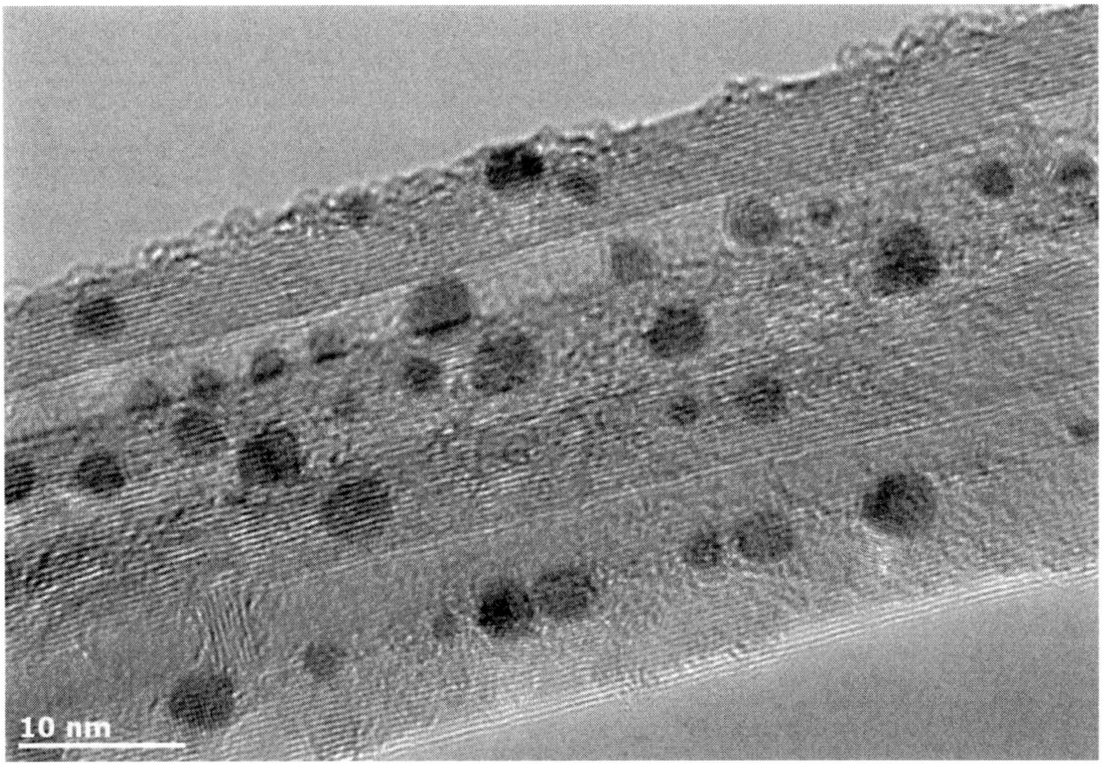

Figure 3 Typical HRTEM image of gold nanoparticles on multi-walled carbon nanotubes after in situ annealing

From the post-anneal tilt-series analysis, it was observed that gold nanoparticles encapsulated within the multi-walled carbon nanotube had a size similar to that of the original particles, while those on the outside had undergone significant ripening. Particles on the exterior were also more commonly located in three lines on the exterior of the carbon nanotube, indicating that they are positioned on an exterior feature of the MWNT.

4. Conclusions

From post-heating tilt-series analysis, it can be seen that particles on the exterior of the nanotube exhibit more significant ripening than those encapsulated within the tube, and are located in a linear orientation on the exterior of the nanotubes. Graphene oxide support films are advantageous for this work, providing improved thermal transport in comparison to conventional amorphous carbon films, while also producing a lower background signal.

References

[1] Turba T, Grant Norton M, Niraula I, McIlroy D N, 2009 J Nanopart Res 11, 2137-2143
[2] Brust M, Walker M, Bethell D, Schiffrin D J, Whyman R, 1994 J Chem Soc Chem Commun, 801-804
[3] La Torre A, Giménez-López M C, Fay M W, Rance G A, Solomonsz W A, Chamberlain T W, Brown P D, Khlobystov A N, 2012 ACS Nano 6 2000-2007
[4] Kang S Y, Kim K, 1998 Langmuir **14** 226-230
[5] Yonezawaa T, Arai S, Takeuchi H, Kamino T, Kuroda K, 2012 Chem Phys Rev **537**, 65-68

Synthesis and characterization of a mixed phase of anatase TiO_2 and TiO_2(B) by low pressure chemical vapour deposition (LPCVD) for high photocatalytic activity

Y. Chimupala[1], G. Hyett[2], R. Simpson[1] and R. Brydson[1]

[1] Institute of Materials Research, SPEME, University of Leeds, Leeds, LS2 9JT,UK.
[2] Department of Chemistry, University of Southampton, Southampton, SO17 1BJ Highfield, UK

Email: pmyc@leeds.ac.uk

Abstract. This project is concerned with enhancing photocatalytic activity by preparing a mixed phase of nano-sized TiO_2. TiO_2 thin films were synthesized by using Low Pressure Chemical Vapour Deposition (LPCVD). Titanium isopropoxide and N_2 gas were used as the precursor and carrier gas respectively. The effects of reaction temperature, carrier gas flow rate and deposited area were studied. TiO_2 thin films with nano-sized TiO_2 particles were obtained under suitable conditions and SEM, TEM, powder XRD and Raman spectroscopy were employed to characterize the phase and physical appearance of synthesized materials. Preliminary results show that a dual phase (TiO_2(B) and anatase) thin film nanopowder was successfully prepared by LPCVD with needle- and polygonal plate-shape crystallites respectively. This thin film deposit produced a preferred orientation of TiO_2(B) needles in the [001] direction of average crystallite size 50-80 nm in length and 5-10 nm in width, whilst the crystallite size of anatase polygonal-plates was around 200 nm. The optimal LPCVD condition for preparing this mixed phase of TiO_2 was 600°C with a 1 mL/s N_2 flow rate.

1. Introduction

Two of the most important world issues are currently energy consumption and environmental pollution. TiO_2 semiconductor photocatalysts are one of the potential solutions to these problems owing to their potential ability for substantial energy production which is cost effective, highly stable and environment friendly [1]. Photocatalytic water-splitting to produce hydrogen using TiO_2 has been investigated however generally the efficiency is low. Currently improvement strategies focus on the use of ultrafine particle sizes and modification of the semiconductor band gap.

Engineering of the TiO_2 band gap by mixing two different phases of TiO_2 (such as anatase/rutile and antase/TiO_2(B) [2]), directly affects the charge transfer process between the different phases reducing recombination of photo-generated electrons and so enhancing the photocatalytic activity [2,3]. TiO_2 (B) has been of significant interest over the last decade because of the distinctive crystal structure which exhibits bigger channels and voids along the c axis in its monoclinic crystal structure as compared with anatase and rutile tetragonal structures [4]. Commonly, alkaline metal titanates have been used as precursors for TiO_2(B) preparation in many synthetic methods such as sol-gel, hydrothermal and wet chemical process [5]. In this research, the LPCVD method was employed to synthesis (for the first time) dual phase anatase and TiO_2(B) thin films for use as a photocatalyst for the water splitting reaction.

2. Experimental Procedure

2.1. Preparation procedure for dual phase TiO_2

Dual phase TiO_2 thin films were prepared by chemical vapour deposition under low pressure at 25 mbar. Titanium tetra-isoproproxide (TTIP, 97%) was used as the chemical vapour precursor and a bubbling chamber settled in an oil bath at 90°C was filled with 5-10 mL TTIP. The TTIP vapour generated was carried into aquartz tube reaction by N_2 at a flow rate of 1mL/s. Titanium oxides were deposited onto a silica glass slide of dimensions 12.5 mm × 150 mm × 1 mm (width × length × thickness) via thermal decomposition of TTIP. The reaction temperature in the tube furnace was varied between 450 and 600°C with a reaction time of 15 min.

2.2. *Materials Characterization*

The deposited microscope glass slide was divided into 12 square samples each of dimensions 12.5 mm × 12.5 mm × 1 mm before being characterized. The samples represented different reactor positions relative to the gas entry point (square 1). The samples were characterised by both in-plane and out-of-plane X-ray diffraction (XRD) and also Raman spectroscopy. The thin film surface morphology and film thickness were studied by scanning electron microscopy (SEM) both in plan view and in cross-section. For transmission electron microscopy (TEM) the deposited materials were scraped off the glass substrate with a scalpel onto a holey carbon film supported on a Cu TEM grid.

3. Results and Discussion

The most interesting synthesis condition was found to be 600°C for 15 min of 5 mL TTIP. All of 12 square deposited glass slides were studied by out of plane XRD as partially shown in figure 1.

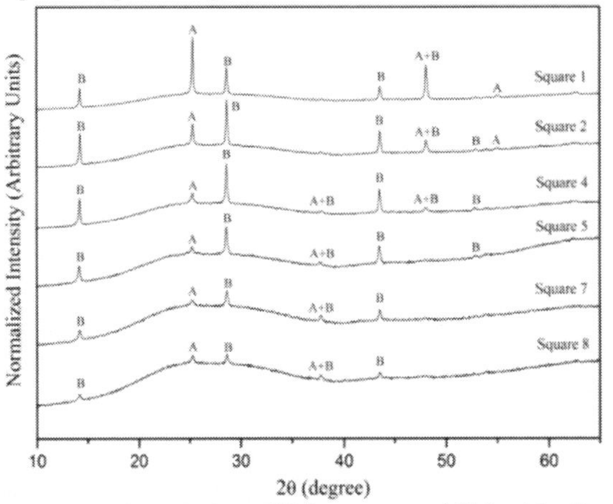

Figure 1. Out of plane XRD patterns of TiO_2 thin film prepared by LPCVD at 600°C: A and B refer to the assignment of the anatase and TiO_2(B) phases respectively.

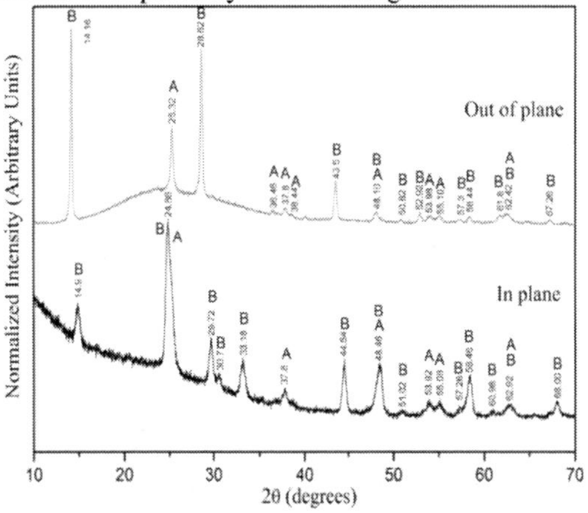

Figure 2. In plane and out of plane XRD patterns of Square 4: A and B refer to anatase and TiO_2(B) phases respectively.

XRD patterns illustrate that all of prepared materials are a mixed phase combination of both TiO_2 anatase and TiO_2(B). In Square 1 the majority phase is anatase, however in all other reactor positions, the main component is the TiO_2(B) phase. It can be seen that two XRD peak positions of anatase at around 2θ = 38° and 48° which corresponds to (112) and (200) peaks respectively, are nearly at the same peak positions as the (401) and (020) peaks of TiO_2(B). Both in-plane and out-of-plane XRD were used to confirm the existence of the mixed phase by focusing on the Square 4 sample, as shown in figure 2.

The in-plane XRD pattern shows different peaks of TiO_2(B) to the out-of-plane pattern which can also be indexed as TiO_2(B) phase. Not only does this confirm the existence of the TiO_2(B) phase it reveals a preferred [001] crystallographic orientation of TiO_2(B) as illustrated at 2θ = 14.16, 28.62 and 43.5 degrees.

Raman spectroscopy was employed to confirm the dual phase (figure 3). The intense peak at 144.96 cm^{-1} and four other peaks at 197.86, 394.4, 515.6 and 638.6 cm^{-1} are the Raman active modes for anatase. The remaining peaks, especially two strong peaks at 121.95 cm^{-1} and 144.96 cm^{-1} indicate the Raman active modes of the TiO_2(B) phase [6].

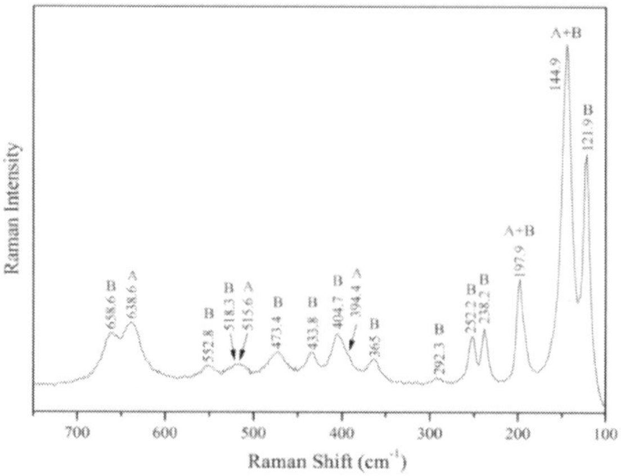

Figure 3. Raman spectrum of Square 4: A and B indicates the anatase and TiO_2(B) phases respectively.

A SEM plan view image of Square 4 is shown in figure 4a and illustrates that there are two kind of particle morphology, one is a large plate with particle size about 100-200 nm, whilst the other one is a nearly round small particle with a diameter of around 20 nm. The average primary particle size of this bimodal distribution at the film surface is 72.95 (S.D =41.87) nm as elucidated from the Particle Size Distribution in figure 4b. SEM/EDX results, shown inset in figure 4a, confirm the presence of Ti and O in the thin film. However, Na, Mg, Si, Ca and excess atomic percentage of O are also present which may come from the glass-slide substrate owing to the large electron beam interaction volume which is bigger that the TiO_2 film thickness of 200 nm as shown in the SEM cross-section in figure 4c. It is also worth nothing that it appears that the thin film again contains two kinds of particles; again a smaller, more equiaxed particles predominantly occurring at the glass substrate/film boundary and the other plate-like particles predominantly deposited on the film surface. This may suggest that the small particles are acting as a seeding layer on the substrate.

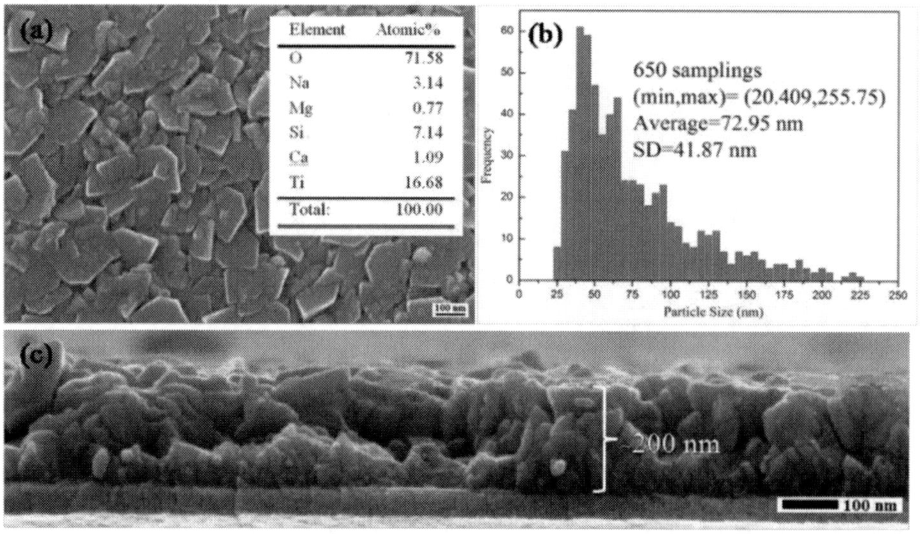

Figure 4. (a) SEM secondary electron image of TiO_2 thin film from Square 4; (b) Primary Particle Size Distribution measured from measurement of 650 particles; (c) SEM cross-sectional image of thin film thickness after fracturing the glass slide.

Both XRD and Raman results indicate the coexistence of anatase and TiO_2(B) therefore the prepared thin film should provide two type of particle morphology as suggested by the SEM images. A typical TEM image, figure 5a, also shows two distinct types of particle morphology which are large polygonal plate-like and smaller needle-shaped. The crystallite size of the needles is in the range 50-80 nm in length and 5-10 nm in width, very much smaller than the polygonal plate of size around 200 nm.

Both of the morphologies were investigated by HRTEM and the relationship between nanostructure and physical appearance determined as shown in figures 5b and 5c. In Figure 5b, a needle shape clearly shows a lattice spacing of 3.18Å and 2.91Å which can be tentatively assigned to the lattice spacings of ($\bar{1}11$) and (400) of TiO_2(B). The measured angle between two lattice planes is 88° which is close to the theoretical angle of 83°. Moreover, the TEM selected area diffraction pattern of a group of needles confirms the existence of TiO_2(B). When combined with the XRD results, this suggest the needles must be growing preferentially along their long axis ([001] direction), i.e oriented vertically. Figure 5c shows that the polygonal plate is anatase as it reveals lattice spacings of 3.47Å, 2.38Å and 2.28Å which correspond to (101), ($\bar{1}03$) and (112) spacings respectively.

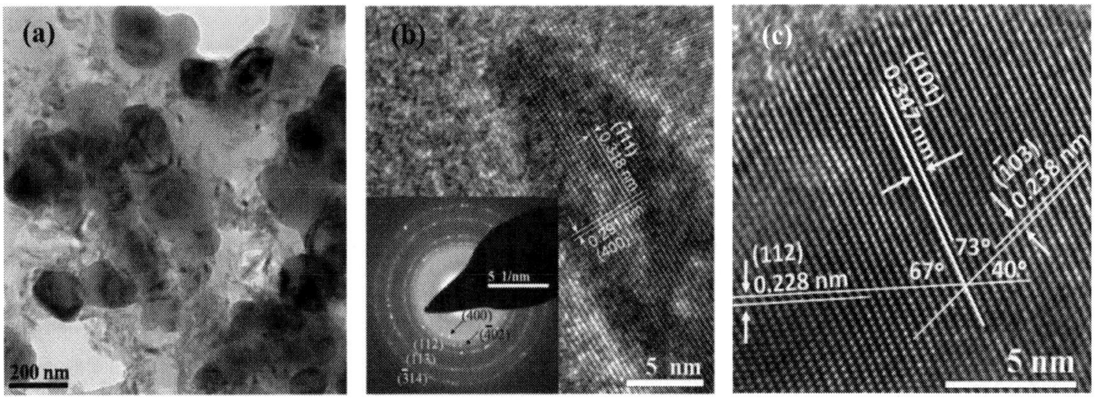

Figure 5. (a) A TEM image taken from Square 4, showing both anatase polygonal plates and TiO_2(B) needle-shaped particles; (b) HRTEM of a TiO_2(B) needle and SADP of a group of the needles; (c) HRTEM image of an anatase polygonal plate.

4. Conclusions

200 nm thick TiO_2 thin films consisting of dual phase monoclinic TiO_2(B) needles and larger anatase polygonal plates have been prepared by an LPCVD method at 600°C for 15 min of 5 mL TTIP precursor. To our knowledge, this is the first time the monoclinic TiO_2(B) phase has been prepared by CVD. The small nano-needles of the TiO_2(B) phase exhibit a preferred orientation along [001].

Acknowledgements

Acknowledgements to the Royal Thai Government for a scholarship to YC.

References

[1] Hoffmann M R , Martin S T, Choi W and Bahnemann D W 1995 *Chem. Rev.* **95** 69
[2] Jiefang Z, Jinlong Z, Feng C and Masakazu A 2005 *Mater. Lett.* **59** 3378
[3] Meng N, Michael K H L, Dennis Y C L and Sumathy K 2007 *Renew. Sust. Energ. Rev.* **11** 401
[4] Marc E, Concepcion D and Jose A A 2010 *Mater. Lett.* **64** 2357
[5] Xiang G L, Li T Y, Zhuang J and Wang X 2010 *Chem. Commun.* **46** 6801
[6] Mouna B Y, Frederic L, Thomas B, Jean S F, Mireille R P, Luc B and Marie L D 2009 *J.Chem. Phys.* **130** 204501

Field emission properties of aligned ZnO nanorods on Ti substrate

H. Algarni[1,a,b], **M. M. El-Gomati**[a], **M. S. Al-Assiri**[c,d]
[a]Department of Electronics, The University of York, York, United Kingdom (UK)
[b]Department of Physics, Faculty of Science, King Khalid University, P.O. Box 9003 Abha, Kingdom of Saudi Arabia.
[c]Promising Center for Sensors and Electronic Devices (PCSED), Najran University, P.O. Box 1988, Najran 11001, Kingdom of Saudi Arabia
[d]Faculty of Sciences and Arts, Department of Physics, Najran University, P.O. 1988, Najran-11001,Kingdom of Saudi Arabia

E-mail: hmha500@york.ac.uk

Abstract. Aligned ZnO nanorods were grown on titanium (Ti) substrate by simple and facile thermal evaporation process using high purity metallic zinc powder in the presence of oxygen. The grown nanorods were studied in terms of their morphological, structural and field emission properties by using different analytical tools such as field emission scanning electron microscopy (FESEM) and X-ray diffraction (XRD). The field emission properties of the as-grown aligned ZnO nanorods were also investigated which resulted in a turn-on field of ~2.65 kV. This research demonstrates that simply prepared ZnO nanorods can be used for field emission device applications.

1. Introduction

Research on 1D semiconductor nanomaterials has received a great deal of attention due to their unique morphologies and wide applications in various high-technological areas including photovoltaics, electronics, sensors and actuators, to name but few such areas [1]. Among the various semiconductor nanomaterials, the II-VI wurtzite hexagonal phase ZnO is particularly interesting due to its own properties and wide applications in various areas of science and technology. Moreover, the 1D nanomaterials of ZnO such as nanorods, nanowires, nanofibers, nanoneedles, etc are very attractive due to their tunable electronic and optoelectronic properties and various high-technological applications, for instance, nanoscale electronic and optoelectronic devices, sensors and actuators, solar cells, nanogenerators, laser operating at room-temperature and so on [2]. Recently, the use of various substrates for the growth of 1D ZnO nanomaterials has received great attention as particular substrates are needed for specific applications. The substrates used for the growth of ZnO nanorods/nanowires are various orientations of silicon chips [3-5], stainless steel grid [6], indium tin oxide substrates [7], nylon fibers [8], alumina [9], anodized aluminum oxide [10] , CaF_2 [11], steel alloy substrates [12], and so on.

[1] To whom any correspondence should be addressed.

In this paper, attempt has been made to grow well-crystalline ZnO nanorods on Ti substrate which were examined in detail in terms of their morphological, structural, and field emission properties. To the best of the author's knowledge, this is the first report in which vertically aligned and well-crystalline hexagonal ZnO nanorods were grown on Ti substrate by facile thermal evaporation of metallic zinc powder in the presence of oxygen at moderate temperatures.

2. Experimental details

For the growth of aligned ZnO nanorods, commercially available high purity (99.9%) Ti substrates were used. Prior to the growth, the substrates were ultrasonically cleaned with DI water, alcohol and acetone, sequentially. The growth of nanorods on several pieces of 1cmx1cm Ti substrates was carried out by a facile thermal evaporation process by using metallic zinc powder in the presence of oxygen. After loading the source materials and substrates, the chamber pressure was evacuated down to 2 mbar using a rotary vacuum pump and the same pressure was maintained during the whole reaction. During the reaction, the furnace was rapidly heated up to 750 °C in the presence of high purity nitrogen (20 sccm) and oxygen (100 sccm) gases. The reaction was completed in 1.5 hr. The furnace was naturally cooled to room-temperature and the deposited structures were analyzed in terms of their morphological and structural properties using a field emission scanning electron microscope (FESEM) and X-ray diffraction (XRD) with Cu-Kα radiation. The field emission properties of the grown structures were also investigated.

3. Results and discussion

3.1 Morphological, compositional and structural properties of as-prepared ZnO nanorods on Ti Substrate

The general morphology of the as-deposited ZnO sample on Ti was imaged by field emission scanning electron microscopy (FESEM). Figure 1 (a) exhibits the typical secondary electron (SE) images of the as-synthesized ZnO nanorods on Ti substrate. The SE image reveals that the synthesized nanorods are grown in very high density and vertically aligned to the substrate surface (figure 1 (a)).

To examine the crystallinity and crystal phases, the as-prepared ZnO nanorods were examined by X-ray diffraction (XRD). Figure 1 (b) exhibits the typical XRD pattern of the as-prepared ZnO nanorods. All the observed diffraction reflections in the XRD pattern are well matched with the wurtzite hexagonal phase of pure well-crystalline ZnO and hence confirm that the prepared nanorods are well-crystalline and possessing wurtzite hexagonal phase of ZnO. Some small diffraction reflections at 38.4°, 40.3°and 53.1° are also seen in the observed XRD pattern of ZnO nanorods which are related with the Ti substrate as was confirmed by measuring the XRD pattern of pure Ti substrate.

Figure 1 (c) exhibits the typical low-magnification TEM image of the as-grown ZnO nanorods which shows clean and smooth surfaces and uniform diameter throughout the length of the ZnO nanorod in the range of 15-20nm. The inset of figure 1 (c) shows the corresponding SAED pattern of the nanorod shown in figure 1 (c). The observed SAED pattern shows very well defined, ordered diffraction pattern which confirmed that the grown nanorods are well-crystalline and preferentially grown along the c-axis direction. Figure 1 (d) exhibits the typical HRTEM image of the nanorods shown in figure (c). The HRTEM image shows the well-defined lattice fringes with the inter-planar distance of 0.52 nm, equal to the lattice constant of crystalline ZnO. This fact clearly confirms that the prepared nanorods are well-crystalline ZnO.

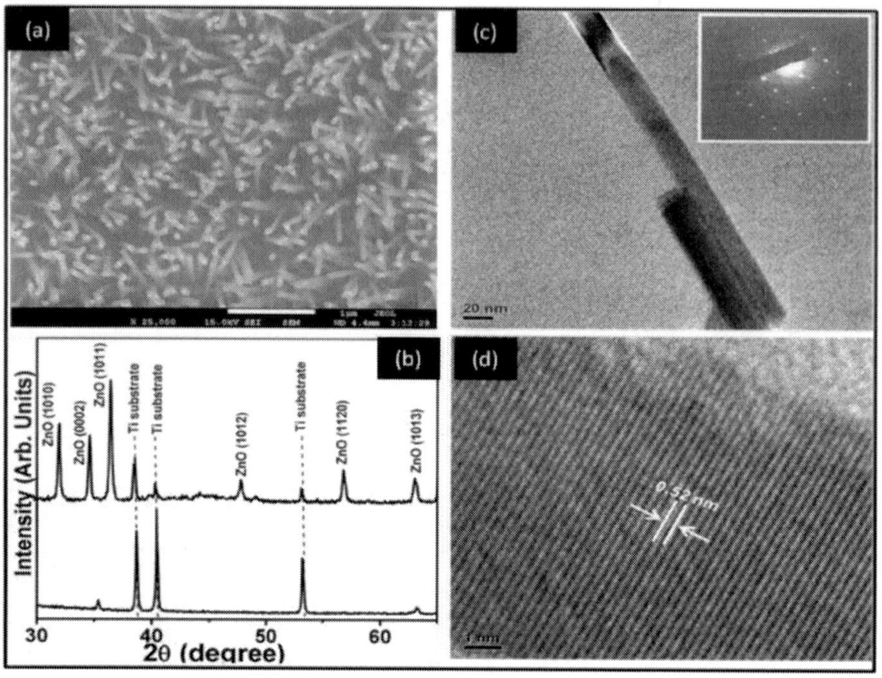

Figure 1. Typical (a) FESEM image, (b) XRD pattern and (c and d) low-magnification and high-magnification TEM images of as -grown ZnO nanorods on Ti substrate and with diameters in the range of 15-20nm.

3.2 Field emission properties of the as-prepared ZnO nanorods on Ti Substrate

The field emission (FE) properties of ZnO nanorods were also investigated using a triode configuration under a base pressure of ~5x10^{-8} mbar. For the FE measurements, two experimental scans were carried out for each sample. Figure 2 (a) shows the typical I-V behaviour (measured current as a function of applied voltage) of the ZnO nanorods. It is seen that in the initial scan, referred to as first scan, the emission current increased with increasing the applied voltage. The initial turn-on voltage is 2.55 kV with the observed emission current of ~100 nA. A maximum emission current of ~13 μA was recorded at an applied voltage of 4.56 kV. In the first scan, the emission current was quite unstable. Moreover, the I-V curve of the first scan usually appears to be noisy and this is attributed to surface contamination of the emitting sites [13]. Therefore, repeated scans were necessary to achieve stable and reproducible emission current.

A much stable and less noisy I-V curve was obtained in case of the second FE measurement which can be attributed to the removal of the surface's contaminants during first FE scan. However, slightly higher voltage was required for the same emission current. The interpolated value of the turn-on voltage for an emission current ~100 nA is ~2.65 kV.

The Fowler–Nordheim equation was employed to analyse the behaviour of the emission current versus the applied voltage. The F-N equation can be expressed as

$$J = 1.54 \times 10^{-6} \frac{(\beta E)^2}{\Phi} \exp\left(-6.83 \times 10^{9} \frac{\Phi^{2/3}}{\beta E}\right)$$

Where J is the current density, E is the macroscopic value of the applied field (E=V/d where V and d are the applied voltage and the distance between the anode and the sample, respectively. β refers to the field enhancement factor and Φ is the work function which is 5.3 eV for ZnO [14]. Fitting the data using the Fowler–Nordheim (F-N) plots showed linear behaviour indicating that the field emission process is dominant. The (I-V) characteristics, a representative image of the field emission pattern from the sample and the F-N relationship are shown in figure 2 (a) and (b) respectively.

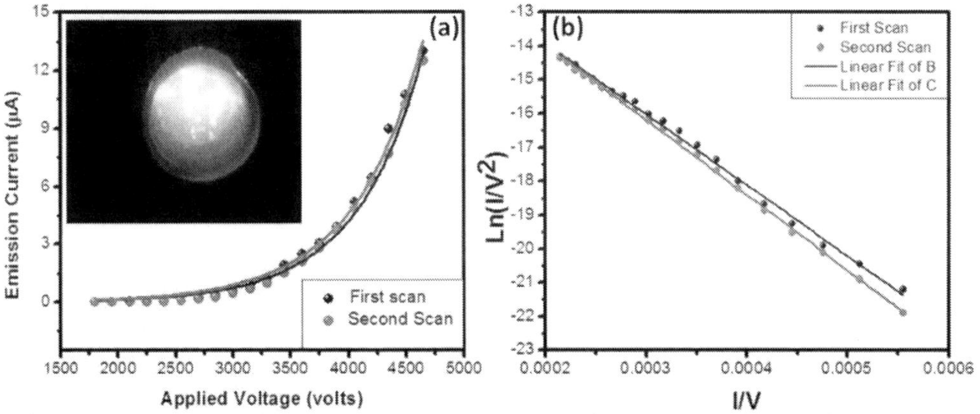

Figure 2. FE measurements of ZnO nanorods; (*a*) current-voltage curve and the typical FE pattern image (inset of (*a*)) and (*b*) corresponding F–N relationship.

4. Conclusion

In summary, a simple and facile approach has been adopted to grow aligned ZnO nanorods on Ti substrate and their morphological, structural and field emission properties have been explored. The detailed morphological and compositional properties of the as-grown products revealed that the prepared nanomaterials are "ZnO nanorods" which are aligned to the substrate surface while the structural studies revealed that the prepared nanorods are well-crystalline and possessing wurtzite hexagonal phase. Finally, the field emission performance of the vertically aligned ZnO nanorods on Ti substrate was investigated and presented in this paper.

Acknowledgements:

This work was supported through a grant (PCSED-008-11) under the Promising Centre on Sensors and Electronic Devices at Najran University, Saudi Arabia.

References

[1] Ed. Umar A and Hahn YB, 2010 "*Metal Oxide nanostructures and their applications*", American Scientific Publishers (ASP), Los Angeles, USA
[2] Ed. Nalwa H S, 2011 "*Encyclopedia of Nanoscience and Nanotechnology*, 25-Volume Set, American Scientific Publishers (ASP), Los Angeles, USA
[3] Ladanov M, Ram MK, Matthews G, Kumar A, 2011 *Langmuir*, **27**, 9012
[4] Yang P, Yan H, Mao S, Russo R, Johnson J, Saykally R, Morris N, Pham J He R and Choi H -J 2002 *Adv. Funct. Mater.* **12**, 323
[5] Wan Q, Li Q H, Chen Y J, Wang T H, He X L, Li J P and Lin C L, 2004 *Appl. Phys. Lett.* **84**, 3654.
[6] Xu X Y, Zhang H Z, Zhao Q, Chen Y F, Xu J and Yu D P, 2005 *J. Phys. Chem. B* **109**, 1699.
[7] H.K. Lee, M.S. Kim, J.S. Yu, 2011 *Nanotechnology*, **22**, 445602
[8] Xue CH, Wang RL, Zhang J, Jia ST, Tian LQ, 2010 *Mater. Lett.* **64**, 327
[9] Wang X D, Summers C J and Wang Z L 2004 *Adv. Mater.* **16**, 1215.
[10] Fang Z, Wang Y, Peng X, Liu X and Zhen C, 2003 *Mater. Lett.* **57**, 4187
[11] Liu Y C, Xu X Y, Mu R, Henderson D O, Lu Y M, Zhang J Y, Shen D Z, Fan X W and White C W, 2003 *Appl. Phys. Lett.* **83**, 1210.
[12] Umar A, Hahn Y B, 2006 *Appl. Phys. Lett.* **88**, 173120
[13] Martin E E, Trolan J K and Dyke W P, 1960, *J. Appl. Phys.* **31**, 782
[14] Lee C J, Lee T J, Lyu S C, Zhang Y, Ruh H and Lee H J 2002 *Appl. Phys. Lett.* **81**, 3648

Journal of Physics: Conference Series **522** (2014) 012076 doi:10.1088/1742-6596/522/1/012076

Indium-doped ZnO nanopencils: structural and field emission properties

H. Algarni[1,a,b], M. M. El-Gomati[a], M. S. Al-Assiri[c,d]
[a]Department of Electronics, The University of York, York, United Kingdom (UK)
[b]Department of Physics, Faculty of Science, King Khalid University, P.O. Box 9003 Abha, Kingdom of Saudi Arabia.
[c]Promising Center for Sensors and Electronic Devices (PCSED), Najran University, P.O. Box 1988, Najran 11001, Kingdom of Saudi Arabia
[d]Faculty of Sciences and Arts, Department of Physics, Najran University, P.O. 1988, Najran-11001,Kingdom of Saudi Arabia

E-mail: hmha500@york.ac.uk

Abstract. This paper reports on the direct growth of In-doped ZnO nanopencils on Si substrate by facile thermal evaporation process using metallic zinc and In powders in the presence of oxygen. The as-grown nanostructures were examined in detail in terms of their morphological, structure and field emission properties. The morphological studies, carried out by FESEM, exhibited that the grown structures possess homogeneous morphologies with pointed tips similar to pencils. The typical tip diameters of as-synthesized nanopencils are in the range of ~13 ± 3 nm which reflects that these nanopencils could be a promising candidate for field emission applications. The detailed structural properties revealed that the prepared nanopencils are well-crystalline and possessing wurtzite hexagonal phase. For application view point, the field emission properties of as-grown In-doped ZnO nanopencils were examined which exhibited a very low turn-on voltage of 1.3 kV.

1. Introduction

Recently, the fabrication, characterization and utilization of 1D nanomaterials are receiving great interest due to their peculiar, fascinating physical, chemical and mechanical properties which are different from their bulk counterparts. Among the various possible applications, the utilization of 1D nanomaterials for field emission (FE) devices is one of the most active research areas because of its practical applications in FE displays, x-ray sources and microwave devices, etc [1-4]. Therefore, many studies have been carried out in this area in the past decades. In this direction, the carbon nanotubes (CNTs) were extensively studied for their potential as field emitters and particularly their very high aspect ratio. However, CNTs as field emitters were found to be sensitive to the working environment (especially, H_2O and O_2 adsorbates) which leads to high instability in the field emission performance [5, 6]. To overcome these problems, researchers have recently tried to use metal oxide nanomaterials for the fabrication of field emission devices [7]. Amongst the various metal oxide materials, the II-VI semiconductor zinc oxide (ZnO) possesses a special place as a "smart material" due to its own properties and wide applications [7]. The attractive properties of ZnO include its wide band gap (3.37

---

[1] To whom any correspondence should be addressed.

eV), high exciton binding energy (60 meV) at room-temperature, low electron affinity, high mechanical strength and chemical and thermal stability. Hitherto, a variety of ZnO nano- and microstructures have been synthesized using various fabrication techniques and their optical and FE properties have been checked and demonstrated in the literature [7]. Moreover, to enhance the field emission properties of ZnO nanomaterials, various metal doping have also been performed and reported in the literature [7]. By detailed studies, it was found that the shape and size of the doped and undoped ZnO nanomaterials have strong influence on the field emission properties of the material. To examine this fact, in the previous study the, various morphologies of undoped ZnO, such as nanosheets networks and nanorods and Ga-doped ZnO nanoneedles have been investigated for field emission device applications. In this paper, the growth, characterizations and field emission applications of In-doped ZnO nanopencils is presented.

2. Experimental details

In-doped ZnO nanopencils were grown on a silicon substrate by facile thermal evaporation process by using metallic indium and zinc powders in the presence of oxygen. Silicon wafers (1 x 1.5 cm) were used as substrates for the deposition which were ultrasonically cleaned with DI water and isopropyl alcohol and finally rinsed with acetone. In a typical reaction process, high-purity metallic Indium and zinc powders were thoroughly mixed in 1:10 ratio and transferred into a quartz boat and placed at the centre of the quartz tube furnace. Consequently, the cleaned substrates were placed adjacent to the source material. After this arrangement and prior to starting the heating, the quartz tube furnace was evacuated to 10-1~ 10-3 mbar and kept under vacuum throughout the synthesis process. After evacuation of the quartz tube, the furnace was heated up to 800 °C using a halogen lamp heating system in the presence of high purity nitrogen and oxygen gases. Once the temperature of the furnace reached over 500 °C, indium and zinc vapors were generated (as the melting point of Indium and zinc powders are less than 500 °C) which were transferred to the substrate by nitrogen carrier gas. The oxygen gas was reacted with the generated zinc and indium vapors and finally the growth was carried out. The reaction was completed in 1 hr. After completing the reaction, the furnace was very slowly cooled down (15 °C/min) to room-temperature and the deposition of a grayish white colored product was observed onto the whole substrate surface. The deposited material was characterized in terms of its morphological, structural and field emission characteristics by using various analytical tools.

3. Results and discussion

3.1 Morphological, compositional and structural properties of as-prepared In-doped ZnO nanopencils

The general morphologies of as-prepared material was examined by field emission scanning electron microscopy (FESEM) and presented in figure 1 (a) which confirm that the prepared material possess pencil-shaped morphologies and that these nanopencils are grown in very high density as. It is also interesting to see that the nanopencils are made of two parts; i.e. a hexagonal nanorod and a nanoneedle which are connected in such a manner that the base of the nanoneedle is attached to the head of hexagonal nanorod. The typical diameters of the base hexagonal nanorods in the nanopencils are ~80 ± 10 nm while the lengths are in the range of ~3 ± 0.5 μm. The diameters for the tips of the nanopencils are ~13 ± 3 nm. The hexagonal morphology of the synthesized nanopencils reveals that the synthesized products possess wurtzite hexagonal phase. It is worth mentioning that such nanostructured morphologies are very promising for the fabrication of field emission devices.

Figure 1 (b) exhibits the typical EDS spectrum of as-prepared In-doped ZnO nanopencils which confirmed that the nanopencils are made of zinc, indium and oxygen as spectrum exhibits peaks only related with these elements. No other peak related with any other element is seen in the EDS spectrum further reveals that the nanopencils are In-doped ZnO without any significant impurity, up to the detection limit of EDS method.

The crystallinity and crystal phases of the as-prepared ZnO nanopencils were examined by x-ray diffraction (XRD) pattern and shown in figure 2. Several well defined diffraction reflections are seen in the observed XRD pattern which are mostly related with the wurtzite hexagonal phase of ZnO. A small diffraction reflection originated at 54.5° is related with In and can be assigned as In (112). In addition to zinc and In reflections in the diffraction pattern, a small reflection appeared at 32.9° which is due to the silicon substrate and can be attributed as Si(002). Finally, due to the presence of a small In peak with other wurtzite ZnO reflections in the pattern reveals that the In atoms are efficiently incorporated into the lattices of grown nanopencils. Moreover, no characteristic diffraction reflection related with any impurity was detected in the pattern within the detection limit of the x-ray diffractometer, which confirms that the synthesized nanopencils are well-crystalline and are pure In-doped ZnO.

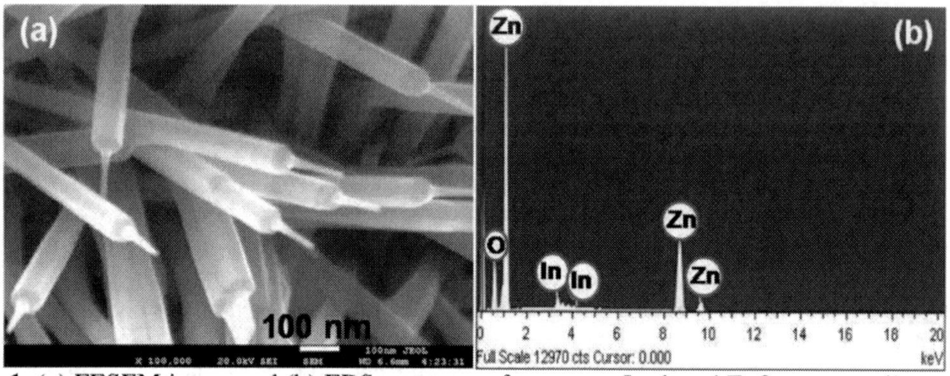

Figure 1. (a) FESEM image and (b) EDS spectrum of as-grown In-doped ZnO nanopencils on silicon substrate prepared by thermal evaporation process

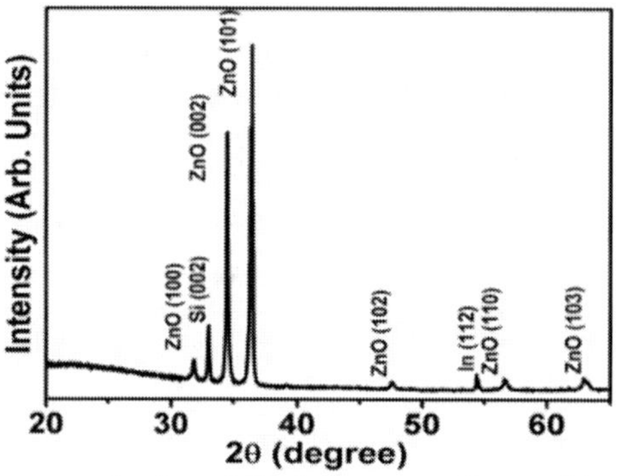

Figure 2. XRD pattern of as-grown In-doped ZnO nanopencils on silicon substrate.

3.2 Field emission (FE) properties of as-prepared ZnO nanorods on Ti Substrate

The determination of FE property of In-doped ZnO nanopencils was carried out in a vacuum chamber with pressure of $3x10^{-8}$ mbar. The field emission curve was obtained by increasing the applied voltage from 0 to 3kV with steps of 0.1 kV. The I-V characteristics curve and FE pattern of the In doped ZnO nanopencils along with the corresponding FN plot are shown in figure 3 (a) and (b) respectively.

The turn on and threshold voltages were found to be ~1.3 kV with emission current of 80 nA and 3kV with emission current of 17.15 μA, respectively. It is well known that the distribution of the field around an emitter depends on its shape, therefore, the low turn on voltage of the In-doped ZnO nanopencils can be related to their distinctive geometrical configuration [8]. The In-doped ZnO nanopencils are excellent candidates for FE due to these reasons: i) the ultra-fine part of the tip intensify and strengthen the electric field around the apex of the emitter giving rise to huge enchantment of the local electric field leading to electrons being emitted at low electric field [9]. ii) The two step configuration localizes the electric field on the ultra-fine part of the emitter and therefore avoids emission from unwanted sites specially the thick base of the emitters [9]. The impact of the doping is equally important to the geometry effect and cannot be ignored. Doping of indium into the ZnO leads to improvement in the electrical properties (i.e. conductivity and carrier concentration) and therefore lowering the work function which gives rise to enhancement of the field emission performance of these emitters [10].

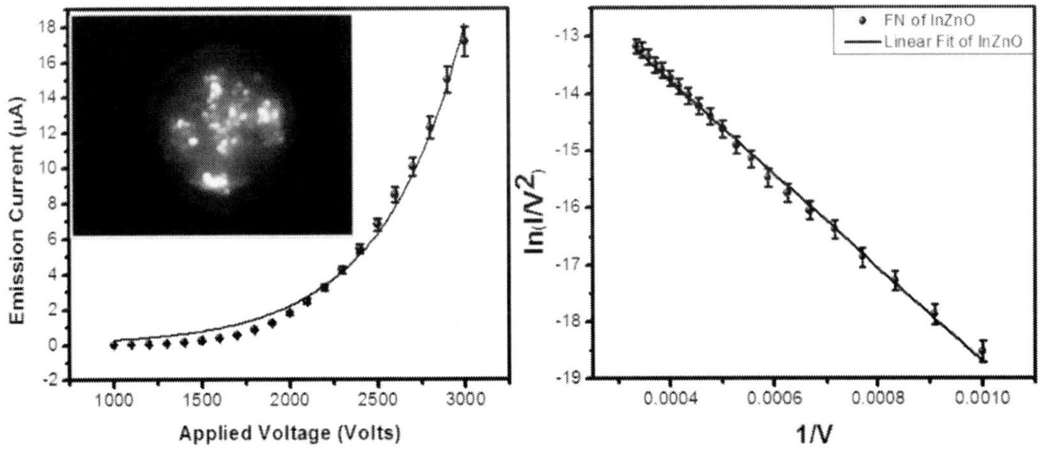

Figure 3. FE measurements of ZnO nanopencils; (*a*) current-voltage curve and the typical FE pattern image (inset of (*a*)) and (*b*) corresponding F–N plot.

4. Conclusion

In summary, In-doped ZnO nanopencils were grown on silicon substrate by facile thermal evaporation process using metallic In and zinc powders in the presence of oxygen. The detailed morphological and structural properties revealed that the prepared nanomaterials are nanopencils, grown in high density and possessing wurtzite hexagonal phase. The field emission properties of the as-prepared nanopencils grown on silicon substrate were also investigated which exhibited a very low turn-on voltage of 1.3 kV. The field emission results show that the In doped ZnO nanopencils emitters can be very efficient candidate in applications such as flat panel displays and high brightness electron sources.

Acknowledgements:
This work was supported through a grant (PCSED-008-11) under the Promising Centre on Sensors and Electronic Devices at Najran University, Saudi Arabia.

References

[1] Milne W I, Teo K B K, Amartunga G A J. Legagneux P, Gangloff L, Schnell J P, Semet V, V. Thien Binhc V and Greening O, 2004 *J. Mater. Chem.* **14**, 933
[2] Bai X, Wang MS, Zhang GM, Yu J, Zhang ZX, Guo DZ, Zhao XY and Xue ZQ, 2007 *J. Vac. Sci. Technol. B*, **25**, 561
[3] Yang YH, Wang CY, Chen US, Hsieh WJ, Chang YS and Shih HC,2007 *J. Phys. Chem. C*, **111**, 1601
[4] Dean K, Allmen P V and Chalamala B, 1999 *J. Vac. Sci. Technol. B*, **17**, 1959
[5] Sarrazin P, *Carbon nanotubes science and applications, 2005 ed.* M. Meyyappan, (CRC press LLC) pp.195
[6] Kim C, Choi YS, Lee SM, Park JT, Kim B and Lee YH, 2002 *J. Am. Chem. Soc.*,**124**, 9906
[7] Ahmad Umar, Y. B. Hahn, "Metal oxide nanomaterials and their applications" Vol. 5, American Scientific Publishers, Los Angeles, USA (2010)
[8] Gomer R, 1961 *Field Emission and Field Ionization*, Harvard University, Cambridge, Mass.
[9] Pan N, Xue H, Yu M, Cui X, Wang X, Hou JG, Huang J and Deng SZ. 2010 *Nanotechnology* **21**, 225707
[10] Huang Y, Zhang Y, Gu Y, Bai X, Qi J, Liao Q and Liu J, 2007 *J. Phys. Chem. C*, **111**, 9039

High Angle Dark Field Imaging of Two-Dimensional Crystals

R Zan[1,2], Q M Ramasse[3], R Jalil[2], T Georgiou[2], U Bangert[1] and K S Novoselov[2]

[1] School of Materials, University of Manchester, Manchester, UK
[2] School of Physics and Astronomy, University of Manchester, Manchester, UK
[3] SuperSTEM, STFC Daresbury Laboratory, Warrington, UK

recep.zan@manchester.ac.uk

Abstract. Two-dimensional (2D) materials, graphene, hexagonal boron nitride (h-BN) and transition metal dichalcogenides (TMD) have been investigated by means of Scanning Transmission Electron Microscopy (STEM), in particular via High Angle Annular Dark Field (HAADF) imaging technique. They are compared in terms of their structure and durability under intense electron beams.

1. Introduction

Graphene and other 2D nanomaterials have attracted considerable attention since their isolation due to their unique physical, chemical, and structural properties as well as their great potential for applications [1]. Single-layer h-BN is one such material whose wide band gap of ~5.9 eV arguably makes it the thinnest possible insulator and therefore a perfect candidate for incorporation into complex heterostructures consisting mostly of low-dimensional materials [2]. Owing to the unique properties of h-BN, much effort has been put into exploring its potential use in electronic and optoelectronic devices such as field emitters and ultraviolet or radiation detectors [3]. It has also been recently found that h-BN is a much better substrate (inert) for graphene studies than SiO_2 as it improves the graphene electronic properties [4]. Other technologically important 2D materials, TMDs e.g. molybdenum disulphide (MoS_2) and tungsten disulphide (WS_2), are also under heavy investigation as a new class of materials with remarkable properties [5]. Single layer TMDs with a thickness of 6.5Å have a direct band gap contrary to their bulk forms which have an indirect band gap. By implementing these layered materials into graphene based devices (by stacking them on top of each other), the device performance can be enhanced [6]. Although electrical, magnetic and mechanical studies of 2D materials and their combination as new material systems are established, there is a lack of direct atom-by-atom visualizations, thus limiting our understanding of these highly exciting material systems.

Here we present a STEM study, in particular via HAADF imaging of graphene, h-BN, MoS_2 and WS_2, which we have used extensively in the past years. The location and nature of individual atoms, defects and peeling thick regions of 2D crystals layer by layer under an intense electron beam will be discussed.

2. Experimental

Two different sample preparation techniques have been employed to fabricate our 2D crystals. The chemical vapor deposition (CVD) technique for graphene and the micromechanical cleavage technique for h-BN, MoS_2 and WS_2 have been used. To prepare CVD-graphene films for the microscopy measurements, the films were initially coated with PMMA followed by etching of the copper substrate, on which graphene was grown, in ammonium persulphate solution. The other 2D materials were 'peeled' several times from their bulk crystals using adhesive tape and then placed onto an oxidized silicon wafer (Si/SiO_2). Thin flakes consisting of single up to many layers were identified under the optical microscope, coated with PMMA and transferred to a quantifoil TEM grid, after having etched-off the Si/SiO_2 substrate. The PMMA layer was then dissolved by dipping the grid into an acetone for about 15 minutes and then drying the samples in a critical point dryer to avoid possible sample damaging due to surface tension [7].

Preliminary studies (location and identification of the layers) of 2D crystals have been performed in a conventional TEM, however, detailed structural and topographic properties have then been studied in an aberration corrected STEM (Daresbury SuperSTEM). The structure, individual atoms and defects in the samples have been directly revealed in HAADF images, which show ~Z^2 sensitivity. The STEM was operated at 60 keV, which is known to be 'safe' (below the knock-on threshold damage) for all studied materials according to the ab-initio calculations for all B, C, N, S, Mo and W[8,9].

3. Results and Discussion

Figure 1 shows raw STEM-HAADF images of single layer graphene at different magnifications. The reasonably large area of graphene in fig. 1a) shows the single layer is not perfectly clean. Residue free (clean) graphene regions are limited in size (from few to few hundred nm^2) and surrounded by contamination, which was found to consist of C, O and Si in our EDX analysis. These hydrocarbon contaminations appears bright due to higher scattering with heavier atoms and larger thickness, however, clean graphene areas are dark. Moreover, due to different thickness of hydrocarbon layers, their brightness in HAADF image appears from grey to very bright, shown in fig. 1b). On the other hand, to be able to see individual carbon atoms and periodicity a closer look is required. Although it is an unprocessed image, the carbon atoms and hexagonal periodicity of the graphene are clearly visible in fig. 1c). Carbon atoms are bright and holes dark in the image in fig. 1c) and benzene rings are overlaid in red highlighting the hexagonal periodicity.

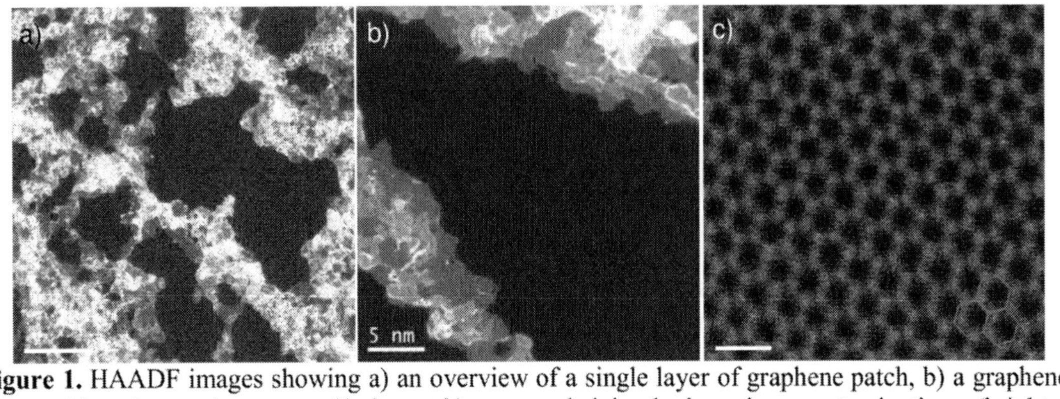

Figure 1. HAADF images showing a) an overview of a single layer of graphene patch, b) a graphene flake with a large clean area (dark patch) surrounded by hydrocarbon contaminations (brighter patches), c) atomic resolution detail in pristine single layer graphene. All images are raw.

Due to its weak contrast on oxidized silicon wafers compared to graphene and other 2Ds, obtaining and identification of single layer h-BN via optical microscope is quite challenging. Double layer h-BN STEM-HAADF images are presented in fig. 2a). The intensity of both, B and N atoms is low similar to carbon in graphene but they are discernible in HAADF image due to the $\sim Z^2$ sensitivity. Although theoretical calculations predict that the knock-on threshold for h-BN should lie above the beam energy used for this study, 60 kV, we found that there is considerable damage inferred to pristine h-BN. As a result, thicker regions of the h-BN flakes can be peeled layer-by-layer under the beam, which conveniently exposes a different number of layers and helps in determining the stacking configuration. An HAADF image of a 'beam-exfoliated' region is shown in fig. 2b), where a single layer patch with triangular shape appears in the middle of the image, surrounded by the original double layer sheet. It is known that h-BN damaging leads to the formation of triangular holes in single layer h-BN sheets[10]. Fig. 2c) is the noise filtered image of fig. 2b) and shows improved visibility of the B and N sites in the h-BN sheet. This allows better differentiation between B and N, and is further demonstrated by drawing an intensity profile (inset) along the line in the triangular feature. The intensity ratio between N and B in fig 2c) is found to be 1.91, which is compatible with the Z^2 ratio of 1.96.

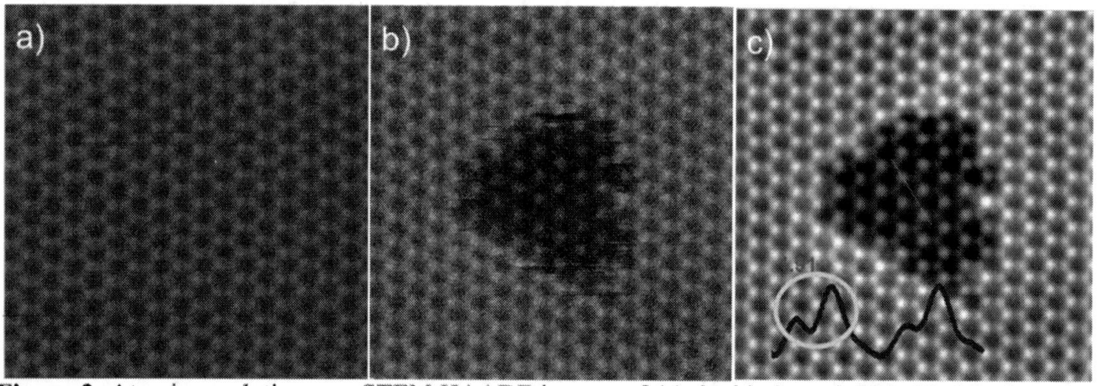

Figure 2. Atomic resolution raw STEM-HAADF images of (a) double layer h-BN, (b) the triangular appearance of single layer h-BN as a result of electron beam etching of the double layer, (c) filtered image of (b). B and N can be distinguished by their intensity in the profile (inset) across along the line in the 'triangle'(c).

Single layer MoS_2 consists of two distinct sub-lattices: Mo (metal) atoms are trigonal-prismatically bonded to pairs of S atoms and arranged hexagonally in a plane. The resulting S-Mo-S slabs are then stacked in various sequences, 2H stacking being the most common. Similar to graphite and h-BN, chemical bonds within the plane are covalent (strong) while weak van der Waals bonds are established between stacked layers to form a bulk (3D) crystal. Figure 3 illustrates how damage initiates immediately: the image in fig. 3b acquired approximately 30 s after that in fig. 3a) demonstrates the ease with which damage is introduced. The single-layer MoS_2 is perforated by losing first a single S atom and then the other Mo-bonded S atom resulting in the formation of a vacancy [9]. As can be seen in fig. 3c), the vacancy is soon turned into a hole (after ~ 1min), which enlarges easily under the scanning electron beam. This is mainly due to the atoms at the edge of the hole being less coordinated with fewer/weaker atomic bonds thus facilitating both ionization and knock-on damage. Furthermore, the sputtered atoms during the damaging behave differently from those in graphene and h-BN in which atoms were knocked-out and lost from the area. However, here the sputtered Mo atoms tend to accumulate at the edge of the hole revealing themselves by extra brightness in the HAADF image in fig 3c). This is probably due to higher atomic weight of Mo, which requires higher displacement energies. EDX analysis (not shown) confirms that the brighter edges are rich in Mo rather than S. This is also the case for single layer WS_2, as is revealed by the decorated edge around the hole in fig. 3d), which is rich in W. The electron dose used in these images was ~$5x10^6$ e/Å^2, which is similar to what we used to acquire other 2Ds.

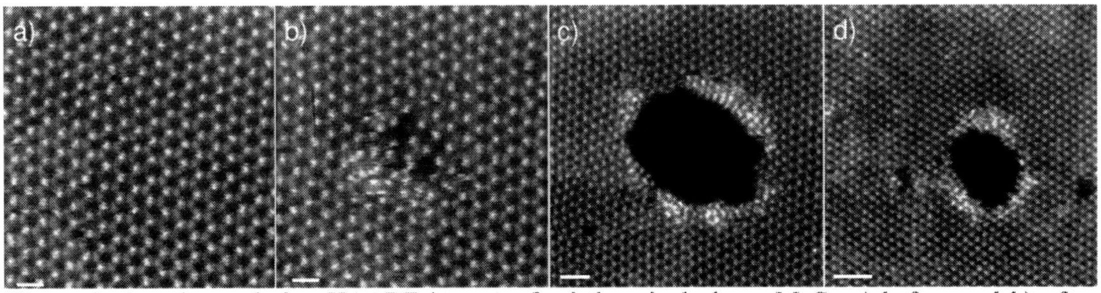

Figure 3. Atomic resolution HAADF images of pristine single layer MoS_2, a) before and b) after consecutive scans, showing vacancy formation; c) expansion of the damaged region; d) damaged single layer WS_2. All images are raw.

4. Conclusions

In conclusion, we have prepared samples of free standing graphene, h-BN, MoS_2 and WS_2 via different techniques and analyzed by aberration corrected STEM at 60 kV. We were able to observe individual carbon, boron and nitrogen, molybdenum and sulfur as well as tungsten and sulfur atoms in the graphene, h-BN, MoS_2 and WS_2 lattices, respectively in HAADF images. However, employing a 60-keV acceleration voltage of the electron beam and near-UHV conditions (<$5x10^{-9}$ torr in the column) did not prevent occurrence of damage in free-standing 2D crystals, except for graphene, either due to knock-on or ionization damage. This indicates the need for application of protective measures to 2Ds during prolonged microscopy analysis, such as graphene encapsulation.

References

[1] Novoselov, K. S.; Falko, V. I.; Colombo, L.; Gellert, P. R.; Schwab, M. G.; Kim, K. *Nature* **2012,** 490, (7419), 192-200.
[2] Gorbachev, R. V.; Riaz, I.; Nair, R. R.; Jalil, R.; Britnell, L.; Belle, B. D.; Hill, E. W.; Novoselov, K. S.; Watanabe, K.; Taniguchi, T.; Geim, A. K.; Blake, P. *Small* **2010,** 7, (4), 465-468.
[3] Kubota, Y.; Watanabe, K.; Tsuda, O.; Taniguchi, T. *Science* **2007,** 317, (5840), 932-934.
[4] Xue, J.; Sanchez-Yamagishi, J.; Bulmash, D.; Jacquod, P.; Deshpande, A.; Watanabe, K.; Taniguchi, T.; Jarillo-Herrero, P.; LeRoy, B. J. *Nat Mater* **2011,** 10, (4), 282-285.
[5] Wang, Q. H.; Kalantar-Zadeh, K.; Kis, A.; Coleman, J. N.; Strano, M. S. *Nat Nano* **2012,** 7, (11), 699-712.
[6] Georgiou, T.; Jalil, R.; Belle, B. D.; Britnell, L.; Gorbachev, R. V.; Morozov, S. V.; Kim, Y.-J.; Gholinia, A.; Haigh, S. J.; Makarovsky, O.; Eaves, L.; Ponomarenko, L. A.; Geim, A. K.; Novoselov, K. S.; Mishchenko, A. *Nat Nano* **2012,** 8, (2), 100-103.
[7] Zan, R.; Ramasse, Q. M.; Bangert, U.; Novoselov, K. S. *Nano Letters* **2012,** 12, (8), 3936-40.
[8] Zobelli, A.; Gloter, A.; Ewels, C. P.; Seifert, G.; Colliex, C. *Physical Review B* **2007,** 75, (24), 245402.
[9] Komsa, H.-P.; Kotakoski, J.; Kurasch, S.; Lehtinen, O.; Kaiser, U.; Krasheninnikov, A. V. *Physical Review Letters* **2012,** 109, (3), 035503.
[10] Jin, C.; Lin, F.; Suenaga, K.; Iijima, S. *Physical Review Letters* **2009,** 102, (19), 195505.

Plasmonic Enhancement at Metal Atoms on Graphene Edges revealed by EFTEM

W Pierce[1], R Zan[1], Q M Ramasse[2], C B Boothroyd[3] and U Bangert[1]

[1]School of Materials, The University of Manchester, Manchester, UK
[2]SuperSTEM, STFC Daresbury Laboratory, Warrington, UK
[3]Ernst Ruska-Centre for Microscopy and Spectroscopy with Electrons and Peter Gruenberg Institute, Juelich Research Centre, D-52425 Juelich, Germany

william.pierce@postgrad.manchester.ac.uk

Abstract. Palladium atoms have been deposited onto graphene where they catalyse etching processes in conjunction with the e-beam, during/after which they reside at the edges of the holes, which have formed in the graphene. Energy filtered imaging reveals that the low loss feature at 2-4 eV constituting the shoulder of the graphene π-plasmon, is up to 10 times enhanced at Pd-decorated graphene edges compared to clean monolayer graphene, rendering it a useful feature for electric field enhancement applications in the optical regime

1. Introduction

Graphene is an extremely versatile material for applications in photodetectors, solar cells and plamonics, as transparent electrode, light enhancer and wave guide. In all these cases graphene needs to be nano-sculpted, i.e., hole and functional impurity atom patterns need to be controllably introduced. Following work by Zhou et al [1], who have shown that individual impurity atoms (in their case Si) can act as atomic antennae, enhancing the graphitic σ-plasmon in graphene, we investigate the possibility of enhancement and energy red-shifting of the π-plasmon and thereto coupled single particle excitations in the visible wavelength regime. For this purpose we introduce metal atoms, which mediate graphene nano-etching under the electron beam [2, 3], and are also expected to enhance absorption and emission within the visible EM spectrum. We concentrate on the case of Pd-adatoms, which induce electron absorption features in the energy regime 2-4 eV, especially at single and double layer graphene edges, scrolls or kinks of one layer above the other.

2. Experimental

We employ two different techniques for graphene preparation, exfoliation from HOPG graphite [4] and the CVD method, in which graphene is grown on copper [5]. Metal impurities have been introduced using electron beam and thermal evaporators, depositing 2 Å of various metals at room temperature. Graphene layer numbers have been identified from electron diffraction patterns by comparing the first and second ring intensities; atomic resolution imaging of metal atoms and clusters on the graphene surface has been carried out in stationary transmission electron microscopy mode (TEM) as well as in high angle annular dark field (HAADF) scanning mode (STEM), the former in a Tecnai F30, the latter in an aberration corrected dedicated STEM (Daresbury SuperSTEM). Atomic sites of singular metal atoms and clusters on graphene have been directly revealed in HAADF imaging mode by their approximate Z^2 dependence, and by relating HAADF with simultaneously obtained electron energy loss (EEL) spectrum images. Further lattice resolution phase contrast images in conjunction with low loss/ plasmon images were obtained in a monochromated and triple aberration corrected Titan PICO. For this purpose stacks of TEM images were obtained, where the same region was successively imaged with an electron beam that has lost increasing amounts of energy. The stacks then consist of a number of energy filtered images (EFTEM images) taken at energy losses typically ranging from -2 eV (energy gain) to 8 eV, in steps of 0.2 eV (in this case the number of images in the stack is 51). A stack can be viewed as a 3-D array, where the x- and y-axis assign the 2-D image dimensions and the z-axis is the energy. Each x-y slice (image) consists of a certain number of pixels (e.g., 512x512); EEL spectra can then be extracted by plotting the intensities of vertically stacked pixels against energy (projection of the pixel intensity onto the z- or energy-axis).

3. Results and Discussion

Metal atom mediated graphene nano-etching [2] is catalised by an electron beam in a (S)TEM (however, knock-on and ionization effects are not believed to be the mechanisms for this). The HAADF image in fig. 1a shows hole formation for the case of Pd, the elemental nature of which was established by single-atom EELS (fig. 1b). The Pd atoms, due to the larger binding energy at graphene edges [6], decorate the forming hole. Pd- atoms are expected to enhance absorption and emission

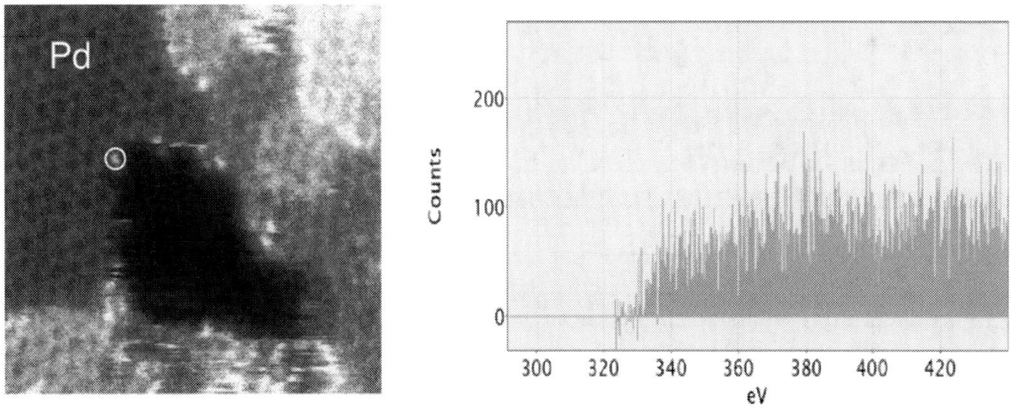

Figure 1. a) HAADF image of 2Å Pd evaporated single layer graphene (false colour: graphene purple, Pd yellow, vacuum dark blue); b) EEL spectrum of the encircled atom in (a) revealing the fingerprint of the Pd L-edge.

within the visible EM spectrum. As can be seen from the EEL spectrum in fig. 2, which was extracted from an EFTEM stack in positions indicated by the arrows, Pd atoms induce electron absorption features at 2-4 eV, especially at single and double layer graphene edges, scrolls or kinks of one layer above the other; judging from the phase contrast in the BF images (figs 2a and b) there appear to always be Pd atoms present (often in chains) when there is an edge-intensity enhancement in fig. 2c. In greater detail: figure 2a was obtained before the acquisition of the EFTEM stack, from which the plasmon map (fig. 2c) comprising of the sum of EFTEM images with losses 3, 3.2, 3.4, 3.6, 3.8 and 4 eV was extracted. The image in fig. 2b was taken after the acquisition of the EFTEM stack. It becomes clear that during the long electron exposure damage to the grapheme occurred resulting in the expansion of the three original holes (fig. 2a) to merge into one (fig.2b). The EFTEM image in fig. 2c shows the intermediate stage. It can nevertheless be seen that Pd-atoms (dark dots and dotty lines in the underfocus phase contrast images) are lining the graphene edges. Fig. 2c shows a clear enhancement of the EEL spectrum feature at 2-4 eV, i.e., at uv/vis energies, at Pd atoms along the hole. Spectra (green) extracted from the EFTEM stack at the hole's edge (arrowed positions) show enhancement by a factor of ~10 compared to spectra (blue) obtained from pure graphene (arrowed position). The top edge of the hole in fig. 2b consists of a double layer; the enhancement for mono- and double-layer edges is similar.

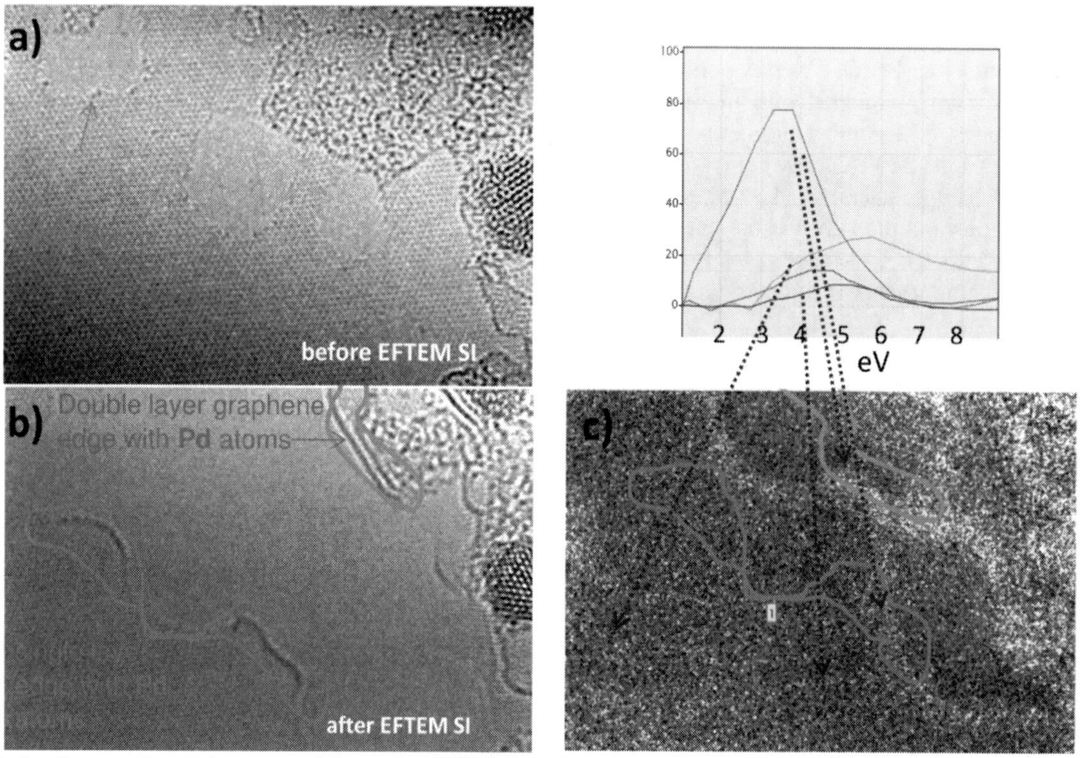

Figure 2. a) 2 Å Pd evaporated on graphene predominantly present as nanocrystals in the ubiquitous contamination but also as individual atoms at the edges of holes; b) changes in the graphene topography under exposure to the electron beam after EFTEM: the 3 separate holes in (a) have merged into one large hole; c) residual intensity of absorption features at 3-4 eV after subtracting the zero loss peak and the π-plasmon at 4.7 eV from an EFTEM stack acquired of the area as in the BF images (a) and (b) showing enhancement in the UV/vis regime at graphene edges; this is further proven by the spectra extracted from the merely zero loss peak subtracted EFTEM stack. Images and energy loss measurement were obtained in a monochromated and triple aberration corrected Titan PICO at 80 keV

The existence of plasmons in graphene is still disputed, hence the issue of whether the observed feature is a 'true' plasmon or an exciton albeit coupled to a plasmon, or even a collective exciton, i.e., an interband transition, requires further investigation. However, since the absorption feature is a strong one, this issue does not defer from its usefulness for potential light -coupling and -enhancing applications.

4. Conclusions

In conclusion, it appears that individual Pd atoms at edges of holes in graphene, the latter being a consequence of metal-mediated etching, give rise to considerable enhancement of absorption features in the uv/vis wavelength regime. Together with the nano-topographic aspect of these features the observed phenomenon could lead to real applications in nano-photonics and –plasmonics.

References

[1] Zhou, W., Lee, J., Nanda, J., Pantelides, S. T., Pennycook, S. J. and Idrobo, J.-C., *Nature Nanotech* **2012,** 7, 161.
[2] Ramasse, Q. M., Zan, R., Bangert, U., Boukhvalov, D. W., Son, Y.-W. and Novoselov K. S., *ACS Nano* **2012**, 6 (5), 4063
[3] Zan, R.; Bangert, U.; Ramasse, Q.; Novoselov, K. S., *J. Phys. Chem. Lett.*, **2012**, *3* (7), 953.
[4] Novoselov, K. S.; et al. A. *Science* **2004,** 306 (5696), 666.
[5] Li, X.; et al. *Science* **2009,** 324 (5932), 1312.
[6] Hardcastle T., Seabourne, C. R., Zan, R., Brydson, R. M. D., Bangert, U., Ramasse, Q. M., Novoselov, K. S. and Scott, A. J., *Phys. Rev.B* **2013,** 87, 195430.

Thermal induced structural transformation of bimetallic AuPd nanoparticles

A. Bruma and Z. Y. Li

The Nanoscale Physics Research Laboratory, School of Physics and Astronomy, The University of Birmingham, B15 2TT, UK

Z.Li@bham.ac.uk

Abstract. High Angle Annular Dark Field Scanning Transmission Electron Microscope (HAADF-STEM) has been employed for the study of thermal effects of structural transformation of AuPd nanoparticles produced by physical vapour deposition. Depending on the duration of annealing at a temperature of 500 K, atomic resolved imaging analysis reveals the formation of various structure morphologies from the ordered $L1_2$ superlattice to the core-shell structure. The effects of Pd-oxides are also discussed.

1. Introduction

Bimetallic AuPd nanoparticles have received considerable attention in recent years partly due to their potential applications in nanocatalysis and fuel cells. In order to design and produce nanoparticles with tailored properties, understanding of the interplay of thermodynamics and kinetics in nanoparticle formation is essential. Theoretically, the structure and chemical ordering of AuPd nanoparticles have been investigated [1-3]. Due to the lower surface energy of Au against Pd, in general surface segregation of Au is favoured, resulting in $Pd_{core}Au_{shell}$ chemical structure. However, depending on how AuPd nanoparticles are synthesized, experimental observations of $Au_{core}Pd_{shell}$ as well as mixed phases have been reported in the literature [4-7]. Here, kinetics may play an important role in structure formation, apart from other external factors such as molecule adsorption. In this paper, we report our recent study on effects of thermal annealing on size, structure and chemical ordering of bimetallic AuPd nanopartilces prepared by physical vapour deposition (PVD). Our goal is to gain a better understanding of the formation mechanisms of various AuPd nanoalloys.

2. Experimental details

The AuPd nanoparticles were produced via PVD on amorphous carbon thin film-covered transmission electron microscopy (TEM) grids under high vacuum conditions (~10^{-7} mbar). Pd atoms were deposited first, followed by deposition of Au. The substrate temperature was held at 300 K during the deposition. Subsequent annealing was carried out at a temperature of 500 K. A 200 kV JEM 2100F scanning transmission electron microscope (STEM), fitted with a CEOS probe corrector for spherical aberration, was employed to image the nanoparticles. All images were recorded using a camera length of 10 cm, which corresponds to an inner collection angle of the high angle annular dark field (HAADF) detector of 62 mrad.

3. Results and discussions

Figure 1 (a) and (b) show typical HAADF-STEM images of AuPd nanoparticles deposited onto amorphous carbon at 300 K. It was observed that the nanoparticles do not show any clear evidence of phase segregation. This is in line with expectations from most of the previous calculations [1-2]. An analysis of the size distribution of AuPd nanoparticles in Figure 1 (c) shows a mean diameter of 1.70±0.34 nm.

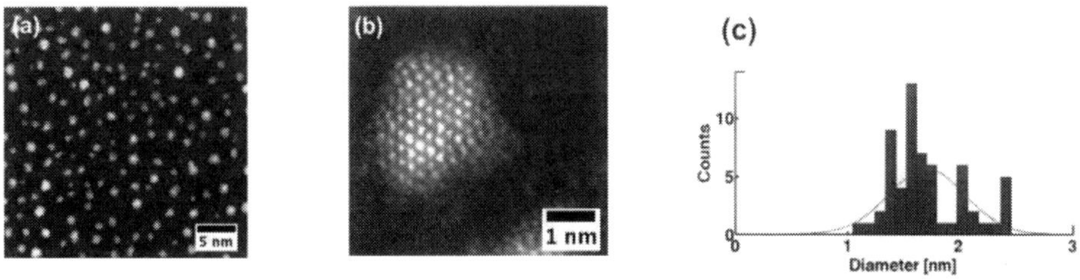

Figure 1 (a, b) Low and high magnification HAADF-STEM image of as-deposited AuPd nanoparticles onto amorphous carbon substrate maintained at 300 K. (c) Diameter distribution of AuPd nanoparticles in (a).

Type	60 minutes	120 minutes	180 minutes
Mixed	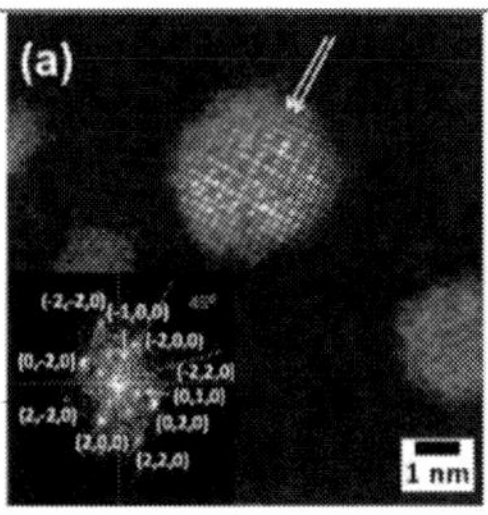	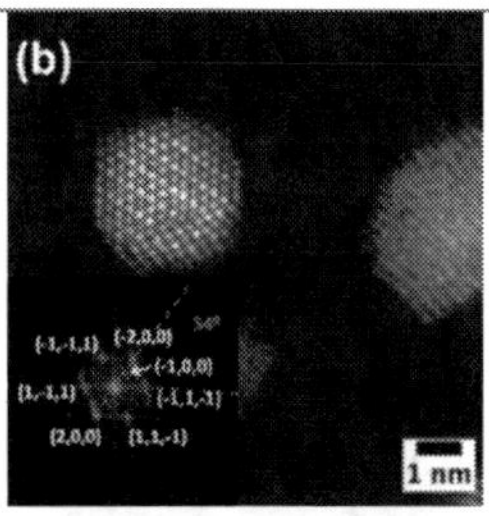	N/A
Layered	N/A	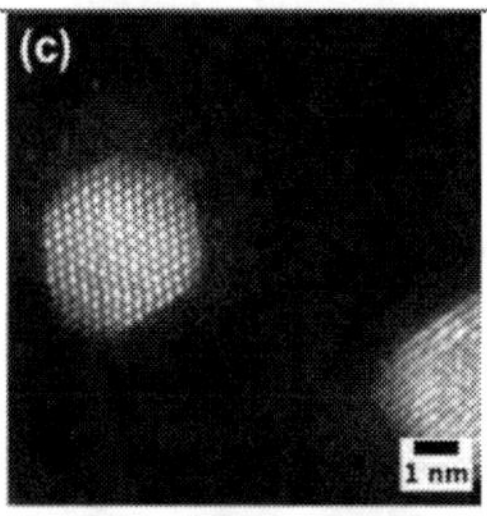	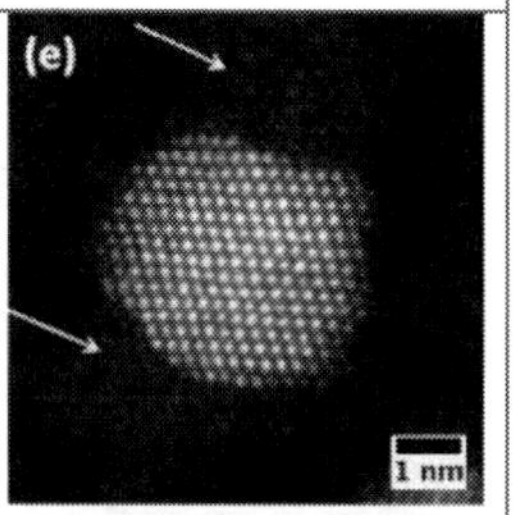
Core-shell	N/A	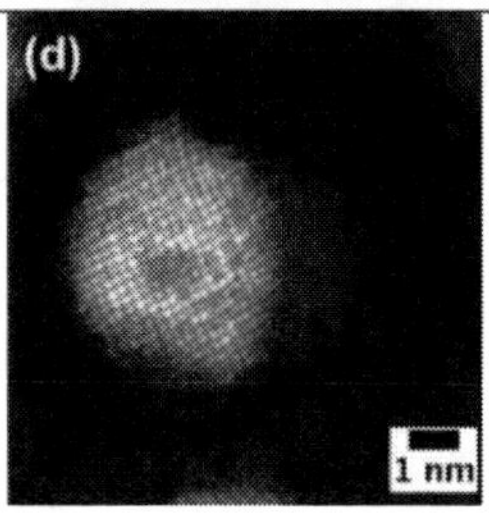	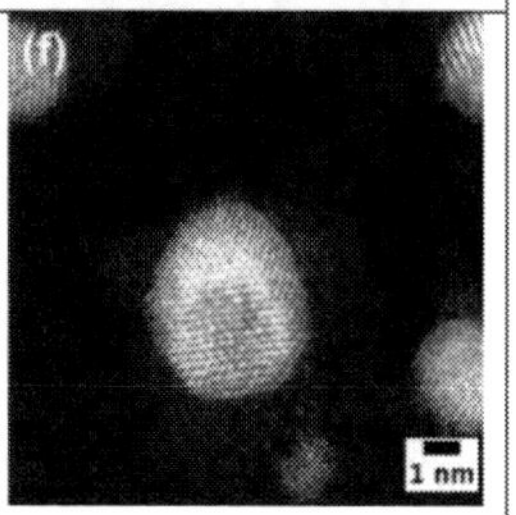

Figure 2 HAADF-STEM images of AuPd nanoparticles annealed at 500 K in situ, for: (a) 60 minutes; (b-d) 120 minutes, and (e-f) 180 minutes.

Journal of Physics: Conference Series **522** (2014) 012079 doi:10.1088/1742-6596/522/1/012079

When the sample was annealed at a temperature of 500 K for 60 mins, a well-ordered alloyed phase of the AuPd nanoparticles could be observed, as shown in Figure 2 (a). Examining the fast fourier transform (fft) pattern of the HAADF-STEM (shown in the inset) reveals the existence of secondary spots in the forbidden positions. The structure can be attributable to $L1_2$ superlattice [7,8]. The particle in Fig. 2 (a) is aligned along the <100> zone axis, with pure Au columns alternating with Au and Pd columns in their two-dimensional projection, as marked by the yellow arrow pointers. The image contrast can be attributed to the difference in the atomic numbers of the constituent elements (Z_{Au}=79, Z_{Pd}=46) [9].

Further annealing of the sample at 500 K up to 120 mins resulted in other nanoparticles morphologies, as shown in Fig. 2 (b-d). Fig. 2 (b) shows a nanoparticle with the same $L1_2$ superlattice as in Fig.2 (a), but aligned in the <110> zone axis. In the layered morphology case, a thin layer can be seen surrounding the exterior of the AuPd nanoparticles, as marked by the arrow pointers in Fig.2(c). The height of this layer is about 2-3 atomic Pd layer using isolated single Pd atoms as mass standard. We attribute this layer to the formation of Pd-oxide layer. A closer inspection of the lattice spacing of Pd revealed a value of 0.30±0.01 nm, which is larger than the nearest neighbour distance of Pd. Possible causes for the dilation of the Pd lattice include the incorporation of either oxygen or carbon into the lattice during the annealing process [10]. No such ordered thin layers were observed for Pd-only particles that had been annealed under the same condition.

For the 120 min.-annealed sample, nanoparticles with core-shell structures were also observed as shown in Fig. 2 (d). The darker contrast core in the images indicate that the structure has a Pd-rich core and Au-rich shell, in agreement with the energetic consideration that Pd has a lower surface energy and slightly higher cohesive energy than that of Au.

Table 1. Diameter distribution for AuPd nanoparticles annealed in situ at 500 K for (a) 60 minutes; (b-d) 120 minutes and (e-f) 180 minutes, respectively.

Type	60 minutes	120 minutes	180 minutes
Alloyed	**(a)** 2.53±0.28 nm 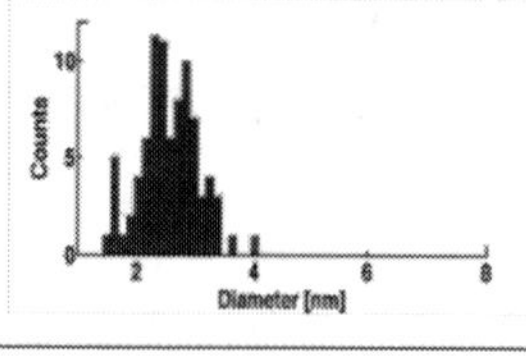	**(b)** 3.43±0.27 nm 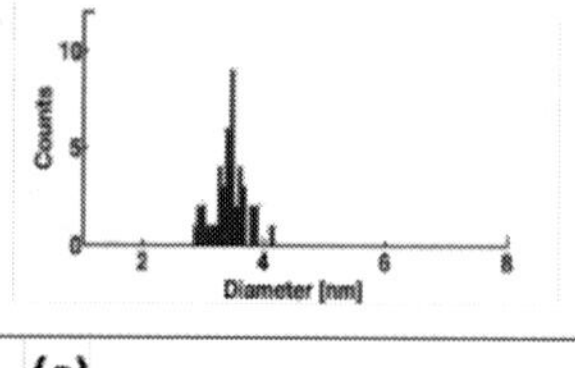	N/A
Layered	N/A	**(c)** 4.69±0.35 nm 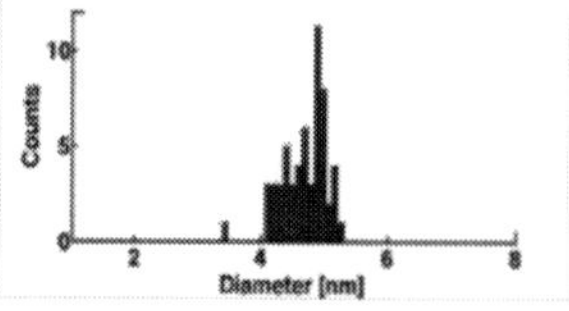	**(e)** 5.96±0.24 nm 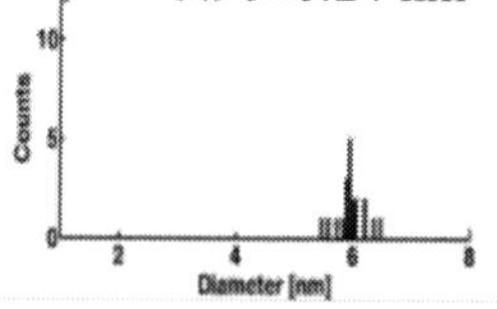
Core- shell	N/A	**(d)** 5.56±0.31 nm 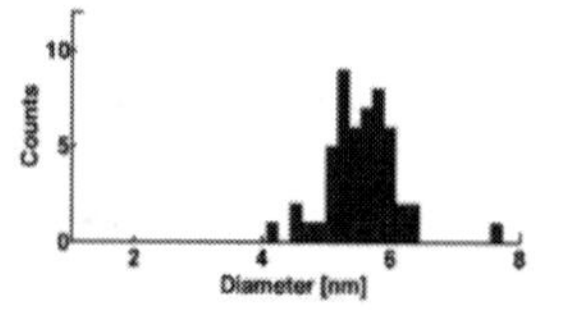	**(f)** 6.71±0.23 nm 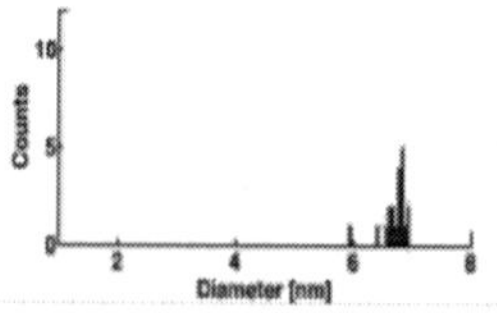

Table 2. Evolution of nanoparticles density as a function of annealing time

Time of annealing	As deposited	60 minutes	120 minutes	180 minutes
Density (particles/μm²)	5.18x10⁴	3.95x10⁴	1.69x10⁴	1.39x10⁴

Previously, Marical *et al.* showed that a very small concentration of CO molecules in the air could be sufficient enough to lead a structural transformation from $Pd_{core}Au_{shell}$ to $Au_{core}Pd_{shell}$ [11]. In the present study, the annealing was conducted under the vacuum condition before the nanoparticles were exposed to air prior imaging. No structure reversal was observed during e-beam irradiation. This could be because the prolonged annealing has already driven the nanoparticles to a thermodynamic stable configuration that has trapped the Pd in the core. We also observed that the core-shell structure was only observed for those particles with large sizes (see Table 1).

To correlate the thermal induced structure transform with size, we analysed nanoparticle size distribution. The results are shown in Table 1. It is apparent that not only had the annealing process caused the structure transformation, but it also resulted in an increase in the nanoparticles' size. As the annealing time increased from 60 to180 minutes, the mean diameter increased from 2.53±0.28 nm to 6.71±0.23 nm. At the same time, the density of the nanoparticle decreased from 5.18×10^4 particles/μm^2 to 1.39×10^4 particles/μm^2 (see Table 2). These values suggest that nanoparticles growth during annealing took place via coalescence or Osward ripening. It is possible that the ordered $L1_2$ superlattice is only possible for small size nanoparticles. It is also interesting to note that, for the prolonged 180 minutes annealing, no nanoparticles with the ordered $L1_2$ superlattice had been observed, indicating that the structure may have been trapped into a thermodynamic metastable state.

4. Conclusions

In summary, we have performed a systematic investigation of atomic structure of AuPd using HAADF-STEM. Upon annealing at an elevated temperature, we show that the AuPd nanoparticles undergo a rich structural transformation, from randomly mixed phase as-deposited to an ordered superlattice and Pd-rich core/Au-rich shell configuration. We have also established a close correlation between the observed structures and the nanoparticle sizes. Our results highlight the paramount importance of understanding the intricate relationships between energetics and kinetics in bimetallic nanoparticle formation for exploring their potential novel applications.

Acknowledgments

We thank Dr. Q. Guo for technical assistance during the evaporation process. We acknowledge financial support from the UK EPSRC.

References

[1] Liu H B, Pal U, Perez R and Ascensio J A 2006 *J Phys Chem* B **110** 5191
[2] Atanasov I and Hou M 2009 *Surf Sci* **603** 2639
[3] Bruma A, Ismail R, Paz-Borbon L O, et al. 2013 *Nanoscale* **5** 646
[4] Deepak FL, Casillas-Garcia G, Esparza R, et al. 2011 *J Crys Growth* **325** 60
[5] Pérez-Tijerina E, Mejía-Rosales S, Inada H et al. 2010 *J Phys Chem C* **114** 6999
[6] Akita T, Hiroki T, Tanaka S et al. 2008 *Catalysis Today* **131** 90
[7] Ding Y, Fan F, Tian Z and Wang Z L 2010 *J Am Chem Soc* **132** 10480
[8] Okamoto H, *Phase Digrams for Binary Alloys*, ASM International, 2nd Ed, Novelty, 1985
[9] Wang Z W, Li Z Y, Park S J, Abdela A, Tang D and Palmer R E 2011 *Phys Rev* B **84** 073408
[10] Mayoral A, Deepak, FL, Esparza R, et al. 2012 *Micron* **43** 557
[11] Mariscal MM, Mayoral A, Olmos-Asar JA, et al. 2011 *Nanoscale* **3** 5013

In-situ electrical and structural characterization of individual GaAs nanowires

V T Fauske[1], D C Kim[2], A M Munshi[2], D L Dheeraj[2], B-O Fimland[2], H Weman[2] and A T J van Helvoort[1]

[1] Department of Physics, Norwegian University of Science and Technology, NO-7491, Trondheim, Norway
[2] Department of Electronics and Telecommunications, Norwegian University of Science and Technology, NO-7491, Trondheim, Norway

E-mail: vidar.fauske@ntnu.no

Abstract. A method for probing the electrical and structural characteristics of individual as-grown III-V nanowires was studied. *In-situ* electrical characterization was performed in a focused ion beam / scanning electron microscopy system by using a fine nano-manipulator and ion beam assisted deposition. Transmission electron microscopy specimens of probed nanowires are prepared afterwards. This method would potentially allow the correlation of electrical and structural characteristics (e.g. crystal faults such as twinning) of the nanowire-substrate system. The challenge is in contacting the nanowires so that the electrical characteristics of the nanowire-substrate system can be extracted correctly.

1. Introduction

Several interesting optoelectronic applications are envisaged for III-V nanowires (NWs), amongst others for solar-cells and light emitting diodes [1]. To move towards devices, precise, homogenous positioning of high quality NWs on low-cost substrates is desired. This requires optimization of both the positioned growth, and the NW crystal structure. To use the growth substrate as an electrode for devices, it is also necessary to optimize the electrical properties of the combined NW-substrate system.

There exist several methods to synthesize NWs, but the most commonly used are based on the vapour-liquid-solid mechanism [2]. Here, a liquid catalyst droplet adsorbs precursor vapour, and facilitates the growth of a solid, crystalline NW. If grown on an appropriate substrate, the crystal growth will nucleate in such a manner that epitaxial growth is achieved. For GaAs NWs, as studied here, common catalysts include Au and Ga, the latter of which is termed self-catalyzed growth. For reproducibility and implementation into devices, positioned growth is required, by either controlling where the catalyst droplets are, or where the crystal can nucleate. Specifically, one way of doing this is by patterning a silicon oxide mask on the substrate [3]. The Ga droplets only form in the holes of the oxide mask, and thereby a position controlled NW growth is achieved.

In general, to measure the electrical properties of a system, it is important to be able to achieve good ohmic contacts. Particularly for GaAs, making ohmic contacts is challenging due to its high level of surface states, requiring proper pre-surface treatment and choice of best contact materials [4].

Here we studied a method to measure the electrical properties of as-grown GaAs NWs in a focused ion beam (FIB) system, and to subsequently characterize the same as-grown NW by (scanning) transmission electron microscopy ((S)TEM) for different NW-substrate systems. This should allow for the correlation of electrical and structural data of individual NWs, and to verify any effects of e.g. crystal defects, on device performance [5].

2. Materials and methods

The NWs used in this study were grown using the self-catalysed growth technique in a solid source Varian Gen II Modular molecular beam epitaxy system. The substrates used in this study were p-type Si(111) and Kish graphite. In total, four GaAs NW samples were studied. In sample A, intrinsic GaAs NWs were grown on a Si substrate for a duration of 25 min with an As flux of 5.6×10^{-6} Torr, and at a substrate temperature of 640 °C. The nominal planar growth rate was 2 Å/s. Sample B was grown with the same growth conditions as sample A, but the NWs were Be-doped (p-doped). The temperature of the Be effusion cell was set to yield a nominal Be concentration of 3.5×10^{18} cm^{-3}. Sample C was grown as sample B, but on a patterned Si substrate (figure 1(a)). The substrate pattern was prepared by depositing a thin layer of SiO_2 (thickness < 60 nm) on the substrate by plasma-enhanced chemical vapour deposition method. Nano-hole array patterning (hole diameter < 200 nm) was done using standard e-beam lithography. Sample D had Be-doped GaAs NWs grown on Kish graphite, by a two-temperature technique [6].

After growth, the samples were mounted on aluminium scanning electron microscopy (SEM) stubs with conductive silver paint. The NWs were characterized in a FEI DualBeam FIB/SEM system (Helios NanoLab 600), equipped with an Omniprobe tungsten (W) nanomanipulator (AutoProbe 200), hereafter called W-probe, and a gas injection system with C-, Pt- and W-sources. The FIB has a Ga source and was operated at 30 kV. Electrical measurements were performed by connecting a Keithley 2636A source-measurement unit to the sample stage (ground) and W-probe. Before each electrical measurement, the W-probe was cleaned by ion sputtering to remove any contaminants and oxides that might interfere with the measurement.

TEM specimen preparation of as-grown NWs was performed *in-situ* in the same FIB/SEM. The as-grown NWs were extracted using the W-probe as a lift-out finger, before the specimen was thinned to electron transparency by ion sputtering. The final sputtering was done with a 5 keV ion beam. Before extraction, the NW was surrounded by carbon deposited by electron beam assisted deposition (EBAD) at 10 kV, to protect the area of interest against ion beam damage. The TEM characterization was performed on a JEOL 2010F operated at 200kV.

3. Results and discussion

To measure the electrical characteristics of the as-grown NW and substrate, current-voltage (I-V) measurements were performed, using the substrate as one electrode, and the W-probe as the other (figure 1). Different approaches to contact the Be-doped NW samples were studied. Initially, the W-probe was simply put in contact with the NW or the Ga droplet on the top, but such a contact was mechanically unstable due to drift and vibrations, leading to fluctuating and irreproducible results. The measurements also indicated a diode like behaviour of the system. A local deposition of W or Pt on the contact area between the probe and the NW by EBAD (a W or Pt weld) gave sufficient mechanical stability, but the signal was again diode-like. As EBAD deposited materials can have high carbon incorporation (from the precursor gas), ion beam assisted deposition (IBAD) was considered as a way to improve the contact. By attaching the W-probe to the NW tip using a local deposition of W by IBAD, (semi-) linear I-V characteristics could reliably be obtained (see figure 2(a)), which implies ohmic contacts. However, while the contact is ohmic, the system still has a significant resistance (~20-30 MΩ). A control measurement was performed in which intrinsic NWs were used (sample A). Intrinsic GaAs wires have a much higher resistivity (~10 GΩ by two-probe measurements [6]), which should give a very low current. However, these control measurements consistently showed similar behaviour as the doped NWs (compare figure 2(a) and (b)). This indicates that the electrical characteristics do not originate from the NW itself. Most likely, W has been deposited along the length of the NW by deposition overspray, which is caused by the wide tails of the ion-beam profile. Consequently, the deposition method would need to be optimized with regards to spatial precision.

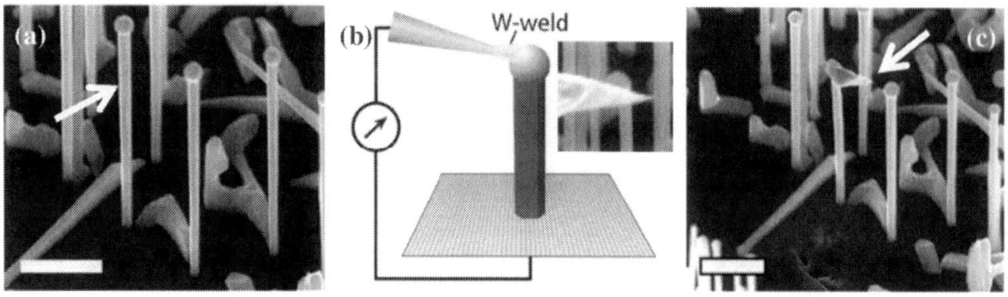

Figure 1. Principle of NW-contacting. (**a**) NW grown on Si with patterned oxide mask (sample C). NW to be contacted indicated by arrow. (**b**) Illustration showing how the NW is contacted (inset: SEM of attached probe to NW). (**c**) NW with part of the W-probe (indicated by arrow) still attached after the probe was cut free by ion sputtering prior to TEM preparation. Scalebar 1 μm in all images.

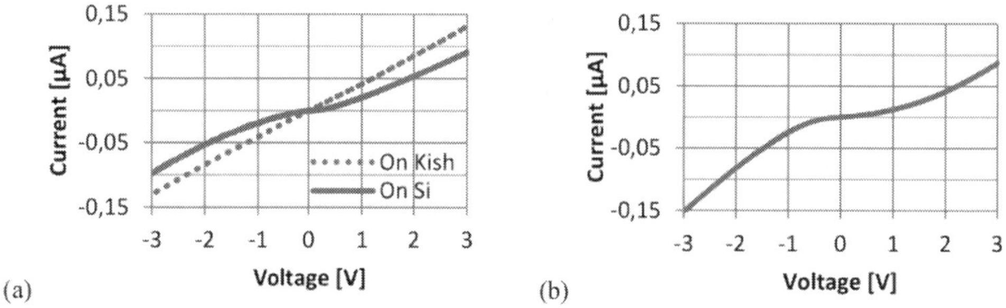

Figure 2. I-V measurements of NWs. (**a**) p-doped NW on p++ (highly p-doped) Si substrate (solid line) and on Kish graphite (dotted line), samples B and D respectively. (**b**) Intrinsic NW on p++ Si substrate (sample A). The W-probe was welded to the NW tip by W-IBAD.

Even if it is the case that material is deposited along the exterior of the NWs, the doped NWs are expected to have a low enough resistivity that the electrons would preferentially pass through it, giving a lower resistance than what was measured. As this is not the case, it is likely that the contact to the NW is blocked by the high surface states of the GaAs. To fix this, the contact quality would need to be improved, e.g. by annealing it *in-situ* with a heating sample holder, or the contact would need to be made with a different material. The preliminary conclusion of the present study is that contacting W-probe to GaAs NWs by FIB is limited to Schottky-like contacts using EBAD.

To correlate the SEM and I-V data with the NW crystal structure, TEM specimens were made from the probed NWs within the same set-up. To do this, the tip of the W-probe is cut off by ion sputtering, and then the TEM specimen is prepared, as outlined in the methods section. This allows the detailed study of the NW, the NW-substrate interface, and the contact between probe and NW.

STEM of the same NW (sample C) as depicted by SEM in figure 1, is shown in figure 3. The high angle annular dark field (HAADF) overview image (figure 3(a)) shows that the NW is about 3.6 μm long, and both the interface and the tip with W-probe remnant have been conserved. Diffraction patterns from the centre of the NW (not shown) confirm zinc blende GaAs with growth direction along the [111] crystal direction. Bright field TEM imaging (not shown) shows that the NW is mostly stacking fault free, except near the substrate interface and near the Ga-droplet. In figure 3(b), a HAADF image of the tip of the NW is shown. From energy dispersive x-ray (EDX) measurements it was confirmed that the W-deposition continues down the NW, approximately down 1/3 of its length, as can be seen as bright areas on NW in figure 3(a) and (b). This W-deposition was also visible by SEM (not shown).

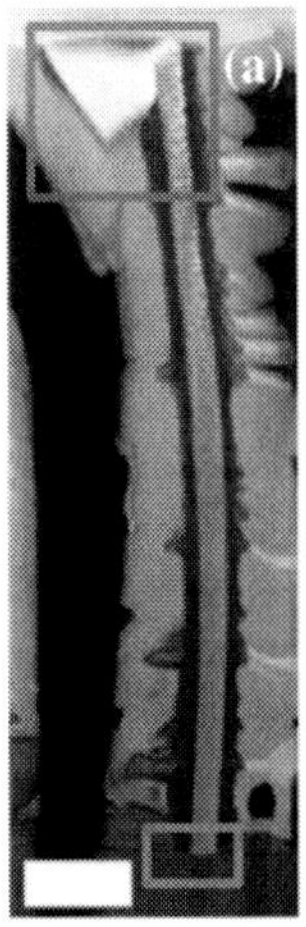

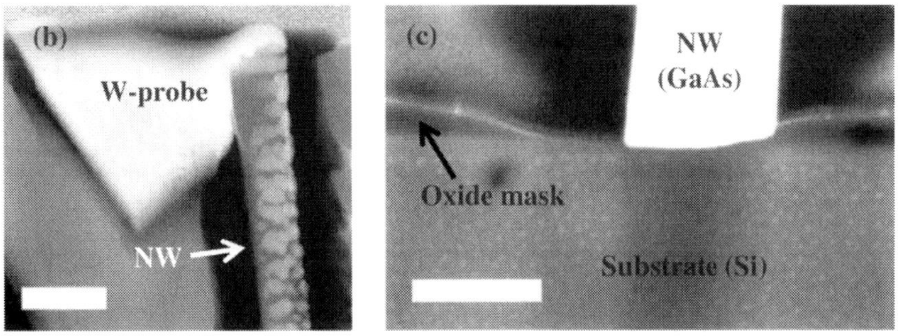

Figure 3. STEM HAADF images of as-grown NW on patterned Si substrate. **(a)** Overview of the entire NW. Boxes indicate the position of figures (b) and (c). **(b)** HAADF of NW tip with the free-cut W-probe. **(c)** HAADF showing the interface region of the NW. The profile of the NW and oxide mask can be seen clearly. Scalebar 500 nm in (a) and (b), and 100 nm in (c).

Finally, in figure 3(c), a higher magnification HAADF image of the NW-substrate interface is shown. The image shows the position of the NW base in relation to the oxide mask, and it can be seen that the NW is growing over the oxide mask in one direction, along the etch profile. This effectively reduces the NW-substrate contact area, which might be detrimental to its function as a contact. A way to avoid this would be to achieve straighter etch profiles, possibly by reactive ion etching. In figure 3(c), a bright layer can be seen on top of the oxide mask. EDX analysis reveals this to be W, further corroborating the earlier suspicion that W overspray cover the length of the NW.

4. Conclusions

Electrical characterization of individual, as-grown GaAs NWs was performed *in-situ* in a FIB-SEM system by using a fine W-probe. The observed I-V curves depend on the way the W-probe is contacted to the NW. For reproducibility, welding the W-probe to the NW is required. Contacting by EBAD is limited to Schottky-like contacts. W-IBAD gives apparent ohmic contacts; however, a conducting layer might have been deposited along the NW, forming a short circuit. Future work should focus on improving the *in-situ* contacting of the probe to the NW, either by improving the deposition process, changing the deposited material and/or the probe material.

The method combines well with TEM, since a TEM specimen can be made directly of the probed, as-grown NWs. The TEM data acquired shows that the NWs grow epitaxially on the Si substrate, and that the NWs are mostly defect free zinc blende, but with a concentration of stacking faults near the substrate interface and NW tip. The NW base is found to expand over the oxide mask. If a low resistance, ohmic contact can be made, the combination of TEM and probing should allow the correlation of electrical and structural data.

References

[1] Wallentin J *et al.* 2013 *Science.* **339** 1057–60.
[2] Wagner R S and Ellis W C 1964 *Applied Physics Letters.* **4** 89–90.
[3] Plissard S, Larrieu G, Wallart X and Caroff P 2011 *Nanotechnology.* **22** 275602.
[4] Howes M J and Morgan D V 1985 *Gallium arsenide: materials, devices, and circuits* Chichester Wiley.
[5] Shimamura K, Yuan Z, Shimojo F and Nakano A 2013 *Applied Physics Letters.* **103** 022105–4.
[6] Munshi A M, Dheeraj D L, Fauske V T, Kim D-C, Van Helvoort A T J, Fimland B-O and Weman H 2012 *Nano Letters.* **12** 4570–6.

AUTHOR INDEX

AUTHOR INDEX

AUTHOR INDEX

Institute of Physics
Dirac House, Temple Back
Bristol BS1 6BE UK

CURRAN ASSOCIATES INC
proceedings.com